Lecture Notes in Computer Science   11157

Commenced Publication in 1973
Founding and Former Series Editors:
Gerhard Goos, Juris Hartmanis, and Jan van Leeuwen

More information about this series at http://www.springer.com/series/7409

Juan C. Trujillo · Karen C. Davis
Xiaoyong Du · Zhanhuai Li
Tok Wang Ling · Guoliang Li
Mong Li Lee (Eds.)

Conceptual Modeling

37th International Conference, ER 2018
Xi'an, China, October 22–25, 2018
Proceedings

 Springer

Editors
Juan C. Trujillo (iD)
Lucentia
University of Alicante
Alicante
Spain

Karen C. Davis
Miami University
Oxford, OH
USA

Xiaoyong Du
Renmin University of China
Beijing
China

Zhanhuai Li
Northwestern Polytechnical University
Xian
China

Tok Wang Ling
Department of Computer Science
National University of Singapore
Singapore
Singapore

Guoliang Li
Department of Computer Science
  and Technology
Tsinghua University
Beijing, Beijing
China

Mong Li Lee
National University of Singapore
Singapore
Singapore

ISSN 0302-9743   ISSN 1611-3349   (electronic)
Lecture Notes in Computer Science
ISBN 978-3-030-00846-8   ISBN 978-3-030-00847-5   (eBook)
https://doi.org/10.1007/978-3-030-00847-5

Library of Congress Control Number: 2018955147

LNCS Sublibrary: SL3 – Information Systems and Applications, incl. Internet/Web, and HCI

This Springer imprint is published by the registered company Springer Nature Switzerland AG
The registered company address is: Gewerbestrasse 11, 6330 Cham, Switzerland

Preface

On behalf of the ER 2018 general co-chairs and program co-chairs, we are pleased to welcome you to the 37th International Conference on Conceptual Modeling (ER 2018), held during October 22–25, 2018, in Xi'an, China. This conference provides an international forum for technical discussion on conceptual modeling of information systems among researchers, developers, and users. This was the second time that the conference was held in China. The first time was in Shanghai in 2004.

The city of Xi'an, located in central-northwest China, is regarded as one of the four ancient capitals in the world. It was the capital of 13 dynasties in Chinese history, reaching its peak of renown in the Tang Dynasty. Xi'an has numerous historical attractions, making it one of the best locations for congresses and tourists in China. In a single day, you can explore the Emperor Qinshihuang's Terracotta Warriors and learn about these amazing artifacts for yourself. Also, you can go to the Drum Tower, the Bell Tower, ancient City Walls, and the Big Wild Goose Pagoda, which are all within minutes of the city center. Nightlife in Xian is equally interesting with neon-lit shops and streets, evening food market, pubs, cafes, a music fountain square, singing and dancing shows, to name but a few.

Since the first version of the entity-relationship (ER) model by Peter Chen appeared in *ACM Transactions on Database Systems (TODS)* in 1976, both the ER model and conceptual modeling have been key success factors for modeling computer-based systems. The International Conference on Conceptual Modeling is an important venue for the presentation and exchange of ideas and concepts that relate to traditional and emerging issues in conceptual modeling of information systems. Work on conceptual modeling has continued to evolve as the ER model has been applied, modified, and extended to research in database management systems, business process management, and management information systems. Conceptual modeling is continuing to play a vital role in the emerging, new data era where the correct design and development of mobile or sensors analytics, big data systems, non-SQL databases, smart cities, and biomedical systems will be crucial. The 37th International Conference on Conceptual Modeling served as a forum where some of these novel areas as well as their fundamental and theoretical issues that are directly related to conceptual modeling were discussed. In this year's edition, we placed special emphasis on research focused on machine and deep learning and how conceptual modeling can be successfully be applied soon in these areas in order to increase the success rates of the application of these artificial intelligence techniques.

The ER conference continues to attract some of the best researchers and keynote speakers, from both academia and industry, who work on topics in traditional and emerging areas of conceptual modeling. This year, 151 full papers were submitted to the conference. Each paper was reviewed by at least three reviewers and, based upon these reviews, 30 full papers and 13 short papers were selected for publication in the proceedings and presentation at the conference. The acceptance rate for regular papers was 19.87%, and for regular and short papers together, 28.48%. These papers were organized into 13 sessions that represent leading research areas in conceptual modeling, including topics related to fundamentals of conceptual modeling, ontologies, semi-structured and spatio-temporal modeling, language and models, and conceptual modeling for machine learning. The scientific program also featured three interesting keynote presentations by Ernesto Damiani, Kyu-Young Whang, and Anqun Pan, each of whom has shared some of their thoughts and insights in these proceedings. Let us also take this opportunity to congratulate Veda C. Storey for her recent Peter Chen award in 2018, which culminates her excellent academic career, being one of the most important influencers in the conceptual modeling area.

We wish to thank the 175 members of the Program Committee and the external reviewers who provided insightful reviews and discussions on the papers. We also appreciate the diligence of the senior reviewers, who provided guidance and recommendations, and for the selection of the best paper awards. Most importantly, we thank the authors who submitted high-quality research papers on a wide variety of topics, thus making this conference possible. We hope you enjoy the proceedings.

We would also like to thank the honorary conference chair, Prof Shan Wang, organization chair, Xuequn Shang, Steering Committee liaison, Il-Yeol Song, workshop co-chairs, Carson Woo and Jiaheng Lu, tutorial co-chairs, Gillian Dobbie and Ernest Teniente, panel co-chairs and demo co-chairs, Qun Chen and Zhifeng Bao, treasurer and registration chair, Wenjie Liu, proceedings chairs, Guoliang Li and Mong Li Lee, publicity chairs, Selmin Nurcan, Bin Cui, Moonkun Lee, and Mengchi Liu, and conference webmaster, Jialu Hu.

We thank the organizers of the individual workshops, Symposium on Conceptual Modeling Education (SCME): Isabelle Comyn-Wattiau and Hui Ma; Doctoral Symposium: Xavier Franch and Chaokun Wang; ER Forum: Heinrich C. Mayr and Oscar Pastor; Empirical Methods in Conceptual Modeling (Emp-ER): Sotirios Liaskos and Jennifer Horkoff; Modeling and Management of Big Data (MoBiD): Il-Yeol Song, Jesús Peral, and Alejandro Maté; Conceptual Modeling in Requirements and Business Analysis (MREBA): Jennifer Horkoff, Renata Guizzardi, and Jelena Zdravkovic; Quality of Models and Models of Quality (QMMQ): Samira Si-said Cherfi, Beatriz Marin, and Oscar Pastor; Corralling the Field of Conceptual Modeling (CCM): Lois Delcambre, Oscar Pastor, Steve Liddle, and Veda C. Storey; and Data Science & Blockchain (DSBC): Peter Chen and Heinrich C. Mayr.

We hope that you will find the proceedings of ER 2018 interesting and beneficial to your research.

August 2018

Juan C. Trujillo
Karen C. Davis
Xiaoyong Du
Zhanhuai Li
Tok Wang Ling

Organization

Honorary Chair

Shan Wang — Renmin University of China, China

Conference General Co-chairs

Zhanhuai Li — Northwestern Polytechnical University, China
Tok Wang Ling — National University of Singapore, Singapore

Organization Chair

Xuequn Shang — Northwestern Polytechnical University, China

Steering Committee Liaison

Il-Yeol Song — Drexel University, USA

Program Committee Co-chairs

Juan C. Trujillo — University of Alicante, Spain
Karen C. Davis — Miami University, USA
Xiaoyong Du — Renmin University of China, China

Workshop Co-chairs

Carson Woo — University of British Columbia, Canada
Jiaheng Lu — University of Helsinki, Finland

Tutorial Co-chairs

Gillian Dobbie — University of Auckland, New Zealand
Matteo Golfarelli — University of Bologna, Italy

Panel Co-chairs

Carlo Batini — University of Milano Bicocca, Italy
Ernest Teniente — Polytechnic University of Catalonia, Spain

Demo Co-chairs

Qun Chen	Northwestern Polytechnical University, China
Zhifeng Bao	RMIT, Australia

SCME (Symposium on Conceptual Modeling Education) Co-chairs

Isabelle Comyn-Wattiau	ESSEC Business School, France
Hui Ma	Victoria University of Wellington, New Zealand

Doctoral Symposium Co-chairs

Xavier Franch	Polytechnic University of Catalonia, Spain
Chaokun Wang	Tsinghua University, China

Forum Co-chairs

Heinrich C. Mayr	Alpen-Adria-Universität Klagenfurt, Austria
Oscar Pastor	Universidad Politècnica de Valencia, Spain

Treasurer and Registration Chair

Wenjie Liu	Northwestern Polytechnical University, China

Proceedings Co-chairs

Guoliang Li	Tsinghua University, China
Mong Li Lee	National University of Singapore

Publicity Co-chairs

Bin Cui	School of EECS, Peking University, China
Moonkun Lee	Chonbuk National University, South Korea
Mengchi Liu	Carleton University, Canada
Selmin Nurcan	Université Paris 1, France

Conference Webmaster

Jialu Hu	Northwestern Polytechnical University, China

Program Committee

Dippy Aggarwal	Intel, USA
Jacky Akoka	CNAM and TEM, France
Jose F. Aldana	University of Malaga, Spain
Raian Ali	Bournemouth University, UK

Joao P. Almeida	Federal University of Espirito Santo, Brazil
Yuan An	Drexel University, USA
Joao Araujo	Universidade NOVA de Lisboa, Portugal
Alessandro Artale	Free University of Bolzano-Bozen, Italy
Claudia P. Ayala	Universitat Politècnica de Catalunya, Spain
Fatma B. Aydemir	Utrecht University, The Netherlands
Doo-Hwan Bae	KAIST, South Korea
Fernanda A. Baiao	UNIRIO, Portugal
Zhifeng Bao	RMIT University, Australia
Judith Barrios Albornoz	University of Los Andes, Chile
Ladjel Bellatreche	LIAS/ENSMA, France
Kawtar Benghazi	University of Granada, Spain
Sandro Bimonte	Irstea, France
Carlos Blanco Bueno	Universidad de Cantabria, Spain
Mokrane Bouzeghoub	UVSQ/CNRS, France
Shawn Bowers	Gonzaga University, USA
Stephane Bressan	National University of Singapore, Singapore
Cristina Cabanillas	Vienna University of Economics and Business, Austria
Diego Calvanese	Free University of Bozen-Bolzano, Italy
Roger Chiang	AIS, New Zealand
Dickson K.W. Chiu	University of Hong Kong, SAR China
Byron Choi	Hong Kong Baptist University, SAR China
Isabelle Comyn-Wattiau	ESSEC Business School, France
Nelly Condori-Fernández	Universidade da Coruña, Spain
Dolors Costal	Universitat Politècnica de Catalunya, Spain
Bin Cui	Peking University, China
Fabiano Dalpiaz	Utrecht University, The Netherlands
Karen C. Davis	Miami University, USA
Valeria De Antonellis	University of Brescia, Italy
Sergio De Cesare	University of Westminster, UK
José Palazzo M. de Oliveira	Federal University of Rio Grande do Sul, Brazil
Adela Del Río Ortega	University of Seville, Spain
Lois Delcambre	Portland State University, USA
Xiaoyong Du	Renmin University of China, China
Marlon Dumas	University of Tartu, Estonia
Johann Eder	Alpen-Adria-Universität Klagenfurt, Austria
Vadim Ermolayev	Zaporizhzhya National University, Ukraine
Ricardo A. Falbo	Federal University of Esprito Santo, Portugal
Eduardo Fernández-Medina	University of Castilla La Mancha, Spain
Hans-Georg Fill	University of Bamberg, Germany
Xavier Franch	Universitat Politècnica de Catalunya, Spain
Ulrich Frank	University of Duisburg-Essen, Germany
Frederik Gailly	Ghent University, Belgium
Aldo Gangemi	Université Paris 13 and CNR-ISTC, France
Jun Gao	Peking University, China
Yunjun Gao	Zhejiang University, China

Faiez Gargouri	Institut Supérieur d'Informatique et de Multimédia de Sfax, France
Marcela Genero	University of Castilla La Mancha, Spain
Aurona Gerber	CAIR, University of Pretoria, South Africa
Sepideh Ghanavati	Texas Tech University, USA
Mohamed Gharzouli	Constantine 2 University, Algeria
Aditya Ghose	University of Wollongong, Australia
Giovanni Giachetti	Universidad Tecnológica de Chile, Chile
Paolo Giorgini	University of Trento, Italy
Cesar Gonzalez-Perez	Incipit, CSIC, Spain
Georg Grossmann	University of South Australia, Australia
Nicola Guarino	ISTC-CNR, Italy
Esther Guerra	Universidad Autónoma de Madrid, Spain
Giancarlo Guizzardi	Federal University of Espírito Santo, Brazil
Renata Guizzardi	Federal University of Espírito Santo, Brazil
Maria Teresa Gómez	University of Seville, Spain
Maria Hallo	Escuela Politècnica Nacional, Ecuador
Sven Hartmann	Clausthal University of Technology, Germany
Martin Henkel	Stockholm University, Sweden
Jennifer Horkoff	Chalmers University of Gothenburg, Sweden
Hao Huang	Wuhan University, China
Sergio Ilarri	University of Zaragoza, Spain
Arantza Illarramendi	Basque Country University, Spain
Matthias Jarke	RWTH Aachen University, Germany
Manfred Jeusfeld	University of Skovde, Sweden
Ivan Jureta	University of Namur, Belgium
Gerti Kappel	Vienna University of Technology, Austria
Dimitris Karagiannis	University of Vienna, Austria
Kamalakar Karlapalem	CDE, IIIT Hyderabad, India
David Kensche	SAP, USA
Vijay Khatri	Indiana University Bloomington, USA
Agnes Koschmider	Karlsruhe Institute of Technology, Germany
John Krogstie	Norwegian University of Science and Technology, Norway
Moonkun Lee	Chonbuk National University, South Korea
Mong Li Lee	National University of Singapore, Singapore
Julio Cesar Leite	PUC-Rio, Brazil
Guoliang Li	Tsinghua University, China
Zhixu Li	Soochow University, China
Stephen Liddle	Brigham Young University, USA
Tok Wang Ling	National University of Singapore, Singapore
Sebastian Link	University of Auckland, New Zealand
Mengchi Liu	Carleton University, USA
Pericles Loucopoulos	University of Manchester, UK
Jiaheng Lu	University of Helsinki, Finland
Hui Ma	Victoria University of Wellington, Australia

Additional Reviewers

Amador Durán
Antonio Manuel Gutierrez
Carlos Müller
Cristina Palomares
Dominik Bork
Faten Atigui
Faycal Hamdi
Guohui Xiao
Ismael Navas-Delgado
Jihae Suh
José A. Galindo
José García-Nieto

Juan Manuel Vara
Karolin Winter
Lidia Lopez
Lingxiao Li
Maciej Rybinski
Maria Del Mar Roldan-Garcia
Nabila Berkani
Pablo Trinidad
Petar Jovanovic
Selma Khouri
Taekyung Kim
Zhuo Peng

Contents

Cloud-Based Modeling

Schema and View Modeling

Languages and Models

NoSQL Modeling

Conceptual Modeling for Machine Learning and Reasoning I

Conceptual Modeling for Machine Learning and Reasoning II

Applications of Conceptual Modeling

Keynotes

Towards Conceptual Models for Machine Learning Computations

Ernesto Damiani[1,2]([ICON]) [iD] and Fulvio Frati[2] [iD]

[1] Centre on Cyber-Physical Systems, Khalifa University, Abu Dhabi, UAE
ernesto.damiani@unimi.it
[2] Computer Science Department, Università degli Studi di Milano, Milan, Italy

Abstract. We make the case for conceptual models that give the human designer full visibility and control over key aspects of ML applications, including input data preparation, training and inference of the ML models. Our models aim to: *(i)* achieve better documentation of ML analytics *(ii)* provide a foundation for a chain of trust in the ML analytics outcome *(iii)* provide a lever to enforce ethical and legal constraints within the ML pipeline. Representational models can dramatically increase reusability of large-scale ML analytics, while decreasing their roll-out time and cost. Also, they will support novel solutions to time-honored issues of analytics like non-uniform data veracity, privacy and latency profiles.

Keywords: Machine learning · Big data analytics
Artificial Intelligence

1 Introduction

Recently, interest in Artificial Intelligence has been triggered by the impressive performance shown by computationally intensive Machine Learning (ML) models. However, most investments in Artificial Intelligence and Big Data analytics so far have been addressed to algorithm development and AI technology acquisition, disregarding engineering of ML applications.

Still, ML is no silver bullet: it is widely acknowledged that ML analytics can achieve satisfactory performance only if training and execution strategy matches the specific application's latency and resource constraints. Enforcing this can be sometimes straightforward, but more often it requires skills, time and experience. For instance, when some real estate (say, a geographical area) is monitored via a single multispectral camera on a satellite, training a classifier involves only the independent training of multiple layer pairs acting as statistical learners, each training requiring a substantial number of matrix multiplications. One may be tempted to say that there is no need for modeling such computation, as the code invoking appropriate ML libraries will be self-documenting enough. However, if the same area is monitored by a "sand-dust" of heterogeneous distributed sensors not all of which are operational at any given time, putting together the feature

© Springer Nature Switzerland AG 2018
J. C. Trujillo et al. (Eds.): ER 2018, LNCS 11157, pp. 3–9, 2018.
https://doi.org/10.1007/978-3-030-00847-5_1

vectors (e.g. via the computation of semantic-driven joins) needed to perform the training. Feature partitioning should be modeled to ensure the application reusability and validation. Other aspects to be modeled include inputs' estimated veracity: complex AI models can be very sensitive to the uncertainty in the input data, due to the very nature of the measurement/acquisition process. This problem is exacerbated by pre-training techniques, aiming to increase the efficiency of the model-training procedures, and acquires a high impact in the case of data from one place/setting used for predictions related to another place/setting.

Besides choosing the ML model to use and the data sources, there are subtler decisions that may impact on the success of ML projects. Today, organizations that wish to use ML for large-scale applications need to solve two major problems:

(a) Enforcing division of labor: Distribute/integrate as needed the data preparation/ pre-processing and analytics computations to handle latency training and execution pipeline of the ML model,
(b) Mapping ML models' topology to execution architecture: Deploy the ML computational tasks on suitable execution targets according to the required performance and to other computational SLAs.

Another fundamental concern is privacy. In the past few years, much research has addressed privacy of ML models *per se*, while less attention has been devoted to possible disclosures via cracks at the seams of ML models' implementations. Indeed, ML pipeline modules can be owned and managed by multiple operators, each with its own interests and agenda; therefore, we believe that disclosure control should be addressed at design time (privacy-by-design). Conceptual models will clarify legal caveats, e.g. pointing out the need for legal advice on whether a given activity (e.g. data randomization) is appropriate to the specific application in case of litigation.

In this paper we advocate an approach based on conceptual models to represent and validate ML models' computations[1]. Such conceptual models are formed after a conceptualization or generalization process based on a proper ontology of concepts. Our main objective is not to impose a new modelling vocabulary or syntax, but to support a methodology for systematic annotation, analysis and validation of code implementing ML models. Also, our conceptual models aim to support generation of code written in programming languages commonly used in ML applications, like Python. Our models will correspond to code skeletons to be completed with configuration information regarding the target execution platform (e.g. core numbers, CPU/GPU capabilities, RAM size and so on). In the fullness of time, ML model computations will be representable in a machine-readable, auditable format, increasing the transparency of the whole process and building a chain of trust on their outcome.

---

[1] Note that the word "model" has been used twice in the same sentence, but with two different meanings. Our conceptual (process) models represent activities needed to train and deploy (mathematical) ML models; the latter play much the same role as algorithms in software design. We will drop the adjective "conceptual" when there is no fear of confusion.

2   State of the Art

The ideas in this paper are a development of the approach taken in the TORE-ADOR project, where activities within Big Data analytics are modeled as services belonging to a OWL/S ontology. We will not try to review our own work in this Section; the interested reader is referred to [4]. Here, we briefly review modelling support made available by other design environments for Big Data Analytics. We claim that none of the approaches reviewed below is based on a conceptual model of the computation, where individual activities belong to a shared ontology. This prevents model-level validation and certification of ML applications, as well as cross-compilation of models across platforms [2,3].

Currently, the main language available for documenting ML applications is the Predictive Model Markup Language (PMML). PMML enables sharing predictive models between applications. It was developed by DMG (Data Mining Group, www.dmg.org), an independent consortium supported by all major ML libraries vendors. PMML uses XML syntax to represent mining models and enables interoperability among analytics tools. The PMML XML schema contains a mechanism for extending the content of a model; however, its target is custom model serialization rather than validation or code generation. A perspective closer to ours is the one of PFA (Portable Format for Analytics), also developed by DMG, which aims to provide interoperability between the steps of the data analytics development process. Developer tools that "speak" PFA (such as Pandas or R) can deploy their engines (e.g. classifiers, predictors, smoothers, filters) on scalable production environments that "understand" PFA (such as Hadoop, Spark or Storm). This way, PFA provides a common interface to deploy analytic workflows across environments, from embedded systems to distributed data centers. Its ease of portability across systems is combined also with algorithmic flexibility: models, pre-processing, and post-processing can be arbitrarily composed, chained, or built into complex workflows described as a JSON or YAML configuration file. PFA is becoming an emerging standard for statistical models and data transformation engines. Still, its use is mainly oriented to serialization and not to compilation or validation. Big Machine Learning (BigML) is oriented to machine learning models' provisioning. It provides a user-friendly interface and a pay-as-you-go business model, exposing traditional machine learning algorithms as Web services integrated via a graphical interface. DataRPM provides a natural language processing interpreter used to understand user's requests and convert it into analytics. It is related to classic SQL conversational interfaces, inasmuch questions and answers are used to generate a visual representation of a query to be executed on the dataset. Another approach towards abstracting the management of data analytics processes is EpiC [6], a system to address the problem of managing Big Data variety. It proposes a solution aimed at defining parallel computations, which is based on an Actor-like concurrent programming model and is independent of the data processing models.

Datameer (www.datameer.com) is a big data analytics platform aiming at building an agile data pipeline, from data preparation to analytics. It allows to define and refine a visual representation of the pipeline. Datameer allows to ingest and integrate more than 70 data formats.

DataRobot (www.datarobot.com) provides a predictive analytics platform to rapidly build and deploy predictive models on the cloud. It aims at automating the entire modeling lifecycle: once the data have been ingested, the platform uses massively parallel processing to train and evaluate models in R, Python, Spark MLlib, H2O and other open source libraries. It searches through many combinations of algorithms, pre-processing steps, features, transformations and tuning parameters to deliver the best models for the dataset and prediction target. The DataRobot process is structured in data ingestion, target variable selection, automatic construction of the available models, and exploration of the top performing models and deployment of the best ones. This pipeline however has limited flexibility and does not addresses the issues of varying veracity and the enforcement of ethical and legal constraints.

Another interesting approach is the one taken by SKYMIND, which allows to translate a ML model into a set of tasks executable on the cloud (https://docs.skymind.ai/docs/welcome). SKYMIND tasks are executed at the core, i.e. on the cloud, with a statically defined division of labor between center and periphery. They assume flat sets of dimensions and fully trusted data feeds.

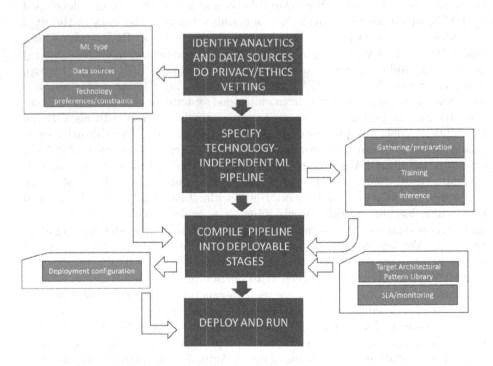

Fig. 1. Model-based design process of ML applications.

3 Design Problems for ML Pipelines

In this Section, we will use the term ML pipeline to designate instances of our conceptual models, i.e. orchestrated process that include all the activities needed to perform the computation of a ML model. Of course, activities of the ML pipeline do not come necessarily in sequence. Instead, the orchestration defined in the conceptual model schedules them in the order and with the repetitions required by the application. In other words, our conceptual models will give to the human designer full visibility of (and control over) all pipeline stages, from input preparation to training and inference, describing ML pipelines as a composition of services enacted by distinct agents on different platforms. This abstract representation can be compiled into deployable computations on target edge, cloud and HPC architectures. It will be tunable to match the application's latency and resource constraints and will assure the desired non-functional properties in terms of data privacy, integrity and protection.

The overall design process we envision for ML pipelines is depicted in Fig. 1.

The compilation step will take as input the technology-independent conceptual process model of the pipeline. The model is represented as a service composition and patterns for efficient parallelization of training and inference algorithms on the target many-core or cluster architecture. The compiler uses the information in the model to fully specify the pattern obtaining an executable computation ready for deployment.

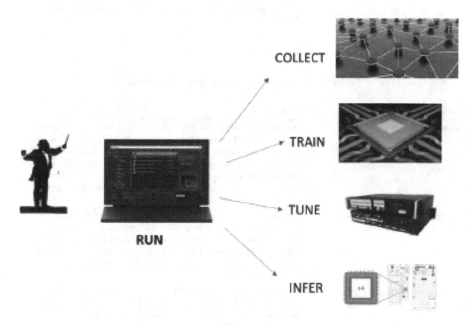

Fig. 2. The execution of a ML pipeline.

Below, in Tables 1 and 2, we discuss in detail some examples of non-trivial decisions during design and deployment of ML models that correspond to variation points in the resulting pipeline composed of data preparation, training and analytics stages. Our methodology will help the designer to handle flexibly such variation points.

Table 1. ML pipeline design decisions.

Design decision	Description
Choose the privacy support	There is a need to choose the level of privacy to address legal and ethical concerns on the training set [1]. The designer needs to decide to manage interactively or non-interactively the right amount of noise to the training set at the periphery (i.e., under the control of the data owner) or in a trusted environment before pushing the trained ML model into production
Choosing the training strategy	There are many training strategies for multi-layer ML models: training may be done independently or joining (some of) the layers in monolithic components. When training time is not an issue, the choice between training strategies will depend mostly on the desired accuracy. However, if there is a deadline for deploying the ML model, the designer should tune depth to achieve an acceptable training time even at the expense of accuracy
Choosing the input partitioning	A feature set collected by many different sensors or mobile devices will have natively a faceted structure. The designer has to decide whether facets (a.k.a. views) can be processed separately, for instance, using multiple kernels when learning classification models or coordinating training of multiple models (co-training)

Table 2. ML pipeline deployment decisions.

Deployment decision	Description
Choosing the data gathering pattern and target	The designer needs to choose the target for data gathering. For distributed cloud processor, data gathering can be split into multiple steps: (1) a commitment step to bound latency, (2) a data transfer step, where periphery push their data to intermediate nodes (3) a ML model feeding step, where a destination node pulls data from intermediate ones and feeds the model with uniform input latency
Choosing the training pattern strategy	The designer needs to identify the target architecture for training. Patterns include Cloud Processors (virtual multi-computers with ultra-band Fabric), ML processors (GPUs or Multi-core CPU with computation-ready RAM) and Wetware-in-Firmware/Silicon, like FPGAs and Bio-Inspired ASIC
Choosing the execution pattern and target	The designer has to identify the target architecture for inference. Well-known patterns include Cloud Processors (virtual multi-computers with ultra-band Fabric), ML processors (GPUs or Multi-core CPU with computation-ready RAM) and Wetware-in-Firmware/Silicon, like FPGAs and Bio-Inspired ASIC

4   Conclusions

In this paper we advocated an approach based on conceptual models to represent and validate ML models' computations. Such conceptual models are formed after a conceptualization or generalization process based on a proper ontology of concepts. The overall execution process we envision for our orchestrated ML pipelines is depicted in Fig. 2.

We argue that orchestrated executions will support establishing Service Level Agreements specific to ML models' computation, allowing users to move away from the use of generic cloud-oriented SLAs [5].

References

1. AlMahmoud, A., Damiani, E., Otrok, H., Al-Hammadi, Y.: Spamdoop: a privacy-preserving big data platform for collaborative spam detection. IEEE Trans. Big Data, 1–1 (2017). https://doi.org/10.1109/TBDATA.2017.2716409
2. Anisetti, M., Ardagna, C., Damiani, E., Ioini, N.E., Gaudenzi, F.: Modeling time, probability, and configuration constraints for continuous cloud service certification. Comput. Secur. **72**, 234–254 (2018). https://doi.org/10.1016/j.cose.2017.09.012, http://www.sciencedirect.com/science/article/pii/S0167404817302018
3. Ardagna, C.A., Asal, R., Damiani, E., Dimitrakos, T., Ioini, N.E., Pahl, C.: Certification-based cloud adaptation. IEEE Trans. Serv. Comput. 1–1 (2018). https://doi.org/10.1109/TSC.2018.2793268
4. Ardagna, C.A., Bellandi, V., Bezzi, M., Ceravolo, P., Damiani, E., Hebert, C.: Model-based big data analytics-as-a-service: take big data to the next level. IEEE Trans. Serv. Comput. 1–1 (2018). https://doi.org/10.1109/TSC.2018.2816941
5. Hussain, W., Hussain, F.K., Hussain, O.K., Damiani, E., Chang, E.: Formulating and managing viable SLAS in cloud computing from a small to medium service provider's viewpoint: a state-of-the-art review. Inf. Syst. **71**, 240–259 (2017). https://doi.org/10.1016/j.is.2017.08.007, http://www.sciencedirect.com/science/article/pii/S0306437917302697
6. Jiang, D., Wu, S., Chen, G., Ooi, B.C., Tan, K.L., Xu, J.: Epic: an extensible and scalable system for processing big data. VLDB J. **25**, 3–26 (2016). https://doi.org/10.1007/s00778-015-0393-2

Recent Trends of Big Data Platforms and Applications

Kyu-Young Whang$^{(\boxtimes)}$

School of Computing, KAIST, Daejeon, South Korea
kywhang@kaist.ac.kr

Abstract. Big data refers to large amounts of data beyond acceptable limits of commonly-used data collection, storage, management, and analysis software. Big data has become a new trend and culture in academia and industry from the beginning of this decade. The importance of big data technology is being widely recognized and getting higher owing to recent technology development. In particular, popular social media services as well as devices connected via Internet-of-Things are accelerating generation of big data. Then, the cloud service improves accessibility of such big data by allowing us to access it everywhere. Furthermore, computing power has also improved rapidly with the introduction of new CPU and GPU hardware technologies. On the basis of these environmental changes, MapReduce and Hadoop significantly contributed to making big data processing prevalent in these days. Hadoop, which is an open-source implementation of MapReduce, enables us to achieve high-performance computing with only commodity machines, but without requiring expensive mainframe computers.

This keynote consists of two parts. The first part introduces the recent trends of big data platforms originated from Hadoop. Then, the second part addresses a few interesting big data applications enabled by such big data platforms.

In the first part, I would like to present the concept of the MapReduce paradigm and its significance in the history of big data processing. Then, I will discuss advantages as well as limitations of Hadoop and systematically review research efforts to overcome the limitations in three catagories: supports of iterative processing, stream processing, and the SQL language. As a solution for the third category, NewSQL, such as Googles spanner and F1, has emerged as a new paradigm. I will also elaborate on ODYS, a massively-parallel search engine which has been developed at KAIST, as an example NewSQL system.

In the second part, I will address the effort of combining AI with big data since big data technology serves as an enabler of artificial intelligence (AI). For example, IBM Watson learned from 200 million pages including Wikipedia and

A joint work with Jae-Gil Lee, Department of Industrial and Systems Engineering, KAIST(email: jaegil@kaist.ac.kr).

© Springer Nature Switzerland AG 2018
J. C. Trujillo et al. (Eds.): ER 2018, LNCS 11157, pp. 10–11, 2018.
https://doi.org/10.1007/978-3-030-00847-5_2

news articles. With this rich knowledge base, IBM Watson surprisingly beat quiz-show human champions. IBM is expanding its application to medical science where Watson for Oncology collaborates with human doctors to diagnose cancers. As another example, smartphone vendors are developing intelligent personal assistants, such as Google Now, Apple Siri, and Amazon Alexa. These assistants benefit from big data because they are getting smarter by learning from huge amounts of user feedback and queries. I will overview some of these data-driven services.

In summary, the keynote will address the characteristics of big data, recent trends of big data platforms, and emerging applications for big data intelligence.

Conceptual Modeling on Tencent's Distributed Database Systems

Anqun Pan(✉), Xiaoyu Wang, and Haixiang Li

Tencent Inc., Shenzhen, China
{aaronpan, xiaoyuwang, blueseali}@tencent.com

Abstract. Tencent is the largest Internet service provider in China. Typical services include WeChat, games, payment, cloud storage and computing. Tencent serves billions of users and millions of enterprises, and some services like WeChat, are required to have the ability to handle more than 208,000 TPS requests at peak time. To do this, Tencent has built an elastic and scalable database service system, namely TDSQL, which can efficiently support their ever-growing service requests. TDSQL is deployed and runs on top of more than ten thousands of compute nodes. In this paper, we present the main challenges that we have encountered, and give our practice of conceptual modeling on TDSQL. First, failures of compute nodes often occur in an X86-based large-scale distributed system architecture. To address this issue, we introduce a fault tolerance model to guarantee the high availability of the services. Second, Tencent serves a huge number of requests, while different types of requests require different storage and compute resources. To improve the resource utilization, we propose a resource scheduling model that enables TDSQL to serve the requests elastically. Third, TDSQL provides a hybrid data modeling to support various data models, and develops DBaaS services to serve 100,000 + DB instances. Finally, we present how to fast develop applications in terms of conceptual modeling on top of TDSQL.

Keywords: TDSQL · Conceptual modeling · AI

1 Introduction

Tencent billing service platform (a.b.a. TBSP) is one of the most advanced financial clouds in the world, which mainly provides three levels of services: SaaS, PaaS and IaaS. In many years of financial service practice, we construct high-performance systems including Midas, Cloud Store and TDSQL (Tencent Distributed Database System) [2] at the SaaS layer to provide reliable and stable financial services. Taking Tencent's business as an example, thousands of online applications such as mobile games, digital business offerings, and advertising services, that require payment transactions all integrate payment modules of TBSP for unified billing management. TBSP deposits approximately 28 billion accounts, with daily transaction volume of more than 10 billion. TBSP includes

© Springer Nature Switzerland AG 2018
J. C. Trujillo et al. (Eds.): ER 2018, LNCS 11157, pp. 12–24, 2018.
https://doi.org/10.1007/978-3-030-00847-5_3

four major modules: account module, order module, risk-control module and settlement module. Reconciliations, audits, risk control data analysis, and construction of user portraits are used to perform comprehensive management and monitoring of payment scenarios.

Due to the sensitivity of financial business data and the stringent requirements for data reliability, we have encountered the following challenges in the process of supporting elastic and scalable 7*24 services to ensure the data reliability:

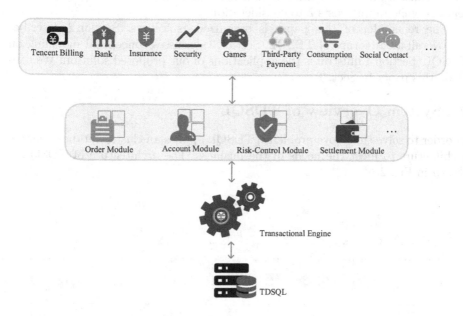

Fig. 1. Overall architecture of TBSP.

Business Scale Varies. For example, order module is mainly responsible for tracking user trading orders, data can be cleared regularly, so the data volume can be controlled. However, the settlement center records all the historical transaction data, which should be capable of storing large amount of data.

Various Data Models. The system needs to support different application scenarios, and the data characteristics of these scenarios are different. For example, personal accounts are mainly stored in relational database. Corporate accounts are stored in key-value storage for the amount of accounts is quite stable, but modifications on a specific account in a short period are extensive.

The Importance of Data is Different. The core account data is frequently accessed and needs high availability. It requires strong data consistency, multiple replications, and disaster recovery across data centers. On the other hand, account data can be defined as hot data. Some historical data like orders created one years ago just need to be archived, which is relatively insignificant than account data.

For the above three problems, it is very important to give a solution to provide effective resource management and to perform unified management of multi-model data. In this paper, we introduce a TDSQL-based solution which contains a novel system architecture, proposes and apply a number of conceptual models including fault recovery modeling, resource scheduling modeling, hybrid data modeling and AI + Database to provide more flexible and highly available services while taking costs into consideration.

The remainder of this paper is organized as follows. Section 2 describes the overall architecture of TDSQL. Section 3 discusses conceptual modeling on TDSQL. Section 4 illustrates the applications of our proposed solution. We conclude this paper in Sect. 6.

2   System Overview of TDSQL

In order to solve the above problems, TDSQL applies an elastic database system architecture to meet the needs of the business. The architecture of TDSQL is shown in Fig. 2.

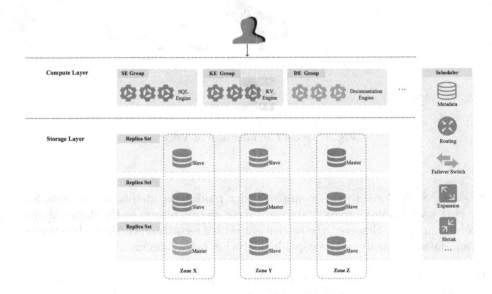

Fig. 2. TDSQL architecture.

Compute Layer. It is charge of protocol analysis, query execution and other data capture and calculation tasks. At present, we can provide multiple data model engines, such as the SQL engine, key-value engine like Redis, document engine like MongoDB, etc.

Scheduler. It is mainly responsible for metadata management, failover recovery management, routing information management and system scaling management.

Replica Set. The data is organized according to Replica Set, and there may exist multiple replications in one Set, which conduct strong synchronous replication protocol based on Raft or asynchronous replication protocol. Multiple replications can be separated in geographically dispersed data centers.

Horizontal Scalability. A logical table may correspond to one or more physical Replica Sets, the logical table shields the physical storage rules of the physical layer, the service does not care about how the data layer stores, nor does it need to integrate the split schemes or buy the middleware in the business code, just like using a centralized (single machine) database [7]. At the same time, it supports the real-time online scaling, the scaling process is completely transparent to the business. No business halt is needed, and only some of partitions are read-only in seconds (for data verification). The whole cluster will not be affected.

Distributed Transactions. TDSQL provides an algorithm to improve 2PC [16] based on Paxos protocol, i.e., Paxos submission algorithm [1,5,6,12,15], which solves the 2PC blocking problem in the distributed environment and achieves better fault tolerance. In addition, based on decentralization of distributed transaction mechanisms [4,8–11,13,14,17,19], a full robust test for all kinds of disaster fault handling and deadlock detection is done. As shown in Fig. 3.

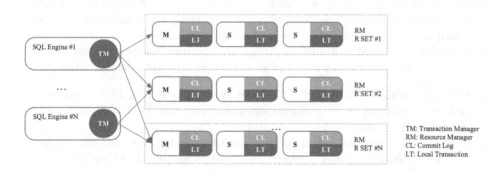

Fig. 3. TDSQL, decentralization of distributed transaction.

3   Conceptual Modeling on TDSQL

In this section, we mainly presents our conceptual modeling from the system level to support scalable and elastic 7*24 services. The models are mainly includes: (a) Fault tolerance. (b) Resource scheduling modeling. (c) Hybrid data modeling. (d) AI + Database.

3.1   Fault Tolerance

According to business requirements for data consistency and latency, we can set single-master or multi-master. Single-master provides strong consistency, which is suitable in financial systems; Multi-master takes the regional latency factors into consideration. We provide the eventually consistency in multi-master, which can be used in social networking systems.

With the cost taken into consideration, different customers and even the same customer may have different network architectures at different stages. For example, some customers may not be able to construct geographically dispersed parallel data centers in the early stage, and they will not possess as many data centers as Tencent does when they become mature companies. Therefore, we ensure the consistency and data availability, and make trade-offs in usability to achieve flexible deployment according to customers different network architectures. We achieve RPO (Recovery Point Objective) as 0 and RTO (Recovery Time Objective) as 40 s in the two deploy models in Sect. 4.

3.2   Resource Scheduling Modeling

We provide two kinds of resource management architectures: Shared-Nothing and Shared-Everything. The Shared-Nothing architecture is mainly used in the scenario of extensive large-scale services, while Shared-Everything is applied to small and medium-scale scenarios. Based on the x86 servers Shared-Nothing distributed storage architecture, customers can freely customize the memory, CPU, and disk parameters under the DBaaS cloud circumstances, which causes a lower utilization of cluster resources. Shared-Everything architecture can be implemented on distributed file system, which can achieve the separation between the data computing and data storage and greatly improve the resource utilization. Moreover, for Shared-Everything architecture, we provide three-level storage strategy which takes different access frequencies into consideration:

1. Standard storage. Its applied in on-line systems requiring high availability and high frequency access, which can basically ensure 7*24 availability even system faults happen.
2. Infrequent access storage. It is suitable for infrequently accessed data storage, generally used for disaster recovery and ensures medium-level availability.
3. Archive (Planned visit). For the purpose of archiving, we apply this storage level for long-term retention, where long-term database backups and regulatory compliance retention data can be achieved.

Differences among these three kinds of storage technology solutions are mainly about the number of copies and storage hardware. For example, in standard storage, we store hot data in 3 replications with SSD disks. In infrequent access storage, cold data is stored in ultra-large SATA disks with relative low speed. Archiving data is stored under erasure code technology instead of to traditional multi-replication strategy, which achieves the same persistence rate when reducing the number of physical storage copies. We usually use 1.3 data replications, combined with hardware customization, custom high-density storage cabinets and low-speed disks. We reduce power costs by applying programmable power-off disks with low-power board, which hibernate after being fully written. We also provide automatic conversion mechanisms for data at different storage levels under different lifecycles to ensure the lowest storage cost.

3.3   Hybrid Data Modeling

The storage layer mainly provides key-value access mode, while the compute layer implements multiple storage engines. We allow customers to specify the use of relational databases (such as MySQL), NoSQL (such as Redis) and Documents databases (such as MongoDB) when creating tables according to their business needs. Free transformation between multiple data models can be conducted based on the provided data schema. Users can access data using MySQL API, Redis API and MongoDB API respectively.

3.4   AI + Databases

With the number of database instances over one million, data security and database tuning become quite complex. We collect a large amount of operation data and use AI to enhance our database security and database intelligent diagnosis optimization. In the future, AI technology [18] will be applied in TDSQL further (Fig. 4).

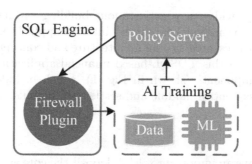

Fig. 4. AI-Security solution.

AI-Security. The SQL Engine includes a Firewall Plugin which restricts access in accordance to the specified type of query, SQL mode, user, client IP, etc. In business, access restriction rules can be pre-configured according to the usage of SQL, or the restriction relies on the malicious SQL model identified by the AI-Security. The AI-Security mainly acquires all queries recorded by the SQL Engine, and identifies whether a specific SQL is a normal business request or a malicious one, like SQL injection, drag, etc., after training. Then AI-Security extracts SQL models, converts them into rules, and appends the rules into Firewall Plugin.

AI-DBA. In terms of database maintenance, we provide intelligent diagnosis services that include rule-based diagnosis system, intelligent parameter tuning based on AI [3], etc.

In the rule-based diagnosis system, we mainly provide instance performance monitoring and overall rating, which includes:

1. Instance basic information, including instance size, location, etc.
2. Instance running status, including active thread, slow query, lock information, SQL time consumption, SQL classification and number of requests, synchronization delay, etc.
3. Resource status, such as CPU, memory, network, disk usage, etc.
4. Database parameter configuration.
5. DB and table information, including table structure, index, etc.

Based on the above information and the long-term accumulation system adjustment strategies, we can make an overall scoring assessment of the specific instance and give corresponding suggestions.

In addition, based on the collected information, the AI-DBA system performs AI training, while the DBAs need to manually label a large amount of data and perform parameter validation after training to ensure that early-stage training is a positive process.

4   Applications

In this section, we provide our practice about building various applications on the basis of TDSQL. We mainly introduce a financial application ecology built on TDSQL to show our presented system architecture and conceptual modeling are robust and scalable. In this TDSQL-based financial application solution, there are mainly four components (shown in Fig. 1), which includes order module, account module, risk-control module and settlement module.

4.1   Deployment

Account module plays an important role to deposit the core account data such as basic information, account balance, etc. Account data is highly sensitive, requires strong data consistency, multiple backups, and disaster recovery across data centers. Also, a specific user may have multiple levels of account. For example, a user will have an overall account and several distinguished accounts in different service systems. Take the overall data requirements into consideration, we designed two deploy strategies which are capable of consistent and high-available data storage (Fig. 5).

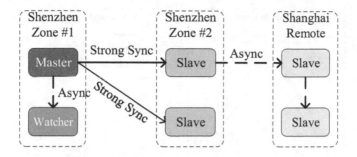

Fig. 5. Geographically dispersed Tri-data-centers architecture.

Geographically Dispersed Tri-data-centers

1. When the master data center is out of service, the cluster will be automatically switched to the standby data center to provide service, which ensures the data consistency and availability. At that time, single data center with strong synchronization is risky.
2. If there are faults in a single server, after comparing the two standby nodes and the watcher node in the same city, the system will switch to the node with the newest data to provide service. The watcher node within the same data center will be selected preferentially to minimize cross data center switching operation.
3. When the standby data center breaks down, the scheduler in another city can automatically make an election.
   (a) Once the failover of standby data center is confirmed, the watcher node in the master data center can be automatically upgraded to a slave node. And the master data center keeps providing service without disturbance during this procedure.
   (b) The same procedure will be conducted when losing network connection between master and standby data centers.

Geographically Dispersed Quad-data-centers. This strategy is much more adaptable, but also requires more data centers.

1. Deploying tri-data-centers in the same city can greatly simplify the synchronization strategy and provide high data availability and consistency.
2. The failure of single data center does not affect data services.
3. The production cluster with tri-data-centers provides multi-active service.
4. The entire city can be manually switched.

4.2   Resource Scheduling

Order module is mainly responsible for tracking user trading orders, data can be cleared regularly, so the data volume can be controlled. Settlement module is

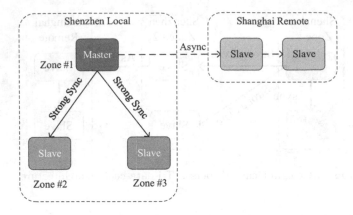

Fig. 6. Geographically dispersed Quad-data-centers architecture.

responsible for recording all historical transaction data, which should be retained for a very long time. However, there are rarely requests to access data that was produced three months or one year ago. The data in order module needs to be accessed frequently, such as querying the order information by order No. and querying all order information by the user identification. The payment data can be divided into three levels by the access frequency (Fig. 6).

Therefore, as is shown in Fig. 7, we apply three-level storage strategy to this two applications. We deposit all the order data in the standard storage and separate the payment data into three storage levels. In this way, we reduce the storage cost without impacting system performance.

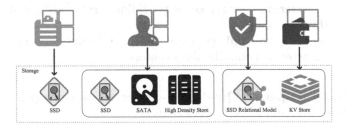

Fig. 7. Geographically dispersed Quad-data-centers architecture.

4.3   Hybrid Data Analysis

The accounts in account module are divided into personal accounts and corporate accounts. Personal accounts are mainly stored in relational model for the account volume is relatively large. Corporate accounts are stored in key-value storage for the amount of accounts is quite stable and in a medium size, but modifications on a specific account in a short period are extensive.

Hence, risk-control module needs to analyze data stored both in relational data model and key-value data model. A core risk control strategy here is based on the property transfer relationship diagram. For example, A personal account obtains illegal property from a corporate account, it can be traced unencumberedly in risk-control module based on the hybrid data access protocol that supports both relational data and key-value data.

5 Experiments

In this section, we compare TDSQL against widely-deployed, open-source databases. We perform two experiments: (a) TDSQL against CockroachDB with benchmark TPCC[1], (b) TDSQL against MySQL with benchmark sysbench[2].

5.1 TDSQL Against CockroachDB

Environment. We deploy TDSQL (TDSQL 5.7.17, with strong-sync and thread pool) and CockroachDB (v1.1.2) on a cluster of 3 servers, of which Table 1 shows the hardware configuration. The network latency among these 3 nodes is about 0.05 ms. For TDSQL, InnoDB Buffer Pool is set as 30 G, and for CockroachDB, we set cache as 90 G and max_sql_memory as 25 G.

Table 1. TDSQL vs CockroachDB, hardware configuration.

Device	Specs	Number
CPU	Intel(R) Xeon(R) CPU E5-2680 v4 @ 2.40 GHz, 14 cores	2
Memory	128 G	
Hard disk	300 G SAS	12
Network card	Intel Corporation Ethernet Controller 10-Gigabit X540-AT2	2
OS	Linux TENCENT 3.10.107-1-tlinux2-0046	

Data Sets. We run benchmark TPCC with 1000 warehouses, warm up for 500 s, and perform the experiment for 1000 s.

Figure 8 shows the result. TDSQL outperforms CockroachDB by about 70%, and as the number of connections grows, TDSQL performs increasingly better than CockroachDB.

---

[1] http://www.tpc.org/tpcc.
[2] https://github.com/akopytov/sysbench.

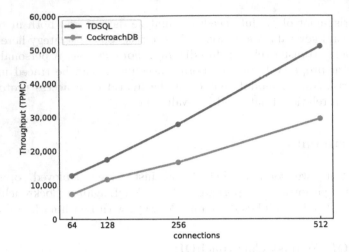

Fig. 8. TDSQL vs CockroachDB, TPCC.

5.2 TDSQL Against MySQL

Environment. We deploy MySQL (MySQL 5.7.17, community edition with semi-sync, without thread pool) and TDSQL (TDSQL 5.7.17, with strong-sync and thread pool) across 2 zones, i.e., 1 server in each zone, and Table 2 shows the hardware configuration of each server. MySQL and TDSQL run on 2 nodes in a master-slave mode, respectively. The network latency between these 2 zones is 3 ~5 ms, and the network response time is about 0.8 ms. InnoDB Buffer Pool is set as 30 G for both MySQL and TDSQL. For TDSQL, we set (a) thread_pool_max_threads as 2000, (b) thread_pool_oversubscribe as 10, (c) thread_pool_stall_limit as 50, (d) thread_handling as 2.

Table 2. TDSQL vs MySQL, hardware configuration

Device	Specs	Number
CPU	Intel(R) Xeon(R) CPU E5-2670 v3 @ 2.30 GHz, 12 cores	2
Memory	512 G	
Hard disk	1.8T SAS (RAID 5) × 4, 1.8T NVMc SSD (RAID 0) × 4	8
Network card	Intel Corporation Ethernet Controller 10-Gigabit X540-AT2	2
OS	Linux TENCENT64.site 3.10.102-1-tlinux2-0400.tl2	

Data Sets. We generate the data sets using sysbench prepare. The data sets consist of 10 tables, each of which contains 2180000 rows.

We perform benchmark sysbench (OLTP), and Fig. 9 shows the result. TDSQL outperforms MySQL by about 80%.

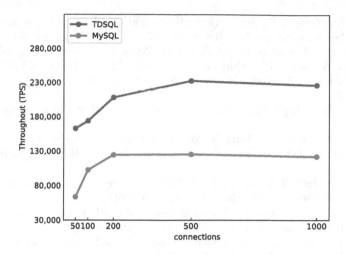

Fig. 9. TDSQL vs MySQL, sysbench (OLTP).

6   Conclusion and Future Work

With the intensification of informatization in various industries, the demand for databases in various scenarios and data sizes is different, which makes it difficult to have a universal product for all scenarios. Therefore, the multi-model and AI + database should be the trend of future development.

References

1. 3PC. https://en.wikipedia.org/wiki/Three-phase_commit_protocol
2. Tencent Distributed Database System(TDSQL). http://tdsql.org
3. Aken, D.V., Pavlo, A., Gordon, G.J., Zhang, B.: Automatic database management system tuning through large-scale machine learning. In: SIGMOD Conference, pp. 1009–1024. ACM (2017)
4. Batra, R.K., Rusinkiewicz, M., Georgakopoulos, D.: A decentralized deadlock-free concurrency control method for multidatabase transactions. In: ICDCS, pp. 72–79. IEEE Computer Society (1992)
5. Gray, J.N.: Notes on data base operating systems. In: Bayer, R., Graham, R.M., Seegmüller, G. (eds.) Operating Systems. LNCS, vol. 60, pp. 393–481. Springer, Heidelberg (1978). https://doi.org/10.1007/3-540-08755-9_9
6. Gray, J., Lamport, L.: Consensus on transaction commit. ACM Trans. Database Syst. **31**(1), 133–160 (2006)
7. Li, H., Yi Feng, P.F.: The Art of Database Transaction Processiong: Transaction Management and Concurrency Control. China Machine Press, Beijing (2017)
8. Haller, K., Schuldt, H.: Towards a decentralized implementation of transaction management. In: Grundlagen von Datenbanken, pp. 57–61. Fakultät für Informatik, Universität Magdeburg (2003)
9. Hwang, B., Son, S.H.: Decentralized transaction management in multidatabase systems. In: COMPSAC, pp. 192–198. IEEE Computer Society (1996)

10. Kang, I.E., Keefe, T.F.: Supporting reliable and atomic transaction management in multidatabase systems. In: ICDCS, pp. 457–464. IEEE Computer Society (1993)
11. Kulkarni, S.S., Demirbas, M., Madappa, D., Avva, B., Leone, M.: Logical physical clocks. In: Aguilera, M.K., Querzoni, L., Shapiro, M. (eds.) OPODIS 2014. LNCS, vol. 8878, pp. 17–32. Springer, Cham (2014). https://doi.org/10.1007/978-3-319-14472-6_2
12. Lampson, B., Sturgis, H.E.: Crash recovery in a distributed data storage system (1979)
13. Lomet, D.B., Fekete, A., Wang, R., Ward, P.: Multi-version concurrency via timestamp range conflict management. In: ICDE, pp. 714–725. IEEE Computer Society (2012)
14. Pang, G.: Scalable Transactions for Scalable Distributed Database Systems. Ph.D. thesis, University of California, Berkeley, USA (2015)
15. Skeen, D., Stonebraker, M.: A formal model of crash recovery in a distributed system. IEEE Trans. Softw. Eng. **SE-9**(3), 219–228 (1983)
16. Skeen, D., Stonebraker, M.: A formal model of crash recovery in a distributed system. IEEE Trans. Softw. Eng. **9**(3), 219–228 (1983)
17. Veijalainen, J., Wolski, A.: Prepare and commit certification for decentralized transaction management in rigorous heterogeneous multidatabases. In: ICDE, pp. 470–479. IEEE Computer Society (1992)
18. Wang, W., Zhang, M., Chen, G., Jagadish, H.V., Ooi, B.C., Tan, K.: Database meets deep learning: challenges and opportunities. SIGMOD Record **45**(2), 17–22 (2016)
19. Yu, X., Pavlo, A., Sánchez, D., Devadas, S.: Tictoc: time traveling optimistic concurrency control. In: SIGMOD Conference, pp. 1629–1642. ACM (2016)

Conceptual Modeling Studies I

A Reference Framework for Conceptual Modeling

Lois M. L. Delcambre[1], Stephen W. Liddle[2]([✉]), Oscar Pastor[3],
and Veda C. Storey[4]

[1] Portland State University, Portland, OR, USA
lmd@pdx.edu
[2] Brigham Young University, Provo, UT, USA
liddle@byu.edu
[3] Universitat Politècnica de València, Valencia, Spain
opastor@dsic.upv.es
[4] Georgia State University, Atlanta, GA, USA
vstorey@gsu.edu

Abstract. With decades of contributions and applications, conceptual modeling is very well-recognized in information systems engineering. However, the importance and relevance of conceptual modeling is less well understood in other disciplines. This paper, through an analysis of existing research and expert opinions, proposes a reference framework for conceptual modeling to help researchers and practitioners position their work in the field, facilitate discussion among researchers in the field, and help researchers and practitioners in other fields understand what the field of conceptual modeling has to offer as well as contribute to its continued, extended influence in multiple domains.

Keywords: Conceptual modeling · Entity-relationship model
Ontology · Constructs · Conceptual modeling languages
Implementation · Reference framework for conceptual modeling

1 Introduction

The impressive capability of conceptualizing is what make humans different from any other species on our planet. Conceptual modeling focuses on capturing the products of human conceptualization and managing them effectively by identifying which modeling primitives should be used to directly express relevant "real world" concepts. Representing reality (the real-world domain of interest) accurately in a computer-based context is critically important for developers of information systems and for other stakeholders pursuing a variety of goals. Conceptual models seek to represent relevant conceptualizations and abstractions of the real world in such a manner that it is possible to support communication, discussion, analysis, and related activities.

© Springer Nature Switzerland AG 2018
J. C. Trujillo et al. (Eds.): ER 2018, LNCS 11157, pp. 27–42, 2018.
https://doi.org/10.1007/978-3-030-00847-5_4

The field of conceptual modeling emerged in the early era of databases and software engineering. Since then it has broadened and matured, with contributions from requirements engineering [56], knowledge representation, philosophy, ontology development [29,51], and applications (e.g. [54,64]). There are many kinds of conceptual models (e.g. [13,18,33,34,53]) with different notations, used for many purposes (e.g. [14,20,48–50,52]).[1]

Analyzing the different definitions that have been proposed for conceptual modeling reveals that there is a reasonably common intuition about what conceptual modeling is. However, there is a lack of a standard, widely-accepted definition of conceptual modeling [80]. There are also different roles that conceptual modeling plays in informations systems engineering [23]. A conceptual model is often referred to as a simplification of a system [4]; an abstraction of a system to support reasoning [8,43]; a specification of a system and its environment (e.g. [56,73]); or a model of concepts (e.g. [62]). Even the distinction between "model" and "conceptual model" is not as clear and precise as it should be; different, sometimes-conflicting perspectives abound [40,78,79].

Although there is little description about the relationships among modeling artifacts [55] and no widely accepted single definition [47], we do not attempt here to provide a universal definition. Rather, we propose a framework to characterize the field of conceptual modeling to:

- Acknowledge ways in which various disciplines have contributed to the field.
- Allow researchers to articulate how their contributions fit within the field.
- Promote debate and discussion within the conceptual modeling community regarding contributions to the field and to identify future research directions.
- Invite researchers in other fields to benefit from conceptual modeling and work to further enrich it.
- Invite practitioners new to conceptual modeling to apply and extend the field's rich contributions.
- Promote the recognition of conceptual modeling as an active field of research within the computing disciplines.

Our framework was derived by: (1) analyzing a number of published definitions of conceptual modeling, and (2) inviting experts with differing backgrounds (computer science, information systems, ontology, philosophy) to answer four questions[2] both by e-mail and in direct conversation at an interactive session where approximately 30 participants engaged in lively discussions.[3] The contribution of this paper is to propose and evaluate our framework to characterize conceptual modeling.

---

[1] Because we seek to give a broad characterization of the field of conceptual modeling, we cite an unusually large and diverse number of papers. Even so, we can only cite a fraction of the high quality related work done by influential researchers.

[2] These questions were: Who are the developers of conceptual models? Who are the users of conceptual models? What is represented in a conceptual model? and What is the purpose of a conceptual model?.

[3] Interactive Session on Definitions of Conceptual Modeling, held at ER 2017.

Section 2 gives an overview of how various research fields have contributed to conceptual modeling. Section 3 summarizes the input obtained from the researchers who participated in our interactive session. In Sect. 4, we introduce our reference framework for conceptual modeling which we evaluate using examples in Sect. 5. Section 6 concludes the paper, presenting areas for future work.

2 Historical Context for Conceptual Modeling

Early work in conceptual modeling was centered on designing a database schema at a more abstract level than the data models used in database management systems (e.g. [1,3,21,32]). Among the first and most widely acknowledged conceptual modeling languages is the Entity-Relationship (ER) model [13]. Many extensions to the ER model were proposed (e.g. [17,35,77]), including extensions for temporal/spatial aspects (e.g. [19,38,61]). Others developed systematic mappings from a conceptual model to lower-level database models such as the relational model (e.g. [74,76]). Conceptual modeling was further developed when database, artificial intelligence, and programming language researchers observed interconnections among their fields [6,7,16]. Others observed a connection between NLP and conceptual modeling [24].

Software engineering (SE) is concerned with producing quality software, typically based on conceptual models of the system to be built. In a typical description, "Conceptual modeling is about describing the semantics of software applications at a high level of abstraction" [23]. In information systems engineering (ISE), conceptual modeling is "the activity that elicits and describes the general knowledge a particular information system needs to know," which is also called a *conceptual schema* [58] that consists of objects, relationships between them, and events. Note that ISE focuses on correctly representing the real world under consideration even when generating software is not the only (or primary) goal.

From a business perspective, conceptual modeling can be viewed as business process modeling (e.g. [20,42,52,66,68,82]), goal modeling (e.g. [37,83]), or enterprise modeling. Enterprise modeling attempts to make explicit knowledge of an enterprise so that it can be used in strategic decision making, developing information technologies to support the organization, communicating, and so forth [5,25,26,46,71]. In artificial intelligence, conceptual modeling can be viewed as the problem of capturing and representing expert knowledge in a form that supports reasoning (e.g. in a Semantic Web context using RDF, OWL, and XML to capture data semantics in an interchangeable way; see e.g. [41]). From a programming language perspective, conceptual modeling (even though it is implementation-independent) can be viewed as describing a system in sufficient detail so that the model can be automatically compiled into an executable system [22,63].

An oft-cited definition describes conceptual modeling as the "activity of formally describing some aspects of the physical and social world around us for purposes of understanding and communication. Conceptual modeling supports structuring and inferential facilities that are psychologically grounded. After all, the descriptions that arise from conceptual modeling activities are intended to be used by humans, not machines" [56]. This definition serves to broaden the perspective of conceptual modeling beyond database and information systems development to the general task of describing and understanding a domain.

Conceptual modeling also has foundations in mathematics (especially set theory and logic) and philosophy (particularly ontology) including the formal definition of conceptual modeling languages as foundational ontologies [9,28, 30,51]. Olivé [59,60] introduces the vision of a universal ontology intended to provide a shared solution to the semantic integration problem in the field of conceptual modeling and to the understandability problem in the field of the Semantic Web.

Over the years, researchers from various communities have offered many definitions of *conceptual modeling* depending upon their distinct disciplinary perspectives [10,12,68–70,72,79]. Thalheim [79] seeks to provide an appropriate definition for conceptual modeling (given its diversity) and to identify research opportunities in the field.

3   A View from the Community of Researchers

Our ER 2017 session featured lively discussion and much interest in this work. Participants defined a *conceptual model* as:

- a representation of reality (but most experts found that definition to be too broad), or
- a representation of what someone has in their mind for a domain; someone's conceptualization of a domain

Some participants expressed the view that to be "conceptual," a model must capture "concepts," which necessarily exist in a human mind, and therefore conceptual modeling cannot be done without human involvement. Participants also stated that a conceptual model:

- should be at a high enough level of abstraction;
- must have explicit semantics (provided in a variety of ways);
- is created and used by people; and
- requires a purpose.

That is, conceptual modeling is an intentional activity that has scope, granularity, and (perhaps multiple) levels of abstractions.

Participants spent most of their time discussing the purpose of a conceptual model. The purpose of a conceptual model is to:

- answer questions about a domain; improve understanding; and promote knowledge sharing;
- expose (the author of a conceptual model's) assumptions about a domain (that can then be confirmed or refuted by others);

- promote communication among people developing a conceptual model, or among people who (later) use a conceptual model; promote dialog or negotiation between two people with different views of the domain (or of "reality");
- support intervention among people when they extend an existing conceptual model (as part of an existing system);
- document the conceptual model for an existing or future system;
- standardize a conceptual model to enable interoperability of systems;
- prescribe the conceptual model for a system (to be built);
- provide normative content; and
- support action.

The final purpose (to support action) suggests that conceptual model development is not done in a vacuum. Nor is it simply to promote communication and understanding. Conceptual modeling is done with a specific goal such as beginning to build or modify a system or support brainstorming.

Participants agreed that language designers and researchers are the developers of conceptual modeling languages, although nearly anyone can develop or use a conceptual model, including: users who want to modify the current conceptual model, people engaged in a dialog about the current conceptual model/system, anyone with whom one communicates using a conceptual model, anyone comparing/evaluating alternative conceptual models, stakeholders, developers, and, ultimately, everyone who uses the system once it is built because they implicitly are exposed to and must learn the conceptual model inherent in the system. When asked what can be done with a conceptual model, responses included: validate it, simulate it, throw it away, and generate code and/or a database schema from it as part of building a system.

4   A Framework to Characterize the Field of Conceptual Modeling

We assume that conceptual modeling research projects/papers are related in some manner to one or more conceptual modeling languages and/or conceptual model(s) that are expressed using a conceptual modeling language.

Definition 1. A *conceptual model* is an artifact (also called a schema, script, document, diagram, ...) that is written using a conceptual modeling language.

Definition 2. A *conceptual modeling language* (sometimes confusingly called a conceptual model, but also called a data model, foundational ontology, ...) defines constructs that can be instantiated and named in any conceptual model defined using this conceptual modeling language as well as the grammar describing how these constructs can be combined. The conceptual modeling language may allow the instantiated, named constructs to be annotated with labels or constraints (e.g. role, cardinality constraints, weak entity).

We use the term *scenario under consideration* for the scope of a conceptual model which may be large/small, simple/complex, associated with (a perhaps future) implementation or not. For a scenario under consideration, a conceptual model may describe:

- **Information of interest**; concepts or classes in the real world (i.e. the scenario under consideration). The conceptual model of information of interest describes and names the type/structure of data instances that will arise if the conceptual model is populated. The conceptual model of information of interest may include instances/values and may describe constraints on the data instances that are to be created, modified, or deleted.
- **Events of interest** that occur in the scenario under consideration and abstract the most important aspects of a process that need to be captured and represented.
- **Processes of interest** that describe the flow of processing steps/event generation that comprise the behavior of the (intended) system. Process descriptions in a conceptual model can be instantiated whenever the appropriate process occurs in the scenario under consideration. Information that tracks or records such process instances may be included in the information of interest of the conceptual model (described just above).
- **Interactions of interest** between humans and the systems being modeled. End-user interaction requirements state user preferences and contexts of use. The interaction perspective complements the well-understood perspectives of data and behavior (events and processes) modeling [2].

A conceptual modeling language may include a means to delineate the boundary/scope of parts of the conceptual model to support modularity.

A conceptual model (whether describing information, processes, events, or interactions) may describe any or all of the following:

- the scenario of interest
- the context (of the scenario of interest)
- the goals (of the scenario of interest)

In the past few decades, the conceptual modeling field has benefited greatly from the application/adoption of ideas and models from work in philosophy on ontology/ies. We do not attempt to summarize the range of efforts that have taken place nor attempt to define the term ontology. We observe that in the conceptual modeling literature, the term ontology is sometimes used to describe a conceptual modeling language (e.g. a foundational ontology [31]) and is sometimes used to describe a conceptual model (e.g. a domain-specific ontology [81]).

Activities. There are a number of activities associated with conceptual modeling languages and the conceptual models written in these languages, as Fig. 1 summarizes. In Fig. 1, gold shading represents conceptual modeling languages, orange shading represents conceptual models, and gray shading represents implemented systems.

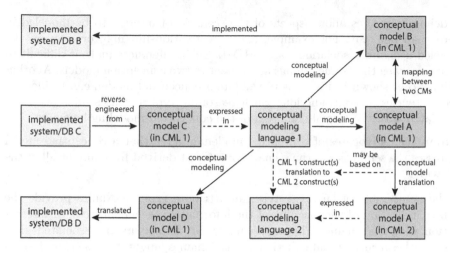

Fig. 1. Activities involving conceptual models

The most basic activity is conceptual modeling: defining a conceptual model using a conceptual modeling language as shown in the figure (conceptual models A, B, and D are defined using conceptual modeling language 1). Another common activity is for a conceptual model to be implemented, as shown at the top of the figure with conceptual model B. Conversely, a conceptual model can be reverse-engineered from an implemented system as shown in the figure on the left where implemented system/DB C is reverse engineered to create conceptual model C [15,39,44]. Another activity is defining or extending a conceptual modeling language (not shown in the figure).

Conceptual model translation is when a conceptual model in one conceptual modeling language is translated into a conceptual model in another conceptual modeling language; Fig. 1 shows conceptual model A in conceptual modeling language 1 being translated into conceptual model A in conceptual modeling language 2. Such a translation, when algorithmic, is often specified at the conceptual modeling language construct level. An entity in the entity-relationship conceptual modeling language might be translated into a relation in the relational data model, for example. A translation may be lossy; there may be constructs in the source conceptual modeling language for which there are no counterparts in the other. Automatic translation of a conceptual model to an implementation is called conceptual-model programming or model-driven software development [22,45,57,63]. Figure 1 shows this in the lower left where conceptual model D is translated into an implementation. As an example, a conceptual schema can be reverse engineered from a database and then automatically translated to a data warehouse schema [65].

It is also interesting to map from one conceptual model to another, e.g. to support information integration. In Fig. 1, conceptual model A is mapped to conceptual model B (both expressed in conceptual modeling language 1). Note that translation and mapping are both difficult because most conceptual modeling languages allow aspects of the scenario of interest to be modeled in more than one way. For example, a binary relationship may or may not be modeled using an association class in UML. Such differences make it difficult to recognize when the "same" things are present in two conceptual models. Another activity (not shown in Fig. 1) is to validate a conceptual model, e.g. to identify inconsistencies or contradictions. Such reasoning typically uses inference rules associated with the formal definition of the conceptual modeling language.

Semantics. Advocates of conceptual modeling agree that a conceptual model expresses the semantics of the scenario of interest derived from any or all of the following:

- the names, e.g. of an entity type, an attribute type, ... Names provide the first indication of the meaning of the information, event, process, or interaction with that name. The names might be defined in an associated glossary, thesaurus, or data dictionary; the names might be considered standard/normative, e.g. because they appear in a standard conceptual model; or they might be (just) names.
- the structure, i.e. the constellation of related, named constructs that appear with this named construct
- the constraints on data (that may populate a conceptual model)
- comments or other documentation that may appear in the conceptual model

One important way to convey the semantics of a conceptual model is to map it to a second conceptual model (e.g. ontology) that is considered to be standard/normative [11,27,75]. When attempting to map from one conceptual model to another it is possible for conflict/differences to occur in all three aspects of semantics: the names, the structure (e.g. leading to what is called structural conflict in the schema mapping literature), and constraints (e.g. conflicting cardinalities). A conceptual model may be developed from a standard or idealized conceptual model and thus have a mapping to this standard conceptual model from the outset. Such standard conceptual models have been called (simply) conceptual models, domain-specific ontologies, domain-related ontology patterns [67], (data model) patterns [36], and seed (data) models.

Dimensions. Conceptual models can be characterized by what is being modeled: information, events, processes, or interactions. In addition, we define several dimensions to characterize conceptual modeling languages (and accordingly, conceptual models written in those languages). Figure 2 illustrates these dimensions.

(1) Timing of Conceptual Modeling: At one end of the spectrum (the most common case) is when a conceptual model is defined first—ahead of any implementation of the model. At the other end of the spectrum is when a conceptual model is extracted from an implementation, i.e. using reverse engineering.

(2) Support for Translation: Regardless of the timing of conceptual modeling, there may or may not be support for automated/algorithmic translation. For conceptual-modeling-first approaches, translation is from a conceptual model to a more implementation-oriented model. For reverse-engineering, there may exist tools to assist with the extraction of a conceptual model from an implementation

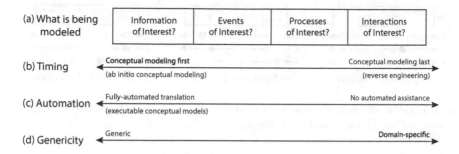

(a) What is being modeled	Information of Interest?	Events of Interest?	Processes of Interest?	Interactions of Interest?

Fig. 2. Dimensions of the reference framework for conceptual modeling

or to translate from a lower-level conceptual model to a higher-level conceptual model. The closer the translation (in either direction) is to an algorithmic process, the more likely there will be a well-defined correspondence or mapping between the source conceptual model and the target conceptual model. When a conceptual model is developed first and there is no support for translation to lower-level models, then the conceptual model serves as documentation and there is a risk that it may not be kept up to date with the implemented system.

(3) Genericity: Conceptual modeling languages may be generic and thus intended for use in any application domain or they may be domain- or application-specific.

Conceptual modeling languages and conceptual models can be characterized according to their purpose and by their intended users, as discussed in Sect. 3 above.

5   Evaluations of the Framework

To evaluate our initial framework, we examine how three scenarios (of many possible scenarios) can be described in the framework.

Scenario 1. Typically, database design begins with requirements analysis followed by conceptual modeling such as drawing an ER (or extended ER) diagram. The developer then generates a relational schema from the conceptual model. In this scenario, what is being modeled is entities, their attributes, and relationships. Events, processes, and interactions are usually ignored. Timing is mostly *ab initio*, with conceptual modeling as a very early activity. However, when the relational schema is updated in later phases of the systems development life cycle (e.g. for performance tuning or to support new features), conceptual modeling can also be done late if the developer chooses to keep the conceptual model in sync with the implemented system. The conceptual modeling language is generic, not specific to any domain. Figure 3(a) illustrates the framework dimensions used in this scenario. Solid dots in Fig. 3 represent primary activities, and dashed dots represent secondary activities, along each dimension.

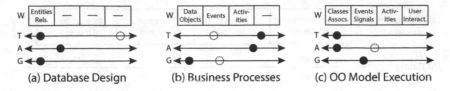

(a) Database Design   (b) Business Processes   (c) OO Model Execution

Fig. 3. Framework dimensions as used in evaluation scenarios (W = what, T = timing, A = automation, G = genericity, see Fig. 2)

Scenario 2. A business analyst may model existing processes and design new (or re-design existing) processes using BPMN as the conceptual modeling language. The resulting business process model includes the events, activities, gateways, and connections that describe the organization's actual or proposed practices. In this scenario, user/system interactions are not generally modeled in detail, though some processes may be designated as manual. The conceptual modeling is usually done relatively late, after an organization has already been operating for some time. Also, often only larger organizations invest the considerable effort required to document their processes in detail, e.g. to support process improvement. Business process modeling generally involves a lot of manual work. Tools support BPMN diagramming and the simulated execution of BPMN models so that analysts can evaluate various design alternatives. The conceptual modeling language in this case is generic. However, there are a number of repositories of existing process models, and it is common for analysts to consult and reuse prior models. Figure 3(b) illustrates these dimensions.

Scenario 3. An object-oriented conceptual model with sufficient detail regarding data, behavior, and interaction can be used to generate a running system directly. For example, OO-Method and the Integranova Model Execution Engine let developers specify an information system in an object-oriented conceptual model and then translate the model into one of several possible system architectures such as client/server or web-based [63]. OO-Method models use classes, associations, events, triggers, activities, and, notably, end-user interaction specifications. The resulting system can be compiled and executed as-is, but more commonly developers tweak the translated output before deploying the system.

This scenario essentially requires that conceptual modeling come first. Translation is fully automatic, though there is manual updating of the generated code base, usually in user interface code, because graphical user interface and user experience design vary significantly from one platform to another. The conceptual modeling language (often UML) is generic, but the approach narrowly suits business information systems rather than the broad field of computing in general. Figure 3(c) illustrates these dimensions graphically.

6   Conclusion

This paper presents an initial framework to describe the field of conceptual modeling with the goals of allowing researchers to place their work in the field, further articulating conceptual modeling as a field of study, and inviting researchers and practitioners in other fields to extend and apply conceptual modeling.

The work could be extended by considering the relationship between general vs. conceptual models/modeling. Conceptual modeling is likely a universal process for conceptualization and its representation that could be used in any research methodology for models and modeling. The framework could be further developed by including a discussion of the features/constructs in the range of conceptual modeling languages that have been proposed. Conceptual modeling languages could also be characterized according to how foundational they are (e.g. to what extent they offer constructs for a conceptual modeling language vs. a conceptual model). We expect that researchers in cognitive science and philosophy will continue their work in investigating which conceptualizations are appropriate for conceptual modeling languages. Such work may investigate the role of purely mathematical constructs (e.g., from set theory) in addition to the use of cognitive abstractions (e.g., generalization or meronomy).

Acknowledgments. We express deep gratitude to colleagues who responded (often with extensive material) to our email requests for their views on questions regarding conceptual modeling. Alphabetically by first name they include: Antoni Olivé, Avi Gal, Bernhard Thalheim, Carlo Batini, Carson Woo, Chris Partridge, David Embley, Dennis McLeod, Dov Dori, Eric Yu, Giancarlo Guizzardi, Heinrich Mayr, Hui Ma, Isabelle Comyn-Wattiau, Jacky Akoka, Jean-Luc Hainaut, Jeff Parsons, Jennifer Horkoff, John Mylopoulos, Jordi Cabot, José Palazzo, Juan Trujillo, Karen Davis, Klaus-Dieter Schewe, Leah Wong, Motoshi Saeki, Nicola Guarino, Peter Scheuermann, Ramez Elmasri, Rick Snodgrass, Roger Chiang, Roland Kaschek, Ron Weber, Sal March, Samira Cherfi, Silvana Castano, Stefano Rizzi, Susan Urban, Terry Halpin, Toby Teorey, Valeria De Antonellis, Yair Wand, and Zoubida Kedad. We are also grateful to the participants in our ER 2017 interactive session.

References

1. Antonellis, V.D., Leva, A.D.: DATAID-1: a database design methodology. Inf. Syst. **10**(2), 181–195 (1985)
2. Aquino, N., Vanderdonckt, J., Panach, J.I., Pastor, O.: Conceptual modelling of interaction. In: Embley, D., Thalheim, B. (eds.) Handbook of Conceptual Modeling-Theory, Practice, and Research Challenges, pp. 335–358. Springer, Heidelberg (2011). https://doi.org/10.1007/978-3-642-15865-0_10
3. Batini, C., Ceri, S., Navathe, S.B.: Conceptual Database Design: An Entity-Relationship Approach. Benjamin/Cummings, Redwood City (1992)
4. Bézivin, J., Gerbé, O.: Towards a precise definition of the OMG/MDA framework. Proc. ASE **2001**, 273–280 (2001)
5. Bjekovic, M., Sottet, J., Favre, J., Proper, H.A.: A framework for natural enterprise modelling. In: IEEE 15th Conference on Business Informatics, CBI 2013, Vienna, Austria, July 15–18, 2013, pp. 79–84 (2013)
6. Brodie, M., Mylopoulos, J., Schmidt, J. (eds.): On Conceptual Modelling, Perspectives from Artificial Intelligence, Databases, and Programming Languages. Topics in Inf. Systems. Springer, New York (1984). https://doi.org/10.1007/978-1-4612-5196-5

7. Brodie, M., Zilles, S.: Proceedings of the workshop on data abstraction, databases and conceptual modelling, pingree park, colorado, june 23–26, 1980. SIGMOD Record 11(2) (1981)
8. Brown, A.W.: Model driven architecture: principles and practice. Softw. Syst. Model. 3(4), 314–327 (2004)
9. Burton-Jones, A., Recker, J., Indulska, M., Green, P.F., Weber, R.: Assessing representation theory with a framework for pursuing success and failure. MIS Q. 41(4), 1307–1333 (2017)
10. Cabot, J., Gómez, C., Pastor, O., Sancho, M., Teniente, E. (eds.): Conceptual Modeling Perspectives. Springer, Cham (2017). https://doi.org/10.1007/978-3-319-67271-7
11. Castano, S., Ferrara, A., Montanelli, S.: Matching ontologies in open networked systems: techniques and applications. In: Spaccapietra, S., Atzeni, P., Chu, W.W., Catarci, T., Sycara, K.P. (eds.) Journal on Data Semantics V. LNCS, vol. 3870, pp. 25–63. Springer, Heidelberg (2006). https://doi.org/10.1007/11617808_2
12. Chen, P.P., Thalheim, B., Wong, L.Y.: Future directions of conceptual modeling. In: Goos, G., Hartmanis, J., van Leeuwen, J., Chen, P.P., Akoka, J., Kangassalu, H., Thalheim, B. (eds.) Conceptual Modeling. LNCS, vol. 1565, pp. 287–301. Springer, Heidelberg (1999). https://doi.org/10.1007/3-540-48854-5_23
13. Chen, P.: The entity-relationship model–toward a unified view of data. ACM Trans. Database Syst. 1(1), 9–36 (1976)
14. Cherfi, S.S.-S., Comyn-Wattiau, I., Akoka, J.: Quality patterns for conceptual modelling. In: Li, Q., Spaccapietra, S., Yu, E., Olivé, A. (eds.) ER 2008. LNCS, vol. 5231, pp. 142–153. Springer, Heidelberg (2008). https://doi.org/10.1007/978-3-540-87877-3_12
15. Chiang, R.H.L., Barron, T.M., Storey, V.C.: Reverse engineering of relational databases: extraction of an EER model from a relational database. Data Knowl. Eng. 12(2), 107–142 (1994)
16. Organick, E.I. (ed.) Proceedings of the 1976 conference on Data Abstraction, definition and structure. ACM, New York, NY 22–24 March 1976. (1925–1985, Elliott Irving) held at Salt Lake City, UT
17. Dietrich, S.W., Urban, S.D.: Fundamentals of Object Databases: Object-Oriented and Object-Relational Design, Synthesis Lectures on Data Management. Morgan & Claypool Publishers (2010)
18. Dori, D.: Model-Based Systems Engineering with OPM and SysML. Springer, New York (2016). https://doi.org/10.1007/978-1-4939-3295-5
19. Edelweiss, N., de Oliveira, J.P.M., Pernici, B.: An object-oriented approach to a temporal query language. In: Karagiannis, D. (ed.) DEXA 1994. LNCS, vol. 856, pp. 225–235. Springer, Heidelberg (1994). https://doi.org/10.1007/3-540-58435-8_187
20. Elahi, G., Yu, E.S.K.: Modeling and analysis of security trade-offs - a goal oriented approach. Data Knowl. Eng. 68(7), 579–598 (2009)
21. Elmasri, R., Navathe, S.B.: Fundamentals of Database Systems, 7th edn. Addison-Wesley-Longman, Boston (2016)
22. Embley, D.W., Liddle, S.W., Pastor, O.: Conceptual-Model Programming: A Manifesto. In: Embley, D., Thalheim, B. (eds.) Handbook of Conceptual Modeling—Theory, Practice, and Research, pp. 3–16. Springer, Heidelberg (2011). https://doi.org/10.1007/978-3-642-15865-0_1
23. Embley, D., Thalheim, B. (eds.): Handbook of Conceptual Modeling-Theory, Practice, and Research Challenges. Springer, Heidelberg (2011). https://doi.org/10.1007/978-3-642-15865-0

24. Fliedl, G., Kop, C., Mayr, H.C.: From textual scenarios to a conceptual schema. Data Knowl. Eng. **55**(1), 20–37 (2005)
25. Frank, U.: Multi-perspective enterprise modeling: foundational concepts, prospects and future research challenges. Softw. Syst. Model. **13**(3), 941–962 (2014)
26. Frank, U., Loucopoulos, P., Pastor, Ó., Petrounias, I. (eds.): PoEM 2014. LNBIP, vol. 197. Springer, Heidelberg (2014). https://doi.org/10.1007/978-3-662-45501-2
27. Gonçalves, B., Guizzardi, G., Filho, J.G.P.: Using an ecg reference ontology for semantic interoperability of ECG data. J. Biomed. Inform. **44**(1), 126–136 (2011)
28. Guarino, N.: Formal ontology and information systems. In: Proceedings of FOIS 1998, pp. 3–15 (1998)
29. Guizzardi, G.: On ontology, ontologies, conceptualizations, modeling languages, and (meta)models. In: Databases and Information Systems IV - Selected Papers from the Seventh International Baltic Conference, DB&IS 2006, July 3–6, 2006, Vilnius, Lithuania, pp. 18–39 (2006)
30. Guizzardi, G., Halpin, T.A.: Ontological foundations for conceptual modelling. Appl. Ontol. **3**(1–2), 1–12 (2008)
31. Guizzardi, G., Wagner, G., Almeida, J.P.A., Guizzardi, R.S.S.: Towards ontological foundations for conceptual modeling: the unified foundational ontology (UFO) story. Appl. Ontol. **10**(3–4), 259–271 (2015)
32. Hainaut, J.: Entity-relationship models: Formal specification and comparision. In: Proceedings of the 9th International Conference on Entity-Relationship Approach (ER 1990), 8–10 October, 1990, Lausanne, Switzerland, pp. 53–64 (1990)
33. Halpin, T.: Conceptual Schema & Relational Database Design, 2nd edn. Prentice-Hall of Australia Pty. Ltd., Sydney (1995)
34. Halpin, T.A., Morgan, T.: Information modeling and relational databases, 2nd edn. Morgan Kaufmann, San Francisco (2008)
35. Hammer, M., McLeod, D.: The semantic data model: A modelling mechanism for data base applications. In: Proceedings of the 1978 ACM SIGMOD International Conference on Management of Data, Austin, Texas, May 31 - June 2, 1978, pp. 26–36 (1978)
36. Hay, D.C.: Data Model Patterns: Conventions of Thought. Dorset House Publishing, New York (1995)
37. Horkoff, J., et al.: Strategic business modeling: representation and reasoning. Softw. Syst. Model. **13**(3), 1015–1041 (2014)
38. Jensen, C.S., Soo, M.D., Snodgrass, R.T.: Unifying temporal data models via a conceptual model. Inf. Syst. **19**(7), 513–547 (1994)
39. Johannesson, P.: A method for transforming relational schemas into conceptual schemas. In: Proceedings of the Tenth International Conference on Data Engineering, February 14–18, 1994, Houston, Texas, USA, pp. 190–201 (1994)
40. Kaschek, R.: 20 years after: what in fact is a model? EMISA J., pp. 28–34, February 2018
41. Kellou-Menouer, K., Kedad, Z.: Schema discovery in RDF data sources. In: Johannesson, P., Lee, M.L., Liddle, S.W., Opdahl, A.L., López, Ó.P. (eds.) ER 2015. LNCS, vol. 9381, pp. 481–495. Springer, Cham (2015). https://doi.org/10.1007/978-3-319-25264-3_36
42. Krogstie, J.: Quality in Business Process Modeling. Springer, Cham (2016). https://doi.org/10.1007/978-3-319-42512-2
43. Kühne, T.: Matters of (meta-)modeling. Softw. Syst. Model. **5**(4), 369–385 (2006)
44. Lammari, N., Comyn-Wattiau, I., Akoka, J.: Extracting generalization hierarchies from relational databases: a reverse engineering approach. Data Knowl. Eng. **63**(2), 568–589 (2007)

45. Liddle, S.: Model-driven software development. In: Embley, D.W., Thalheim, B. (eds.) Handbook of Conceptual Modeling-Theory, Practice, and Research Challenges, pp. 17–54. Springer, Heidelberg (2011). https://doi.org/10.1007/978-3-642-15865-0

46. Lin, Y., Strasunskas, D., Hakkarainen, S., Krogstie, J., Solvberg, A.: Semantic annotation framework to manage semantic heterogeneity of process models. In: Dubois, E., Pohl, K. (eds.) CAiSE 2006. LNCS, vol. 4001, pp. 433–446. Springer, Heidelberg (2006). https://doi.org/10.1007/11767138_29

47. Ludewig, J.: Models in software engineering. Softw. Syst. Model. **2**(1), 5–14 (2003)

48. Lukyanenko, R., Parsons, J.: Lightweight conceptual modeling for crowdsourcing. In: Ng, W., Storey, V.C., Trujillo, J.C. (eds.) ER 2013. LNCS, vol. 8217, pp. 508–511. Springer, Heidelberg (2013). https://doi.org/10.1007/978-3-642-41924-9_46

49. Ma, H., Noack, R., Schewe, K., Thalheim, B.: Using meta-structures in database design. Informatica (Slovenia) **34**(3), 387–403 (2010)

50. Ma, H., Schewe, K.-D., Thalheim, B.: Geometrically enhanced conceptual modelling. In: Laender, A.H.F., Castano, S., Dayal, U., Casati, F., de Oliveira, J.P.M. (eds.) ER 2009. LNCS, vol. 5829, pp. 219–233. Springer, Heidelberg (2009). https://doi.org/10.1007/978-3-642-04840-1_18

51. March, S.T., Allen, G.N.: Toward a social ontology for conceptual modeling. CAIS **34**, 70 (2014)

52. Maté, A., Trujillo, J., Mylopoulos, J.: Key performance indicator elicitation and selection through conceptual modelling. In: Comyn-Wattiau, I., Tanaka, K., Song, I.-Y., Yamamoto, S., Saeki, M. (eds.) ER 2016. LNCS, vol. 9974, pp. 73–80. Springer, Cham (2016). https://doi.org/10.1007/978-3-319-46397-1_6

53. Mellor, S., Balcer, M.: Executable UML: A Foundation for Model-Driven Architectures. Addison-Wesley Longman Publishing Co. Inc, Boston (2002)

54. Michael, J., Mayr, H.C.: Conceptual modeling for ambient assistance. In: Ng, W., Storey, V.C., Trujillo, J.C. (eds.) ER 2013. LNCS, vol. 8217, pp. 403–413. Springer, Heidelberg (2013). https://doi.org/10.1007/978-3-642-41924-9_33

55. Muller, P., Fondement, F., Baudry, B.: Modeling modeling. In: Proceedings of the 12th International Conference, Model driven engineering languages and systems. MODELS 2009, Denver, CO, USA, October 4–9, 2009, pp. 2–16 (2009)

56. Mylopoulos, J.: Conceptual modeling and Telos. In: Conceptual Modelling, Databases and CASE: An Integrated View of Information Systems Development, pp. 49–68. Wiley, New York (1992)

57. Olivé, A.: Conceptual schema-centric development: a grand challenge for information systems research. In: CAiSE, pp. 1–15 (2005)

58. Olivé, A.: Conceptual Modeling of Information Systems. Springer, Berlin (2007). https://doi.org/10.1007/978-3-540-39390-0

59. Olivé, A.: The universal ontology: a vision for conceptual modeling and the semantic web (Invited Paper). In: Mayr, H.C., Guizzardi, G., Ma, H., Pastor, O. (eds.) ER 2017. LNCS, vol. 10650, pp. 1–17. Springer, Cham (2017). https://doi.org/10.1007/978-3-319-69904-2_1

60. Olivé, A.: A universal ontology-based approach to data integration. EMISA J. **13**, 110–119 (2018)

61. Parent, C., Spaccapietra, S., Zimányi, E.: Conceptual Modeling for Traditional and Spatio-Temporal Applications - The MADS Approach. Springer, Heidelberg (2006). https://doi.org/10.1007/3-540-30326-X

62. Partridge, C., Gonzalez-Perez, C., Henderson-Sellers, B.: Are conceptual models concept models? Proc. ER **2013**, 96–105 (2013)

63. Pastor, O., Molina, J.: Model-Driven Architecture in Practice: A Software Production Environment Based on Conceptual Modeling. Springer, New York (2007). https://doi.org/10.1007/978-3-540-71868-0
64. Pastor, O.: Conceptual modeling of life: beyond the homo sapiens. In: Comyn-Wattiau, I., Tanaka, K., Song, I.-Y., Yamamoto, S., Saeki, M. (eds.) ER 2016. LNCS, vol. 9974, pp. 18–31. Springer, Cham (2016). https://doi.org/10.1007/978-3-319-46397-1_2
65. Phipps, C., Davis, K.C.: Automating data warehouse conceptual schema design and evaluation. In: Design and Management of Data Warehouses 2002, Proceedings of the 4th International Workshop DMDW 2002, Toronto, Canada, May 27, 2002, pp. 23–32 (2002)
66. Pichler, H., Eder, J.: Business process modelling and workflow design. In: Embley, D., Thalheim, B. (eds.) Handbook of Conceptual Modeling - Theory, Practice, and Research Challenges, pp. 259–286. Springer, Heidelberg (2011). https://doi.org/10.1007/978-3-642-15865-0_8
67. Presutti, V., Lodi, G., Nuzzolese, A., Gangemi, A., Peroni, S., Asprino, L.: The role of ontology design patterns in linked data projects. In: Comyn-Wattiau, I., Tanaka, K., Song, I.-Y., Yamamoto, S., Saeki, M. (eds.) ER 2016. LNCS, vol. 9974, pp. 113–121. Springer, Cham (2016). https://doi.org/10.1007/978-3-319-46397-1_9
68. Recker, J., Rosemann, M., Indulska, M., Green, P.: Business process modeling-a comparative analysis. J. Assoc. Inf. Syst. 10(4), 1 (2009)
69. Rizzi, S.: Conceptual modeling solutions for the data warehouse. In: Database Technologies: Concepts, Methodologies, Tools, and Applications, vol. 4 , pp. 86–104. IGI Global (2009)
70. Saeki, M.: Object-oriented meta modelling. In: OOER'95: Object-Oriented and Entity-Relationship Modelling, 14th International Conference, Gold Coast, Australia, December 12–15, 1995, Proceedings, pp. 250–259 (1995)
71. Sandkuhl, K., et al.: From expert discipline to common practice: a vision and research agenda for extending the reach of enterprise modeling. Bus. Inf. Syst. Eng. 60(1), 69–80 (2018)
72. Schewe, K., Thalheim, B.: Conceptual modelling of web information systems. Data Knowl. Eng. 54(2), 147–188 (2005)
73. Soley, R., Group, O.S.S.: Model driven architecture (2003). http://www.omg.org/cgi-bin/doc?omg/00-11-05
74. Storey, V.: Relational database design based on the entity-relationship model. Data Knowl. Eng. 7, 47–83 (1991)
75. Storey, V.C.: Conceptual modeling meets domain ontology development: a reconciliation. J. Database Manag. 28(1), 18–30 (2017)
76. Teorey, T., Yang, D., Fry, J.: A logical design methodology for relational databases using the extended entity-relationship model. ACM Comput. Surv. 18(2), 197–222 (1986)
77. Thalheim, B.: Entity-Relationship Modeling: Foundations of Database Technology. Springer, Berlin (2000). https://doi.org/10.1007/978-3-662-04058-4
78. Thalheim, B.: The science and art of conceptual modelling. In: Hameurlain, A., Küng, J., Wagner, R., Liddle, S.W., Schewe, K.-D., Zhou, X. (eds.) Transactions on Large-Scale Data- and Knowledge-Centered Systems VI. LNCS, vol. 7600, pp. 76–105. Springer, Heidelberg (2012). https://doi.org/10.1007/978-3-642-34179-3_3
79. Thalheim, B.: Conceptual model notions-a matter of controversy: conceptual modelling and its lacunas. EMISA J. 13, 9–27 (2018)
80. Wand, Y., Weber, R.: Research commentary: information systems and conceptual modeling - a research agenda. Inf. Syst. Res. 13(4), 363–376 (2002)

81. Weber, R.: Ontological Issues in Accounting Information Systems, pp. 13–33. American Accounting Association, Sarosota (2002)
82. Woo, C.: The role of conceptual modeling in managing and changing the business. In: Jeusfeld, M., Delcambre, L., Ling, T.-W. (eds.) ER 2011. LNCS, vol. 6998, pp. 1–12. Springer, Heidelberg (2011). https://doi.org/10.1007/978-3-642-24606-7_1
83. Yu, E.S.K., Mylopoulos, J.: From E-R to "a-r" - modelling strategic actor relationships for business process reengineering. In: Proceedings of ER 1994, pp. 548–565 (1994)

Empirical Comparison of Model Consistency Between Ontology-Driven Conceptual Modeling and Traditional Conceptual Modeling

Michaël Verdonck[1](✉), Robert Pergl[2], and Frederik Gailly[1]

[1] Faculty of Economics and Business Administration, Ghent University,
Ghent, Belgium
{michael.verdonck,frederik.gailly}@ugent.be
[2] Faculty of Information Technology, Czech Technical University,
Prague, Czech Republic
perglr@fit.cvut.cz

Abstract. This paper conducts an empirical study that explores the differences between adopting a traditional conceptual modeling (TCM) technique and an ontology-driven conceptual modeling (ODCM) technique with the objective to understand how these techniques influence the consistency between the resulting conceptual models. To determine these differences, we first briefly discuss previous research efforts and compose our hypothesis. Next, this hypothesis is tested in a rigorously developed experiment, where a total of 100 students from two different Universities participated. The findings of our empirical study confirm that there do exist meaningful differences between adopting the two techniques. We observed that novice modelers applying the ODCM technique arrived at higher consistent models compared to novice modelers applying the TCM technique. More specifically, our results indicate that the adoption of an ontological way of thinking facilitates modelers in constructing higher consistent models.

1 Introduction

Conceptual modeling constitutes an important role in representing and supporting complex human design activities. Conceptual models were introduced to increase understanding and communication of a system or domain among stakeholders. Some commonly used conceptual modeling techniques and methods include: Business Process Model and Notation (BPMN), entity relationship modeling (ER), object-role modeling (ORM), and the Unified Modeling Language (UML). We refer to these techniques and methods as *traditional conceptual modeling* (TCM). In order to arrive at high quality conceptual models, the framework of Lindland et al. [1] identifies several quality goals for conceptual models and also discusses the means for achieving them. Two of such major goals – corresponding to semantic quality – are validity and completeness. The former guarantees that all the statements made by the model are correct and relevant to the problem, while the latter assures that the model contains all the

© Springer Nature Switzerland AG 2018
J. C. Trujillo et al. (Eds.): ER 2018, LNCS 11157, pp. 43–57, 2018.
https://doi.org/10.1007/978-3-030-00847-5_5

required statements about the domain which are correct and relevant. In order to achieve a feasible level of validity, consistency is considered as a principal semantic aspect of the conceptual models since it allows verifying the internal correctness of specifications. Consistency can be defined as a state in which two or more elements, which overlap in different models of the same system, have a satisfactory joint description [2]. A failure in consistency can be related to the use of synonyms, inconsistencies between diagrams that represent a certain domain or inconsistencies between models at different development life cycle stages [3]. Inconsistent models can lead to duplication of information, diffuse interpretations and contradictory specifications – resulting in faulty system design and numerous errors in software development [4].

One approach to counter inconsistency is through the adoption of an ontology. Ontologies provide a foundational theory, which articulate and formalize the conceptual modeling grammars needed to describe the structure and behavior of the modeled domain [5]. More specifically, we can describe the utilization of ontological theories, coming from areas such as formal ontology, cognitive science and philosophical logics, to develop engineering artifacts – e.g. modeling languages, methodologies, design patterns and simulators – for improving the theory and practice of conceptual modeling, as *ontology-driven conceptual modeling* (ODCM) [6]. Furthermore, since ODCM provides unambiguous definitions for terms used in a domain, it plays a crucial role in the communication between modelers, as such maintaining consistency between conceptual models and integrating different user perspectives [7]. However, while several research efforts [8, 9] have adopted ontological theories to perform analyses or develop techniques that aim to increase the consistency of conceptual models, there exists little research that actually demonstrates that ontologies are capable of increasing consistency.

Therefore, it is the goal of this paper to investigate the effect of adopting ontological theories on the consistency of conceptual models. More specifically, we will examine the difference in adopting an ontology-driven modeling technique compared to a traditional conceptual modeling technique (without the application of an ontology), in order to verify if the adoption of ontologies can truly lead to increased consistency. As such, we will differentiate between modelers that are trained in a TCM approach and modelers that have been taught an ODCM approach. Through our study, we will then compare the two modeling approaches by investigating the consistency of the resulting conceptual models. To properly measure these effects, we conducted an empirical study. This study is part of an overall research project, that not only intends to investigate the consistency between an ODCM and TCM approach, but also aims to measure the difference in model quality between both approaches. This article however, will focus solely on the difference in consistency. As the foundation for the further development of this paper, we formulate our **research question** as follows: *Are there meaningful differences in the consistency of the resulting conceptual model between novice modelers trained in an ontology-driven conceptual modeling technique and novice modelers trained in a traditional conceptual modeling technique.* In Sect. 2 of this paper, we formulate our hypothesis. Next, we will draft our experimental design to test these hypotheses in Sect. 3. We will then present the results of our experiment in Sect. 4 and discuss their outcome on the hypothesis. Next, in Sect. 5, we will interpret the results of our experiment, and discuss their consequences and implications. Finally, we will present our conclusion and discuss the validity of the study in Sect. 6 of this paper.

2 Hypothesis Development

A substantial amount of research has been performed to identify the issues corresponding to the consistency of conceptual models, and consequently the proposal of techniques to overcome these issues [2]. Similarly, ontological theories have been applied as a means to improve consistency. Applying ODCM requires a modeler to identify and categorize phenomena according to the concepts, rules and patterns that correspond to the specific ontology – which is argued to facilitate a higher degree of consistency between the conceptual models. An example of adopting an ODCM technique can be found in the research of [9], where they propose a novel approach to model-driven software development based on three different types of ontologies, allowing the verification of the developed models both for consistency and ontological adequacy Another way of adopting ODCM to improve consistency is the study of [8], where they aim to overcome the lack of consistency between business processes and their underlying business models by performing an ontological analysis to evaluate the expressiveness of the business models in terms of ontological coverage and overlap.

Although the examples above clearly demonstrate that research efforts have been performed to develop different techniques or perform analyses in order to improve the consistency of conceptual models, they often lack an empirical evaluation. Moreover, the studies that have performed empirical research concerning the application of ODCM have focused more on demonstrating that the adoption of an ontological way of thinking will lead for instance to fewer modeling errors, lower cognitive difficulties or higher quality models [10, 11]. It is then often implied that since ODCM provides unambiguous definitions for terms used in a domain, they will consequently enhance the design and development of higher consistent models. However – as to the knowledge of the authors – no explicit study exists that empirically validates that adopting an ODCM technique will lead to higher consistent models compared to adopting a TCM technique. Therefore, we will empirically investigate the impact of adopting an ODCM technique compared to a TCM technique on the consistency of the resulting conceptual models. Moreover, it is our intention to properly train subjects in either one of these techniques, and then let them develop a conceptual model with the respective technique. More specifically, we will train novice modelers that have no prior knowledge in either of the techniques – enabling us to measure the full influence of the modeling approach that is being taught. We will then assess the impact of each technique by evaluating the consistency between the models that have been constructed by the respective technique. Since inconsistencies are often caused by ambiguities and complex descriptions corresponding to a certain scenario, we will provide our subjects with a sufficiently complex modeling task.

Hence, the **objectives** of our empirical study are: (1) to compare the adoption of an OCDM and a TCM technique; (2) to compare subjects that have been properly trained in both techniques – over a period of several months – in order to guarantee a thorough understanding; (3) to require subjects to apply the technique in order to construct a conceptual model and (4) to examine if models developed by the ODCM technique encompass a higher degree of consistency compared to models developed by the TCM technique. Thus, based upon the assumptions given above, we formulate our **hypothesis** as follows: *Novice modelers applying an ODCM technique will arrive at higher consistent models compared to novice modelers applying a TCM technique – given a thorough understanding of the respective technique and a sufficiently complex modeling task.*

3   Experimental Design

In order to rigorously test our hypothesis, we will carefully plan and design the experiment as to arrive at validated experimental results. Due to the lack of a random assignment of subjects between our testing groups – infra Experimental Design Type – we would like to emphasize that we will perform a quasi-experiment, since key characteristics may differ. As such, when referring to the term 'experiment' in the further development of this paper, we refer to a quasi-experiment. We base ourselves upon the experimental design described in [12], where an experiment can be divided into several steps. First, we will select the independent and dependent variables, based upon our hypothesis, and the instruments that we will apply to measure these variables. Next, the selection of subjects is carried out. Further, we describe the experiment design type of the experiment. After the planning process is iterated, we can conduct the actual experiment, and collect the data in order to either accept or reject the hypothesis. We would like to note that all materials – the assignments per treatment, the case description, knowledge assessments, competency questions etc. –can be found at our online repository at Open Science Framework (OSF)[1].

3.1   Variable Development

Independent Variable
The independent variable constitutes of the two different modeling techniques or approaches our subjects can apply to construct a conceptual model. More specifically, we will compare the enhanced entity relationship (EER) modeling technique with the ontology-driven OntoUML modeling technique. The entity-relationship (ER) approach still remains the principal model for conceptual design [13]. It allows the representation of information in terms of entities, their attributes, and associations among entity occurrences referred to as relationships. The ER modeling technique can be applied in combination with several notations, such as the UML notation – more specifically class diagram notation. By enhancing the ER approach with the UML notation – also known as the EER technique – the conceptual model gains substantial benefits, including easier communication and a more truthful representation of a particular domain.

   Likewise, OntoUML is a well-known technique in the domain of ODCM and has been frequently adopted for various purposes. Moreover, OntoUML also applies the

---

[1] osf.io/vs2bx.

UML notation – i.e. class diagrams – with the UFO ontology as an underlying foundational theory. The purpose of OntoUML is to enhance the truthfulness to reality by constructing conceptual models supported by ontological concepts [14]. Therefore, both techniques have been primarily developed to deliver conceptual models that offer faithful representations of a particular domain. In addition, both techniques use the same UML notation, but are grounded in two different underlying theories – the EER approach and the UFO ontology.

Dependent Variable
The purpose of our experiment is to measure the difference in consistency of the resulting conceptual models of our subjects when applying either a traditional modeling technique or an ontology-driven modeling technique. More specifically, every subject will be assigned with the same scenario that describes a certain domain. We will thus compare the consistency of conceptual models that represent the same exact scenario. As mentioned above, consistency can be described as the state in which two or more elements have a satisfactory joint description of the same system. Therefore, we will determine the consistency between the resulting conceptual models by comparing the different elements and their interdependent relationships for their likeness.

In order to objectively measure and assess the consistency of a model, we will apply a rather novel approach and rely on the use of *competency questions* while measuring the underlying deviations between subjects. Originally, competency questions were applied in ontology development [15], where a particular ontology was found adequate to represent a certain domain providing that the ontology could represent and answer a specific set of competency questions. In our experiment, we will relate the competency questions to specific concepts and relationships that should be included in the model in order to be deemed a good representation of the domain. Thus, by applying the competency questions we are measuring the validity of each conceptual models, as well as their degree of completeness. For instance, the case that will be given to the subjects consists of the description of the university domain. A specific aspect of this case describes that a professor can only work at one department of a faculty. We consequently have a competency question that specifically focuses on this aspect and verifies if a subject has included both the concepts professor and department, and if the existential relationship is fulfilled in means of a composition relationship.

However, while the competency questions are an indication that a certain treatment was more effective at identifying the essential elements corresponding to a domain, it does not necessarily mean that the models are more consistent. It is possible that a treatment includes many of the necessary elements but displays a high degree of discrepancy between the different models of the subjects. Therefore, in order to measure the consistency of each treatment, we are going to measure the standard deviation that exists between the different treatments in order to examine the variance that exists between the different models of each subject in the respective treatment. This approach is similar to the research of [16], where consistency was also measured by regarding the percentage differences of each participant compared to the mean of the group.

Furthermore, we will differentiate between two sets of competency questions. One set of questions will measure if subjects adequately represented the domain as described in the assignment. The second set of questions will measure how the models of the subjects represented certain 'complications' described in the case. As mentioned above, inconsistencies are often caused by ambiguities and complex descriptions corresponding to a certain scenario. Therefore, by having a scenario that encompasses certain complexities, we increased the difficulty for obtaining an overall consistency between the models. The adoption of an ODCM technique should nonetheless achieve higher consistency despite these difficulties – which is the purpose of this experiment. Hence, we will make a distinction between competency questions that measure Content Interpretation (CI) and Content Sophistication (CS), corresponding to the work of [17, 18]. While the former is defined as the identification of the entities that exists in the domain by an applicant or modeler, the latter can be seen as the process of gradually improving the model such that it provides a more precise representation of the world.

Tailored to our experiment, participants will receive a case that describes the university domain. More specifically, the scenario describes a company that desires to develop a student information system for universities. As part of the development process, a conceptual model is required that should be applicable to multiple universities. As means of a reference case, a description is given of a university. Subjects are given specific instructions that the concepts and entities of the university should be modeled, but that their model should also be accessible for representing the structure of other universities. The purpose of the task is thus of a rather businesslike nature, with the objective to deliver a 'complete' representation of the case, and which should at the same time be adaptive enough to apply to the structure of other universities. Consequently, when modeling the domain, they have to identify the necessary constructs, relationships and cardinalities that govern this domain – i.e. content interpretation. However, as mentioned, the case (deliberately) contains ambiguous descriptions or certain complications. The competency question corresponding to the content sophistication will then check for the concepts and relationships that provide a more precise representation of the university domain. As such, the standard deviation of the competency questions enables us to evaluate the consistency of participants' models in a rather objective way, by comparing the consistency of the completeness of the model (i.e. content interpretation), and the consistency related to the validity of the models (i.e. content sophistication).

Finally, the correction of these competency questions will be reviewed by several authors of this paper. The given scores of each reviewer are then compared and checked for their correlation in order to enhance the objectivity of the correction of these competency questions. All competency questions can be found in the OSF repository.

Control Variables

Since we will be testing participants modeling with a TCM and an ODCM technique, we need to ascertain that all subjects have an equal understanding of each technique they are modeling with. Therefore, we apply a control variable to test every subject's knowledge and understanding of the modeling technique, before the start of the experiment. More specifically, we evaluate each subjects' understanding with several written statements. Each of these statements describe a certain phenomenon or scenario, to which the subject has to choose the correct corresponding element of the modeling technique. In total, six statements were given for each treatment (see OSF repository). Each of these statements was derived from examples from existing literature or exercises related to the techniques. The results from the subjects that failed the knowledge test will not be incorporated into the results of the experiment.

Next, to provide a complex enough case description as required in our hypotheses, we have selected a modeling case that served as an assignment of a modeling course given at the University Ghent. The feedback and the final results of the assignment that applied the modeling case confirmed that the modeling case is of a rather complex degree. Additionally, we have presented the modeling assignment at the OntoCom workshop at the 36th International Conference on Conceptual Modeling. During this workshop, the case has been given to several experts in the domain of conceptual modeling and ontology. Each of these experts have then created a conceptual model, according to their interpretation of the case. Afterwards, the different models were discussed for their completeness and how they dealt with the challenges or ambiguities that could be found in the case. During this workshop, many of the competency questions were derived from the models of the workshop and the feedback from the different experts. Additionally, the experts who have modeled the case themselves also have labeled the case as sufficiently complex to be applied in an experimental setting.

3.2   Subject Selection

The subjects in our study all were novice conceptual modelers and were attending two different courses on conceptual modeling at Ghent University (Belgium) and the Technical University of Prague (Czech). While the subjects at Ghent University were taught how to adopt a TCM technique to construct a conceptual model, the course at the Technical University of Prague taught their students the ODCM technique. Both courses focused on software engineering, where each of the respective techniques were taught in a similar fashion. As stated by [19], using students as participants remains a valid simplification of reality needed in laboratory contexts. It is an effective way to advance software engineering theories and technologies but, like any other aspect of study settings, should be carefully considered during the design, execution, interpretation, and reporting of an experiment. Consequently, we decided to select students as our test subjects since they have no prior knowledge of conceptual modeling and can thus be seen as novice modelers who can be trained in either TCM or ODCM. Hence, our selection of students enabled us to train subjects without having prior experience in another modeling technique. Consequently, we could measure the full impact of the modeling technique that is being taught.

At Ghent University, students have been taught the EER conceptual modeling technique through both theoretical classes and practical sessions. In these practical sessions, students were required to solve modeling assignments of certain scenarios. Additionally, students were required to submit a rather extensive group assignment, where they had to design and implement an information system. An important aspect of this assignment was to develop a sound EER conceptual model that forms the foundation of their database. Similarly, students at the Technical University of Prague received both theoretical classes as well as practical sessions on a weekly basis. Furthermore, they also had to complete a work assignment that required them to create sound OntoUML models, to serve as a foundation for a software system. Moreover, all subjects have the same age (i.e. early-twenties) and the majority of our subjects have a business/technical-oriented background. Concerning motivation, students were asked to participate with the experiment out of self-interest and as an opportunity to improve their skills in conceptual modeling. There was no reward-based incentive. As such, students that participated in our experiment were essentially self-motivated based on the inclination to learn more and to improve their skillset. Thus, the specific selection and the education program leads to a controlled sample of subjects, all being novice modelers, that are properly trained in the respective modeling technique and with no prior knowledge of any other modeling technique. Finally, in order to determine the number of subjects for our empirical study, we base ourselves on the differences in the averages in the model comprehension scores from another empirical study that also compared two different modeling techniques [20]. Based upon the sample size formula below [21], assuming a Type I error (α) of 5% and a Power ($1 - \beta$, where β is Type II error) of 0.8, we require a total number of 43 subjects per treatment group. In total, 100 subjects participated in the study, of which 50 in each treatment. Hence, the number of participants in our experiment is sufficient in regard to the required statistical minimum.

3.3 Experimental Design Type

An experiment consists of a series of tests of different treatments [12]. To get the desired results to answer our research question, the series of tests must be carefully planned and designed. Based on our hypotheses, we can derive two treatments: an UML and an OntoUML treatment. The assignment in each treatment constitutes of a case study that has to be modeled by the participants of the respective treatment. We have assigned the participants to these treatments according to the balancing design principle. By balancing the treatments, we assign an equal number of subjects to each separate treatment, to arrive at a balanced design. Balancing is desirable since it both simplifies and strengthens the statistical analysis of the data. However, due to practical limitations we could not balance the students of the two different universities between the two treatments, e.g. half of the students of Ghent University being trained in TCM and ODCM and vice versa for the students at the University of Prague. As such, one group may differ from the other – e.g. due to the students' specific profile or the teaching method of the respective professor. As such, our type of experiment is a quasi-experiment. The most important consequence of this quasi-experimental design is that our study may suffer from increased selection bias, meaning that other factors instead of our dependent variable may have influenced the outcome of our results. As a result, this also impacts the internal validity of our study, which is again emphasized below in the conclusion section. The design type of our quasi-experiment is a one-factor with two treatments design, meaning that we compare the two treatments against each other with one dependent variables – the consistency of the conceptual models. Each subject also takes part in only one treatment. Most commonly, the means of the dependent variables for each treatment are compared. We will thus assign scores to the different measures of the dependent variables in order to compare our two different treatments objectively.

4 Results

When calculating the arithmetic mean of the total results for each treatment, we observed that the ODCM treatment scores substantially higher (66,75%) compared to the TCM treatment (53,70%). More specifically, we observed a rather large difference in means between the Content Interoperation (CI) and the Content Sophistication (CS) questions. The ODCM treatment (87,5%) scored somewhat higher for the CI question compared to the TCM treatment (83,4%). However, regarding the CS question, the ODCM treatment achieves almost double the total score (46%) compared to the TCM treatment (24%), strongly indicating that the ontology-driven technique aids subjects in better identifying the required concepts and relationships when dealing with complex and ambiguous case descriptions.

However, as mentioned above, in order to determine the degree of consistency between the conceptual models of each treatment, we consider the variance that exists between the competency questions. More specifically, Table 1. displays the standard deviation related to the results of the competency questions, and also distinguished between the scores of the CI questions and the CS questions. As we can observe, the total scores vary substantially more for the TCM treatment (14,84) than for the ODCM treatment (8,93). This observation also holds when we make the distinction between the CI and the CS questions. It would thus appear that the deviation between the subjects' models in the TCM treatment was overall higher compared to the ODCM treatment.

Table 1. Descriptive results Competency Questions

Scores per Treatment		Std. Deviation
Total score competency questions	TCM	14,84
	ODCM	8,93
Total score content interpretation questions	TCM	15,50
	ODCM	10,06
Total score content sophistication questions	TCM	16,57
	ODCM	11,61

While the results above seem to confirm our hypothesis, we will perform additional testing – more specifically the Levene's test for homogeneity of variances– to examine if the difference in variance between the treatments is also significantly different. We would like to remark that besides the Levene's test, we have also performed the Brown Forsythe test for the equality of group variances – a more robust variant of the Levene's test – which arrived at exactly the same results as the Levene's test described below.

Table 2. displays the results concerning the Levene's test, where df1 and df2 correspond to the degrees of freedom of the Levene statistic. Furthermore, since we are performing a directional test – i.e. the variance of the TCM treatment is higher than the variance of the ODCM treatment – we will regard the one-tailed significance of the Levene's test. More specifically, regarding the p-values corresponding to the test, we notice that for every score, the p-value is below 0,05. Meaning that we cannot assume homogeneity of variances between the two treatments and that the variance of the TCM treatment is significantly higher compared to the variance of the ODCM treatment on a 5% significance level. As such, we **accept our hypothesis**, and confirm that novice modelers applying an ODCM technique will arrive at higher consistent models compared to novice modelers applying a TCM technique.

Table 2. Levene's test for homogeneity of variances of the scores per treatment

Test of homogeneity of variances	Levene statistic	df1	df2	Sig. (one-tailed)
Total score competency questions	23,106	1	98	0,000
Total score content interpretation questions	8,014	1	98	0,003
Total score content sophistication questions	13,277	1	98	0,000

5 Discussion

The results of our empirical study seem to acknowledge that meaningful differences in the consistency of the resulting conceptual model between novice modelers trained in an ODCM technique and in a TCM technique do exists. More specifically, the results indicate that novice modelers applying ODCM arrived at more consistent models compared to novice modelers applying TCM. Below, we will discuss two derivations that are based upon these results. We would like to note that we intentionally call these derivations instead of findings or conclusions, since we are of the opinion that one experiment does not suffice in order to confirm that all ODCM techniques lead to a higher degree of consistency. More empirical research with different sets of subjects are required to support this hypothesis.

Derivation 1: *Novice modelers applying an ODCM technique will arrive at higher consistent models compared to novice modelers applying a TCM technique.*

When regarding the descriptive statistics in Table 1, we observed that the standard deviation for the competency questions was substantially higher for the ODCM technique (14,84) compared to the TCM technique (8,93). Moreover, this difference was found to be significant, confirming our hypothesis that the ODCM technique results in more consistent models than the TCM technique. Furthermore, we also distinguished between the Content Interpretation and the Content Sophistication questions, in order to identify how well the subjects per treatment coped with the more complex and ambiguous descriptions of the assignment. The standard deviation for both type of questions was found to be significantly different in favor of the ODCM technique. Consequently, our research results confirm that the ODCM technique facilitates a higher degree of consistency, both in identifying the basic concepts and relationships as well as the more complex and ambiguous requirements of the case description.

Derivation 2. *Adopting an ontological way of thinking leads to more consistent models.*

A likely explanation to the observed differences in the results discussed above can perhaps be found in the manner how modelers of the ODCM treatment are adopted to an ontological way of thinking when learning and applying an ODCM technique. Idiosyncratically, these modelers have to interpret and recognize the domain or scenario that they wish to model in the ontological concepts and rules that correspond to this technique. These ontological concepts and rules are governed by a fixed set of axiom's, constraints and patterns corresponding to the underlying ontology. In other words, we believe that since the modelers of the ODCM technique are committed to develop a conceptual model according to a shared set of ontological rules and patterns, consequently they arrive at more consistent conceptual models.

An example of such a pattern is displayed in Fig. 1. In this figure, a typical pattern of the UFO ontology is displayed. Without going into much details about the specific structure of the UFO ontology, a Kind can be seen as a 'rigid type', meaning that it is an existentially independent concept, that contains its own principle of identity. A Role is always a specialization of a rigid type – in our case a Kind – where the specialization condition is a relationally dependent one. For instance, husband or wife can be seen as Roles, which are specializations of the concepts Man and Woman or more generally Person. Hence, modelers adopting OntoUML idiosyncratically model concepts such as Husband and Wife as Roles of a Person. Similar to the case description of our empirical study, modelers applying the OntoUML technique will have the tendency to model the different actors in the description – i.e. 'Student' and 'Professor' – as Roles of a more general entity such as Person or University Member. Consequently, applying an ODCM technique such as OntoUML leads to more consistent models since all of its modelers are trained to identify and apply the underlying ontological patterns.

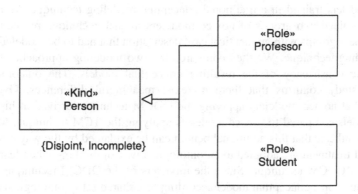

Fig. 1. ODCM pattern - case description example

Another example of such an ontological pattern can be found in the specialization of a Phase from a Kind – where the specialization condition is always an intrinsic one. For instance, a child can be seen as a Phase of a Person, where the specific range of categorizing someone as a child can be specifically determined. Similarly, modelers applying the OntoUML technique have the proclivity to model the different states of a course – i.e. 'Active' and 'Inactive' – as specialized Phases of a course itself. Alternatively, we could assign active/inactive as a property of a course. However, when relating other concepts such as exam or exam date to a course, a conflicting situation can arise where an exam or exam date is scheduled for an inactive course. Therefore, adopting the ontological pattern enables subjects to recognize (in)active states as further specializations of a course, prompting modelers to more carefully consider the structure of their concepts and the intertwining relationships. The impact of such patterns can also be clearly found in the answers corresponding to the competency questions. For instance, when regarding the 10th Content Sophistication question – "Can exams and exam dates be associated only to active courses?" – the ODCM treatment scored a total of 74% on this question, compared to a 45% of the TCM treatment. Thus, we have reason to believe that adopting the concepts, rules and patterns corresponding to an ODCM technique enables modelers to develop conceptual models in a more consistent way, but also prompts modelers to consider the structure and order of their model more carefully.

6 Conclusion

While various research efforts have adopted ontological theories to perform analyses or develop techniques that aim to increase the consistency of conceptual models, there exists little research that actually demonstrates that ontologies are capable of increasing consistency. Moreover, little knowledge exists on how ontologies aid modelers in developing conceptual models of a high degree of consistency. Therefore, this paper conducted an empirical study with the principal research question if there exist meaningful differences in the consistency of the resulting conceptual model between novice modelers trained in an ontology-driven conceptual modeling technique and novice modelers trained in a traditional conceptual modeling technique. More specifically, we trained two groups of novice modelers in each technique respectively and assigned these groups with an identical case description that had to be modeled with the corresponding technique. We then compared the two modeling approaches by investigating the consistency of the resulting conceptual models. The outcome of our empirical study confirms that there do exist meaningful differences. Overall, we observed that novice modelers applying the ODCM technique arrived at higher consistent models compared to novice modelers applying the TCM technique. Moreover, our results indicate that this additional benefit can be explained by the way modelers of the ODCM treatment are adopted to an ontological way of thinking when learning and applying an ODCM technique. Since the modelers of the ODCM technique are committed to develop a conceptual model according to a shared set of ontological rules and patterns, consequently they arrive at more consistent conceptual models. Finally, we would like to acknowledge that one experiment does not suffice in order to confirm that all ODCM techniques lead to a higher degree of consistency. Hence, we encourage future research into comparing different techniques with one another. Below we also discuss the validity of this study.

Internal Validity

We have carefully designed and monitored the conduct of this experiment. Several experimental standards were also implemented to strengthen the validity of the experiment: (1) we applied the balancing design principle in order to balance our treatments, since balancing between the treatments was not possible due to practical limitations, we emphasize that this is a quasi-experiment; (2) subjects were selected from a 'controlled' environment, meaning that they all share a similar background and were novice modelers in the field of conceptual modeling; (3) neither of the subjects had any prior knowledge of either of the modeling techniques that were applied in the treatments; (4) we inserted a control variable in the experiment to assert that subjects had a similar understanding of the techniques before commencing the experiment; (5) we have presented the modeling case at the OntoCom workshop at the 36th International Conference on Conceptual Modeling in order to evaluate our case and the related competency questions by several experts in the domain of conceptual modeling and ontology; and finally (6) the correction of the competency questions – although already rather objective by themselves – has been conducted by several authors of this article.

External Validity

We are well aware that by conducting our experiment on students, we limit the overall generalizability of our results. However, as stated by [19], using students as participants remains a valid simplification of reality needed in laboratory contexts. It is an effective way to advance software engineering theories and technologies but, like any other aspect of study settings, should be carefully considered during the design, execution, interpretation, and reporting of an experiment. Consequently, we decided to select students as our test subjects since they have no prior knowledge of conceptual modeling and can thus be seen as novice modelers who can be trained in either TCM or ODCM. Furthermore, although we have balanced our number of subjects across our treatments, an even better approach would have been to also balance subjects of the different universities over each treatment. In our current setup, only one type of technique was taught at each university. This was due to the practical organization of the classes given at the universities. We therefore acknowledge that dividing students over the different treatments per university would have increased the external validity of this study. We would like to remark however, that the nature of our results quite accurately follows the distinctions that exist between the techniques that have been applied in this study. For instance, the results of some competency questions can be clearly attributed to the existence of the ontological patterns that exist in the ODCM technique. Finally, we deliberately also chose our assignment to deal with the university domain since students are well aware of this domain and so that there would not exist an additional advantage in modeling between the students.

References

1. Lindland, O.I., Sindre, G., Solvberg, A.: Understanding quality in conceptual modeling. IEEE Softw. **11**, 42–49 (1994)
2. Lucas, F.J., Molina, F., Toval, A.: A systematic review of UML model consistency management. Inf. Softw. Technol. **51**, 1631–1645 (2009)
3. Finkelstein, A., Gabbay, D.M., Hunter, A., Kramer, J., Nuseibeh, B.: Inconsistency handling in multperspective specifications. Tse. **20**, 569–578 (1994)
4. Muskens, J., Bril, R.J., Chaudron, M.R.V: Generalizing consistency checking between software views. In: Proceedings of the 5th Work. IEEE/IFIP Conference on Software Architecture WICSA 2005, pp. 169–180 (2005)
5. Wand, W.R.: On the ontological expressiveness of information systems analysis and design grammars. Inf. Syst. J. **3**, 217–237 (1993)
6. Guizzardi, G.: Ontological foundations for conceptual modeling with applications. In: Ralyté, J., Franch, X., Brinkkemper, S., Wrycza, S. (eds.) CAiSE 2012. LNCS, vol. 7328, pp. 695–696. Springer, Heidelberg (2012). https://doi.org/10.1007/978-3-642-31095-9_45
7. Uschold, M., Gruninger, M.: Ontologies: principles, methods and applications. Knowl. Eng. Rev. **11**, 93–116 (1996)
8. Buder, J., Felden, C.: Ontological analysis of value models. In: ECIS 2011 Proceedings (2011)
9. Hoehndorf, R., Ngomo, A.C.N., Herre, H.: Developing consistent and modular software models with ontologies. Front. Artif. Intell. Appl. **199**, 399–412 (2009)
10. Bera, P.: Analyzing the cognitive difficulties for developing and using UML class diagrams for domain understanding. J. Database Manag. **23**, 1–29 (2012)
11. Evermann, W.Y., Evermann, J., Wand, Y.: Ontological modeling rules for UML: an empirical assessment. J. Comput. Inf. Syst. **46**, 14–29 (2006)
12. Wohlin, C., Runeson, P., Host, M., Ohlsson, M.C., Regnell, B., Wesslen, A.: Experimentation in Software Engineering. Springer, Heidelberg (2012). https://doi.org/10.1007/978-3-642-29044-2
13. Fettke, P.: How conceptual modeling is used. Commun. Assoc. Inf. Syst. **25**, 571–592 (2009)
14. Guizzardi, G., Wagner, G.: Towards ontological foundations for agent modelling concepts using the unified fundational ontology (UFO). In: Bresciani, P., Giorgini, P., Henderson-Sellers, B., Low, G., Winikoff, M. (eds.) AOIS -2004. LNCS (LNAI), vol. 3508, pp. 110–124. Springer, Heidelberg (2005). https://doi.org/10.1007/11426714_8
15. Grüninger, M., Fox, Mark S.: The role of competency questions in enterprise engineering. In: Rolstadås, A. (ed.) Benchmarking — Theory and Practice. IAICT, pp. 22–31. Springer, Boston, MA (1995). https://doi.org/10.1007/978-0-387-34847-6_3
16. Moody, D.L.: The method evaluation model : a theoretical model for validating information systems design methods. In: Proceedings of the 11th European Conference of Information Systems (ECIS), pp. 1327–1336 (2003)
17. Daga, A., et al.: An ontological approach for recovering legacy business content. In: Annual Hawaii International Conference on System Sciences 00, pp. 1–9 (2005)
18. De Cesare, S., Partridge, C.: BORO as a foundation to enterprise ontology. J. Inf. Syst. **30**, 83–112 (2016)

19. Falessi, D., et al.: Empirical software engineering experts on the use of students and professionals in experiments. Empir. Softw. Eng. 1–38 (2017)
20. Verdonck, M., Gailly, F.: An exploratory analysis on the comprehension of 3D and 4D ontology-driven conceptual models. In: Link, S., Trujillo, Juan C. (eds.) ER 2016. LNCS, vol. 9975, pp. 163–172. Springer, Cham (2016). https://doi.org/10.1007/978-3-319-47717-6_14
21. Shao, J., Wang, H., Chow, S.-C.: Sample Size Calculations in Clinical Research. Chapman & Hall/.CRC, New York (2008)

A Conceptual Framework for Supporting Deep Exploration of Business Process Behavior

Arava Tsoury[(✉)], Pnina Soffer[(✉)], and Iris Reinhartz-Berger[(✉)]

University of Haifa, Mount Carmel, Haifa 3498838, Israel
{atsoury, spnina, iris}@is.haifa.ac.il

Abstract. Process mining serves for gaining insights into business process behavior based on event logs. These techniques are typically limited to addressing data included in the log. Recent studies suggest extracting data-rich event logs from databases or transaction logs. However, these event logs are at a very fine granularity level, substituting business-level activities by low-level database operations, and challenging data-aware process mining. To address this gap, we propose an approach that enables a broad and deep exploration of process behavior, using a conceptual framework based on three sources: the *event log* that holds information regarding the business-level activities, the *(relational) database* that stores the current values of data elements, and the *transaction (redo) log* that captures historical data operations performed on the database as a result of business process activities. Nine types of operations define how to map subsets of elements among the three sources in order to support human analysts in exploring and understanding the reasons of observed process behavior. A preliminary evaluation analyzes the outcomes for four useful scenarios.

Keywords: Business process · Process mining · Data-aware · Database
Event log · Transaction log

1 Introduction

Business process management is an essential way of operating and managing organizations for achieving business goals. Business processes entail activities and decisions, relying on and manipulating data, which is typically stored in databases. To fully understand a process, a combination of the control (activities, decisions, and their order) and data views is needed.

Process mining [13] has emerged as a promising area for gaining insights into business process behavior. For this purpose, process mining uses event logs which capture events that typically correspond to activities performed in the business. Some event logs are quite basic ("thin") and document only information that relates to the control flow (e.g., event label and timestamp). Other event logs may contain additional data attributes, such as affected data elements and their values.

Many research efforts have been dedicated to data-aware process mining. These kinds of analysis require richer event logs in terms of data. Data-aware process discovery [4], for example, attempts to discover the control flow, the related data flows, and

© Springer Nature Switzerland AG 2018
J. C. Trujillo et al. (Eds.): ER 2018, LNCS 11157, pp. 58–71, 2018.
https://doi.org/10.1007/978-3-030-00847-5_6

decision guards from event logs. Decision mining [10] aims to discover (data-related) decision rules employed in the process. Data-aware conformance checking [3, 11] assesses conformance of an actual process instance with a prescribed process model, considering both control flow and data operations, and diagnosing data exceptions that may be associated with deviations from the normative process.

Typically, when an event log is created, only a subset of data attributes is included. This subset should be indicated by analysts, based on what they expect to be relevant for certain analyses that are of interest for the business goals [8]. However, different analysis goals can be set, and a log that fits one goal may not fit the other. Furthermore, the expectations of relevance can only be based on evidence if the data attributes are included in the log. It turns out that to discover that a data attribute has an impact that is relevant for some analysis, it needs to be included in the log. To be included in the log, the existence of such impact should be assumed beforehand; otherwise, such impact will not be discovered. A possible way to avoid this conflict is generating the "appropriate" event log, e.g., from a database [2] or a transaction log [5]. However, as discussed later, this may hinder the business-level activities and keep the analysis in the lower level of database operations.

This paper aims at overcoming the aforementioned challenges by introducing a conceptual framework that supports a deep exploration of process behavior. The framework is based on three sources: (1) the event log which may include only basic control-related information or additional data attributes; (2) the database that stores the current values of the data attributes, and (3) the transaction (redo) log that bridges between the two other sources by capturing historical data operations performed on the database as a result of business process activities. The approach promotes storing information from the three sources in a big data repository and analyzing it through operations that map corresponding pieces of information. We demonstrate these operations that set the ground for various detailed exploration trails and enable the analysts to drill into the roots of phenomena that can initially be spotted using standard analyses (e.g., conformance checking) and utilizing analysts' domain knowledge. The analysts can further refine or extend this set of operations for improving the search, filtering, and merging of information from the three sources.

The rest of the paper is structured as follows. Section 2 provides background on event and transaction logs, and refers to existing studies that use such logs for various types of analysis. Section 3 presents a running example to illustrate and motivate the need for our approach. Section 4 presents the proposed conceptual approach, including a set of generic mapping operations, while Sect. 5 discusses insights from using the approach in four useful scenarios. Finally, Sect. 6 discusses the strengths and limitations of the approach, and Sect. 7 concludes and outlines future directions.

2 Event and Transaction Logs: Background and Related Work

Event logs are widely available in a variety of information systems, most notably business process management systems [13]. They serve as a primary input to process mining analyses [14]. The content of event logs varies and can be configured by the organizational IT administrators. Basic logs, a.k.a. "thin" logs, contain only an event label (e.g., activity) and a case identifier. The case identifier uniquely ties the event to a specific execution of the process (e.g., a specific customer order). Event logs often contain information about the activity timestamp and the resource who performed it. Sometimes event logs contain data elements attached to events [14] (e.g., the total payment of an order). Table 1 exemplifies a basic event log that records a sales process, including the order identifier, the (event end) timestamp and the activity.

Table 1. An example of a few lines in a basic event log of a sales process

OrderId	TimeStamp	Activity
10259	2018-02-21 10:26:11.387	New order
10259	2018-02-21 10:29:16.760	Allocate product
10259	2018-02-21 10:30:17.840	Allocate product
10259	2018-02-21 10:35:16.887	Set delivery info
10259	2018-02-21 10:42:21.900	Handle payment
10259	2018-02-23 12:06:27.933	Ship and close order

Based on event logs, it is possible to discover a process model, find bottlenecks in the process, predict costs and risks, and more [13]. When a process model is available, it is possible to further check the conformance of the event log with the process model or to check for compliance, detecting behavior which violates some modeled business rules [13]. Over the years, various process mining techniques have been proposed, addressing control flow aspects (e.g., [1, 15]), data flows and resources (e.g., [11, 9]) and the time perspective (e.g., [3]). For applying all these techniques, event logs containing the required data are needed. Even if such event logs are provided, the analysis is limited to the perspective that is recorded in the event log.

While business process management systems generate event logs as a standard, it is possible but sometimes challenging to extract logs in other systems [8, 14]. In general, generating event logs that include data attributes relevant for process mining is difficult and time consuming. In many cases it requires domain knowledge and a profound understanding of what and how to record [14]. Some studies, such as [2, 8], propose extraction of event logs from relational databases, typically used in enterprise systems. These approaches, however, are capable of capturing only events whose history is recorded in the database (e.g., the statuses of an order which are stored as records). For other events, only the last (current) values of the related data attributes can be retrieved (e.g., the up-to-date address of the customer). This limits the possible analyses that can be performed regarding the process behavior.

The existence of transaction logs (also termed redo logs) in relational databases may help address the aforementioned gap. Transaction logs store information about operations that change data in the database, including insertion, update, and deletion of records. Different database systems provide diverse formats of transaction logs, but a typical transaction log records for each transaction an identity, the start and end times, the types of performed operations (insert, update, or delete), and additional information, such as the record to which each change was performed, the attribute that was changed (including the new value), and the success status (commit or rollback).

Table 2 demonstrates a partial transaction log that records database operations of a sales system. Such logs can be easily generated from different database management systems using tools such as MS-SQL change data capture[1], which present database changes in a relational format rather than as a collection of SQL queries.

Table 2. An example of a few lines in a transaction log of a sales system

TranId	BeginTime	EndTime	Operation	orderID	Attribute	NewValue
0 × 000000008A7D	2018-02-22 09:12:23.783	2018-02-22 09:12:23.783	Insert	10473	CustomerID	ISLAT
0 × 000000008A7D	2018-02-22 09:12:23.783	2018-02-22 09:12:23.783	Insert	10473	EmployeeID	1
0 × 000000008A7D	2018-02-22 09:12:23.783	2018-02-22 09:12:23.783	Insert	10473	RequiredDate	2018-03-07 11:43:25.910
0 × 000000008B72	2018-02-24 15:22:20.977	2018-02-24 15:22:25.673	Update	10473	RequiredDate	2018-03-10 11:43:25.910

Although the main aim of transaction logs is to support database recovery and hence their content is typically deleted after recovery points, it is possible to store their content before synchronization. Storing such information into a big data repository may enable effective and efficient manipulation for various analyses.

As can be seen, the information in the two types of logs is different. Particularly, the exact impact of events on the database is not fully shown in the event log and the business-level activities (which typically entail a number of database operations) are not shown in the transaction log. However, these types of information are related. Particularly, the relations can be reflected through two attributes: (1) the case identifier – which explicitly appears in the event log and is captured as the primary key of a table in the database (e.g., the OrderId values in our example logs); and (2) the timestamp – the event (end) timestamp is between the transaction's begin and end times, meaning that we assume writing to the database is not delayed.

The relations between event and transaction logs motivated a recent approach [5], which suggested a systematic extraction of an event log from a transaction log in order to enrich analyses with data. Specifically, the approach treats database changes as events, and supports data-aware process mining based on a meta-model that connects database changes, data values, and events. The resulting event log, however, is at a very fine granularity level, substituting business-level activities by low-level database operations. Using process mining techniques on this kind of log is expected to yield highly complex "spaghetti" models. Yet, the available data makes it possible to filter and mine subsets of events that meet some conditions, overcoming complexity. Similarly to this approach, we propose to utilize transaction logs. However, we propose to

---

[1] https://docs.microsoft.com/en-us/sql/relational-databases/track-changes/about-change-data-capture-sql-server?view=sql-server-2017.

use it *in combination* with an event log to keep the business-level activities that trigger database operations. Our proposed approach allows analysis at varying abstraction levels, as well as makes *all* data attribute available for process mining.

3   Motivating Example

To motivate the need for our approach, consider the following simple sales process in which customers order products that are shipped after payment. Figure 1 depicts a model of this sales process, setting the required flow. The data used in the process is stored in a relational database, which includes tables related to customers, orders, shippers (i.e., shipping companies), employees, and so on.

Assume also the following business rule:

> If the value of the order's *total payment* exceeds 10,000, only a *manager* is allowed to perform *handle payment*.

For checking compliance with this business rule, the event log should contain the relevant data attributes, namely, total payment and the employee's role. If it does not, the values could be extracted from the database and checked [2]. However, the data values of these attributes may change during the process, and their values in the database do not necessarily reflect their values when certain events occurred. Alternatively, the transaction log could be used as a basis for analysis, but this would only reflect updates in the payment value, not necessarily related to the *handle payment* activity.

Assume that we decide after all to include *total payment* and the employee's *role* in the event log. Now, a new business rule is introduced:

> *Handle payment* will not be possible if the requested *delivery date* is not set.

Apparently, the existing event log is not sufficient for checking compliance with this rule, since it does not record information regarding the delivery date. Moreover, assume non-compliance with one of the rules is discovered. Additional questions may be raised for gaining a deeper understanding of the situation. For example, do the non-compliant cases relate to specific VIP customers whose orders should not be delayed? Such investigations require dynamic analyses, using varying sets of data attributes.

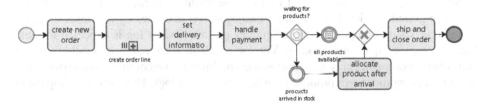

Fig. 1. A model of the sales process

4 The Suggested Approach

The main idea of the suggested approach is to combine information from three resources – the event log, the transaction log, and the database itself, in order to enrich the business process analysis with data-related aspects. The basic conceptual framework of the approach is outlined in Fig. 2. The main building blocks are defined and formalized below.

Event log (E): a multiset of traces, where each trace, marked by a *case id*, describes a particular execution of the process, in the form of a sequence of events. Each event is a tuple *(activity, timestamp)*. For simplicity, we flatten the structure and assume that an event log is a set of tuples of the form *(caseObject, caseID, activity, timestamp)*, where *caseObject* represents the attribute which uniquely identifies a single process execution and *caseID* is its value for the specific event.

As noted, event logs have different formats, but the conversion to our format is quite straightforward. For example, the conversion from Table 1 to our format requires renaming the column *orderID* by *caseID* and adding a column *caseObject* whose value for all rows is *orderID*.

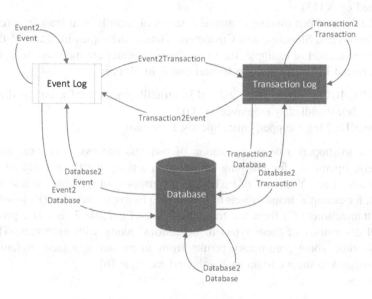

Fig. 2. The suggested conceptual framework

Transaction log (T): a multiset of transactions executed by a database management system, where each transaction, identified by a transaction id and characterized by begin and end times, is a sequence of changes (insert, update, or delete) on specific data elements that modify certain attributes of them. Hence, a transaction log can be represented by a set of tuples of the form *(transactionID, beginTime, endTime, operation, caseObject, caseID, attribute, newValue)*, where:

- *transactionID* uniquely identifies the transaction;
- *beginTime, endTime* are the timestamps when the transaction begins and ends, respectively;
- *operation* ∈ {insert, update, delete} specifies the change type;
- *caseObject* and *caseID* identify the data element that is changed;
- *attribute* and *newValue* specify the change essence.

Table 2 can be easily transformed to this format by repeating the changes suggested above for converting the event log.

Database (D): a multiset of assignments to relations, where each relation is a collection of (uniquely named) attributes (A_{ij}) over certain domains (D_{ij}) [16]. Namely, D is a set of tuples of the form $((A_{i1}, v_{i1}), \ldots (A_{in_i}, v_{in_i}))$, where $v_{ij} \in D_{ij}^2$.

We further assume the existence of relational algebra operations, including projection (π), selection (σ), and natural join ($\bowtie$). Recall that relational algebra is used to support analysis of different types of databases, including relational as well as unstructured ones [13].

The database of our running example consists of records in at least four relations: Orders, Employees, Shippers, and Customers. Hence, the tuples in D are of different sizes and structures. For example, the two following tuples are included in D; the first tuple is a record in Orders and the second one – in Shippers:

1. ((orderID, 10473), (orderDate, 2/2/2017), (totalPayment, 1350), (shippedBy.ShipperID, 73e), (handledBy.employeeID, 1))
2. ((ShipperID, 73e), (shipperName, Speedy Express))

In order to support a deep exploration of business process behaviour, we define nine generic operations for mapping information among the three building blocks, $M_{X \rightarrow Y}$, where X and Y are one of E, T, or D. A mapping of type $M_{E \rightarrow T}$, also denoted as ET(E'), for example, maps subsets of events (E') from the event log (E) to subsets of (related) transactions (T') from the transaction log (T). Table 3 lists and provides a high-level description of these types of operations, along with examples. The next section provides some preliminary results from using our approach in four useful scenarios related to the well-known Northwind example [6].

---

[2] This definition abstracts from low level implementation aspects such as keys and foreign keys.

Table 3. Operations of the suggested approach

#	Type	Description	Notation	Examples
1	Event2Event	Exploring sets of events	$M_{E\rightarrow E}$	All cases in E' in which activity B is executed at least th time units after activity A: $EE_{A\rightarrow^{th}B}(E') = \{c\in E \mid \exists c'\in E'$ such that e'.activity=A $\wedge$ e.activity=B $\wedge$ e.caseID = e'.caseID $\wedge$ e.timestamp - e'.timestamp > th\}
2	Event2Transaction	Exploring sets of transactions that correspond to sets of events	$M_{E\rightarrow T}$	All data attributes which were changed (updated or deleted) by events in E': $ET(E') = \{t\in T \mid \exists e\in E'$ such that (t.operation$\in$\{"update","delete"\} $\wedge$ t.caseObject = e.caseObject $\wedge$ t.caseID=e.caseID $\wedge$ e.timestamp $\geq$ t.beginTime $\wedge$ e.timestamp $\leq$ t.endTime\}
3	Event2Database	Exploring sets of database records that correspond to sets of events	$M_{E\rightarrow D}$	The current values of data attributes associated with cases in E': $ED(E') = \{d\in D \mid \exists e\in E'$, i, j$\in$N such that e.caseObject = d.A_{ii} $\wedge$ e.caseID=d.v_{ij}\}
4	Transaction2Event	Exploring sets of events that correspond to sets of transactions	$M_{T\rightarrow E}$	All activities updating the value of attributes in T': $TE(T')=\{e\in E \mid \exists t\in T'$ such that e.caseObject=t.caseObject $\wedge$ e.caseID=t.caseID $\wedge$ e.timestamp $\geq$ t.beginTime $\wedge$ e.timestamp $\leq$ t.endTime\}
5	Transaction2Transaction	Exploring sets of transactions	$M_{T\rightarrow T}$	All data attributes in T' which are updated following a change in a given attribute att: $TT_{att}(T') = \{t\in T \mid \exists t'\in T'$ such that t'.attribute=att $\wedge$ t'.operation$\in$\{"update", "delete"\} $\wedge$ t.caseObject = t'.caseObject $\wedge$ t.caseID = t'.caseID $\wedge$ t.beginTime > t'.beginTime\}
6	Transaction2Database	Exploring sets of database records that correspond to sets of transactions	$M_{T\rightarrow D}$	The current values of data attributes associated with transactions in T': $TD(T')=\{d\in D \mid \exists t\in T'$, i, j$\in$N such that t.caseObject=d.A_{ij} $\wedge$ t.caseID=d.v_{ij}\}
7	Database2Event	Exploring sets of events that correspond to sets of database records	$M_{D\rightarrow E}$	All cases associated with database records in D': $DE(D') = \{e\in E \mid \exists d\in D'$, i, j$\in$N such that e.caseObject = d.A_{ii} $\wedge$ e.caseID=d.v_{ii}\}
8	Database2Transaction	Exploring sets of transactions that correspond to sets of database records	$M_{D\rightarrow T}$	All updates performed on database records in D': $DT(D') = \{t\in T \mid \exists d\in D$, i, j$\in$N such that t.caseObject=d.A_{ij} $\wedge$ t.caseID=d.v_{ij} $\wedge$ t.operation$\in$\{"update", "delete"\}\}
9	Database2Database	Exploring sets of database records	$M_{D\rightarrow D}$	Querying over the database using relational algebra operations

5   Preliminary Results

In order to evaluate the feasibility of the approach and its ability to obtain useful results, we used a variant of Microsoft Northwind example, which contains the sales data for Northwind Traders, a fictitious specialty foods export-import company [6]. Based on a model of the sales business process, part of which is shown in Fig. 1, we created through simulation a transaction log with 7500 records and a related event log with 7000 records. These corresponded to about 850 different orders. The simulated data was stored in Google Bigquery[3], a well-known big data repository, along with implementations of mapping operations, some of which are demonstrated in Table 3 (the whole set of operations is marked m_i in the sequel). We analyzed four scenarios that went along different trails of the conceptual framework, demonstrating possible useful exploration paths of process behavior that can be taken by an analyst.

Scenario #1 : Investigating "hidden" updates (T→T→E)	
Description:	
Certain updates should be regularly performed through specific activities. This scenario aims to identify cases where the changes were made in other activities, possibly avoiding control measures taken for changes that are done according to the regulations.	
Steps:	

1. Retrieve the transactions which change a certain attribute
    Input: T
    Output: S1
    Operation: m_1 of type $M_{T \to T}$
	Which transactions update the value of the attribute *total payment*?
	S1={t∈T
		t.attribute = "total payment" ∧ t.operation="update"}

2. Search for the events corresponding to the set of transactions retrieved in the previous step
    Input: S1
    Output: S2
    Operation: m_2 of type $M_{T \to E}$
	Which activities update the value of the attribute *total payment*? We expect that the relevant transactions will be executed through the activity *handle payment*
	S2={e∈E
		∃t∈S1 such that[a] e.caseObject=t.caseObject ∧ e.caseID=t.caseID ∧ e.timestamp ≥ t.beginTime ∧ e.timestamp≤t.endTime}

3. From the set of events retrieved in the previous step, select the events in which the activity is unexpected
    Input: S2
    Output: S3
    Operation: m_3 of type $M_{E \to E}$
	Which activities other than *handle payment* update the value of the attribute *total payment*?
	S3= {e ∈ E
		e.activity ≠ "handle payment" }

Example of findings/conclusions	
The analyst found a few cases where the *total payment* value was updated through the *create new order* activity (by directly changing the value in the order record). These cases potentially bypassed the chief accountant who was not automatically informed of the update (as is the case for updates that take place through the *handle payment* activity).	

[a] We assume that writing to the event log occurs within the time interval of the corresponding transaction, namely, t.beginTime ≤ e.timestamp ≤ t.endTime.

---

[3] https://cloud.google.com/bigquery.

Scenario #2: Investigating non-compliance of data-related business rules (E→E→T→T →D)	
Description:	
This scenario attempts to identify cases of non-compliance to a business rule that relates to data values. The scenario starts with retrieving events related to the business rule, continues with identifying corresponding data changes, and ends with retrieving additional information required for the business rule from the database.	
Steps:	

1. Filter the event log
     Input: E
     Output: S1
     Operation: m_4 of type $M_{E→E}$
	Assuming the first business rule from Section 3[b], which events in the event log are related to the activity *handle payment*?
	S1={e∈E │ e.activity ="handle payment"}

2. For the events retrieved in the previous step, search for all related data changes that satisfy the condition in the business rule
     Input: S1
     Output: S2
     Operation: m_5 of type $M_{E→T}$
	For the above events, which transactions changed the attribute *total payment* to a value greater than 10,000?
	S2 = {t∈T │ ∃e∈S1 such that t.caseObject = e.caseObject ∧ t.caseID = e.caseID ∧ e.timestamp ≥ t.startTime ∧ e.timestamp ≤ t.endTime ∧ t.attribute="total payment" ∧ t.newValue>10000}

3. For the set of transactions retrieved in the previous step, search for additional information relevant to the business rule from the database
     Input: S2
     Intermediate: S3'
     Output: S3
     Operation: m_6 of type $M_{T→D}$,
                  m_7 of type $M_{D→D}$
	For the set of transactions retrieved in the previous step, which orders (S3') were performed by non-manager employees (S3)?
	S3'={d∈D │ ∃t∈S2, i, j∈N such that t.caseObject=d.A_{ij} ∧ t.caseID=d.v_{ij}}
	S3= $\pi_{Orders.*}$ ($\sigma_{Employees.role ≠ "manager"}$ (S3' ⋈ Employees))

Example of findings/conclusions	
In some cases, the analyst found that employees exceeded their authority.	

[b] If *total_payment* > 10,000, then *handle_payment* is performed by a *manager*.

Scenario #3: Investigating delays in the process (E→E→D→D→D→T)	
Description:	
This scenario is guided by the question what causes delays in certain activities. It starts with identifying a set of cases in which delays are detected, then additional information that according to domain knowledge may be related to delays is retrieved from the database, and finally past changes in these data elements are examined.	
Steps:	

1. Do performance analysis, e.g. [16]
     Input: E
     Output: S1
     Operation: m_8 of type $M_{E→E}$
	For which cases was the *ship & close order* activity delayed with respect to the *set delivery information* activity over a threshold *th* set by the analyst?
	S1= {e∈E │ ∃e'∈E such that e'.activity="set delivery information" ∧ e.activity="ship & close order" ∧ e.caseID = e'.caseID ∧ e.timestamp - e'.timestamp > th}

2. For cases from the previous step, search additional information from the database
     Input: S1
     Intermediate: S2'
     Output: S2
     Operation: m_9 of type $M_{E→D}$,
                  m_{10} of type $M_{D→D}$
	For the above cases, what are the shipping companies (S2) involved in the corresponding orders (S2')?
	S2'={d∈D │ ∃e∈S1, i, j∈N such that e.caseObject = d.A_{ij} ∧ e.caseID=d.v_{ij}}
	S2 = $\pi_{Shippers.*}$ (S2' ⋈ Shippers)

3. For the data elements addressed in the previous step, search for data records sharing the same value
     Input: S2
     Output: S3
     Operation: m_{11} of type $M_{D→D}$
	Which orders were shipped by the retrieved shipping companies?
	S3 = $\pi_{Orders.*}$ (S2 ⋈ Orders)

4. Search for changes related to the information retrieved from the database
     Input: S3
     Output: S4
     Operation: m_{12} of type $M_{D→T}$
	For the above orders, how many times was the *delivery date* changed on the average?
	S4 = {t∈T │ ∃d∈S3, i, j∈N such that t.caseObject=d.A_{ij} ∧ t.caseID=d.v_{ij} ∧ t.attribute= "delivery date" ∧ t.operation="update"}[c]

Example of findings/conclusions	
The analyst discovers that the delays are commonly caused by a few shipping companies that frequently change their planned delivery date.	

[c] Additional processing of the results in S4 was performed for obtaining the average number of changes per order.

Scenario #4: Investigating non-conformance of control flows (E→E→T→T→T→D→D)	
Description:	
This scenario addresses what underlies control flow related non-conformance, observed using standard process mining. The scenario starts with a specific non-conformant sequence and asks what data attributes are changed in these cases. With the identified attributes, additional drill-down is performed, identifying questionable behavior for a set of cases. For these cases related data values are retrieved.	
Steps:	
1. Perform conformance checking and identify non-conformant control flows Input: E Output: S1 Operation: m_{13} of type $M_{E→E}$	For which cases *set delivery information* follows *handle payment* (violating the model)? S1= {e∈E | ∃e' ∈ E such that e.activity = "set delivery information" ∧ e.caseID = e'.caseID ∧ e'.activity = "handle payment" ∧ e.timestamp > e'.timestamp}
2. Search for all data attributes changed by the set of cases retrieved in the previous step Input: S1 Output: S2 Operation: m_{14} of type $M_{E→T}$	For the above cases, what data attributes were changed (updated/deleted) in all events of these cases (before or after the mentioned activities)? S2 = {t∈T | ∃e∈S1 such that t.operation∈ {"update", "delete"} ∧ t.caseObject=e.caseObject ∧ t.caseID=e.caseID }
3. From the set of data attributes retrieved in the previous step, select a "suspicious" one and search for all data attributes changed afterwards in the same traces[d] Input: S2 Output: S3 Operation: m_{15} of type $M_{T→T}$	The analyst selects the suspicious attribute *quantity* which was changed only in a small number of cases. For these cases, what additional attributes were changed *following* quantity's change? S3 = {t∈T | ∃ t' ∈ S2 such that t'.attribute ="quantity" ∧ t'.operation∈ {"update", "delete"} ∧ t.caseObject=t'.caseObject ∧ t.caseID =t'.caseID ∧ t.beginTime > t'.beginTime}
4. Based on the information in the previous steps, select an unusual combination of attribute changes (e.g., A is changed without any change in B) and retrieve all cases in which such an unusual combination occurs Input: S2 (rather than S3; S3 helped to determine the unusual combination) Output: S4 Operation: m_{16} of type $M_{T→T}$	The analyst selects the unusual combination of quantity change without any change in the total payment. What are the cases in which the quantity is changed without changing the total payment afterwards? S4 = {t ∈ S2 | t.attribute ="quantity" ∧ ¬∃t' ∈T such that t.caseObject=t'.caseObject ∧ t.caseID=t'.caseID ∧ t'.operation∈ {"update", "delete"} ∧ t'.attribute="total payment" ∧ t.beginTime < t'.beginTime}
5. From the set of cases retrieved in the previous step find additional information from the database. Input: S4 Intermediate: S5' Output: S5 Operation: m_{17} of type $M_{T→D}$, m_{18} of type $M_{D→D}$	For the cases where quantity was changed without a following change in *total payment* (S5'), who was the customer (S5)? S5'={d∈D | ∃t∈S4, i, j∈N such that t.caseObject=d.A_{ij} ∧ t.caseID=d.v_{ij}} S5 = $\pi_{Customers.}$• (customers ⋈ S5')
Example of findings/conclusions	
Most of the investigated cases involve one specific customer, for whom special payment agreements exist. Thus, the normative behavior in these cases is different than the one specified in the process model.	

[d] Assuming the following steps are of interest based on domain knowledge.

6   Discussion

The three sources of information and the nine mapping operations suggested by our approach support a dynamic, flexible, human-driven exploration that enables gaining a deep understanding of what underlies observed behavior. This was demonstrated in four plausible scenarios on the Northwind example.

The first scenario, of investigating "hidden" updates, demonstrates a mapping from the transaction log to the event log. With this mapping the importance of the business-level information, appearing only in the event log, becomes apparent, abstracting from the low-level data operations of the transaction log. Such analysis would not be possible with event logs generated from transaction logs (as suggested in [5]), since data updates may be made through different activities due to workarounds. Revealing such cases, as demonstrated in this scenario, can support taking appropriate measures.

The second scenario shows how compliance to business rules which rely on both control flow and data can be checked. Indeed such rules could potentially be checked with a data-enriched log using data aware compliance checking [9]. However, this requires inclusion of all the relevant data in the log in advance, and would not be possible for rules whose relevant data is not in the log. Particularly, when new rules are introduced, an existing log might not support compliance checking with them.

The third and the fourth scenarios start with standard process mining analyses of performance or conformance. They then show how findings in such cases can lead to further investigation. In the third scenario the operations that are used are of mapping event log cases to database values, internal querying of the database, and mapping data values to transaction log changes. The fourth scenario goes from the process mining findings to the transaction log, refining the investigation and turning to the database records. These two scenarios demonstrate the possibility to address different subsets of the available information, focusing, expanding, and refocusing, by following relations among data elements. The investigation steps can be determined by the analyst based on domain knowledge and the specific findings of each step.

Note that in the above scenarios, we did not use the full capabilities of relational algebra that enables aggregation functions, intersections, and more. These can further support the analyst in determining the exploration directions. Moreover, although we refer to three sources of data, these actually stand for three types of sources, so multiple formats of event logs, databases and transaction logs can be used together (once the relations among them are defined). In our proof-of-concept prototype we relied on a specific environment, namely, Google Bigquery to which the event and transaction logs were loaded. However, the suggested approach is not dependent on this specific platform, although detailed implementation changes may be required to adapt our prototype to other platforms.

Finally, it may be argued that providing only a set of atomic operations is not sufficient for the needs of an analyst, and some automation of combinations is missing. However, since, as demonstrated, the exploration is highly dependent on the questions asked and the intermediate results, the use of atomic operations is essential to support various kinds of combinations. Yet, some automated or semi-automated methods are desirable, but their development should follow an extensive study of the common combinations actually used by analysts in such explorations.

7 Conclusion and Future Work

The approach suggested in this paper aims at broadening the scope of possible analyses of process behavior. While existing process mining techniques are capable of indicating unusual or questionable behavior and bring it to attention, further investigations of the causes and roots of observed phenomena are limited. Our basic premise is that human-driven exploration is essential and should be supported. To this end we suggest an approach that combines information from three related data sources, event logs, transaction logs, and databases, in order to support deep, human-driven exploration of business process behavior. We demonstrate possible utilization of the approach and its business value in four useful scenarios using Northwind example.

The proposed approach can open promising directions for future work. It can be used as a basis for the identification of interesting patterns of data-related process behavior, which can then be detected for monitoring and analysis purposes. Furthermore, the actual combinations and exploration trails used by analysts can be studied as a basis for additional automation. For developing scalable environments that utilize the suggested approach, several steps are still needed. First, usability of the approach should be evaluated. Particularly, we intend to examine how analysts can use dynamic querying in order to add and refine mapping operations between data sources and explore different aspects of process behavior. Second, a methodological support for the ETL (extract, transform, and load) phases is needed for extending the supported formats and supporting their preparation for analysis. Finally, additional tools, such as data impact analysis [12] and data mining tools (e.g., generating decision trees or clustering [7]), should be integrated for supporting extended analysis possibilities.

Acknowledgment. This research is supported by the Israel Science Foundation under grant 856/13.

References

1. Bezerra, F., Wainer, J., van der Aalst, W.M.P.: Anomaly detection using process mining. In: Halpin, T., Krogstie, J., Nurcan, S., Proper, E., Schmidt, R., Soffer, P., Ukor, R. (eds.) BPMDS/EMMSAD -2009. LNBIP, vol. 29, pp. 149–161. Springer, Heidelberg (2009). https://doi.org/10.1007/978-3-642-01862-6_13
2. Calvanese, D., Kalayci, T.E., Montali, M., Tinella, S.: Ontology-based data access for extracting event logs from legacy data: the onprom tool and methodology. In: Abramowicz, W. (ed.) BIS 2017. LNBIP, vol. 288, pp. 220–236. Springer, Cham (2017). https://doi.org/10.1007/978-3-319-59336-4_16
3. de Leoni, M., van der Aalst, W.M.P.: Aligning event logs and process models for multi-perspective conformance checking: an approach based on integer linear programming. In: Daniel, F., Wang, J., Weber, B. (eds.) BPM 2013. LNCS, vol. 8094, pp. 113–129. Springer, Heidelberg (2013). https://doi.org/10.1007/978-3-642-40176-3_10
4. De Leoni, M., van der Aalst, W.M.:. Data-aware process mining: discovering decisions in processes using alignments. In: Proceedings of the 28th Annual ACM Symposium on Applied Computing, pp. 1454–1461. ACM, March 2013
5. de Murillas, E.G.L., Reijers, H.A., Van Der Aalst, W.M.: Connecting databases with process mining: a meta model and toolset. Softw. Syst. Modeling (2018). https://doi.org/10.1007/s10270-018-0664-7
6. Dyer, J.N., Rogers, C.: Adapting the access northwind database to support a database course. J. Inf. Syst. Educ. **26**(2), 85 (2015)
7. Hand, D.J.: Principles of data mining. Drug Saf. **30**(7), 621–622 (2007)
8. Jans, M., Soffer, P.: From relational database to event log: decisions with quality impact. In: Teniente, E., Weidlich, M. (eds.) BPM 2017. LNBIP, vol. 308, pp. 588–599. Springer, Cham (2018). https://doi.org/10.1007/978-3-319-74030-0_46
9. Ramezani, E., Fahland, D., van der Aalst, W.M.P.: Where did i misbehave? Diagnostic information in compliance checking. In: Barros, A., Gal, A., Kindler, E. (eds.) BPM 2012. LNCS, vol. 7481, pp. 262–278. Springer, Heidelberg (2012). https://doi.org/10.1007/978-3-642-32885-5_21

10. Rozinat, A., van der Aalst, W.M.P.: Decision mining in ProM. In: Dustdar, S., Fiadeiro, J.L., Sheth, A.P. (eds.) BPM 2006. LNCS, vol. 4102, pp. 420–425. Springer, Heidelberg (2006). https://doi.org/10.1007/11841760_33
11. Taghiabadi, E.R., Gromov, V., Fahland, D., van der Aalst, W.M.P.: Compliance checking of data-aware and resource-aware compliance requirements. In: Meersman, R., Panetto, H., Dillon, T., Missikoff, M., Liu, L., Pastor, O., Cuzzocrea, A., Sellis, T. (eds.) OTM 2014. LNCS, vol. 8841, pp. 237–257. Springer, Heidelberg (2014). https://doi.org/10.1007/978-3-662-45563-0_14
12. Tsoury, A., Soffer, P., Reinhartz-Berger, I.: Towards impact analysis of data in business processes. In: Schmidt, R., Guédria, W., Bider, I., Guerreiro, S. (eds.) BPMDS/EMMSAD - 2016. LNBIP, vol. 248, pp. 125–140. Springer, Cham (2016). https://doi.org/10.1007/978-3-319-39429-9_9
13. Van der Aalst, W.: Process Mining: Discovery, Conformance and Enhancement of Business Processes. Springer-Verlag, Heidelberg (2011). https://doi.org/10.1007/978-3-642-19345-3
14. Aalst, W.M.P.: Extracting event data from databases to unleash process mining. In: vom Brocke, J., Schmiedel, T. (eds.) BPM - Driving Innovation in a Digital World. MP, pp. 105–128. Springer, Cham (2015). https://doi.org/10.1007/978-3-319-14430-6_8
15. Van der Aalst, W., Adriansyah, A., van Dongen, B.: Replaying history on process models for conformance checking and performance analysis. Wiley Interdiscip. Rev.: Data Min. Knowl. Discov. 2(2), 182–192 (2012)
16. Winslett, M., Smith, K., Qian, X.: Formal query languages for secure relational databases. ACM Trans. Database Syst. (TODS) 19(4), 626–662 (1994)

Conceptual Modeling Studies II

An In-Depth Benchmarking Study on Bill of Materials for High-End Manufacturing

Yurui Wang, Shanlei Mu, Feiran Huang, Wei Lu$^{(\boxtimes)}$, and Yueguo Chen

DEKE, MOE and School of Information, Renmin University of China, Beijing, China
{wangyurui,msl,huangfeiran,lu-wei,chenyueguo}@ruc.edu.cn

Abstract. A piece of the bill of materials (a.b.a BOM) data is a hierarchical graph data model showing the assembly structure to manufacture an end product. The BOM is widely used in modeling product structures and is of great importance in high-end manufacturing. Therefore, the conduction of an in-depth benchmarking study on the BOM is of practical necessity for high-end manufacturing. While the state-of-the-art work focuses on a general graph model benchmarking, investigating a hierarchical graph data model benchmarking still remains an open problem. To address this issue, we make three contributions in this paper. First, we propose the BOM data generator by taking into consideration the given application scenario. Second, we abstract a group of queries that are widely used in high-end manufacturing. Third, we conduct an in-depth evaluation over both relational databases and graph databases that can be used to manage BOMs, and establish a unified evaluation benchmark for querying and modifying BOMs stored in different database systems.

Keywords: Database benchmark · Bill of materials · Graph data
High-end manufacturing

1 Introduction

Driven by emerging technologies such as cloud computing, big data, and the Internet of Things, a new round of transformation and upgrading in the manufacturing industry has been started. For high-tech manufacturing industry, in order to keep track of changes in product and maintain accurate lists of required components, the most common data structure being used is bill of materials.

A bill of materials is a list of the raw materials, sub-assemblies, intermediate assemblies, sub-components, parts and the quantities of each needed to manufacture an end product [8]. It is a structure used to guide to manufacture the final product. The BOM data gives instruction to the manufacturing chain on what type of and how many raw materials and components are required, and can generate the procurement plan. It also includes information about packaging and assembling, and is often represented as a tree structure with hierarchical relationship among different components and materials [7]. Here, we give an example of contracted BOMs of three different products in Fig. 1 to illustrate the idea of BOM [7].

© Springer Nature Switzerland AG 2018
J. C. Trujillo et al. (Eds.): ER 2018, LNCS 11157, pp. 75–90, 2018.
https://doi.org/10.1007/978-3-030-00847-5_7

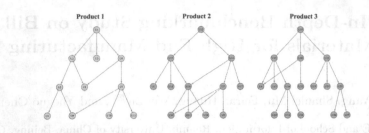

Fig. 1. An example of contracted BOMs of three different products.

Using this structure, the BOM ensures that products are produced according to the correct schema and are finished on time, which is fundamental to a company's profitability.

The BOM can be stored in either relational database or graph database, reaching different performance when accessing and modifying data. Given different data structures and query workloads, choosing the right database is important to enhance the query performance. However, currently, we don't have a unified evaluation benchmark for querying and modifying big BOM data stored in different database systems. A benchmark can give instructions to high-end manufacturing industry about choosing the right database system to store and query BOMs. Also, the manufacturers can make sure their products are tailored to produce good performance according to the benchmark [9]. Therefore, a unified evaluation benchmark is needed in many aspects.

However, to measure the performance of a database system can be a very complex task. Performance evaluation of database system is a non-trivial activity [9], made more complicated by the existence of various data structures and query workloads serving specific requirements. Identify certain key aspects generally desired of all database systems and try to define benchmarks for them are of great importance [9] and are very complex.

To measure the performance, a data generator and a query load generator are needed to simulate real-world application scenarios. BOMs are of hierarchical nature [8] and have parent-child relationships, generating data that has similar structures with BOMs is difficult, and have lots of parameters to configure. Querying the BOM is a common action in many business and industrial environments, and various types of queries and operations are needed to stimulate the actual application.

In this paper, we design a test data generator to create BOM structures of different data distribution. Based on actual demands in manufacturing industry, we devise query workloads to compare the performance and evaluate each database system using the generated data. The unique contribution of this paper is that we establish a unified evaluation benchmark for querying and modifying BOM stored in different database systems, which is of far-reaching significance for the development of manufacturing industry.

2   Related Work

Database benchmark is an important topic for it gives the measurements about overall performance of each database system, and has been studied a lot. Kabangu [2] provides a good survey for TPC benchmarks designed for assessing typical relational database queries in business and industrial applications. Here, we briefly revisit some of the recent work in this area.

The TPC Benchmark H (TPC-H) is a decision support benchmark consisting of a suite of business oriented queries and concurrent data modifications [3]. TPC benchmarks emphasize join query, selections, aggregations, and sort operations. However, because graph databases target different types of queries, the benchmarks on these relational databases are not sufficient to assess their performance [5].

The object-oriented database (OODB) has similarities with the graph database in some ways. Although OODB benchmarks create graphs, the graphs have a very different structure from typical graphs in graph analysis applications [5]. Investigating the real world data, we can find that graphs are irregular. The degree of the nodes exhibit a large variance, nodes are clustered in communities and graphs have small diameters [5]. Therefore, a graph data generator is needed to better characterize the feature of BOM data.

The problem of schema-driven generation of synthetic graph instances and corresponding query workloads for use in experimental analysis of graph database systems has been studied before, and there already exists the design and engineering principles of gMark, a domain- and query language-independent graph instance and query workload generator [13]. However, gMark generates graph instances and query workloads of random distribution, and only controls the diversity of properties [13]. To establish a unified evaluation benchmark for querying and modifying BOM, we need to create BOM structures of different data distribution and devise query workloads that have real-world application.

In summary, there are some benchmarks for general-purpose database management systems, but there is a lack of evaluation benchmarks for high-end manufacturing database management systems, and high-end manufacturing data is quite different and needs to be evaluated in another way. Therefore, based on the evaluation benchmarks of the existing database management system benchmarks and the feature of BOM data, we need a more targeted benchmark for high-end manufacturing that are more suitable for practical use in light of the characteristics of high-end manufacturing graph data.

3   Overview of Benchmark

Based on the above research, the main task of this paper is to develop benchmarks and test tools for high-end manufacturing databases. The specific research steps and contents are as follows:

(1) Data Model Analysis and Design
(2) Data Generator Design and Implementation
(3) Query Load Generator design and implementation
(4) Benchmarking and Iterative Optimization.

The overall idea is shown in Fig. 2. We analyze the structural characteristics of the graph data, and generate multiple types of query load instances based on the query result selectivity. A set of evaluation criteria is formulated to quantitatively evaluate the performance of the different databases.

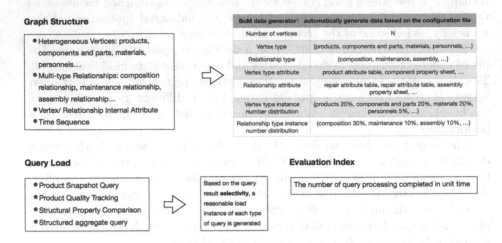

Fig. 2. The overall idea.

Figure 3 shows the framework of this approach.

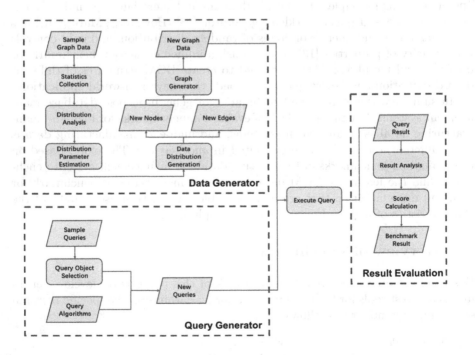

Fig. 3. The framework.

4   Data Generation

4.1   Data Model Analysis

To analyze the data structure of BOM for high-end manufacturing, we use a bill of materials for two cranes as a research case. Figure 4 depicts the BOM data of the two cranes. Each material can be connected with more than 10 types of materials and the BOM is the basis for resource planning (purchasing, inventory, and outsourcing).

Fig. 4. The BOM data of the two cranes in neo4j.

Based on features and requirements, we design hierarchy data structures that comprise of parent child relationship to describe the connection and structure within the product, and can be used as the basis for quality tracking, which is the bridge between the various departments of the factory [6]. To be more specific, we designed the data structures of the BOM's objects and relationships according to the data set of the real application which are shown in Figs. 5 and 6 as below

plm_m_oid	unique identifier, used for the relationship connection
plm_oid	unique identifier
plm_m_id	unique identifier
plm_i_name	part name
plm_i_createtime	create time
plm_i_checkintime	check in time
plm_cailiao	material number
plm_checkintime	product check in time
plm_weight	weight
plm_wllx	material type
plm_wlly	source of material
plm_guige	specification
plm_gylx	process type

plm_oid	relationship (edge) unique identifier
plm_leftobj	left object identifier (parent object)
plm_rightobj	right object identifier (child object)
plm_createtime	create time
plm_order	order number
plm_jianhao	number information
plm_number	number information

Fig. 5. The data structure of objects.   **Fig. 6.** The data structure of relationships.

4.2 Data Statistic

In order to generate data set as close as possible to reality, we do some statistics on the data of the two graphs. We import the BOM data of the two cranes into Neo4j, on which we do some statistics about the graph of BOM.

The steps of the data statistics are:

Step 1. Collect statistics on name, in-degree edge and level of each node in the graph.

Step 2. Get statistics on the level of each type of node.

Step 3. For each type of node in the graph, collect statistics on level of each type of in-degree edge (Get statistics on level of each type of edge of each in-degree node).

4.3 Distribution Analysis

Three Basic Distributions in Graph Data. After some investigation, we found that graph data distribution from reality is usually divided into three categories: normal distribution, uniform distribution and zipfian distribution.

Normal Distribution. The normal distribution are important in statistics and are often used in the natural and social sciences to represent real-valued random variables whose distributions are not known [11].

Uniform Distribution. The uniform is a useful choice when a random variable has definite maximum and minimum values, and there is no basis for assuming that any range of its values is more likely than any other range of the same size [12].

Zipfian Distribution. The zipfian distribution represents a distribution that satisfies Zipf's law. Within a wide area and making appropriate approximations, many natural phenomena conform to Zifp's law.

Data Distribution Analysis Algorithm. The key step in distribution analysis is to determine the parameters in the distribution probability density function. Based on the existing data set, we need to perform recursive iterative algorithm to find the optimal solution of a parameter (Fig.7).

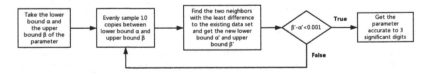

Fig. 7. The algorithm to determine the parameter in the distribution.

4.4   Scalable Data Distribution Generator

The amount of data should be such that the new data not only conforms to the hierarchical distribution of sample data, but also conforms to the newly set level and the amount of data.

We define the size of the sample graph data as S_{old}, the number of layers as L_{old}, the size of the generated target map data as S_{new}, and the number of layers as L_{new}. We introduce a new concept named scaling factor:

Definition 1 (Scaling factor). *Scaling factor is used to scale the distribution probability density function from initial range to the new range. We use the maximum level of the graph to scale.*

After we get the scaling factor, as well as with the parameter of size, upper bounds and lower bounds, we can scale the distribution probability density function to a wider or narrower range.

The Scaling of Nodes/Edges. In high-end manufacturing industry, the application scenario of scale expansion may be to expand the functions of industrial products, such as "adding a new mechanical module" and "manufacturing machinery with larger volume product". In these scenarios, if there is no change in the core design, then the composition of the parts is only multiplied by the number, and the composition between the parts does not change much. Therefore, distribution density does not change, but only increases in number. The size of scaling factor S_{size} is $S_{size} = \frac{S_{new}}{S_{old}}$.

The Scaling of the Nodes' Level. Second, we need to consider the expansion of the hierarchy. The method we use is to scale the distribution density function. The size of the scaling factor S_{level} is $S_{level} = \frac{L_{new}}{L_{old}}$. The resulting distribution is exactly the L_{new} layer graph data. This ensures that our new probability density distribution produces the same node distribution on the L_{new} layer as the sample data set.

The Scaling of the Edges' Level. First, we introduce Theorem 1. From this theorem, we find that the other node pointing to a node cannot be higher than the upper layer of this node. Otherwise, the hierarchical relationship of the entire graph will be disrupted.

Theorem 1. *Given two nodes N_1 and N_2, their levels are L_1 and L_2 respectively. If there is an edge E between N_1 and N_2 and edge E points from N_1 to N_2, then we can conclude that $L_2 - 1 \leq L_1$.*

Proof. Suppose that $L_1 < L_2 - 1$, since the level of each node is determined by the length of the shortest path from the root node to the node, if the level of node N_1 is L_1, since edge E points from N_1 to N_2, then N_2 Level $L_2 = L_1 + 1$, substituting this into hypothesis, we get $L_1 < L_1$, contradict.

4.5   Generate New Nodes

The steps to generate new nodes are:

Step 1. Collect statistics on the existing graph and store it in a dictionary.

Step 2. Perform data distribution analysis on each type of node in the graph.

Step 3. Get the number of layers ($Level_{max\_new}$) of the graph to be generated, compare to the original layer ($Level_{max\_old}$) of the original graph data and get scaling factor (SF).

Step 4. Scale distribution based on SF. Multiply the density distribution for each type of node by SF (equivalent to extending the distribution from $Level_{max\_new}$ to $Level_{max\_old}$).

Step 5. Use the data distribution to generate new data. The data size is the scale of the original data of this type of node multiplied by SF.

After finishing these steps, all nodes are generated and distributed at a certain level. The algorithm can be defined as:

Algorithm 1. Generate new nodes

1: **procedure** GENERATE_NODES($raw\_nodes, size\_raw, size\_target, level_raw, level\_target, nodeid$)
2:     **for** i in raw_nodes **do**
3:         size = round(1.0 * size_target)/(size_raw * len(raw_data))
4:         scale = (1.0 * level_target) /(level_raw)
5:         item = ds.simulate(size, scale)
6:     **end for**
7: **end procedure**

4.6   Generate New Edges

To determine the corresponding original node in the sample data, we use the method of "retracting" the scaling, that is, dividing the number of layers of the new node.

The steps to generate new edges are:

Step 1. Collect statistics on the existing graph.

Step 2. Get the number of layers ($Level_{max\_new}$) of the graph to be generated, compare to the original layer ($Level_{max\_old}$) of the original graph data and get scaling factor (SF).

Step 3. Analyze the data distribution of each type of edge of each node.

Step 4. Scale according to scaling factor.

Step 5. Use the data distribution to generate new edges. At this point, all nodes have a certain incoming degree.

Step 6. Divide the incoming degree by the scaling factor: $S_{level} = \frac{L_{new}}{L_{old}}$

The implementation of this function is as follows:

Algorithm 2. Determine the level distribution of the incoming edges

---

1: **procedure** INCOMING_EDGES_LEVEL(*level, level_new, level_old, raw_outs*)
2:　　in_level = round(1.0 * level / (1.0 * level_new / level_old))
3:　　position = random.randint(0, len(raw_outs[i][outs_level]['number'])-1)
4:　　item_out = DataSimulator(raw_outs[i][outs_level]['level']).simulate(int(round(size_target / size_raw * raw_outs[i][outs_level]['number'][position])), (level_target / level_raw))
5: **end procedure**

Next, we need to obtain each incoming nodes when we determine the incoming distribution. The specific algorithm is as follows:

Algorithm 3. Match the incoming edge with node

1: **procedure** MATCHING_NODES(*outs_level, level_new, level_old, raw_outs*)
2:　　names == raw_outs[i][outs_level]['name']
3:　　**if** names[m] in new_nodes_by_name and item_out_data[count2] in new_nodes_by_name[names[m]] **then** target_nodes = new_nodes_by_name[names[m]][item_out_data[count2]]
　　　　temp_out = target_nodes[(inturn_count1 % len(target_nodes))].id
4:　　**end if**
5: **end procedure**

5　Query Load Generation

We focus on the issues of what query workloads are generated in different environments, how query workloads are implemented, and how query workloads are combined.

5.1　Basic Queries Definition

We divide the query load into four basic queries and define as follows:

Definition 2 (4 Basic Queries)

(1) Generate structure of a product.
(2) Where used query for a component.
(3) Query the differences between the structures of the products.
(4) Product structure aggregation.

5.2   Basic Queries Implementation

(1) Generate Structure of a Product: Oracle handles queries for generating structures by performing several join operations between tables. We extend neo4j by writing user-defined procedures in Java, and call the procedures from Cypher.

Algorithm 4. Gene_struct in Oracle	**Algorithm 5.** Gene_struct in neo4j
1: **procedure** GENER-ATE_STRUCTURE(*rootid, lev*) 2: *cursor* = *select* inform *from* item_table p *join* relation_table r *join* item_table c *start with* rootid *connect by* p.plm_m_oid = *prior* r.plm_rightobj *and* level < *lev* 3: *fetch cursor and return results* 4: **end procedure**	1: **procedure** GENER-ATE_STRUCTURE(*rootid, lev*) 2: *td* = traversalDescription().depthFirst().relationships(rtp, OUTGOING).evaluator(Evaluators.toDepth(lev)) 3: Node node = graphDb.findNode(label, "plm_m_id",rootid) 4: **end procedure**

(2) Where Used Query for a Component: In Oracle, we define where_used function to query the use of a part in a product instance. Provided rootid is used to find the root node. In neo4j, we set the relationships() parameter to Direction.INCOMING.

Algorithm 6. Where used in Oracle	**Algorithm 7.** Where used in neo4j
1: **procedure** WHERE_USED(*rootid*) 2: *cursor* = *select* inform *from* item_table p *join* relation_table r *join* item_table c *start with* rootid *connect by prior* p.plm_m_oid = r.plm_rightobj *order sibling by* p.plm_m_id 3: *fetch cursor and return results* 4: **end procedure**	1: **procedure** WHERE_USED(*rootid*) 2: TraversalDescription *td* = graphDb.traversalDescription().depthFirst() .relationships(rtp, INCOMING).evaluator(toDepth(lev)) 3: Node node = graphDb.findNode(label, "plm_m_id",rootid) 4: **end procedure**

(3) Query the Differences Between the Structures of the Products: In Oracle, we define function structure_diff. In this connection, compare the differences between the source and target. In neo4j, we compare the structural differences at each depth between products.

Algorithm 8. Struct diff in Oracle
1: **procedure**
STR_DIFF(*sourceid, targetid*)
2:    *cursor = select* **when** s.plm_r_oid
is null **then** ' ←' **when** d.plm_r_oid
is null **then** ' →' **when** s.plm_r_oid
<> d.plm_r_oid **then** ' <>'   **else**
' =='
diff_type, other inform *from*
(select info *from* item_table p *join*
relation_table r *join* item_table c
start with p.plm_m_id = source_id
connect by p.plm_m_oid = *prior*
r.plm_rightobj) *s full outer join*
(select info *from* item_table p *join*
relation_table r *join* item_table c
start with p.plm_m_id = target_id
connect by p.plm_m_oid = *prior*
r.plm_rightobj) *d*
3: **end procedure**

Algorithm 9. Struct diff in neo4j
1: **procedure**
STR_DIFF(*sourceid, targetid*)
2:    **for** *i*=1 to *lev* **do**
3:        TraversalDescription   *td* =
graphDb.depthFirst() .relationships(
rtp,   OUTGOING   ) .evaluator(
atDepth(i))
4:        **for**   Path   childPath:
td.traverse(target) **do**
5:            compare relationship *with*
hashmap
6:        **end for**
7:    **end for**
8: **end procedure**

(4) Product Structure Aggregation: In Oracle, we define structure_aggr function to return the aggregated structure as strings. In this query, we use **UNION ALL** keyword to aggregate the attributes. In neo4j, we just add the aggregated attribute to results instead of the real attribute.

Algorithm 10. Struct aggr in Oracle
1: **procedure** STRUCTURE_AGGR(*sourceid*)
2:    *cursor = select* inform *from* item_table p
join relation_table r
join item_table c
where p.plm_m_id = sourceid
union all (select inform *from* full_BOM p
join (select inform *from* item_table p
join relation_table r
join item_table c) *n*
select information *from* full_BOM
3:    *fetch cursor and return results*
4: **end procedure**

Algorithm 11. Struct aggr in neo4j
1: **procedure** STRUCTURE_AGGR(*sourceid*)
2:    TraversalDescription   *td*   =
graphDb.traversalDescription().depthFirst()
.relationships(rtp, OUTGOING).evaluator(
atDepth(lev))
3:    **for** Path childPath: td.traverse(source) **do**
4:        *add relationship to hashmap*
5:    **end for**
6:    **for** Path childPath: td.traverse(target) **do**
7:        *compare relationship with hashmap*
8:    **end for**
9: **end procedure**

5.3   Large-Scale Generation of Complex Query Workloads

Once we define these 4 basic functions, we can generate lots of queries that has real-life application in business and industry. We can:

(1) Generate a product's entire or part structure by using generate_structure.
(2) Check the use of a specific component by using where_used.
(3) Compare structural differences between two products by using structure_diff.

(4) Downward to get the total number of items after aggregation by using structure_aggr.
(5) Downward aggregation to generate purchase plan by using structure_aggr and specifying the source of material (plm.wlly).
(6) Downward structural aggregation to obtain parts borrowing situation by using structure_aggr.
(7) Upward structural aggregation to get total weight by using structure_aggr, same way to get total cost, total purchase plan, total inventory plan, etc.

In industry, the number of various types of queries and operations has a certain percentage and needs to be set according to the actual needs.

6 Experiment

We empirically evaluate the performance of Oracle and neo4j in three aspects: the efficiency to import huge data, to handle basic queries and complex query workloads of real world application.

6.1 Data Insertion

We test the performance for importing data in oracle and neo4j, and Table 1 gives the summary results.

Table 1. performance for data insertion.

DBMS	Data size	Execution time
oracle	196,733 records	140.015 s
neo4j		25.431 s

6.2 Basic Query Evaluation

To evaluate performance of relational database oracle and graph database neo4j dealing with the basic queries for BOM, we generate BOM data of different sizes and levels, execute 4 basic functions on each data set 3 times, and take the average run time (ms). Figures 8 and 9 show the results.

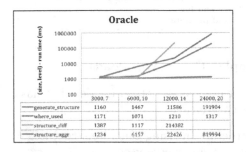

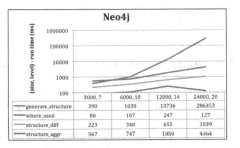

Fig. 8. Oracle running basic functions.   **Fig. 9.** Neo4j running basic functions.

6.3   Complex Query Load Evaluation

In industry, the number of various types of queries and operations has a certain percentage and needs to be set according to the actual needs of the manufacturers or users. To simulate the actual operations in high-end manufacturing industry, we design a query load pattern and test the overall performance for oracle and neo4j (Fig. 10).

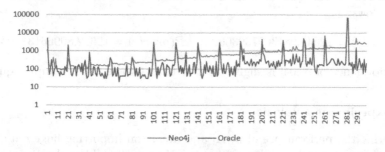

Fig. 10. General performance of oracle and neo4j.

In experiment, we generate 15 graphs of different sizes through the data generator and generated 20 sets of complex queries for each graph. We design a indicator P to measure the performance of the database under complex combine queries. We use the graph scale expression $f(N)$ as the numerator and the query time expression $g(t)$ as the denominator.

$$v = \frac{f(N)}{g(t)}$$

v_{ij} is the performance of database executing query j on graph i.

We assign different weights to the query according to its order. For the same graph, the probability of occurrence of a query should decrease exponentially. At the same time, the basic amplitude of this decline should be greater, so we set the weight according to the probability of its occurrence. w_j represents the weight of the query j for a graph. According to the definition of v_{ij} and w_j, we get the expression of P.

$$w_j = (\frac{1}{j})^{j-1}, \qquad P = \lambda \sum_{i=1}^{n} \sum_{j=1}^{m} w_j \cdot v_{ij}$$

where n is the number of graphs, m is the number of query, and λ is a constant.

Finally, we need to determine $f(N)$ and $g(t)$. In queries, long-time queries require more attention than short-time queries, so when t is small, the performance difference should be reduced. We chose the logarithmic function to solve this problem, so

$$f(N) = N, \qquad g(t) = log(t)$$

The calculation formula of the evaluation on the performance of the database under complex queries is as follows:

$$P = \lambda \sum_{i=1}^{n} \sum_{j=1}^{m} w_j \cdot v_{ij} = \lambda \sum_{i=1}^{n} \sum_{j=1}^{m} \frac{f(N_i)}{g(t_{ij})} \cdot \left(\frac{1}{j}\right)^{j-1} = \lambda \sum_{i=1}^{n} \sum_{j=1}^{m} \frac{N_i}{log(t_{ij})} \cdot \left(\frac{1}{j}\right)^{j-1}$$

According to the above formula, we conducted experiments on Neo4j and Oracle based on the generated data and queries. And the experiments results are as follows:

$$P(neo4j) = 12645.981309730309, \qquad P(oracle) = 12622.809661317044$$

The performance of Neo4j is slightly better than Oracle under complex queries.

6.4 Experiment Analysis

Considering the performance of neo4j and oracle on importing huge data, handling basic queries and complex queries based on real world applications, we set the following evaluation indexes P1, P2, and P3. We have

$$P_1 = \frac{N}{log(t)}, \qquad P_2 = \sum_{i=1}^{n} \sum_{j=1}^{m} \frac{N_i}{t_{ij}}$$

where N represents the number of records to load, t is the time required for loading data, n represents the number of types of graphs, m is the number of query types, N_i is the size of the i_{th} graph, and t_{ij} is the time spent on the i_{th} graph for querying the j_{th} query. We set $P_3 = P$ calculated in Sect. 6.3.

According to the above formulas, we have (Table 2)

Table 2. Performance indexes.

DBMS	P_1	P_2	P_3
oracle	16602	17136	15059
neo4j	19394.4	24457	14834

According to the importance of the three types of experiments, we assign the following proportions: $P = P_1 \times 0.2 + P_2 \times 0.3 + P_3 \times 0.5$. We have $P(neo4j) = 18632.98$, $P(oracle) = 15990.7$.

Because $P(neo4j) > P(oracle)$, the overall performance of neo4j for querying BOM is superior to oracle in this experiment.

7   Conclusion

We first present a data generator that satisfies the key criteria of generating the similar structures of BOM in high-end manufacturing industry, and is extensible and highly configurable by allowing lots of input parameters to control the size and structure of the graph data. We also present a query load generator to stimulate real-life application scenarios in business and industry. The query workloads can be configured in terms of the expected query selectivity of a given workload, so as to fully test the capability of a database. Using these generated data and queries, our experiments highlight the weak parts in recursive query processing for both relational database and graph database.

The novel and important contribution is that we define a unified evaluation benchmark for querying and modifying BOM stored in different database systems for the first time. This benchmark can give instructions to high-end manufacturing industry about choosing the right database system to store and query BOMs, and allows the manufacturers to make sure their products are tailored to produce good performance according to this benchmark.

Acknowledgements. This work is supported by the National Key Research and Development Program of China (No. 2016YFB1000702). The National Natural Science Foundation of China in part supports this work under Grant No. 61502504. Wei Lu is the corresponding author.

References

1. Bagan, G., Bonifati, A., Ciucanu, R.: gMark: schema-driven generation of graphs and queries. IEEE Trans. Knowl. Data Eng. **29**(4), 856–869 (2017)
2. Kabangu, S.: An Investigation of the TPC-H Benchmark Suite and Techniques Used in the Performance Optimization of Decision Support Systems (DSS). http://pppj2012.ru.ac.za/g09k3351/CSHnsThesis.pdf
3. TPC-H. http://www.tpc.org/tpch/default.asp
4. Cattell, R., Skeen, J.: Object operations benchmark. TODS **17**(1), 1–31 (1992)
5. Dominguez-Sal, D., Martinez-Bazan, N., Muntes-Mulero, V., Baleta, P., Larriba-Pey, J.L.: A discussion on the design of graph database benchmarks. In: Nambiar, R., Poess, M. (eds.) TPCTC 2010. LNCS, vol. 6417, pp. 25–40. Springer, Heidelberg (2011). https://doi.org/10.1007/978-3-642-18206-8_3
6. Zhou, C., Cao, Q.: Design and implementation of intelligent manufacturing project management system based on bill of material. Clust. Comput. **5**, 1–9 (2018)
7. Cinelli, M., Ferraro, G., Iovanella, A., et al.: A network perspective on the visualization and analysis of bill of materials. Int. J. Eng. Bus. Manag. (2017, to appear)
8. Bill of Materials. https://en.wikipedia.org/wiki/Bill_of_materials
9. Svensson, M.: Performance Analysis and Benchmarking of a Database with Cached Content. TRITA-NA-E05021. http://citeseerx.ist.psu.edu/viewdoc/download?doi=10.1.1.102.9902&rep=rep1&type=pdf
10. Paul, S.: Database systems performance evaluation techniques (2008)
11. Casella, G., Berger, R.L.: Statistical Inference, 2nd edn. Duxbury, Belmont (2001). ISBN 0-534-24312-6

12. Gailmard, S.: Statistical Modeling and Inference for Social Science (2014). ISBN-13 978–1107003149
13. Bruno, N., Chaudhuri, S., Thomas, D.: Generating queries with cardinality constraints for DBMS testing. IEEE Trans. Knowl. Data Eng. **18**(12), 1721–1725 (2006)
14. Poess, M., Stephens, J.M.: Generating thousand benchmark queries in seconds. In: Proceedings of 13th International Conference on Very Large Data Bases, pp. 1045–1053 (2004)
15. Bagan, G., Bonifati, A., Ciucanu, R., Fletcher, G., Lemay, A., Advokaat, N.: gMark: schema-driven generation of graphs and queries (2015). http://arxiv.org/abs/1511.08386
16. Bagan, G., Bonifati, A., Ciucanu, R., Fletcher, G.H.L., Lemay, A., Advokaat, N.: Generating flexible workloads for graph databases. Proc. VLDB Endow. **9**(13), 1457–1460 (2016)
17. Schmidt, A., Waas, F., Kersten, M., et al.: XMark: a benchmark for XML data management. In: Proceedings of the International Conference on Very Large Data Bases (VLDB), pp. 974–985 (2002)

Towards Data Visualisation Based on Conceptual Modelling

Peter McBrien[1](✉)[iD] and Alexandra Poulovassilis[2][iD]

[1] Department of Computing, Imperial College,
180 Queen's Gate, London SW7 2BZ, UK
p.mcbrien@ic.ac.uk

[2] Birkbeck Knowledge Lab, Birkbeck, University of London,
Malet Street, London WC1E 7HX, UK
ap@dcs.bbk.ac.uk

Abstract. Selecting data, transformations and visual encodings in current data visualisation tools is undertaken at a relatively low level of abstraction - namely, on tables of data - and ignores the conceptual model of the data. Domain experts, who are likely to be familiar with the conceptual model of their data, may find it hard to understand tabular data representations, and hence hard to select appropriate data transformations and visualisations to meet their exploration or question-answering needs. We propose an approach that addresses these problems by defining a set of visualisation schema patterns that each characterise a group of commonly-used data visualisations, and by using knowledge of the conceptual schema of the underlying data source to create mappings between it and the visualisation schema patterns. To our knowledge, this is the first work to propose a conceptual modelling approach to matching data and visualisations.

1 Introduction

Current data visualisation approaches base their visualisations on simple table data presentations, and fail to capture the full schema knowledge when the underlying data source is a structured database, such as a relational database. Furthermore, creating visualisations requires a fresh data mapping effort for each visualisation that is created, be it programmer effort or end-user effort. We propose an approach that addresses these problems by firstly defining **visualisation schema patterns** that characterise each distinct (from a data representation capability) group of commonly-used data visualisations, and secondly that uses the conceptual schema of the underlying data source to create mappings between the data schema and the visualisation schema patterns. The benefits of this approach are firstly that we use the full knowledge of the conceptual model of the underlying data to identify which are feasible visualisations for that data, by matching the data schema with the set of visualisation schemas; and secondly, once this mapping is in place, the creation of actual visual charts can utilise the mapping to extract data, drill-down, roll-up, pivot, switch visualisation *etc.*

© Springer Nature Switzerland AG 2018
J. C. Trujillo et al. (Eds.): ER 2018, LNCS 11157, pp. 91–99, 2018.
https://doi.org/10.1007/978-3-030-00847-5_8

To our knowledge, ours is the first work to propose a conceptual modelling approach to matching data and visualisations. We refer readers to [2] for a review of
related work on visualisation tools, taxonomies, recommendation, and languages
for manipulating graphical data.

2 Motivating Example

A fragment of the Mondial database [1] is illustrated in the ER diagram in
Fig. 1. It describes a schema about countries (including the current population),
the history of a country's population in the weak entity country_population, and
provinces in countries). For some countries, data about the GDP of the country
is recorded in the subset entity economy, the attributes of which are all optional,
indicated by the use of a question mark. Also recorded is which continent or
continents a country belongs to: most countries will belong 100% to one continent; but the cardinality constraint of 1:2 allows some (e.g. Russia, Turkey) to
spread over two continents, with the percent attribute of encompasses recording
the proportion of their land area that belongs to each continent.

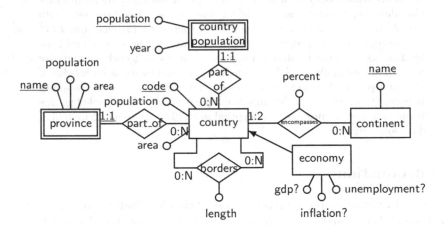

Fig. 1. ER schema of a fragment of the Mondial database

Suppose we wished to explore the relationship between inflation, unemployment, and GDP in countries. We could first extract a table of data with
scheme (country,inflation,unemployment,gdp), where country corresponds to the
key attribute code of the country entity in Fig. 1, and without null values for
inflation, unemployment, and gdp. Importing that table to Tableau, and choosing
to represent countries as a 'dimension', and putting the inflation and unemployment figures on the x and y axis, produces the chart shown in Fig. 2(a).

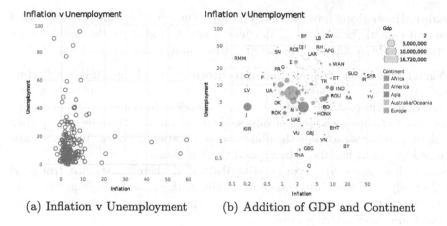

(a) Inflation v Unemployment   (b) Addition of GDP and Continent

Fig. 2. Presentation of country data in Tableau

We see that, because of a few outlying data values, most of the data appears in a small cluster to the bottom left of the diagram and is largely illegible. No use has been made of the fact that the data distribution can easily be determined to be skewed, and hence an alternative scaling could have been used. Furthermore, no suggestion is made on how to include the gdp column of the table, despite the fact that this is numeric-valued, which would suggest displaying its data using a graphical construct suitable for representing ranges of numbers. Figure 2(b) shows the result of a user (manually) determining that a logarithmic scale will better spread the data relating to the relationship between inflation and unemployment, and that the data in gdp can be used to scale the size of the circles, to make a bubble chart. Figure 2(b) also colour-codes countries by their continent—as suggested by the database schema, which connects countries to continent via a relationship with restricted (upper bound 2) cardinality.

3 Visualisation Schema Patterns

Our starting premise is that each instance of an entity in the database is associated with one or more graphic elements, which in visualisation are usually classified [5] as **marks** (points, lines, areas, *etc.*) or **channels** (colour, length, shape, coordinate, texture, orientation, movement, *etc.* of a mark). An attribute value of an entity, or the participation of an entity in a relationship, is associated with a dimension of the visualisation, and the process of visualisation is about choosing the correct graphic elements for a given schema.

Taking an approach similar to Tableau, we identify the following two major types of dimensions (which differ from the discrete and continuous classification found in [4]):

- **discrete dimensions** have a relatively small number of distinct values, that may nor may not have a natural ordering; they are used to choose a mark or to vary a channel of a mark.

- **scalar dimensions** have a relatively large number of distinct values with a natural numeric ordering (e.g. integers, floats, timestamps, dates); these are represented by a channel associated with a mark.

When a dimension is represented by a **colour channel**, then if it is a discrete dimension it lends itself to using a colour key, where each colour represents a discrete value. Alternatively, if it is a scalar dimension, then a spectrum of colours can be used to represent a range of values. Hence, in our descriptions below, when we talk of a colour we assume the ability to automatically choose between these two representations based on the type of the dimension.

Scalar dimensions are **evenly distributed** if their values are (roughly) spread evenly over the entire range of values in the dimension (many visualisations struggle to represent data where most data is in a small range of values and there are some outlying values).

As is well known [5], what we are naming discrete or scalar dimensions may have specific real-world characteristics, and may for example be a **geographical**, **temporal**, or **lexical** dimension. This characterisation then may suggest specific visualisations for their representation (*e.g.* a map, time slider, word cloud, *etc.*). However, in this paper we focus on what assistance can be given to the visualisation process by the knowledge represented in the schema of the data, and hence we only consider these real-world characteristics if required for the use of a particular visualisation. Indeed our work should be viewed as providing assistance to existing visualisation techniques, to be used where data is sourced from a structured database. Our work is therefore complementary to aspects such as task-based visualisation design and interaction during design.

In the following subsections we present successively more complex visualisation schema patterns, and the visualisations that they encompass. Our survey of visualisation techniques has so far not found any visualisations that require more complex schema patterns than those presented here, and in particular none that require a pair of relationships to be considered together.

3.1   Basic Entity Visualisations

An ER entity can be regarded as a conceptual modelling of a relational table. Many visualisations are designed to represent such tabular data, so we begin by identifying a category of visualisations that are suitable for representing an entity with its keys and attributes. This 'basic entity' visualisation schema pattern is illustrated in Fig. 3(a), where it should be noted that the key attribute k might be inherited from a parent, such as economy in Fig. 1 having an inherited key code from country. Many visualisations fit into this category, and we list below a sample to illustrate the way in which different features of each visualisation are represented in our approach (more are given in [2], e.g. choropleths, word clouds).

- Basic **bar charts** represent instances of an entity E (identified by the value of k) as bars, with the length of the bar determined by the value of an attribute a_1. Hence a_1 should be a scalar attribute.

- A **calendar chart** (found in both D3 and Google Charts) represents instances of E according to a date-valued attribute a_1.
- In **scatter diagrams** (such as in Fig. 2(a)), each point represents an instance of E, and two dimensions a_1 and a_2 are used to plot its x and y coordinates. Optionally, a third dimension a_3 can be used to colour it.
- In **bubble charts** (such as in Fig. 2(b)), each bubble denotes an instance of E; two dimensions a_1 and a_2 are used to plot its coordinates, and a third dimension a_3 its size. A fourth dimension a_4 may be used to colour it.

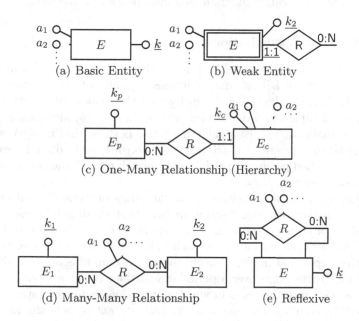

Fig. 3. Visualisation schema patterns for data visualisations

The table below summarises the above analysis, where $|k|$ denotes the number of distinct values of the key k. The upper cardinality of 100 shown in relation to the bar chart is subjective, and aesthetics-driven; it would be user-configurable in any implementation.

Basic Entity Visualisations			
Name	$|k|$	mandatory	optional
Bar Chart	1..100	a_1 scalar	-
Calendar	1..*	a_1 temporal scalar	-
Scatter Diagrams	1..*	a_1, a_2 scalar	-
Bubble Charts	1..*	a_1, a_2, a_3 scalar	a_4 colour

Note that all of the above visualisations (and indeed those listed in the following subsections) may have additional temporal scalars represented by time sliders, and discrete scalars represented by snapshot or paging options.

In our approach, visualisation schema patterns are used in conjunction with the database schema to guide the process of choosing a visualisation, by finding sub-graphs of the database schema that match each visualisation schema. Although this is an instance of the (NP-complete) subgraph isomorphism problem, the query graph (i.e. the visualisation schema) will be small and hence we anticipate fast execution times using state-of-the-art algorithms such as [3].

For example, starting with the schema in Fig. 1 and matching Fig. 3(a) against it, a match is found with the entity country, with k matching code and choices area and population for the scalars a_1 and a_2. The user can therefore be offered a bar chart or scatter diagram as a visualisation of the data.

3.2   Weak Entity Visualisations

A particular form of compound key (often arising from the representation of weak entity data in an ER schema) identifies a family of visualisations where one part of the key, k_1 (the key of the entity that the weak entity is attached to) identifies a set of tuples, and the second part of the key, k_2, identifies a tuple in the set. The visualisation schema pattern for this is shown in Fig. 3(b), where it should be noted that k_1 would match a key relationship in the data; for example, in Fig. 1 if E matched province then k_1 would match part_of and hence be based on the code of country.

The values of k_2 must lie within a similar range of values for all instances of k_1 (so as to make their visualisation in one chart meaningful). Also, we say that the values of k_2 are **complete** with respect to k_1 if it is the case that the same set of values appears for k_2 for each value of k_1. For example, the weak entity country_population in Fig. 1 meets the range requirement since the dates for population figures range over a period of less than 200 years, but it fails the completeness test since the years in which population figures are available vary from country to country. By contrast, the province entity fails the range test, since the names of provinces are almost entirely disjoint with those of countries.

As with the basic entity visualisation, there are many visualisations suited to present the weak entity visualisation, a selection of which are listed below, together with a summary table:

- In a **line chart** each line represents a distinct value of k_1; k_2 represents a scalar dimension to be plotted along the x-axis; and a_1 must be a scalar dimension to be plotted along the y-axis. XY variations allow an additional dimension a_2 to be added to the y-axis.
- In a **stacked bar chart**, distinct values of k_1 are represented by a bar, with one of the elements in the stack representing a value of k_2, and the length of the bar determined by a scalar dimension a_1. Each value of k_1 should appear with the same (or almost same) set of values for k_2 (the completeness property) so that the elements in each stack can be compared.
- In a **spider chart**, each ring represents a value of k_1 and each spoke a value of k_2; the intersection of the ring with a spoke is determined by a_1.

Weak Entity Visualisations					
Name	$|k_1|$	$|k_2|$	complete	mandatory	optional
Line Chart	1..20	1..*	no	k_2, a_1 scalar	a_2 scalar
Stacked Bar Chart	1..20	1..20	yes	a_1 scalar	-
Spider Chart	3..10	1..20	yes	a_1 scalar	-

3.3 One-Many Relationships

Relationships that are one-many (such as part_of in Fig. 1) lend themselves to visualisations that are hierarchical in nature. The visualisation schema for these relationships is illustrated in Fig. 3(c), where the entity that is on the 'many' side of the relationship (such as country for part_of) will be considered the parent entity E_p, and the other entity (province for part_of) the child entity E_c. Visualisations that represent the one-many visualisation schema are less common, but some examples are listed below together with a summary table.

- In a **tree map**, rectangles representing instances of E_p are divided into rectangles representing E_c, the area of which is proportional to the value of a scalar dimension a_1. A selector may be added to alter the proportion to be determined by other scalar dimensions $a_2, a_3, \ldots$
- In a **hierarchy tree**, nodes represent instances of E_p that are connected by lines to circles representing instances of E_c. A discrete dimension a_1 may optionally be used to colour the lines linking the entities.

One-many relationships				
Name	$|k_1|$	$|k_2|$ per k_1	mandatory	optional
Tree Map	1..100	1..100	a_1 scalar	a_2 colour
Hierarchy Tree	1..100	1..100	-	a_2 colour

3.4 Many-Many Relationships

Relationships that are many-many (such as borders in Fig. 1) lend themselves to visualisations that represent networks of data. The visualisation schema pattern for these relationships is illustrated in Fig. 3(d), where it should be noted that the data that governs the visualisation is now present as attributes of the relationship between entities E_1 and E_2. Visualisations that represent the many-many visualisation schema are the rarest, with two being the following:

- In **sankey** diagrams, the left hand elements of the diagram represent instances of E_1, the right hand elements represent instances of E_2, and the width of the flow between the left and right elements represents scalar dimension a_1. Optionally, a second attribute a_2 of the many-many relationship may be represented by varying the colour of the connection.

– In **chord** diagrams, instances of the entities are represented by points on the perimeter of the circle, with the value of a_1 varying the width of the connection between pairs of points. Again a second attribute a_2 of the many-many relationship may be represented by varying the colour of the connection. We note that chord diagrams are particularly suited to **reflexive** relationships, shown in Fig. 3(e), since then the points around the circle represent instances of just one type of entity E, and are not grouped according to which entity type they belong to.

Many-many relationships					
Name	$|k_1|$	$|k_2|$	reflexive	mandatory	optional
Sankey	1..20	1..20	no	a_1 scalar	a_2 colour
Chord	1..100	1..100	yes	a_1 scalar	a_2 colour

4 Conclusions

We have proposed, for the first time, a conceptual modelling approach to matching data and visualisations. Our approach makes use of the conceptual schema associated with the data and automatically matches it against a set of visualisation schema patterns (expressed in the same ER formalism) each of which characterises a group of potential visualisation alternatives. We also propose the use of well-known schema transformations in order to transform the database schema to that required for matching particular visualisation patterns (details of this can be found in [2]).

With this approach, domain experts can interact with conceptual models of their data, rather than lower-level tabular representations. By providing a set of visualisation schema patterns, each of which captures the data representation capabilities of a set of common data visualisations, we make it easier for the user to select a visualisation that is meaningful in relation to their data and their information seeking requirements; and to select from a more focussed set of visualisations. By matching between the visualisation schema patterns and the conceptual database schema, full schema knowledge can be used to automatically map between the data and a range of possible visualisations. By applying, again at the level of the conceptual database schema, a set of well-known schema transformations, it is possible to generate additional matchings between the transformed database schema and the set of visualisation schema patterns.

An implementation of the approach would include also data analysis capabilities to determine whether a dimension is scalar or discrete (or both), and to determine appropriate scaling of numeric dimensions (e.g. linear, logarithmic) by supporting an additional dimension characteristic of 'skew'. Also important is extension of our visualisation schema patterns to include descriptive elements (also populated from attributes of the database schema). Finally, a full implementation would include a second stage of mapping, from a visualisation schema pattern to an actual physical visualisation representation rendered by a target data visualisation tool.

References

1. May, W.: Information extraction and integration with FLORID: the MONDIAL case study. Technical report 131, Universität Freiburg, Institut für Informatik (1999). http://dbis.informatik.uni-goettingen.de/Mondial
2. McBrien, P., Poulovassilis, A.: Towards data visualisation based on conceptual modelling and schema transformations. Technical report No. 39, AutoMed (2018). www.doc.ic.ac.uk/automed
3. Ren, X., Wang, J.: Exploiting vertex relationships in speeding up subgraph isomorphism over large graphs. Proc. VLDB Endow. **8**(5), 617–628 (2015)
4. Tory, M., Moller, T.: Rethinking visualization: a high-level taxonomy. In: Proceedings of Information Visualization, pp. 151–158. IEEE (2004)
5. Ware, C.: Information Visualization: Perception for Design, 3rd edn. Morgan Kaufmann, San Francisco (2013)

Ontological Deep Data Cleaning

Scott N. Woodfield[1], Spencer Seeger[1], Samuel Litster[1], Stephen W. Liddle[1], Brenden Grace[1], and David W. Embley[1,2(✉)]

[1] Brigham Young University, Provo, UT 84602, USA
embley@cs.byu.edu
[2] FamilySearch International, Lehi, UT 84043, USA

Abstract. Analytical applications such as forensics, investigative journalism, and genealogy require deep data cleaning in which application-dependent semantic errors and inconsistencies are detected and resolved. To facilitate deep data cleaning, the application is modeled ontologically, and real-world crisp and fuzzy constraints are specified. Conceptual-model-based declarative specification enables rapid development and modification of the usually large number of constraints. Field tests show the prototype's ability to detect errors and either resolve them or provide guidance for user-involved resolution. A user study also shows the value of declarative specification in deep data cleaning applications.

Keywords: Data quality · Data cleaning
Declarative constraint specification
Conceptual-model-based deep data cleaning

1 Introduction

Data cleaning improves data quality by detecting and removing errors and inconsistencies [9]. *Deep data cleaning* includes data cleaning but adds general constraints that serve to detect application-dependent semantic errors and inconsistencies. These general constraints can be either crisp or fuzzy and are often expressed probabilistically. Ontologies are a natural framework for these semantic constraints. Within an ontology their purpose is to accurately characterize objects and data instances and their interrelationships. As such, they become the basis for *ontological deep data cleaning*.

In this paper we illustrate the principles and possibilities of ontological deep data cleaning in the realm of genealogy—an application not only for family history enthusiasts, but also for studying inherited diseases and for establishing public policy such as for intergenerational poverty.

To illustrate ontologically deep data cleaning, consider the text snippet in Fig. 1 taken from *The Ely Ancestry* [11]. When processed by our ensemble of information extraction engines [2], most of the genealogical information on the page was correctly extracted. The ensemble, however, incorrectly associated Mary Ely and Gerard Lathrop (the first two numbered children in Fig. 1) with four parents: Mary Eliza Warner, Joel M. Gloyd, Abigail Huntington Lathrop,

© Springer Nature Switzerland AG 2018
J. C. Trujillo et al. (Eds.): ER 2018, LNCS 11157, pp. 100–108, 2018.
https://doi.org/10.1007/978-3-030-00847-5_9

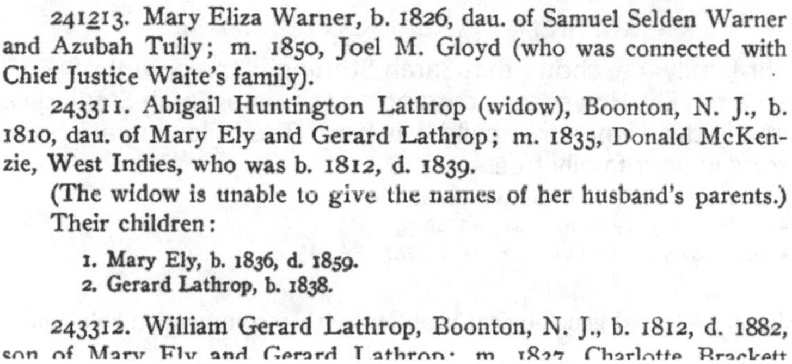

241213. Mary Eliza Warner, b. 1826, dau. of Samuel Selden Warner and Azubah Tully; m. 1850, Joel M. Gloyd (who was connected with Chief Justice Waite's family).

243311. Abigail Huntington Lathrop (widow), Boonton, N. J., b. 1810, dau. of Mary Ely and Gerard Lathrop; m. 1835, Donald McKenzie, West Indies, who was b. 1812, d. 1839.

(The widow is unable to give the names of her husband's parents.) Their children:

1. Mary Ely, b. 1836, d. 1859.
2. Gerard Lathrop, b. 1838.

243312. William Gerard Lathrop, Boonton, N. J., b. 1812, d. 1882, son of Mary Ely and Gerard Lathrop; m. 1837, Charlotte Brackett

Fig. 1. Text Snippet from *The Ely Ancestry* [11], p. 421.

and Donald McKenzie. The data cleaning component of our extraction application not only flagged Mary Ely and Gerard Lathrop as having more than two parents, but also flagged Mary Eliza Warner as likely being too young to be the mother of Mary and Gerard since she would have been only 10–12 years old at the time of their births. After detection, our data cleaning component considers the possibility of retracting assertions to correct problems and in this case can automatically retract the assertions linking Mary Ely and Gerard Lathrop to their incorrect parents.

Preparation of data for import into a target repository is only one part of our genealogical deep cleaning application. Upon import, the first problem is to reconcile the new data with the old. For example, when we imported the data in Fig. 1, a search for duplicates revealed confusion with the name Mary Ely because several different people on the page share the same name. Further, Mary Eli in the repository was another possible duplicate. Aided by the constraint that it is impossible to be one's own ancestor, the data cleaning system arrived at the correct set of assertions. Our data cleaning tool can also generally check assertions in the repository. As Fig. 2 shows, our cleaning component detected that Abigail's third great-grandmother, Sarah Sterling, died before a child of hers was born. When a violation is detected, the tool explains what may be wrong as Fig. 2 shows.

Although our approach to data cleaning corresponds with the central ideas of detecting and removing errors [4,9,10], it differs from other data cleaning research in two fundamental ways: (1) it is deep—centered on ontological semantic constraints; and (2) for applications that of necessity must accommodate dirty data, it applies to cleaning the repository post-import as well as to cleaning pre-import and on-import. Aspects of our approach have similarities with other data cleaning research: ontological reasoning for data cleaning [1], duplicate detection [8], entity resolution [6], and declarative cleaning operators [4].

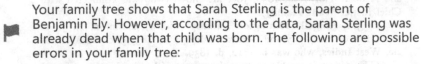

Fig. 2. Detected genealogical problem with recommended solution.

2   Deep Data Cleaning

Discovering, analyzing, and organizing family histories is necessarily investigative, requiring researchers to gather and organize clues from census records,
obituaries, and other sources. Source information is not always accurate, is sometimes unreadable and is often non-existent. Researchers inevitably draw some
conclusions that are incomplete and inaccurate. Of necessity, deep data cleaning
is required both to clean data for import into the repository and to clean data
already residing in the repository.

2.1   Application System

FamilySearch [3] hosts *Family Tree*, a wiki-like repository of genealogical information that allows world-wide collaboration on a single, shared, ancestry of
humankind. *Family Tree* contains over 1.2 billion person records. FamilySearch
also hosts historical records containing 6.2 billion searchable names. Most of
these historical records, typically hand-written, have been indexed for search by
more than a million registered volunteers. Advances in automatic indexing are
being developed to index the vast number of ever growing unindexed documents.

Among several other automated information extraction projects, our Fe6
project [2] is specifically aimed at indexing FamilySearch's growing collection of
over 360,000 family history books that have been scanned and OCR'd. Figure 1,
with which we illustrated our data cleaning tools, is taken from one of the many
millions of pages of these books. Targets for the Fe6 extracted information are
(1) the historical records collection of indexed documents and (2) *Family Tree*.

2.2   Pre-import Data Cleaning

Fe6 stands for **F**orms-based **e**nsemble of extraction tools with **6** pipeline phases.
Briefly explained,[1] it is the pre-import pipeline that extracts information from
the pages of a book and prepares it for import into FamilySearch repositories.
Phases 1, 2, and 3 prepare a book for processing, extract data with an ensemble of extraction engines, and merge and shallow-clean the data, storing it in an

---

[1] An extended version of this short paper can be found at http://deg.byu.edu/papers/.

ontological data model. Phase 5 standardizes the data and infers additional information which is implied but not explicitly stated in the text. Phase 6 prepares the data for import.

Our deep data cleaning component operates in Phase 4. Formally, each object set in the underlying conceptual model is a one place predicate and each n-ary relationship set is an n-place predicate. Thus, we are able to specify deep data cleaning detection rules as Datalog-like inference rules (e.g. those implicit in the discussion in the introduction). To obtain the probability distributions for fuzzy constraints, we define descrete functional predicates. We populate these distributions by sampling the vast store of data in FamilySearch's *Family Tree*.

2.3 Post-import Data Cleaning

We program post-import deep data cleaning with a conceptual-model-equivalent language [7]. Figure 3 shows an example. The first three statements indicate how the conceptual model is specified. Each statement declares a relationship set in the model in which its related object sets are capitalized nouns. Colons specify is-a hierarchies. Participation constraints for a statement's relationship set declaration are in square brackets. General constraints for the model are specified in *ENSURE* statements. The *ENSURE* statement in Fig. 3 is one of the ontological constraints defined for the conceptual model's representation of the ontology. Its name is "I am not my own ancestor" with which it can be referenced. Its body is a Datalog-like statement, defining the constraint. Note the use of relationship set names in these general constraints.

```
Person[1] was born on Birthdate:String[1:*];
Child:Person[2] is a child of Person[0:*];
Person[0:*] married Spouse[1:*] on MarriageDate:String[1:*]
    at MarriagePlace:String[1:*];
...
ENSURE I am not my own ancestor BEGIN
    IF Child(c) is a child of Person(p) THEN
        Person(p) is an ancestor of Descendent:Person(c);
    IF Person(p) is an ancestor of Descendent(d) AND
        Child(c) is a child of Person(d) THEN
            Person(p) is an ancestor of Descendent(c);
    IF Person(p) is an ancestor of Descendent(p) THEN
        Person(p) is an ancestor of Descendent(p) has Probability(prob)
            WHERE prob = PROBABILITY OF PersonIsOwnAncestor(p);
END;
```

Fig. 3. Conceptual-model-equivalent language

The language is fully declarative, which allows a user to easily change both the model and the constraints. Thus, it is easily configurable to accommodate additional concepts such as an occupation that might be of interest or to accommodate a host of GedcomX [5] *FactTypes* (e.g Adoption, BarMitzva, ...). More importantly for ontological deep data cleaning, it provides for ease of adding, deleting, and modifying constraints.

On import into *Family Tree*, the first check is for potential duplicates. Given potential duplicates x and y, the check loads the instance graphs of x and y into the conceptual model defined in Fig. 3. It then conflates the *Person* object identifiers for x and y and runs constraints against the resulting graph.

As an example, consider merging the instance graphs obtained from Fig. 1 for Mary Ely (the first numbered child on the page) and Mary Ely (the mother of Abigail Huntington Lathrop). As *Person*-IDs, let Mary Ely (the child) be P_1, Mary Ely (Abigail's mother) be P_2, and Abigail be P_3. Conflating P_1 and P_2 (as $P_{1.2}$), yields a new graph with many relationships. Figure 4 shows the graph reduced to just those edges connected to $P_{1.2}$ plus the edges connecting person objects to names. Among the many relationships are *Child*(P_3) *is a child of Person*($P_{1.2}$) because Abigail is a child of Mary Ely (her mother) and *Child*($P_{1.2}$) *is a child of Person*(P_3) because Mary Ely (the child) is a child of Abigail. Now, we have a cycle which the *ENSURE* constraint in Fig. 3 will detect and thus will state that the probability of such an occurrence is zero. Hence, "I am not my own ancestor" is false, and the proposed merge should be rejected.

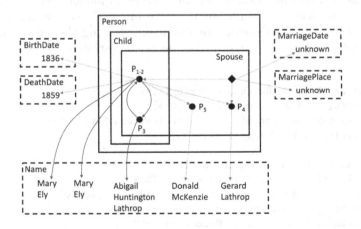

Fig. 4. Instance graph with Mary Elys conflated.

Beyond import, even when the input data is clean and even when duplicates are correctly detected and merged, there is no guarantee that the data in *Family Tree* is clean. (Over 4.3 million users have contributed to the creation of the tree, each in his or her own way based on available data or even just on memories of family lore.) To aid in cleaning *Family Tree*, we have implemented a tool, called *Tree Sweeper*, which runs over the ancestry of a given person in *Family Tree*. It imports into the conceptual model in Fig. 3 the instance graphs of ancestors in the tree up to a specified generation. Then, given the ontological constraints of the model, it detects and reports encountered problems. As Fig. 2 shows, a Tree Sweeper run over the ancestry of Abigail Huntington Lathrop in Fig. 1 raised a red flag about Sara Sterling, Abigail's great-grandmother. Tree Sweeper can also detect fictitious persons in *Family Tree*, most likely created by improper merges of presumed duplicates. An analysis of the person's instance graph is often sufficient to detect bogus persons.

3   Experimental Evaluation

We have conducted several field tests with Tree Sweeper, and we have conducted a user study to assess how well people with varying computer expertise can declare *ENSURE* rules.

3.1   Field Tests

Error Detection: Tree Sweeper vs. FamilySearch. To evaluate Tree Sweeper, we compared its error detection capability with FamilySearch's. We randomly chose four persons in *Family Tree* and collected the records of all their ancestors up to the eighth generation, yielding a sample size of 423 persons. Table 1 shows the results of running Tree Sweeper over these 423 person records and the errors reported by FamilySearch within these records. Tree Sweeper found 17 (31%) more crisp errors than did FamilySearch. In addition, Tree Sweeper found 17 probabilistically unlikely (fuzzy) errors. FamilySearch does not consider fuzzy errors. Thus, for example, a 12-year-old mother is considered impossible, but a 13-year-old mother is not an error. Tree Sweeper uses a distribution when evaluating a mother's age at the birth of her first child. It establishes an unlikely threshold and an error threshold, both of which are percentages. The Tree Sweeper error threshold for a mother's age at the birth of her first child was 0.01% and the unlikely threshold was 1%. This capability is enabled by the use of distributions in Tree Sweeper to measure the probability that a constraint has been violated rather than using a single crisp error threshold.

Merge Problem Detection. Wanting to determine the extent to which merge problems occur in *Family Tree*, we sampled 51 people randomly from the tree and found that 9.8% were formed by merging. Next, we sampled 140 people with reported errors and found that 76% of them had been formed by merging. There is a significant difference in the merge percentages, implying that there is a correlation between people with errors and whether they are merged individuals. This observation, however, does not suggest causality. To investigate further we evaluated the 18 individuals who had three or more reported errors. Of these 18, 15 were merged individuals. Further evaluation of those 15 showed that 6 of the merges were erroneous. Thus, a significant number of errors are caused by improper merging, but there are certainly other factors to be considered.

Table 1. Error detection: Tree Sweeper vs. FamilySearch.

	Crisp	Fuzzy
Number of people in sample	423	423
Number of errors found	72	17
Errors per person	16.9%	4.0%
Errors Found & Percent		
Tree Sweeper	68, 94.4%	17, 100%
FamilySearch	55, 76.4%	0, 0.0%
Errors found by Tree Sweeper but not FamilySearch	17	17
Errors found by FamilySearch but not Tree Sweeper	4	0
Errors found by both	51	0

3.2   User Study: Declarative vs. Imperative Specification

In our Error Detection field test in Sect. 3.1, we found that FamilySearch and Tree Sweeper each had one false positive. In Tree Sweeper's case, it reported that a person was probably too old to get married. The problem occurred because of our misinterpretation of the distribution we were using. We thought it represented the probability of a person marrying at a given age. Instead, it was the probability of a person's age when married for the first time.

The underlying conceptual model of Tree Sweeper, including general constraints, was designed to be easy to write, understand, and modify. To evaluate this claim, we performed an experiment involving the solution of our misunderstanding of Tree Sweeper's marrying age distribution. We conducted an experiment with 31 subjects. Three of the participants were experts in reading, writing, and modifying the text-based conceptual modeling language; the others had never seen the language.

In the experiment each subject was given a one-page tutorial on the language. Examples were included. Next subjects were shown how to modify the syntax of an unrelated constraint. The subjects were then given a second page describing the problem and the erroneous marrying-age constraint. After reading the description, they started a timer and began modifying the constraint to solve the problem. When they thought they were finished, they presented their solution to the proctor who indicated whether their solution was correct or not. If incorrect, they were to modify their solution and resubmit. This submission process repeated until they had correctly modified the constraint. We used two metrics: time to correctly modify the constraint and the number of tries it took to be successful. Table 2 shows the results.

At an α-level of 0.05, we determined that experts were faster than computer scientists, and computer scientists were faster than non-computer scientists. Even the worst subject took only five minutes and three tries.

Table 2. Declarative vs. imperative coding.

Group (Nr. Subjects)	Avg. Min.	Avg. Tries	Max Min.	Max Tries
Experts (3)	1.00	1.00	1	1
Computer Scientists (12)	1.83	1.33	3	2
Non-Computer Scientists (6)	3.16	1.33	5	2
Uncategorized Subjects (9)	2.33	1.56	5	3
Overall	2.18	1.30	5	3

4   Concluding Remarks

Ontological deep data cleaning relies on a plethora of declaratively specified crisp and fuzzy constraints. It is particularly useful for investigatory applications in which the data in the application's repository is never complete, consistent, and error free and in which data cleaning is needed both pre- and post-data import.

Our prototype implementation within a genealogical information-extraction application is declarative—based on a conceptual-model-equivalent programming language. Field tests bespeak its usefulness in discovering and providing guidance for fixing discovered errors. A user study shows that a broad spectrum of users can quickly learn the language and effectively modify its constraint rules.

In future work, (1) we intend to enhance Tree Sweeper to aid users in their quest to clean their ancestry in *Family Tree*. (2) After additional work in which we need to code the temporary conflation of merge-proposed individuals and test and enhance the accuracy of proposed ontological constraints, we plan to offer it as a background sanity check for the merge operation in FamilySearch. (3) Since the Fe6 extraction engines are capable of reading documents [2], an interesting future possibility is to have them read source documents and determine whether the evidence supports the conclusions posted in *Family Tree*.

References

1. Dolby, J., et al.: Scalable cleanup of information extraction data using ontologies. In: Aberer, K., et al. (eds.) ASWC/ISWC -2007. LNCS, vol. 4825, pp. 100–113. Springer, Heidelberg (2007). https://doi.org/10.1007/978-3-540-76298-0_8
2. Embley, D., Liddle, S., Lonsdale, D., Woodfield, S.: Ontological document reading: an experience report. Enterp. Model. Inf. Syst. Arch. Int. J. Concept. Model., 133–181 (2018)
3. FamilySearch. http://familysearch.org
4. Galhardas, H.: Data cleaning and transformation using the AJAX framework. In: Lämmel, R., Saraiva, J., Visser, J. (eds.) GTTSE 2005. LNCS, vol. 4143, pp. 327–343. Springer, Heidelberg (2006). https://doi.org/10.1007/11877028_12
5. Gedcom X. http://www.gedcomx.org/
6. Kang, H., Getoor, L., Shneiderman, B., Bilgic, M., Licamele, L.: Interactive entity resolution in relational data: a visual analytic tool and its evaluation. IEEE Trans. Vis. Comput. Graph. **14**(5), 999–1014 (2008)

7. Liddle, S.W., Embley, D.W., Woodfield, S.N.: Unifying modeling and programming through an active, object-oriented, model-equivalent programming language. In: Papazoglou, M.P. (ed.) ER 1995. LNCS, vol. 1021, pp. 55–64. Springer, Heidelberg (1995). https://doi.org/10.1007/BFb0020520

8. Low, W., Lee, M., Ling, T.: A knowledge-based approach for duplicate elimination in data cleaning. Inf. Syst. **26**(8), 585–606 (2001)

9. Rahm, E., Do, H.: Data cleaning: problems and current approaches. IEEE Data Eng. Bull. **23**(4), 3–13 (2000)

10. Raman, V., Hellerstein, J.: Potter's Wheel: an interactive data cleaning system. In: Proceedings of the 27th International Conference on Very Large Data Bases, VLDB 2001, Rome, Italy, pp. 381–390, September 2001

11. Vanderpoel, G.: The Ely Ancestry: Lineage of RICHARD ELY of Plymouth, England. The Calumet Press, New York (1902)

Automatic Generation of Security Compliant (Virtual) Model Views

Salvador Martínez[1]($\boxtimes$), Alexis Fouche[1], Sébastien Gérard[1], and Jordi Cabot[2]

[1] CEA-LIST, Paris-Saclay, France
{salvador.martinez,alexis.fouche,sebastien.gerard}@cea.fr
[2] ICREA-UOC, Barcelona, Spain
jordi.cabot@icrea.cat

Abstract. The increased adoption of model-driven engineering in collaborative development scenarios raises new security concerns such as confidentiality and integrity. In a collaborative setting, the model, or fragments of it, should only be accessed and manipulated by authorized parties. Otherwise, important knowledge could be unintentionally leaked or shared artifacts corrupted. In this paper we explore the introduction of access-control mechanisms for models. Our approach relies on the definition of a domain specific language tailored to the definition of access-control rules on models and on its enforcement thanks to the automatic generation of security compliant (virtual) views.

1 Introduction

The increased adoption of the model driven engineering (MDE) paradigm in complex collaborative scenarios introduces the need for effective confidentiality and integrity protection mechanisms at the modeling level. Indeed, in such scenarios, a given model (or type of model) may be shared among different stakeholders over possibly untrusted channels. This model, or fragments of it, should only be accessed and manipulated by authorized parties. Otherwise, the collaboration process could lead to the leak of important knowledge, likely triggering reputation and/or economical losses.

Access-control (AC) policies are often the mechanism of choice to implement the security requirements of confidentiality and integrity and thus, they constitute a pervasive mechanism in current information systems. However, while there exist standard access-control languages based on well-defined paradigms (e.g., Role-based access-control (RBAC) [4] and attribute-based access control (ABAC) [12]) the available AC frameworks can not be directly used in an MDE scenario as they either provide a coarse granularity that only allow for managing permissions at the file level (e.g., file systems rights), requiring the manual fragmentation of the model to enforce security, or do not take into account specificities of the modelware technical space such as the existence of metamodels and the conformance relation.

This makes it difficult to define and enforce policies that respect the *least privilege principle* that states that subjects must only have the rights they need

© Springer Nature Switzerland AG 2018
J. C. Trujillo et al. (Eds.): ER 2018, LNCS 11157, pp. 109–117, 2018.
https://doi.org/10.1007/978-3-030-00847-5_10

to perform their assigned duties. Indeed, an effective access-control mechanism for models must (1) provide the means to define and enforce fine-grained access-control rules (i.e., the means to control access to any part of a model, be it a class, an attribute, a relation or an operation); (2) protect both models and their metamodels since metamodel information is also valuable and should be protected; (3) be usable by modelers (which requires a language that uses familiar model concepts as language primitives) and (4) keep the consistency of the secured models, so that they can be viewed and manipulated by existing MDE tools with no adaptation.

To the best of our knowledge, a language satisfying all these constraints does not exist. While modeling has been intensively used as a means of including security and access-control concerns in the early phases of systems design and specification [7,8], very few approaches are specially tailored to the protection of the models themselves, e.g. [3], and none of them consider the modification of a model's metamodel as a requirement and enforcement mechanism.

Thus, we have decided to build a new fine-grained access-control mechanism for models. More specifically, in this paper we present an approach composed of: (1) a role-based access-control language specially designed to work with models allowing for the specification of conditions at the M2 and M1 level and (2) an enforcement mechanism based on the automatic generation of security compliant (virtual) views that protects both the model and metamodel.

2   Concepts

In order to ease the discussion, we give a few notions on the concepts of access-control and model views.

−**Access-Control:** Fig. 1(a) shows the core concepts of access-control: Objects represent the passive resources that can be accessed within a system and that we may want to protect (files in an operation system, tables in a database,...). **Subjects** are the active entities in a system. They represent the actors to which the access to *Objects* is controlled. **Actions** are any kind of access to the *Objects* that may be performed by the *Subjects* in a given system. **Permissions** relate *Actions* with *Objects*. A permission is thus the right to perform a given *Action* (or set of actions) on a given *Object* (or set of objects). These permission are, in turn, granted to *Subjects*.

Nevertheless, directly assigning permissions to Subjects becomes unpractical when the user-base of the applications is large. Hence, in real applications, the definition of the permissions and its assignment is often performed by using the concepts of **Rule** and **Policy**. A rule is the assignment (or denial) of a permission to a given subject. Generally, access control rules have the form: $R_i : \{conditions\} \rightarrow \{decision\}$, where the sub-index i specifies the ordering of the rule, *decision* can be accept or deny and *conditions* is a set of rule matching attributes (e.g., hold roles). An access-control security policy is the set of permission assignments within a given information system, which is composed of a set of Rules. This policy constitutes a mere definition of the security requirements for the system, while the process of implementing the mechanisms to make the system follow the rules it defines is called enforcement.

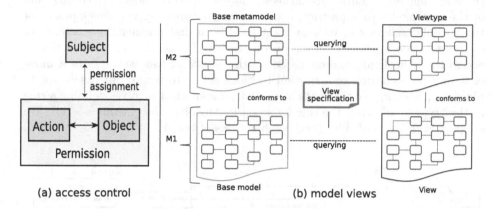

Fig. 1. Core concepts

–Model-Views: The concept of view is very common in the field of databases, and serves as a way to provide a certain perspective tailored for a specific type of use. As such, a database view may be considered as a security enforcement mechanism that filters and restricts the information a certain user can see from the database. There exists several solutions bringing the concept of views to the modelware realm using a number of different strategies. We base our work on the *virtual models* approach where views are not serialized but instead are the result of executing live queries on the original model. In particular, we adopt the terminology described in [2]. The main concepts (depicted in Fig. 1(b)) are: *Base meta-model*: a regular meta-model involved in the definition of a viewtype; *base model*: a regular model used as an input for building a view; *viewtype*: a metamodel which structure is defined through the specification of queries on one (or more) meta-model; *view*: a model conforming to a viewtype, and resulting from the querying of one (or more) base model.

3   Approach

This section introduces the main elements of our approach and a running example to illustrate them.

Running Example: We show in Fig. 2 the metamodel of a medical record, the same example used in the XACML specification [10]. Medical *records*, identified by a *recordid* string, contain five different types of information: (1) information about the insurer of the patient represented by the *Company* metaclass; (2) Information about the *Patient*, including name and *Contact* information; (3) information about the parents of the patient represented by the *Parent-Guardian* metaclass; (4) information about the prescribed treatments, represented by the *Medical* metaclass that aggregates data regarding the *Treatment* and the *Drugs* it uses and regarding the *Result* of medical visits; and finally (5) information about the *physician* assigned to that patient. Given this metamodel, there are different security scenarios we may want to consider: (1) Hiding part of the metamodel to a partner, e.g. if we are outsourcing the development of the "people" subsystem, we may want to hide to that partner the existence of classes to store medical information on treatments and drugs, (2) Restricting access to medical information based on the profile of the user, e.g. in a models@run.time scenario, we may want to block access to record objects except to physicians in charge of that patient. Note that scenario 1 involves defining a rule at the "type-level" while the rule for scenario 2 involves the "instance-level". We support both kinds of rules (and combinations of both).

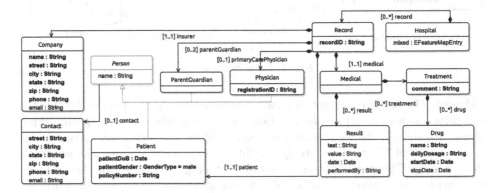

Fig. 2. Medical record running example

Our Solution: Fig. 3 summarizes our approach for providing access-control for models. The process starts with a security engineer that specifies the desired policy. This policy is written using a language specially tailored to define such modeling access control rules. This policy is then transformed to a viewtype specification. The viewtype specification is interpreted by a view engine in order to generate a filtered metamodel (viewtype) and eventually, a filtered model (view). These are the elements the end-user will obtain upon an access request. Note that different filtered artefacts (viewtypes and views) are generated for each accessing role or end-user.

Our approach is thus composed of two main building blocks, an AC language and a view generator in charge of the enforcement of the policy defined with such AC language. We provide tool support[1] for the AC language, its transformation towards a view specification, and the execution of such view specification.

---

[1] https://gitlab.com/smartine/SecureModelViews.

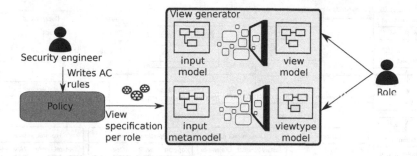

Fig. 3. Approach

4   Access-Control Language for Models

The first step requires providing the means to define access-control rules for models. We do so by creating a domain specific policy language. This language allows security engineers to write rules based on the base metamodel (i.e. filtering access based on certain element types; the types themselves should not even be visible to the users) and/or its instances (i.e. preventing access to model elements with certain values). In the rest of this section we discuss the language's abstract syntax and its execution semantics and we provide a textual concrete syntax to ease its utilization.

4.1   Abstract Syntax

Figure 4 shows the conceptual schema of our RBAC-based policy language. It allows the definition of rules that associate roles to the permissions (or prohibitions) to perform actions on model elements. In this language, a *Policy* contains a number of access-control *Rules*. These rules are composed of a left-hand side and a right-hand side. The left-hand side is meant to be used to express a number of conditions for a given access-control rule to apply to a given access request. We provide our language with three specific condition elements:

(1) *Subject* identified by its reference to a *RoleDeclaration* and representing the subject accessing the protected resource.
(2) *Action* that represents the operation to be performed on the protected resource. The values of the *actionType* attribute of type *ModelAction* can be *Read*, *Write* or *Execute* (this action only applies to model operations). Note that we include the three CRUD operations *Create*, *Update* and *Delete* on the *Write* operation as the granularity of our language permits to obtain the same effects by the combination of *Write* and *Read* operations on the different model elements.
(3) *ObjectCondition* that represents the resource to which the rule applies. Our language allows the definition of *ClassCondition* to express permissions that apply to a class, *AttributeCondition ReferenceConditions* that are meant to represent permissions on specific attributes and references and *OperationCondition*

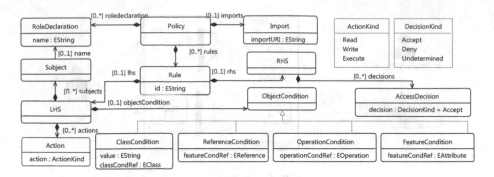

Fig. 4. Access-control policy metamodel

to represent permissions on model operations. Note that all *ObjectCondition* elements hold a reference to the metamodel element they refer to. Besides, the *ClassCondition* holds a *value* attribute of type String meant to be a place holder for model queries to filter model elements w.r.t. complex conditions. Concrete implementations may link this query place holder to concrete query languages such as OCL [11]. Their evaluation is nevertheless delegated to the view framework in charge of generating the view models from a given view specification.

The right-hand side of rules is used to express the effect the application a *Rule* has by means of a *Decision*. *DecisionKind* lists the type of decisions that can be issued, namely *Allow* for granting a permission, *Deny* for a prohibition. Note that the *Policy* holds a *default* attribute of type *DefaultPolicy* used to provided a general decision when no rule applies. This policy language can be easily enhanced so that elements such as roles and actions have attributes as we have done in a previous work [9]. Finally, we consider user management issues (such as role hierarchies or role delegation) and policy constraints (such as Separation of Duty) out of the scope of this paper. Nevertheless, our language may be easily extended to support advanced concepts and integrate contributions where OCL constraints are used to impose constraints on the policy [1].

4.2 Concrete Syntax

In order to ease its use, we also define a textual concrete syntax for the Policy language. Access-control policies may easily become large and thus, a textual syntax would be easier to read and manipulate than a graphical or form-based one. Listing 1.1 shows an example of this concrete syntax. A policy with a default behaviour of *deny* is defined for our Medical Record example introduced in Sect. 2. The policy defines three roles, Physician, Patient and Clerk and then proceeds to define access-control rules for the Clerk role. Rules c1 to c4 grant read access to the Company, Hospital, Patient and Record elements. Rule c3 However restricts the granted access to Records with recordID greater than 100. c5 to c7 deny the access to the Physician, Medical and Parent elements. Finally, rule c8 forbids the access to the *patientGender* attribute of Patient.

As we can see, the access control policy mixes positive with negative permissions. This simplifies the propagation of permissions leading to policies that are more compact and easier to read. Also, it does not list all of the metamodel classes, attributes, references and operations (which would be tedious and error prone). For its correct interpretation we need to provide our language with a set of precise execution semantics. We do so in the following subsection.

Listing 1.1. Policy Example

```
import "platform // Test1 / records . ecore "

DeclareRole  Physician ,  Patient ,  Clerk

rule  c1  (Clerk;  Read;  class  records . Company )-> Accept
rule  c2  (Clerk;  Read;  class  records . Hospital )-> Accept
rule  c3  (Clerk;  Read;  class  records . Record
  WithValue = < "self . recordID .>.100" >)-> Accept
rule  c4  (Clerk;  Read;  class  records . Patient )-> Accept
rule  c5  (Clerk;  Read;  class  records . Physician )-> Deny
rule  c6  (Clerk;  Read;  class  records . Medical )-> Deny
rule  c7  (Clerk;  Read;  class  records . Parent )-> Deny
rule  c8  (Clerk;  Read;  att  records . Patient . patientGender )-> Deny
```

4.3   Execution Semantics

In the general case, the calculation of the permissions for each metamodel and model element requires (1) to match an applicable access-control rule if such a rule exists and (2) to apply permissions propagation policies that may affect the evaluation of that rule. These two mechanisms work as follows:

Rule Matching. As shown in Fig. 4, each rule may list several *Role* and several *Action* elements. However, they only list one metamodel element (*metaclass*, *structuralfeature*, or *operation*). This limitation simplifies the process of matching applicable rules and, more importantly, prevents rule conflicts due to the intersection of rule conditions (and well-formedness rules defined at the policy language level already prevent the existence of two rules on the same metamodel element with an overlapping set of roles). An additional generic query language helps to write arbitrary conditions to precisely define the set of instances of the metamodel elements affected by the rule.

Permission Propagation. In order to clarify the interpretation of access-control policies, we propose a numbers of permission propagation *principles* that simplify the specification of such rules freeing the designers from manually defining in each rule the priorities in case of conflict:

- *SuperClass Propagation.* Permissions defined for a superclass are inherited by the subclasses if no other rule is defined for the subclasses.
- *Containment Relationship Propagation.* Permissions are propagated through the containment relationship as defined in the metamodel.
- *Containment Propagation.* Permissions on a Class are propagated to its contained elements (i.e., its attributes, references and operations).

- *Deny Overrides.* A rule denying a permission on a given model element is propagated to all the subtree of contained elements overriding any rule granting the permission that may be found on the subtree (including its contained attributes, references and operations). This guarantees that the containment hierarchy is preserved, a requirement to have valid models.
- *Default Propagation.* If no access rule is applicable to a given model element, and no permission is inherited from the previous propagation principles, the default policy applies;

As a result of applying the aforementioned execution semantics to the example in Listing 1.1 we will obtain the following list of permissions: Accept Classes: *Hospital, Record (with RecordId greater than 100), Company and Patient*; Deny Classes: *Physician, ParentGuardian, Medical, Treatment, Result, Drug and Contact*; Accept all Attributes and References from Accept Classes apart from patientGender and contact; Deny all other Attributes and References. Note that the rule C1 is redundant, as *Company* inherits the accept permission from *Record*. We detect this kind of anomalies and report them to the user during the transformation process we describe in Sect. 5.

5   Enforcement with Virtual Model Views

As stated in Sect. 3, the second building block of our approach is a model view generator. Indeed, we use *views* as an access-control enforcement mechanism. We adapt the implementation of the virtual model view approach as described in EMFViews [2] for that purpose. A view specification corresponding to an access-control policy conforming to our PolicyDSL language is automatically obtained by the use of a model transformation. This transformation takes as input three models: an access-control policy; the metamodel referred by that policy; and a parameters model indicating for which role the view is to be generated. It produces as output a view definition.

Due to space limitations we do not show here the transformation nor the View definition and Parameter metamodels. They are available in the project website together with a demo showing a view automatically derived from the policy in Listing 1.1.

The biggest advantage of enforcing access-control through the use of virtual model views resides in (1) its capacity to modify the metamodel of the model to be protected thanks to the generation of a specific viewtype for the view; (2) the elimination of the synchronization issues that would appear otherwise between the original model and the filtered parts (synchronization between the views and the original model is also supported at the individual attribute level, while more complex updates fall under the limits of the well-known view update challenge [5]).

As future work we intend to support other modelling artifacts, like OCL queries and model transformations that should be adapted when producing a view so that they continue to be executable and meaningful in the *secure* context. From a tooling perspective, we will complete the integration of the components described into the open source Papyrus UML environment [6].

References

1. Ben Fadhel, A., Bianculli, D., Briand, L.: GemRBAC-DSL: a high-level specification language for role-based access control policies. In: SACMAT 2016, pp. 179–190. ACM (2016)
2. Bruneliere, H., Perez, J.G., Wimmer, M., Cabot, J.: EMF views: a view mechanism for integrating heterogeneous models. In: Johannesson, P., Lee, M.L., Liddle, S.W., Opdahl, A.L., López, Ó.P. (eds.) ER 2015. LNCS, vol. 9381, pp. 317–325. Springer, Cham (2015). https://doi.org/10.1007/978-3-319-25264-3_23
3. Debreceni, C., Bergmann, G., Ráth, I., Varró, D.: Enforcing fine-grained access control for secure collaborative modelling using bidirectional transformations. SOSYM, 1–33 (2017)
4. Ferraiolo, D., Cugini, J., Kuhn, D.R.: Role-based access control (RBAC): features and motivations. In: ACSAC, pp. 241–48 (1995)
5. Foster, J.N., Greenwald, M.B., Moore, J.T., Pierce, B.C., Schmitt, A.: Combinators for bidirectional tree transformations: a linguistic approach to the view-update problem. ACM TOPLAS **29**(3), 17 (2007)
6. Gérard, S., et al.: Papyrus UML, August 2012. http://www.papyrusuml.org
7. Jürjens, J.: UMLsec: extending UML for secure systems development. In: Jézéquel, J.-M., Hussmann, H., Cook, S. (eds.) <<UML>> 2002. LNCS, vol. 2460, pp. 412–425. Springer, Heidelberg (2002). https://doi.org/10.1007/3-540-45800-X_32
8. Lodderstedt, T., Basin, D., Doser, J.: SecureUML: a UML-based modeling language for model-driven security. In: Jézéquel, J.-M., Hussmann, H., Cook, S. (eds.) <<UML>> 2002. LNCS, vol. 2460, pp. 426–441. Springer, Heidelberg (2002). https://doi.org/10.1007/3-540-45800-X_33
9. Martínez, S., García, J., Cabot, J.: Runtime support for rule-based access-control evaluation through model-transformation. In: SLE 2016, pp. 57–69. ACM (2016)
10. Rissanen, E., et al.: eXtensible access control markup language (XACML) 3.0 (2013)
11. OMG, UML 2.0 OCL specification. OMG Adopted Specification (ptc/03-10-14) (2003)
12. Yuan, E., Tong, J.: Attributed based access control (ABAC) for web services. In: ICWS 2005. IEEE (2005)

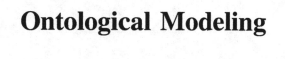

Ontological Modeling

The Common Ontology of Value and Risk

Tiago Prince Sales[1,2(✉)], Fernanda Baião[3], Giancarlo Guizzardi[4],
João Paulo A. Almeida[5], Nicola Guarino[2], and John Mylopoulos[1]

[1] University of Trento, Trento, Italy
{tiago.princesales,john.mylopoulos}@unitn.it
[2] ISTC-CNR Laboratory for Applied Ontology, Trento, Italy
nicola.guarino@cnr.it
[3] Federal University of the State of Rio de Janeiro (UNIRIO), Rio de Janeiro, Brazil
fernanda.baiao@uniriotec.br
[4] Free University of Bozen-Bolzano, Bolzano, Italy
giancarlo.guizzardi@unibz.it
[5] Federal University of Espírito Santo, Vitória, Brazil
jpalmeida@ieee.org

Abstract. Risk analysis is traditionally accepted as a complex and critical activity in various contexts, such as strategic planning and software development. Given its complexity, several modeling approaches have been proposed to help analysts in representing and analyzing risks. Naturally, having a clear understanding of the nature of risk is fundamental for such an activity. Yet, risk is still a heavily overloaded and conceptually unclear notion, despite the wide number of efforts to properly characterize it, including a series of international standards. In this paper, we address this issue by means of an in-depth ontological analysis of the notion of risk. In particular, this analysis shows a surprising and important result, namely, that the notion of risk is irreducibly intertwined with the notion of value and, more specifically, that risk assessment is a particular case of value ascription. As a result, we propose a concrete artifact, namely, the Common Ontology of Value and Risk, which we employ to harmonize different conceptions of risk existing in the literature.

Keywords: Risk · Risk modeling · Value · Enterprise modeling
OntoUML

1 Introduction

Risk analysis is traditionally accepted as a complex and critical activity in various contexts, such as strategic and project planning, finance, engineering of complex systems, and software development. It offers techniques and tools for systematically identifying potential issues, analyzing their impact and designing and evaluating mitigation strategies.

Given the complexity of risk analysis, several modeling approaches have been proposed to help analysts in representing and analyzing risks in different contexts. Examples include the Goal-Risk framework [2], an approach designed to

© Springer Nature Switzerland AG 2018
J. C. Trujillo et al. (Eds.): ER 2018, LNCS 11157, pp. 121–135, 2018.
https://doi.org/10.1007/978-3-030-00847-5_11

support risk analysis in the context of requirements engineering; RiskML [30], an i*-based modeling language tailored for dealing with risks inherent to the adoption of open source software; the CORAS method [22], a model-driven approach focused on the protection of enterprise assets; and Archimate [6], in which risks are analyzed in the context of enterprise architecture models.

Naturally, having a clear understanding of the ontological nature of risk is fundamental for performing risk analysis, and even more for developing modeling languages to support it. Yet, risk is still a heavily overloaded and conceptually unclear notion [5,25], despite the wide number of efforts to properly characterize it [3,7,27]–including several standardization efforts [9,16–18].

In this paper, we address this issue by means of an in-depth ontological analysis, conducted under the principles of the Unified Foundational Ontology (UFO) [13]. As we shall see, our analysis shows an important result: the notion of *risk* is irreducibly intertwined with the notion of *value* and, more specifically, the process of assessing risk is a particular case of that of ascribing value. Indeed, we are not the first to relate value and risk. For example, Boholm and Coverllec [7] defended, in their relational theory of risk, that *"for an object to be considered 'at risk', it must be ascribed some kind of value"*, and Rosa [27] defined risk as *"a situation or event where something of human value [...] has been put at stake"*. Our analysis, however, is (to the best of our knowledge) the first to show and formally characterize the process of ascribing risk as a particular case of the process of ascribing value (in the sense of use value, as we shall discuss in Sect. 2). This opens the possibility of applying methodologies and techniques developed in marketing and economics for value analysis to the case of risk analysis, and vice versa, linking together two historically disconnected bodies of research. As a result of our analysis, we propose a concrete artifact, namely the Common Ontology of Value and Risk, formalized in OntoUML [13].

The remainder of this paper is organized as follows. First, in Sects. 2 and 3 we start with separate characterizations of the concepts of value and risk, contrasting their different interpretations found in the literature. Then, in Sect. 4 we compare the two concepts and discuss how several characteristics, historically ascribed to value, also apply to risk and vice versa. In Sect. 5 we present the common ontology of value and risk resulting from our analysis, and finally we discuss the implications of our findings on the practice of conceptual modeling of risk and value in Sect. 6, adding some further remarks in Sect. 7.

2   On Value and Value Ascription

The term 'value' is heavily overloaded, standing for various meanings in different fields. Thus, it is paramount to this paper to clarify what we mean (and what we do not mean) by value. There is one sense in which value stands for *ethical value* [26], as in "the values of our company are passion, integrity and diversity". In this sense, a value can be some sort of high-level and long term goal an agent is committed to pursuing or a sort of constraint that guides the behavior of an agent. This notion of value is important in the study of Ethics and human behavior, but it is not what we mean by value in this paper.

Another common meaning for value is that of *exchange value* [31], an interpretation that is widely adopted in economics. This meaning of value is exemplified in sentences such as "the value of my bicycle is 100 €" and "the value of my house is equivalent to that of two cars". Exchange value captures how much people are willing to pay for something or, more broadly, the worth of one good or service expressed in terms of the worth of another. This meaning of value is fundamental for economics and has been used in modeling approaches such as e3value [11], but still, it is not the interpretation we adopt in this paper.

Moreover, value may stand for *use value* [1,29,31], as in "my bicycle is valuable to me because I ride it to the office every day" and "the heating system of my car is of little value to me because I live in a city that is warm all year round". In this sense, the value of a thing emerges from how well its affordances match the goals/needs of a given agent in a given context. The notion of use value (or value-in-use) is mostly used in the business literature, in particular in marketing and strategy research, as it is a core part of understanding relevant phenomena such as what motivates customers to buy a particular product, why they choose one offering over another, and how companies differentiate themselves from their competitors. *Use value is the interpretation of value we adopt in this paper.*

In recent works [1,29], some of us have investigated the ontological nature of use value, aiming at understanding foundational questions such as: "What do we ascribe value to?" and "Which factors influence value?". In these works, we were able to identify and formalize various characteristics of use value. Its first characteristic is *goal-dependency*, i.e. things have value to people because they allow them to achieve their goals. This means that value is not intrinsic to anything, and the same object may have different values to different agents, or even according to different goals of the same agent. For example, a winter jacket has value on a cold night because, by wearing it, one is protected from the cold.

A second characteristic of use value is that, ultimately, it is ascribed to experiences, not objects. This may sound counter intuitive at first, as we have mentioned several examples of value seemly being attributed to objects. To clarify this point, let us go back to the winter jacket example. To ascribe some value to a jacket, we need to consider the situations in which we envision ourselves using such a jacket. It could be a snowy day while we go to work, a winter hike on the Italian Dolomites, or a rainy evening when we go to a dinner. In each of these situations, we will have different goals that we expect the jacket to help us fulfill, such as staying warm and dry, looking fashionable, keeping our belongings and so on. The value that we ascribe to the jacket, thus, will be "calculated" from the value ascribed in these envisioned experiences.

Despite its subjective nature and the fact that value is ultimately grounded on experiences, value is directly affected by the intrinsic properties of the objects that participate in these experiences. For instance, if a jacket is worn during a hike, it will be more or less valuable depending on its weight (an intrinsic quality that inheres in the jacket), as lighter jackets facilitate exercise. The same analysis holds for a jacket's waterproof capability on a rainy day: the more it can repel water, the more it satisfies its wearer's goal of staying dry.

By considering the whole experience in which objects are used, we are able to explain that not only intrinsic properties of things affect their value, but also the properties of other objects and of the experience itself. This is useful, for instance, to explain that the value experienced by a user of a movie streaming service is affected by the speed of the internet connection used to access it, as well as the screen resolution of the streaming device.

Note that the conceptualization of value we proposed in our previous works [1,29] is not restricted to the positive dimensions of experiences. As extensively discussed in the literature [19,21], value is a composition of benefits, which emerge from goal satisfaction, and sacrifices, which emerge from goal dissatisfaction. Thus, the value of an airline service is not only taking passengers from one place to another, but doing so minus the price one has to pay for the respective flight ticket, the effort to arrive to the point of departure, and so on.

The types of sacrifices that affect customer value have even been classified in the literature. Kambil and colleagues [19], for instance, propose to distinguish between three types of sacrifices, namely price, risk and effort. Note that the explicit representation of risk as a value reducing factor already suggests the process of ascribing value is strongly related to the process of assessing risk.

3   On Risk and Risk Assessment

The notion of risk has been systematically investigated for over 50 years [25]. Throughout this time, a wide number of definitions have been proposed and, although much progress has been made to clarify the nature of risk, the term remains overloaded and conceptually unclear [4,5,25,27].

One of the definitions that gained significant traction over the years in the risk community was proposed by the sociologist Eugene Rosa [27], who defined risk as "*a situation or event where something of human value (including humans themselves) has been put at stake and where the outcome is uncertain*". Rosa argues that his definition contains the three necessary and sufficient conditions to characterize risk. First, risk relates to some possible state of reality that affects someone's *interest*, either positively or negatively. Second, risk involves *uncertainty* about whether or not such a state will hold in the future; thus, if an event is certain to happen (such as the sun rising tomorrow), one cannot ascribe a risk to it. Third, risk is about a *possible* state of reality (thus ruling out the possibility of talking about the risk of someone turning into a werewolf).

Note that, intentionally, Rosa's definition does not exclude the case of "positive risks", i.e., risks related to events that can exclusively affect one's interests in a positive way. This idea that risks are not necessarily "bad", however, is in fact much older, dating back to at least the 1960's, when the distinction of speculative and pure risks was already being discussed [32]. In this context, pure risk stands for uncertain events that exclusively lead to negative outcomes (such as the risk of being in a car accident or the risk of being robbed), while speculative risk stands for the possibility of getting either a positive or a negative outcome, such as when investing in a company or playing the lottery.

More recently, Boholm and Corvellec proposed the so-called *relational theory of risk* [7], which defines risk as a triple composed of a *risk object*, an *object at risk*, and a *risk relationship* connecting the former two. In this theory, *risk objects* are said to be the source of risks, such as a drunk driver that poses a threat to the wellbeing of pedestrians, or a blizzard that puts car drivers in risk of an accident. Note that, even though the authors use the term *object*, they also include events and states as possible risk "objects". *Objects at risk* are the things of value[1] that are at stake because of a risk object. In the former examples, the objects at risk could be the pedestrians, the car, the driver and so on. The *risk relationship* is what connects risk objects to objects at risk. The authors adopt a cognitive approach towards the nature of risk, arguing that these relationships do not just occur, but instead they must be crafted or imagined by some agent. What follows from this position is that being a risk object or an object at risk is neither an intrinsic nor a necessary property of anything. Thus, an object may be a risk object to one person and an object at risk to another.

In [5], Aven and colleagues compared eleven definitions of risk from different sources, categorizing them in three groups, each capturing a particular sense in which risk is used. A first group refers to risk as a quantitative concept "attached" to an event. This interpretation is fundamental to make sense of sentences such as "it is riskier to drive when it is snowing than when it is not". The second group refers to risks as if they were the actual events, defined in terms of a chain of causality leading to consequences to some agent. This perspective is fundamental to explain what we assess risk for, and where risk comes from. The last group refers to risks as people's perceptions, equating objective risk to assessed risk. In this sense, risk is not just "out there", but, as argued in the relational theory of risk, it must be necessarily assessed by someone.

This plethora of risk definitions led to a number of standardization efforts [9,16–18] that aimed to provide an ultimate definition for those working on risk management. One of these efforts resulted in the ISO 3100:2018 [17,18] standard, which defines risk as the *"effect of uncertainty on objectives"*. This very abstract and concise definition is further explained in the standard by a number of commentaries, including that risks might refer to positive or negative impact on objectives (in line with Rosa's proposal [27]) and that risks are often explained in terms of events, consequences and likelihood.

In summary, what can be extracted from these different definitions is that to conceptualize risk, one must refer to: (i) agents and their goals; (ii) events and their triggers, and events' impacts on goals; and (iii) uncertainty.

4   Similarities Between Value and Risk

In this section, we elaborate on the evidences that motivated our pursue of a common ontology of value and risk. In particular, we explore the role of goals, context, uncertainty and impact in the conceptualization of both risk and value.

---

[1] Boholm and Corvellec [7] do not explicitly state which notion of value they use, but, through their argumentation and examples, we inferred that they mean use value.

4.1   Goal Dependency

The first similarity between value and risk is that they are both goal-dependent notions, in the sense that nothing is intrinsically valuable and nothing is intrinsically at risk. Things do not just have value, they have value *for* someone, and in case their *affordances* enable certain happenings that positively contribute to the achievement of one's goals. Analogously, things are only at risk from one's perspective, in case their *vulnerabilities* enable happenings that hurt one's goals. Just as "beauty is in the eye of the beholder", so are value and risk.

Take a pack of cigarettes, for instance. It has a high positive value for a smoker, as it enables him to satisfy his addiction. The same pack of cigarettes, however, would have arguably no value for a non-smoker, as its affordances would not help such a person to make any progress towards her goals. Similarly, if one drops her wallet on the street, we would claim that the wallet is at risk of being stolen, as it is unattended and the owner probably wants to keep her money and documents. However, from the perspective of an alert thief, such an unattended wallet is not at risk, but it presents an opportunity for an easy theft.

Note that even though risk and value are subjective, they still depend on the intrinsic properties of things or, to put it more precisely, on their *dispositions* (or dispositional properties). Note that when dispositions are perceived as beneficial, i.e., they enable the manifestation of events desired by an agent, they are usually labeled as *capabilities*[23], as in the capability of a smartphone to make calls. Conversely, when dispositions enable undesired events, as in "the fragility of my phone's screen material makes it susceptible to breaking", they are referred to as *vulnerabilities* [6]. That is why conceptualizations of value, which usually focus on positive outcomes, refer to capabilities, and those of risk, which usually focus on negative outcomes, refer to vulnerabilities.

Still, some argue that risks can be absolute in some situations [5], such as in the risk of dying. This argument is built upon the claim that some things, such as human lives, are universally valuable, thus any death is an event that necessarily "destroys" value. We argue against this position because, from an utilitarian perspective, value always emerges from goal achievement, which makes it necessarily relative. Take for instance the extreme case of suicidal terrorist attacks. From the perspective of the attacker, his death does not destroy value, but creates it for his terrorist organization.

4.2   Context Dependency

Another similarity between value and risk comes from the process we follow to "calculate" them. The value/risk ascribed to an object is always derived from the value/risk ascribed to events (or experiences) "enabled" by their dispositions, regardless if these events are intentional or not, and if they affect one's goals positively or negatively. Ascribing value to a notebook, for instance, means ascribing value to a number of different experiences enabled by the notebook, such as streaming a movie, giving a presentation, using a social media platform, working on a paper while traveling for a conference in another country, or playing a computer game. The ascribed value could be even high for some of these cases and low for others. Nonetheless, *the value of the notebook* cannot be computed without considering the different scenarios in which it will be used. Analogously, ascribing risk to an object means ascribing risks to different events involving this object. For instance, the risk of a car being stolen is ascribed based on the risk of it being stolen when parked on a private garage, when parked on the street, when being driven in a city with high criminality rates, and so on.

Stating that risk and value are contextually dependent means that they emerge not only from intrinsic properties of an object, but also from contextual properties. The value of watching a film on Netflix indeed depends on the properties of the Netflix service, but it also depends on the properties of the other objects involved in the experience, such as the resolution of the streaming device and the speed of the internet connection. In an analogous manner, the risk of being involved in a car accident when driving on a highway is certainly affected by the car's properties, such as how reliable the breaking system is, but it is also affected by the properties of the other participants of the driving event, including the highway's physical conditions and the traffic intensity.

4.3 Uncertainty and Impact

A further evidence that the processes of conceptualizing risk and value are similar regards the role played by uncertainty and impact. According to the popular risk equation, risk is equal to the likelihood of an event times its impact.

To understand why impact is positively correlated with risk, consider the following example. Two business angels invested in the same startup. One invested a hundred thousand euros and the other a million. If everything else but the invested amount is the same, the bigger investment is said to be riskier, since the impact of an eventual bankruptcy of the startup would be ten times worse.

But what about value? Is the value ascribed to an experience also positively correlated with its impact? As we have previously discussed, value emerges from achieving goals. Thus, the more an event makes progress towards achieving one, the more valuable it is. For instance, imagine that a traveler wants to fly from Rome to Brussels and that there are only two flights available. If the only significant difference between them is that one takes two hours and the other takes four hours, the shorter flight would be more valuable to the traveler, assuming that she has the goal of minimizing the duration of her trip.

The other parameter in the risk equation is likelihood, often referred to as probability or frequency, which states that the more likely an event is to happen, the riskier it is. To understand this correlation, consider two trips. The first takes place in a highway during a bright sunny day, whilst the second takes place during a snowstorm. The risk of an accident in the latter scenario is greater simply because an accident is more likely to happen in conditions that included reduced visibility and reduced adhesion of the car tires to the road.

Note that the likelihood parameter also applies to value. To see how, let us consider a mobile app that works as a compass. It is very unlikely that urban smartphone users would ever need such an app for guidance. Thus, having it in their phones is of very little value to them, even though it could be useful in a theoretical scenario. If we then compare it to other apps, such as a calendar, a camera, or an alarm, the value of the compass app seems to be even lower, as these other apps are often used in a daily basis.

In summary, the computation of the likelihood of an event times its impact on one's objectives and preferences fits the quantitative analysis of both risk and value. The differences between them rely on the kind of event one usually analyzes (unwanted for risks, expected and desired for value) and the nature of the expected impact on goals (negative for risks and positive for value).

5   The Common Ontology of Value and Risk

In this section, we present a well-founded ontology that formalizes the assumptions on value and risk discussed in the previous sections. Given the polysemic nature of these terms [20,31], we aim to disentangle three perspectives: (i) an *experiential perspective*, which describes value and risk in terms of events and their causes, (ii) a *relational perspective*, which identifies the subjective nature of value and risk, and (iii) a *quantitative perspective*, which projects value and risk on measurable scales. In the OntoUML diagrams depicting this ontology, we adopt the following color coding: events are represented in yellow, objects in pink, qualities and modes in blue, relators in green, situations in orange, and powertypes in white. Additionally, in the models represented in these diagrams, we use the OntoUML semantics of non-sortals proposed in [15].

The *experiential perspective* is depicted in Figs. 1 and 2. As argued in the previous section, value and risk can be ascribed to both objects and events. Still, whenever they are ascribed to an object, one must always consider all the relevant events involving it, which will ultimately ground value and risk. These events, named VALUE and RISK EXPERIENCE, have some agents as key participants, deemed VALUE and RISK SUBJECT respectively. These identify the perspective from which the judgment is made and whose INTENTIONS are considered.

Note that, as we argued in [29], value can be ascribed to past, actual or envisioned experiences. Risk, however, is only ascribed to envisioned experiences that may (but are not certain to) happen. We are aware that there is a controversy concerning the ontological nature of future events. The classical view of events assumes that they are immutable entities and that only past events truly exist as genuine perdurants (occurrences) [10]. However, accounting for future events (which is the case for envisioned experiences) seems to be unavoidable for any theory of risk, as uncertainty and possibility are core aspects of this concept. This means that we need to refer to future events – whose expected temporal properties are not completely fixed – as first-class citizens in our domain of discourse. As bold as this assumption may seem (see [12] for details), conceptualizing risk with no reference to the future would sound as an oxymoron to us, given the explanatory purposes of our paper. So, we shall talk of expected events as regular entities of our domain, not differently from, say, a planned air trip in a flight reservation system. In order to use this non-classical notion of events in our analysis while maintaining its ontological rigour, we employ the formulation of events as proposed in [12], which was already successfully employed in our work on value propositions [29].

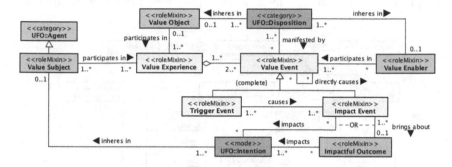

Fig. 1. Value experiences, their parts and participants.

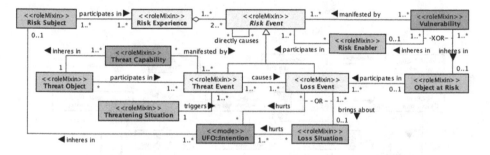

Fig. 2. Risk experiences, their parts and participants.

VALUE and RISK EXPERIENCES are commonly decomposed into "smaller" events to clarify their internal structure and how they affect multiple goals. One component type of a VALUE EXPERIENCE is a TRIGGER EVENT, which is defined by causing, directly or indirectly, events of gain or loss. A second component type is an IMPACT EVENT, which is defined by its impact on INTENTIONS. Note that such an impact might be direct or indirect, and positive or negative. An example of an event with a *direct positive* impact is that of eating which directly satisfies a goal of being fed, while an example of an event with a *direct negative* impact is that of being robbed, which directly hurts the goal of feeling safe. An example an event with an *indirect positive* impact is that of taking a bus, which, upon its completion, will satisfy the goal of arriving at a destination. Lastly, an event with an *indirect negative* impact would be that of having your phone stolen, which puts one in a phone-less situation, which in turn hurts one's goals of

contacting people. The difference for RISK EXPERIENCES is that the focus is on
unwanted events that have the potential of causing losses. Thus, its components
are restricted to THREAT and LOSS EVENTS. A THREAT EVENT is one with the
potential of causing a loss, which might be intentional, such as a hacker attack,
or unintentional, such as an accidental liquid spill on a computer. LOSS EVENTS
are simply *Impact Events* that necessarily impact intentions in a negative way.

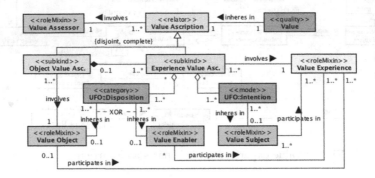

Fig. 3. Modeling value ascriptions.

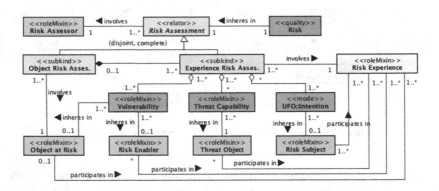

Fig. 4. Modeling risk assessments.

In the ontology, we differentiate between several roles played by objects
in VALUE and RISK EXPERIENCES. In the value case, we distinguish between
VALUE OBJECTS and VALUE ENABLERS. These are the objects whose dispositions "enable" the occurrence of a VALUE EXPERIENCE (or of one of its parts).
Their difference is that the former is the focus of a given value ascription (e.g.
a car, a music streaming service), whilst the latter plays an ancillary role in
VALUE EXPERIENCES (e.g. the fuel in car, the device used for streaming). In the
risk case, we distinguish between three roles: (i) the THREAT OBJECT, as that

which causes a threat; (ii) the OBJECT AT RISK, as that which is exposed to potential damage; and (iii) the RISK ENABLER, as that which plays an ancillary role in RISK EXPERIENCES. To exemplify this latter distinction, consider a situation in which a factory worker gets hurt while operating a machine. In this case, the worker is both the THREAT OBJECT and the OBJECT AT RISK, but she only got hurt because her equipment, the THREAT ENABLER, was not sturdy enough. Analogously to the value case, the dispositions of all these objects are manifested in risk experiences. Those of THREAT OBJECTS, however, are labeled THREAT CAPABILITIES (e.g. the skill of a pick-pocketer to swiftly grab a wallet), whilst those of OBJECTS AT RISK and RISK ENABLERS are labeled VULNERABILITIES (e.g. the flammability of a house manifested in a fire, a security flaw in an information system which allows hackers to steal sensitive data).

The *relational perspective* is depicted in Figs. 3 and 4. We capture it by means of objectified relationships labeled VALUE ASCRIPTION and RISK ASSESSMENT, which involve: (i) an agent responsible for the judgment, deemed the VALUE and RISK ASSESSOR respectively; and (ii) the target of the judgment, either an object or an event. Judgments made for objects are labeled OBJECT VALUE ASCRIPTION and OBJECT RISK ASSESSMENT and involve, respectively, exactly one VALUE OBJECT and OBJECT AT RISK. Judgments on events are deemed EXPERIENCE VALUE ASCRIPTION and EXPERIENCE RISK ASSESSMENT, and involve, respectively, one VALUE and RISK EXPERIENCE.

The *quantitative perspective* is also depicted in Figs. 3 and 4. We represent it by means of the VALUE and RISK qualities inhering in the aforementioned relationships. In UFO, a quality is an objectification of a property that can be directly evaluated (projected) into certain value spaces [13]. Common examples include a person's weight, which can be measured in kilograms or pounds, and the color of a flower, which can be specified in RGB or HSV. Thus, representing value and risk as qualities means that they can also be measured according to a given scale, such as a simple discrete scale like $<Low, Medium, High>$ or a continuous scale (e.g. from 0.0 to 100.0).

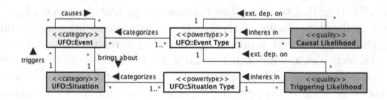

Fig. 5. Representing likelihood in UFO.

Lastly, as discussed in Sect. 4, conceptualizing value and risk requires accounting for the likelihood of events, which is typically expressed by probability measures. However, as noted by Aven et al. [5], there are two conflicting interpretations on the ontological nature of probability: (i) the *frequentist* interpretation, in which the probability of an event is the fraction of times an event of that *type* occurs. For example, the likelihood of the Brazil beating Germany in the 2018 World Cup may be calculated based on the number of times it happened in the past; and (ii) the *subjective* interpretation, in which the probability of an event expresses the assessor's uncertainty (degree of belief) of an event to occur, conditioned on some background knowledge. In the World Cup example, the likelihood of a Brazilian victory for a sports analyst depends on her knowledge about the physical conditions of players, the teams' tactics, etc.

Discussing the ontological nature of probability is out of the scope of this paper. For us, it suffices that, in both perspectives, likelihood is a quantitative concept that inheres in *types*, not in individuals. Thus, we need to include types of events and situations in our domain of discourse. We do that by employing the notion of *powertype* incorporated into OntoUML. This means taking its ontological interpretation as proposed in [14] and following the modeling guidelines proposed in [8]. In particular, following the latter, we employ the relation of *categorization* between a *powertype* t and its base type t' such that: a type t categorizes a type t' iff all instances of t are proper specializations of t'.

Figure 5 illustrates the excerpt of our ontology w.r.t. the concept of likelihood. We distinguish between a TRIGGERING and CAUSAL LIKELIHOOD. The former inheres in a SITUATION TYPE and represents how likely a SITUATION TYPE will trigger an EVENT TYPE once a situation of this type becomes a fact. The latter inheres in an EVENT TYPE and captures that, given the occurrence of an event e and a certain EVENT TYPE t, how likely e will – directly or indirectly – cause another event of type t to occur. In the value case, how likely a TRIGGER EVENT of a VALUE EXPERIENCE will cause an IMPACT EVENT, whilst on the risk case, how likely a THREAT EVENT of a RISK EXPERIENCE will cause a LOSS EVENT.

6   Implications for Conceptual Modeling

The ontology we proposed in this paper has a number of implications for research on the conceptual modeling of value and risk. First, the ontology can provide well-founded real world semantics for existing risk modeling languages, such as CORAS [22], RiskML [30], Goal-Risk Framework [2] and Archimate [6].

Second, following existing methods for ontology-based language evaluation and (re)design [13,28], our proposal can serve as a reference model to assess how well these modeling approaches stand w.r.t to the risk domain in terms of *domain appropriateness* and *comprehensibility appropriateness*[13]. More concretely, these methods can be systematically employed for the identification of a number of types of deficiencies that can occur in language design (e.g., *construct deficit* - when there is concept in the domain that does not have a representation in terms of a construct of the language, and *construct overload* - when a construct in the language represents more than a domain concept). For example, such an analysis of RiskML [30] would identify a construct deficit with respect to the representation of vulnerabilities, whilst one in the CORAS approach [22] would identify a construct deficit regarding the explicit representation of goals. A case of construct overload is also found in ArchiMate [6], in which the RISK construct collapses: (i) a complex event, (ii) the overall risk an asset is exposed to, and (iii) an assessment regarding what to do about an identified risk.

Third, the ontology we propose allows for the comparison and integration of risk modeling approaches by means of semantic interpretation of the languages' constructs–after all, language integration is a semantic interoperability problem. For instance, the constructs of Threat Scenario and Unwanted Event in CORAS seem to be equivalent to those of Threat Event and Loss Event in Archimate [6], respectively, which in turn, are all specializations of RiskML's Event.

Languages for modeling use value are much less developed than those for risk. Thus, a relevant impact of this work is to demonstrate that a lot of the effort that has been done on risk modeling could be fruitfully leveraged in developing tools for value modeling. This is a noteworthy impact, given that value modeling approaches are in high demand, as evinced by the increasing popularity of tools such as the Business Model Canvas [23] and the Value Proposition Canvas [24].

7   Final Remarks

In this paper, we have presented an ontological analysis of risk which explicits the deep connections between the concepts of value and risk. The ontology that resulted from this analysis formally characterizes and integrates three different perspectives on risk: (i) risk as a quantitative notion, which we labeled simply as RISK in our ontology; (ii) risk as a chain of events that impacts on an agent's goals, which we labeled as RISK EXPERIENCE; and (iii) risk as the relationship of ascribing risk, which we labeled as RISK ASSESSMENT.

Moreover, this paper further extends the ontological analysis on use value initiated in [1] and revisited in [29]. In particular, we improved these works by (i) discussing how likelihood influences value, (ii) refining the internal structure of value experiences and their participants, (iii) clarifying the role of dispositions in value creation, and (iv) distinguishing between value objects and value enablers.

We are aware, however, that the current ontology does not fully describe the domain of risk management, as it lacks security-related concepts such as *mitigation* and *control strategies*. These are recurrently found in risk modeling languages (e.g. [6,22]), as analysts do not just need to identify and model risks, but also decide on how to address them. As future work, we plan to extend and further validate our ontology with risk analysis experts from different domains, such as finance and software development. We also plan to validate it by means of systematic comparisons with other theories and formalizations of risk. Then, we can leverage it to analyze the domain adequacy of existing risk modeling languages and, if needed, redesign them so that they are clearer and more expressive to model risks.

Acknowledgement. This work is partially supported by CNPq (grants number 407235/2017-5, 312123/2017-5 and 312158/2015-7), CAPES (23038.028816/2016-41) and FUB (OCEAN Project).

References

1. Anderson, B., Guarino, N., Johannesson, P., Livieri, B.: Towards an ontology of value ascription. In: 9th International Conference on Formal Ontology in Information Systems (FOIS), vol. 283, p. 331. IOS Press (2016)
2. Asnar, Y., Giorgini, P., Mylopoulos, J.: Goal-driven risk assessment in requirements engineering. Requir. Eng. 16(2), 101–116 (2011)
3. Aven, T.: On how to define, understand and describe risk. Reliab. Eng. Syst. Saf. 95(6), 623–631 (2010)
4. Aven, T.: Misconceptions of Risk. Wiley, New York (2011)
5. Aven, T., Renn, O., Rosa, E.A.: On the ontological status of the concept of risk. Saf. Sci. 49(8), 1074–1079 (2011)
6. Band, I., et al.: Modeling enterprise risk management and security with the Archi-Mate language - W172 (2017)
7. Boholm, Å., Corvellec, H.: A relational theory of risk. J. Risk Res. 14(2), 175–190 (2011)
8. Carvalho, V.A., Almeida, J.P.A., Fonseca, C.M., Guizzardi, G.: Multi-level ontology-based conceptual modeling. Data Knowl. Eng. 109, 3–24 (2017)
9. Committee of Sponsoring Organizations of the Treadway Commission (COSO): Enterprise Risk Management - Integrated Framework (2004)
10. Diekemper, J.: The existence of the past. Synthese 191(6), 1085–1104 (2014)
11. Gordijn, J., Akkermans, J.M.: Value-based requirements engineering: exploring innovative e-commerce ideas. Requir. Eng. 8(2), 114–134 (2003)
12. Guarino, N.: On the semantics of ongoing and future occurrence identifiers. In: Mayr, H.C., Guizzardi, G., Ma, H., Pastor, O. (eds.) ER 2017. LNCS, vol. 10650, pp. 477–490. Springer, Cham (2017). https://doi.org/10.1007/978-3-319-69904-2_36
13. Guizzardi, G.: Ontological foundations for structural conceptual models (2005)
14. Guizzardi, G., Almeida, J.P.A., Guarino, N., de Carvalho, V.A.: Towards an ontological analysis of powertypes. In: 1st Joint Ontology Workshops (JOWO) (2015)
15. Guizzardi, G., Fonseca, C.M., Benevides, A.B., Almeida, J.P.A., Porello, D., Sales, T.P.: Endurant types in ontology-driven conceptual modeling: Towards OntoUML 2.0. In: 37th International Conference on Conceptual Modeling (ER) (2018)
16. Institute of Risk Management (IRM): A Risk Management Standard (2002)
17. ISO: Risk Management - Vocabulary, ISO Guide 73:2009 (2009)
18. ISO: Risk Management - Guidelines, ISO 31000:2018 (2018)
19. Kambil, A., Ginsberg, A., Bloch, M.: Re-inventing value propositions. In: Information Systems Working Papers Series (1996)
20. Kjellmer, G.: On the awkward polysemy of the verb 'risk'. Nord. J. Engl. Stud. 6(1), 57–70 (2007)
21. Lanning, M.J., Michaels, E.G.: A business is a value delivery system (1988)
22. Lund, M.S., Solhaug, B., Stølen, K.: Model-Driven Risk Analysis: The CORAS Approach. Springer Science & Business Media, Heidelberg (2010)
23. Osterwalder, A., Pigneur, Y.: Business Model Generation: A Handbook for Visionaries, Game Changers, and Challengers. Wiley, Hoboken (2010)
24. Osterwalder, A., Pigneur, Y., Bernarda, G., Smith, A.: Value Proposition Design: How to Create Products and Services Customers Want. Wiley, Hoboken (2014)
25. Renn, O.: Three decades of risk research: accomplishments and new challenges. J. Risk Res. 1(1), 49–71 (1998)
26. Rokeach, M.: The Nature of Human Values. Free Press, New York (1973)

27. Rosa, E.A.: Metatheoretical foundations for post-normal risk. J. Risk Res. **1**, 15–44 (1998)
28. Rosemann, M., Green, P., Indulska, M.: A reference methodology for conducting ontological analyses. In: Atzeni, P., Chu, W., Lu, H., Zhou, S., Ling, T.-W. (eds.) ER 2004. LNCS, vol. 3288, pp. 110–121. Springer, Heidelberg (2004). https://doi.org/10.1007/978-3-540-30464-7_10
29. Sales, T.P., Guarino, N., Guizzardi, G., Mylopoulos, J.: An ontological analysis of value propositions. In: 21st IEEE International Enterprise Distributed Object Computing Conference (EDOC), pp. 184–193 (2017)
30. Siena, A., Morandini, M., Susi, A.: Modelling risks in open source software component selection. In: Yu, E., Dobbie, G., Jarke, M., Purao, S. (eds.) ER 2014. LNCS, vol. 8824, pp. 335–348. Springer, Cham (2014). https://doi.org/10.1007/978-3-319-12206-9_28
31. Vargo, S.L., Maglio, P.P., Akaka, M.A.: On value and value co-creation: a service systems and service logic perspective. Eur. Manag. J. **26**(3), 145–152 (2008)
32. Williams, C.A.: Attitudes toward speculative risks as an indicator of attitudes toward pure risks. J. Risk Insur. **33**(4), 577–586 (1966)

Endurant Types in Ontology-Driven Conceptual Modeling: Towards OntoUML 2.0

Giancarlo Guizzardi[1,2]([⊠]), Claudenir M. Fonseca[1],
Alessander Botti Benevides[2], João Paulo A. Almeida[2], Daniele Porello[1],
and Tiago Prince Sales[3]

[1] Conceptual and Cognitive Modeling Research Group (CORE),
Free University of Bozen-Bolzano, Bolzano, Italy
{giancarlo.guizzardi,daniele.porello,cmoraisfonseca}@unibz.it
[2] NEMO, Federal University of Espírito Santo, Vitória, Brazil
abbenevides@inf.ufes.br
[3] DISI, University of Trento, Trento, Italy
tiago.princesales@unitn.it

Abstract. For over a decade now, a community of researchers has contributed to the development of the Unified Foundational Ontology (UFO)
- aimed at providing foundations for all major conceptual modeling constructs. This ontology has led to the development of an Ontology-Driven
Conceptual Modeling language dubbed OntoUML, reflecting the ontological micro-theories comprising UFO. Over the years, UFO and OntoUML
have been successfully employed in a number of academic, industrial and
governmental settings to create conceptual models in a variety of different domains. These experiences have pointed out to opportunities of
improvement not only to the language itself but also to its underlying
theory. In this paper, we take the first step in that direction by revis-
ing the theory of types in UFO in response to empirical evidence. The
new version of this theory shows that many of the meta-types present
in OntoUML (differentiating Kinds, Roles, Phases, Mixins, etc.) should
be considered not as restricted to Substantial types but instead should
be applied to model Endurant Types in general, including Relator types,
Quality types and Mode types. We also contribute a formal characterization of this fragment of the theory, which is then used to advance a
metamodel for OntoUML 2.0. Finally, we propose a computational support tool implementing this updated metamodel.

Keywords: OntoUML · UFO · Ontology-Driven Conceptual Modeling

1 Introduction

In recent years, there has been a growing interest in the use of foundational
ontologies (i.e., ontological theories in the philosophical sense) to evaluate and

© Springer Nature Switzerland AG 2018
J. C. Trujillo et al. (Eds.): ER 2018, LNCS 11157, pp. 136–150, 2018.
https://doi.org/10.1007/978-3-030-00847-5_12

(re)design conceptual modeling languages. For over a decade now, a community of researchers has contributed to the development of a foundational ontology (termed the Unified Foundational Ontology – UFO) aimed at providing foundations underlying all conceptual modeling major constructs. This ontology has also been systematically used to design an ontology-driven conceptual modeling (ODCM) language termed OntoUML [8,9]. UFO and OntoUML have been successfully employed in academic, industrial and governmental settings to create conceptual models in a number of different domains, including Geology, Biodiversity Management, Organ Donation, Petroleum Reservoir Modeling, Disaster Management, Context Modeling, Datawarehousing, Enterprise Architecture, Data Provenance, Measurement, Logistics, Complex Media Management, Telecommunications, Heart Electrophysiology, among many others [9]. In fact, research shows that they are among the most used foundational ontology and modeling language in the ODCM literature, respectively [13]. Moreover, empirical evidence shows that OntoUML significantly contributes to improving the quality of conceptual models without requiring an additional effort to produce them. For instance, the work of [12] reports on a modeling experiment conducted with 100 participants in two countries showing the advantages (in these respects) of OntoUML when compared to a classical conceptual modeling language (EER).

The observation of the application of OntoUML over the years conducted by several groups in a variety of domains also amounted to a fruitful empirical source of knowledge regarding the language and its foundations[1]. In particular, we have managed to observe a number of different ways in which people would slightly subvert the syntax of the language, ultimately creating what we could call *"systematic subversions"* of the language [9]. These "subversions" would (purposefully) produce models that were grammatically incorrect, but which were needed to express the intended characterization of their underlying conceptualizations that could not be expressed otherwise. Moreover, they were "systematic" because they would recur in the works of different authors that would, independently of each other, subvert the language in the same manner and with the same modeling intention. One of these "language subversions" led us in this article to reconsider some of the theoretical foundations underlying the language, i.e., it led us to rethink and evolve a core theory in UFO, namely, its theory of *Endurant Types and Taxonomic structures* [8].

Structural conceptual modeling languages (e.g., ER, UML, ORM but also OntoUML) are designed to model types whose instances are endurants (object-like entities). These can be of the regular independent sort but also *dependent* endurants (e.g., weak entities/objects and objectified relationships). In ODCM approaches, such as OntoUML (but also OntoClean [7]), there are finer-grained distinctions among types, for instance, differentiating endurant types according to meta-types such as *Kinds*, *Phases*, *Roles* and *Mixins*.

Following the original version of UFO's theory of types, many of these meta-types present in OntoUML are currently restricted to the modeling of Substantial types (independent objects). For instance, we can represent that "Person" is the

---

[1] Several dozens of these models are available at http://www.menthor.net/.

Kind of entity Mick Jagger is, but that he is also in a "Senior Citizen" *Phase*, that he plays the *Role* of "Singer" and "Knight of the British Empire" in the scope of certain relations to The Rolling Stones and the Order of the British Empire, respectively. But, now, how can we model that the relationship between Giovanni and UN is of the *Kind* "Employment", that it is currently in a "Tenured" *Phase*, and that it can play the *Role* of "Legal Grounds" for his visa application?

Consciously ignoring this restriction, users of the language started to systematically employ these meta-type distinctions to other types of endurants, in particular, to *existentially dependent* endurants such as qualities (e.g., the perceived value of the experience, the color of the apple), modes (e.g., Paul's Dengue Fever, Matteo's capacity of programming in Scratch) and relators (e.g., John and Mary's Marriage, Giovanni's Employment at the UN). These "subversions" were needed to capture subtle aspects of domains such as value, service and economic exchange (among many others) [1,2,6,14]. This called our attention to the fact that, like full-fledged endurants, qualities, modes and relators are also subject of both essential and accidental properties, and as such they can also instantiate contingent types such as phases and roles and that, in complex domains, their types can also be involved in sophisticated taxonomic structures. In other words, meta-types such as phases, roles, role mixins, mixins, categories, etc. are meta-types of Endurants, in general, and not only of substantials.

The contributions of this paper are three-fold. First, we propose a new formal theory of endurant types and taxonomic structures for UFO. Although developed in the framework of UFO, this theory amounts to a contribution to ODCM, more broadly. In particular, we can influence approaches such as OntoClean and ORM, which are sensitive to these matters. Second, following the same ontology-based language engineering approach that was used to create the original version of OntoUML [8], we employ this new formalized version of the theory to advance an enhanced metamodel for OntoUML 2.0. Finally, we employ this metamodel to implement a software tool for OntoUML 2.0 supporting model verification.

The remainder of this paper is organized as follows: Sect. 2 presents the background on OntoUML and UFO; Sect. 3 discusses and formalizes the changes on the underlying theory, UFO; Sect. 4 introduces a new version of OntoUML, presenting its constructs and syntactic constraints; Sect. 5 concludes the paper with our final considerations.

2   Background: UFO and OntoUML

OntoUML, as all structural conceptual modeling languages is meant to represent type-level structures whose instances are *endurants* (object-like entities), i.e., they are meant to model *Endurant Types* and their type-level relations. Figure 1 depicts the hierarchy of Endurant Types in UFO.

UFO distinguishes Endurant Types into Substantial Types and Moment Types. Naturally, these are sorts of types whose instances are Substantials and Moments [8], respectively. Substantials are existentially independent objects such as John Lennon, the Moon, an organization, a car, a dog. Moments, in contrast,

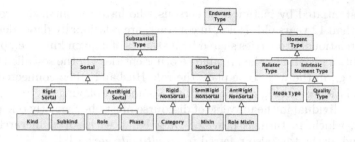

Fig. 1. A taxonomy of endurant types in UFO.

are *existentially dependent* individuals such as (a) Sofia's capacity to speak Italian (which depends on her) and (b) the marriage between John and Yoko (which depends on both John and Yoko). Moments of type (a) are termed modes; those of type (b) are termed relators. Relators are individuals with the power of connecting entities. For example, an Enrollment relator connects an individual playing the Student role with an Educational Institution. Every instance of a relator type is existentially dependent on at least two distinct entities. Moreover, relators are typically composed of modes, for example, in the way that the marriage between John and Mary is composed of their mutual commitments and claims. Furthermore, there is a third sort of moments termed qualities. Qualities are individual moments that can be mapped to some quality space, e.g., an apple's color which may change from green to red while maintaining its identity [8].

Concerning the substantial type hierarchy, sortal types are the ones that either provide or carry a uniform *principle of identity* for their instances. A principle of identity regarding a sortal S makes explicit the properties that no two instances of S can have in common, because such properties uniquely identify S instances. In particular, it also informs which changes an individual can undergo without changing its identity, i.e., while remaining the same. Within the category of sortals, we can further distinguish between rigid and anti-rigid. A rigid type is one that classifies its instances necessarily (in the modal sense), i.e., the instances of that type cannot cease to be so without ceasing to exist. Anti-rigidity, in contrast, characterizes a type whose instances can move in and out of its extension without altering their identity [8]. For instance, contrast the rigid type Person with the anti-rigid types Student or Husband. While the same individual John never ceases to be an instance of Person, he can move in and out of the extension of Student or Husband, depending on whether he enrolls in/finishes college or marries/divorces, respectively.

Kinds are sortal rigid types that provide a uniform principle of identity for their instances (e.g., Person). Subkinds are sortal rigid types that carry the principle of identity supplied by a unique Kind (e.g., a kind Person can have the subkinds Man and Woman that carry the principle of identity provided by Person). Concerning anti-rigid sortal types, we have the distinction between roles and phases. Phases are relationally independent types defined by contingent but intrinsic instantiation conditions [8]. For example, a Child is a phase of

Person, instantiated by instances of persons who have the intrinsic property of being less than 12 years old. Roles, in contrast, are relationally dependent types, capturing relational properties shared by instances of a given kind, i.e., putting it baldly: entities play roles when related to other entities via the so-called material relations (e.g., in the way some plays the role Husband when connected via the material relation of "being married to" with someone playing the role of Wife). Since each individual in the universe of discourse must obey exactly one principle of identity, which, in turn, is provided by a Kind, each sortal hierarchy has a unique Kind at the top, also referred to as *ultimate sortal* [8].

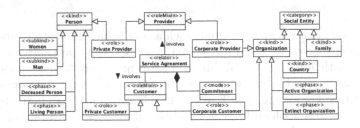

Fig. 2. OntoUML example.

Non-Sortals (also called *dispersive types* [8]) are types that aggregate properties that are common to different sortals, i.e., that ultimately classify entities that are of different Kinds. Non-sortals do not provide a uniform principle of identity for their instances; instead, they just classify things that share common properties but which obey different principles of identity. Furniture is an example of non-sortal that aggregates properties of Table, Chair and so on. Other examples include Work of Art (including paintings, music compositions, statues), Insurable Item (including works of arts, buildings, cars, body parts) and Legal Entity (including people, organizations, contracts, legislations). The meta-properties of rigidity and anti-rigidity can also be applied to distinguish different types of Non-Sortals. A Category represents a rigid and relationally independent non-sortal, i.e., a dispersive type that aggregates essential properties that are common to different rigid sortals [8] (e.g., Physical Object aggregates essential properties of tables, cars, glasses). A Role Mixin represents an anti-rigid and relationally dependent non-sortal, i.e., a dispersive type that aggregates properties that are common to different Roles (e.g., the type Customer that aggregates properties of individual customers and corporate customers) [8]. Although not prescribed in the original set of UFO endurant types, over the years, the notion of a Phase Mixin emerged as a useful notion that was found missing by different authors [4,11]. A *Phase Mixin* represents an anti-rigid and relationally independent non-sortal, i.e., a dispersive type that aggregates properties that are common to different Phases (e.g., the type Active Agent that aggregates properties of living people and active organizations). Finally, a Mixin is a non-sortal that represents properties shared by things of different kinds but which are essential to some of these instances and accidental to some others. For example, the type Insured Item can be essential to cars (suppose all cars must be insured) while being accidental to houses (e.g., houses can be insured but are not necessarily insured).

The leaf ontological distinctions represented in Fig. 1 as well as their corresponding axiomatization are reflected as modeling constructs in OntoUML [8]. An example of a model illustrating these notions is presented in Fig. 2.

3 A New Formal Theory of Endurant Types

In this section, we present a first-order modal theory of endurant types, in which types and their instances are both in the domain of quantification (i.e., first-order citizens). Types and their instances are connected by instantiation relations (symbolized as ::). For our purposes, the first order modal logic QS5 plus the Barcan formula and its converse suffices [5]. That means that we assume a *fixed* domain of entities for every possible world, what is traditionally associated to a *possibilistic* view of the entities of the domain, i.e., the domain includes all the *possibilia*. In the following formulas, we drop both the universal quantifier and the necessity operator in case their scope takes the full formula. In what follows, (a*) and (t*) patterns refer to axioms and theorems, respectively, and the modal operators of *necessity* ($\square$) and *possibility* ($\lozenge$) are used with their usual meaning. This theory has been specified and verified in TPTP (http://www.tptp.org), being automatically proven with provers SPASS 3.9 and Z3 4.4.1.[2]

Firstly, *types* are implicitly defined as those entities that are possibly instantiated (a1), while *individuals* are those necessarily not instantiated (a2). Since we are only concerned with first-order types, the domain of :: is Individual and the codomain is Type (a3). From (t1), (t2), entities are partitioned into individuals and types . We introduce the *specialization* relation between types ($\sqsubseteq$) defining it in terms of necessary extensional inclusion (a4), i.e., inclusion of their instances. By means of (a4), it follows that the specialization relation is quasi-reflexive (t3) and transitive (t4). Whenever two types have a common instance, they must share a supertype or a subtype for this instance (a5).

a1 $\quad$ Type$(x) \leftrightarrow \lozenge(\exists y(y :: x))$ $\qquad$ **a3** $\quad x :: y \rightarrow$ Individual$(x) \wedge$ Type(y)

a2 $\quad$ Individual$(x) \leftrightarrow \square(\neg \exists y(y :: x))$ $\qquad$ **t1** $\quad$ Individual$(x) \vee$ Type(x)

t2 $\quad \neg \exists x($Individual$(x) \wedge$ Type$(x))$

a4 $\quad x \sqsubseteq y \leftrightarrow$ Type$(x) \wedge$ Type$(y) \wedge \square(\forall z(z :: x \rightarrow z :: y))$

t3 $\quad x \sqsubseteq y \rightarrow (x \sqsubseteq x \wedge y \sqsubseteq y)$ $\qquad$ **t4** $\quad x \sqsubseteq y \wedge y \sqsubseteq z \rightarrow x \sqsubseteq z$

a5 $\quad \forall t_1, t_2, x((x :: t_1 \wedge x :: t_2 \wedge \neg(t_1 \sqsubseteq t_2) \wedge \neg(t_2 \sqsubseteq t_1)) \rightarrow (\exists t_3(t_1 \sqsubseteq t_3 \wedge t_2 \sqsubseteq t_3 \wedge$
$\quad x :: t_3) \vee \exists t_3(t_3 \sqsubseteq t_1 \wedge t_3 \sqsubseteq t_2 \wedge x :: t_3)))$

---

[2] For the formal specification, see https://github.com/nemo-ufes/ufo-types.

We implicitly define *rigidity* of types as rigid (a6), semi-rigid (a8) and anti-rigid (a7), concluding that every type is either one of the three ((t5) and (t6)) and rigid and semi-rigid types cannot specialize anti-rigid ones ((t7) and (t8)).

a6 $\mathsf{Rigid}(t) \leftrightarrow \mathsf{Type}(t) \land \forall x(\Diamond(x :: t) \rightarrow \Box(x :: t))$

a7 $\mathsf{AntiRigid}(t) \leftrightarrow \mathsf{Type}(t) \land \forall x(\Diamond(x :: t) \rightarrow \Diamond(\neg x :: t))$

a8 $\mathsf{SemiRigid}(t) \leftrightarrow \mathsf{Type}(t) \land \neg\mathsf{Rigid}(t) \land \neg\mathsf{AntiRigid}(t)$

t5 $\mathsf{Type}(t) \leftrightarrow \mathsf{Rigid}(t) \lor \mathsf{AntiRigid}(t) \lor \mathsf{SemiRigid}(t)$

t6 $\neg\exists x((\mathsf{Rigid}(x) \land \mathsf{AntiRigid}(x)) \lor (\mathsf{Rigid}(x) \land \mathsf{SemiRigid}(x)) \lor (\mathsf{SemiRigid}(x) \land \mathsf{AntiRigid}(x)))$

t7 $\neg\exists x, y(\mathsf{Rigid}(x) \land \mathsf{AntiRigid}(y) \land x \sqsubseteq y)$

t8 $\neg\exists x, y(\mathsf{SemiRigid}(x) \land \mathsf{AntiRigid}(y) \land x \sqsubseteq y)$

On *sortality*, our basic assumption is that every individual necessarily instantiates a kind (a9), and everything necessarily instantiates at most one kind (a10). We implicitly define *sortals* as those types whose instances necessarily instantiate the same kind (a11); while a *non-sortal* is a type that is necessarily not a sortal (a12). As theorems, we have that kinds are rigid (t9), kinds are necessarily disjoint (t10); a kind cannot specialize a different kind (t11); kinds are sortals (t12); sortals specialize a kind (t13); sortals cannot specialize different kinds (t14); a non-sortal cannot specialize a sortal (t15); and non-sortals do not have direct instances, their instances are also instances of a sortal that either specializes the non-sortal, or specializes a common non-sortal supertype (t16).

a9 $\mathsf{Individual}(x) \rightarrow \exists k(\mathsf{Kind}(k) \land \Box(x :: k))$

a10 $\mathsf{Kind}(k) \land x :: k \rightarrow \neg\Diamond(\exists z(\mathsf{Kind}(z) \land x :: z \land z \neq k))$

a11 $\mathsf{Sortal}(t) \leftrightarrow \mathsf{Type}(t) \land \exists k(\mathsf{Kind}(k) \land \Box(\forall x(x :: t \rightarrow x :: k)))$

a12 $\mathsf{NonSortal}(t) \leftrightarrow \mathsf{Type}(t) \land \neg\mathsf{Sortal}(t)$

t9 $\mathsf{Kind}(k) \rightarrow \mathsf{Rigid}(k)$

t10 $\mathsf{Kind}(x) \land \mathsf{Kind}(y) \land x \neq y \rightarrow \Box(\neg\exists z(z :: x \land z :: y))$

t11 $\mathsf{Kind}(x) \land \mathsf{Kind}(y) \land x \neq y \rightarrow (\neg(x \sqsubseteq y) \land \neg(y \sqsubseteq x))$

t12 $\mathsf{Kind}(t) \rightarrow \mathsf{Sortal}(t)$

t13 $\mathsf{Sortal}(x) \rightarrow \exists k(\mathsf{Kind}(k) \land x \sqsubseteq k)$

t14 $\neg\exists x, y, z(\mathsf{Kind}(y) \land \mathsf{Kind}(z) \land y \neq z \land x \sqsubseteq y \land x \sqsubseteq z)$

t15 $\neg\exists x, y(\mathsf{NonSortal}(x) \land \mathsf{Sortal}(y) \land x \sqsubseteq y)$

t16 $(\mathsf{NonSortal}(t) \land x :: t) \rightarrow (\exists s(\mathsf{Sortal}(s) \land s \sqsubseteq t \land x :: s) \lor \exists n, s(\mathsf{NonSortal}(n) \land \mathsf{Sortal}(s) \land s \sqsubseteq n \land t \sqsubseteq n \land x :: s))$

Regarding the leaves of the taxonomy of types according to their sortality and rigidity, kinds and subkinds are disjoint (a13), and together encompass all rigid sortals (a14). Phases and roles are disjoint (a15), and together encompass all antirigid sortals (a16). Semi rigid sortals are those that are semirigid and sortal (a17). Categories are those types that are rigid and non-sortals (a18). Mixins are those types that are semirigid and non-sortals (a19). Phase-mixins and role-mixins are disjoint (a20), and together encompass all antirigid non-sortals (a21). Let $\mathcal{L}_T$ be the set of the leaf categories of the UFO taxonomy of types {Kind, Sub-Kind, Role, Phase, SemiRigidSortal, RoleMixin, PhaseMixin, Category, Mixin}, it follows that these leaf categories are pairwise disjoint (t17) and complete (t18).

a13 $\neg\exists t(\text{Kind}(t) \wedge \text{SubKind}(t))$

a14 $\text{Kind}(t) \vee \text{SubKind}(t) \leftrightarrow \text{Rigid}(t) \wedge \text{Sortal}(t)$

a15 $\neg\exists t(\text{Phase}(t) \wedge \text{Role}(t))$

a16 $\text{Phase}(t) \vee \text{Role}(t) \leftrightarrow \text{AntiRigid}(t) \wedge \text{Sortal}(t)$

a17 $\text{SemiRigidSortal}(t) \leftrightarrow \text{SemiRigid}(t) \wedge \text{Sortal}(t)$

a18 $\text{Category}(t) \leftrightarrow \text{Rigid}(t) \wedge \text{NonSortal}(t)$

a19 $\text{Mixin}(t) \leftrightarrow \text{SemiRigid}(t) \wedge \text{NonSortal}(t)$

a20 $\neg\exists t(\text{PhaseMixin}(t) \wedge \text{RoleMixin}(t))$

a21 $\text{PhaseMixin}(t) \vee \text{RoleMixin}(t) \leftrightarrow \text{AntiRigid}(t) \wedge \text{NonSortal}(t)$

t17 $\bigwedge\limits_{i,j\in\mathcal{L}_T, i\neq j} i(t) \rightarrow \neg j(t)$ **t18** $\text{Type}(x) \leftrightarrow \bigvee\limits_{i\in\mathcal{L}_T} i(x)$

On the UFO taxonomy of endurants, endurants are individuals (a22), and the Endurant type is partitioned into Substantial and Moment (a23), (a24). Moreover, the Moment type is partitioned into Relator and IntrinsicMoment (a25), (a26). Finally, the IntrinsicMoment type is partitioned into Mode and Quality (a27), (a28). Let $\mathcal{L}_e$ be the set of the leaf categories of endurants {Substantial, Relator, Mode, Quality}, it is a theorem that these leaf categories partition Endurant (t19), (t20).

a22 $\text{Endurant}(x) \rightarrow \text{Individual}(x)$

a23 $\text{Substantial}(x) \vee \text{Moment}(x) \leftrightarrow \text{Endurant}(x)$

a24 $\neg\exists x(\text{Substantial}(x) \wedge \text{Moment}(x))$

a25 $\text{Relator}(x) \vee \text{IntrinsicMoment}(x) \leftrightarrow \text{Moment}(x)$

a26 $\neg\exists x(\text{Relator}(x) \wedge \text{IntrinsicMoment}(x))$

a27 $\text{Mode}(x) \vee \text{Quality}(x) \leftrightarrow \text{IntrinsicMoment}(x)$

a28 $\neg\exists x(\text{Mode}(x) \wedge \text{Quality}(x))$

t19 $\bigwedge\limits_{i,j\in\mathcal{L}_e, i\neq j} i(t) \rightarrow \neg j(t)$

t20 $\text{Endurant}(x) \leftrightarrow \text{Substantial}(x) \vee \text{Relator}(x) \vee \text{Mode}(x) \vee \text{Quality}(x)$

We define a taxonomy of endurant types according to the ontological nature of their instances. Let $\mathcal{P}_E$ be the set of pairs {(EndurantType, Endurant); (SubstantialType, Substantial); (MomentType, Moment); (RelatorType, Relator); (ModeType, Mode); (QualityType, Quality)}. We implicitly define these types in the axiom schema (a29). It follows that these types are pairwise disjoint (t21).

a29 $\bigwedge_{(i,j)\in\mathcal{P}_E} i(t) \leftrightarrow \mathsf{Type}(t) \wedge \Box(\forall x(x :: t \rightarrow j(x)))$

t21 $\bigwedge_{i,j\in\{\mathsf{SubstantialType},\mathsf{RelatorType},\mathsf{ModeType},\mathsf{QualityType}\},i\neq j} i(x) \rightarrow \neg j(x)$

Kinds are also specialized according to the ontological nature of their instances. Let $\mathcal{P}_K$ be the set of pairs {(SubstantialKind,SubstantialType); (RelatorKind, RelatorType); (ModeKind, ModeType); (QualityKind, QualityType)}. We implicitly define these kinds in the axiom schema (a30). It is a theorem that all entities that possibly instantiate an endurant kind are endurants (t22). Moreover, every endurant instantiates one of the specific endurant kinds (a31). It follows that every endurant sortal is a type in $\mathcal{L}_{ES}$ = {SubstantialKind, RelatorKind, ModeKind, QualityKind, SubKind, Phase, Role, SemiRigidSortal} (t23); and that some sortals specialize specific kinds (t24).

a30 $\bigwedge_{(i,j)\in\mathcal{P}_K} i(t) \leftrightarrow j(t) \wedge \mathsf{Kind}(t)$

t22 $\Diamond(\exists k((\mathsf{SubstantialKind}(k)\vee\mathsf{RelatorKind}(k)\vee\mathsf{ModeKind}(k)\vee\mathsf{QualityKind}(k))\wedge x :: k)) \rightarrow \mathsf{Endurant}(x)$

a31 $\mathsf{Endurant}(x) \rightarrow \Diamond(\exists k((\mathsf{SubstantialKind}(k) \vee \mathsf{RelatorKind}(k) \vee \mathsf{ModeKind}(k) \vee \mathsf{QualityKind}(k)) \wedge x :: k))$

t23 $\mathsf{EndurantType}(x) \wedge \mathsf{Sortal}(x) \rightarrow \bigvee_{i\in\mathcal{L}_{ES}} i(x)$

t24 $(\mathsf{Sortal}(t)\wedge(\mathsf{SubstantialType}(t)\vee\mathsf{RelatorType}(t)\vee\mathsf{ModeType}(t)\vee\mathsf{QualityType}(t))) \leftrightarrow \exists k((\mathsf{SubstantialKind}(k) \vee \mathsf{RelatorKind}(k) \vee \mathsf{ModeKind}(k) \vee \mathsf{QualityKind}(k)) \wedge t \sqsubseteq k)$

We have as theorems that the leaves of the taxonomy of endurant types— $\mathcal{L}_{ET}$ = {SubstantialKind, SubKind, RelatorKind, ModeKind, QualityKind, SemiRigidSortal, Category, Phase, Mixin, Role, PhaseMixin, RoleMixin}—are disjoint (t25); and that every endurant type is a type in $\mathcal{L}_{ET}$ (t26).

t25 $\bigwedge_{i,j\in\mathcal{L}_{ET},i\neq j} i(t) \rightarrow \neg j(t)$ **t26** $\mathsf{EndurantType}(x) \rightarrow \bigvee_{i\in\mathcal{L}_{ET}} i(x)$

The results of this section are summarized in Fig. 3. When contrasting this figure with Fig. 1, one can appreciate how in this new theory, the taxonomy reflecting the ontological *nature* of the entity being classified (e.g., whether a substantial, a mode, a relator) is orthogonal to the one reflecting meta-properties such as sortality, rigidity, etc. Although the formal characterization of this theory is generally defined for *Type*, from an ontological perspective, endurants are the natural bearers of modal properties [6]. As a consequence, the interpretation of modal notions such as rigidity and anti-rigidity only makes (ontological) sense when applied to *Endurant Types*. Finally, although a logically possible combination of meta-properties, the category of semi-rigid sortals has been excluded from our ontology, given that it seems to play no role in Conceptual Modeling [8], as also confirmed by the empirical analysis of OntoUML models.

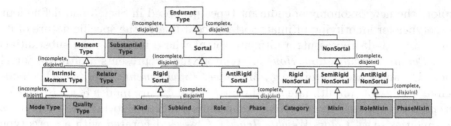

Fig. 3. Proposed taxonomy of endurant types in UFO.

4 Towards a New UML Profile for Modeling Endurant Taxonomic Structures

OntoUML is a ODCM language that extends UML by defining a set of stereotypes in order to reflect UFO ontological distinctions into language constructs. Constructs decorated by OntoUML stereotypes carry precise semantics grounded in the underlying ontology. Additionally, a number of *semantically motivated syntactic constraints* [3] govern OntoUML models driving them to conform to UFO. This combination of stereotypes and constraints enforce this conformance, making every valid OntoUML model compliant to UFO. In this section we provide a new UML profile (*lightweight extension*) for OntoUML that reflects the taxonomy of endurant types previously discussed and presented in Fig. 3. Only the elements in gray of the profile presented in Fig. 4 represent concrete stereotypes.

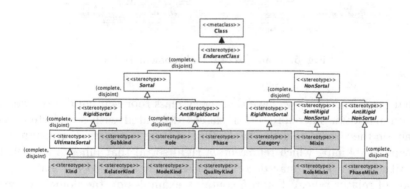

Fig. 4. OntoUML profile.

In the current version of OntoUML, the only stereotype for ultimate sortals (i.e. kinds) is «kind», which decorates substantial types that are ultimate sortals. Since the language was silent about modal properties of other types of endurants the stereotypes «relator», «mode» and «quality» would only carry information about the nature of the instances of that given moment type. In order to account for modal properties of both substantial and moment types, we explore the new taxonomy of endurant types proposed in Sect. 3 and define four stereotypes for identifying ultimate sortals according to the specific nature of its instances: «kind» represents a ultimate sortal whose instances are substantials (e.g., *Person, Organization, House*); «relatorKind» represents a ultimate sortal whose instances are relators (e.g., *Marriage, Employment, Enrollment*); «modeKind» represents an ultimate sortal whose instances are modes (e.g., *Headache, Intention, Goal*); and «qualityKind» represents a ultimate sortal whose instances are qualities (e.g., *Color, Weight, Height*). Classes decorated with a stereotype from this set not only provide the principle of identity to their instances, but also identify their nature. Notice that «kind» keeps the same semantics from the current profile, but «relator», «mode» and «quality» are discontinued in the new one. Furthermore, the additional stereotypes of this profile { «subkind», «role», «phase», «category», «mixin», «roleMixin», «phaseMixin»} have direct counterparts in UFO (see Fig. 3) and each of these stereotypes reflect the properties of its counterpart into the decorated class. In Figs. 5 and 6 we present examples of their application. These examples are not intended to defend the particular modeling choices therein, but rather to elucidate language application.

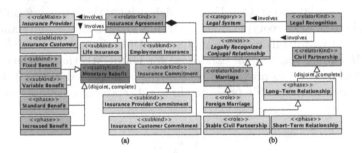

Fig. 5. Examples of sortal moment types.

From (a13) and (a14) we have that every class representing a sortal specializes a unique kind. Therefore, other classes decorated with sortal stereotypes may rely on their kinds to define the nature of their instances. For instance, the stereotype «subkind», which represents a rigid sortal that inherits its principle of identity from a kind, may decorate classes that apply to any nature (e.g., subkinds of relator types). Figure 5 contains examples on the domain of agreements (e.g., insurance agreements, civil partnerships, marriages) and related sortals represented in the proposed profile. In this example, *Life Insurance* and *Employment Insurance* are rigid types that inherit their principle of identity and their nature from *Insurance Agreement*, and thus, are decorated as «subkind»and represent relator types. Likewise, *Insurance Provider Commitment* and *Insurance Customer Commitment*, which represent commitments of the provider and customer involved in an insurance agreement, respectively, are rigid sortal mode types whose instances constitute *Insurance Agreements*. Our conceptualization may also include *Monetary Benefit* as a «qualityKind» that represents the amount of money the customer may receive for her insurance, amount that may be fixed *a priori* or variable according to the conditions of the agreement.

Moreover, «role» and «phase» represent anti-rigid sortals that are externally dependent, in the case of roles, or independent, in the case of phases. Figure 5(b) includes examples of phases and roles of moment types that are covered by the new profile. *Foreign Marriage*, for instance, can be defined as a «role» played by a marriage involved in a legal recognition relationship with a foreign legal system. *Civil Partnership*, another sort of relationship, may be classified into phases of short-term and long-term relationship. In legal systems where long-term relationships are legally recognized, acquiring marriage-like status, we also have the role *Stable Civil Partnership*. As an example of «phase» of quality types, *Standard Benefit* and *Increased Benefit* can represent benefits of an insurance which may change its prize value according to the clauses of the related contract.

Figure 6 presents examples of non-sortal moment types, i.e., types of moments whose instances follow different principles of identity. Here, *Agreement* represents a general category of agreements (e.g., conjugal relationships, insurance agreements). *Involved Part* is a non-sortal role (i.e., role mixin) that classifies entities involved in some agreement. Involved parts that bear some *Commitment* within an agreement are classified as *Party*. An example of non-sortal phases (i.e., phase mixins) are the phase of *Commitment, Fulfilled Commitment, Unfulfilled Commitment* and *Broken Commitment*. Finally, *Legally Recognized Conjugal Relationship* is an mixin that classifies endurants that are involved in some *Legal Recognition*. Instances of it include both entities that are necessarily classified as such (e.g., marriages), as well as entities that are contingently classified (e.g., civil partnerships that are long-term).

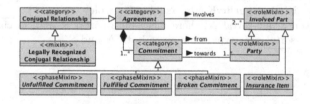

Fig. 6. Examples of non-sortal universals.

Differently from the sortals, this profile does not explicitly capture the nature of instances of non-sortal classes. Accordingly, non-sortal classes may classify not only endurants of distinct principles of identity, but also endurants of different natures. This feature allows the modeler to capture types such as *Insurance Item*. *Insurance Item* is a sort of *Involved Part* that classifies endurants insured by some insurance agreement. This example of role mixin includes as instances both substantials (e.g., cars, houses, machines) and moments (e.g., an employment or a contract), mixing instances of different basic ontological categories.

Table 1 summarizes the *semantically motivated syntactic constraints* regarding this new profile. The constraints discussed here are limited to those related to taxonomies of endurant types. Along with this new OntoUML profile, the revision of the language includes the development of computational support for ODCM. In that spirit, we have developed a plug-in for the Visual Paradigm UML CASE tool[3], which contemplates the aforementioned stereotypes and allows the verifications of the related syntactic constraints over models, thus, supporting the user for consistent representation of the intended conceptualizations. In case of constraint violation, the plug-in presents a log of the inconsistencies identified and the involved entities.

Table 1. OntoUML constraints for taxonomies of endurant universals.

Constraints
From (t25) and (t26), every class representing an endurant type must be decorated with exactly one stereotype from the set {«kind», «relatorKind», «modeKind», «qualityKind», «subkind», «role», «phase», «category», «mixin», «roleMixin», «phaseMixin»}. Semi-rigid sortals are excluded from the profile (see sec.3).
From (t24), every class representing an endurant sortal that is not a kind (including «subkind», «role», «phase») specializes a class decorated with a stereotype from the set {«kind», «relatorKind», «modeKind», «qualityKind»}.
From (t11), a class representing a kind cannot specialize another kind.
From (t14), a class cannot specialize more than one kind.
From (t7), a class representing a rigid type {«kind», «relatorKind», «modeKind», «qualityKind», «subkind», «category»} cannot specialize a class representing an anti-rigid type {«role», «phase», «roleMixin», «phaseMixin»}.
From (t8), a class representing a semi-rigid type {«mixin»} cannot specialize a class representing an anti-rigid type {«role», «phase», «roleMixin», «phaseMixin»}.
From (t15), a class representing a non-sortal {«category», «mixin», «roleMixin», «phaseMixin»} cannot specialize a class representing a sortal one {«kind», «relatorKind», «modeKind», «qualityKind», «subkind», «role» «phase»}.
From (t16), given a non-sortal N, there must be a sortal S that specializes N, or specializes a non-sortal supertype common to both N and S.

5   Final Considerations

In this paper, we make a contribution to the ontological foundations of conceptual modeling by proposing a formal theory of Endurant Types and the taxonomic structures involving them. This theory was developed to address a number of empirically elicited requirements, collected from observing the practice of the OntoUML community while using these notions to model a variety of domains (*claim to relevance*). Despite the empirical origin of these requirements, they are very much in line with the philosophical literature (*claim to ontological adequacy*). For example, Moltmann [10] uses the notion of *variable tropes* (moments

---

[3] The OntoUML plug-in for Visual Paradigm is available at https://github.com/nemo-ufes/OntoUML-2.0-for-Visual-Paradigm.

that can change while maintaining identity) to address a number of fundamental phenomena in cognition and language. Additionally, this formal theory has been checked for its consistency using theorem provers (*claim to consistency*).

The theory proposed here addresses an important fragment of a new version of the Unified Foundational Ontology (UFO) and serves as a foundation for a new version of the Ontology-Driven Conceptual Modeling Language OntoUML. In particular, the precise relation between language and ontology is exercised here via a design process that uses: the ontological distinctions put forth by the theory to derive the metamodel of the language; the formal axioms and ontological constraints of the theory to derive semantic and, ultimately, syntactical constraints for this language. This process is at the core of classical approaches on ontology-based language engineering [8]. As a result of applying this process, we manage to use this theory to construct two artifacts (*claim to realizability*), namely: a new UML profile (a lightweight extension to the UML 2.0 metamodel) capturing the concepts and formal constraint proposed by this theory; a computational tool implementing this profile.

The work developed here focuses exclusively on endurant types and taxonomic relations. In an extension of this work, we intend to address other relations involving endurants and endurant types (e.g., relational dependence for roles and role mixins, existential dependence for moment types, the foundation between relators and events [8]). More broadly, the work presented is part of a research program aimed at addressing a fuller evolution of UFO and OntoUML.

Acknowledgments. We thank Nicola Guarino for the fruitful discussions; and CNPq, CAPES, and FAPES for funding. The third author was supported by the grant N° 71024352.

References

1. Blums, I., Weigand, H.: Financial reporting by a shared ledger. In: 8th International Workshop on Formal Ontologies Meet Industry (FOMI) (2017)
2. do Carmo, A.P., Zamperini, T., de Mello, M.R., de Castro Leal, A.L., Garcia, A.S.: Ontologia das coisas para espaços inteligentes baseados em visão computacional. In: 9th Brazilian Ontology Research Seminar (2017)
3. de Carvalho, V.A., Almeida, J.P.A., Guizzardi, G.: Using reference domain ontologies to define the real-world semantics of domain-specific languages. In: Jarke, M., et al. (eds.) CAiSE 2014. LNCS, vol. 8484, pp. 488–502. Springer, Cham (2014). https://doi.org/10.1007/978-3-319-07881-6_33
4. de Carvalho, V.A., Almeida, J.P.A., Fonseca, C.M., Guizzardi, G.: Multi-level ontology-based conceptual modeling. Data Knowl. Eng. **109**, 3–24 (2017)
5. Fitting, M., Mendelsohn, R.L.: First-Order Modal Logic. Synthese Library, vol. 277. Springer Science & Business Media, Dordrecht (2012). https://doi.org/10.1007/978-94-011-5292-1
6. Guarino, N., Guizzardi, G.: "We Need to Discuss the *Relationship*": revisiting relationships as modeling constructs. In: Zdravkovic, J., Kirikova, M., Johannesson, P. (eds.) CAiSE 2015. LNCS, vol. 9097, pp. 279–294. Springer, Cham (2015). https://doi.org/10.1007/978-3-319-19069-3_18

7. Guarino, N., Welty, C.A.: An overview of ontoclean. In: Staab, S., Studer, R. (eds.) Handbook on Ontologies, pp. 201–220. Springer, Heidelberg (2009). https://doi.org/10.1007/978-3-540-24750-0_8

8. Guizzardi, G.: Ontological foundations for structural conceptual models. Telematica Instituut/CTIT (2005)

9. Guizzardi, G., Wagner, G., Almeida, J.P.A.A., Guizzardi, R.S.: Towards ontological foundations for conceptual modeling: the unified foundational ontology (UFO) story. Appl. Ontol. **10**(3–4), 259–271 (2015)

10. Moltmann, F.: Events, tropes and truthmaking. Philos. Stud. **134**, 363–403 (2007)

11. Rybola, Z.: Towards OntoUML for software engineering: transformation of OntoUML into relational databases. Czech Technical University in Prague (2017)

12. Verdonck, M.: Ontology-driven conceptual modeling: model comprehension, ontology selection, and method complexity. Ph.D. thesis submitted to the Applied Economics Program of Ghent University, Belgium (2018)

13. Verdonck, M., Gailly, F.: Insights on the use and application of ontology and conceptual modeling languages in ontology-driven conceptual modeling. In: Comyn-Wattiau, I., Tanaka, K., Song, I.-Y., Yamamoto, S., Saeki, M. (eds.) ER 2016. LNCS, vol. 9974, pp. 83–97. Springer, Cham (2016). https://doi.org/10.1007/978-3-319-46397-1_7

14. Zamborlini, V., Betti, A., van den Heuvel, C.: Toward a core conceptual model for (Im)material cultural heritage in the golden agents project. In: 1st International Workshop on Understanding Events Semantics in Cultural Heritage (2017)

Reification and Truthmaking Patterns

Nicola Guarino[1]([⊠]), Tiago Prince Sales[1,2], and Giancarlo Guizzardi[3]

[1] ISTC-CNR Laboratory for Applied Ontology, Trento, Italy
`nicola.guarino@cnr.it`
[2] University of Trento, Trento, Italy
`tiago.princesales@unitn.it`
[3] Free University of Bozen-Bolzano, Bolzano, Italy
`giancarlo.guizzardi@unibz.it`

Abstract. Reification is a standard technique in conceptual modeling, which consists of including in the domain of discourse entities that may otherwise be hidden or implicit. However, deciding what should be reified is not always easy. Recent work on formal ontology offers us a simple answer: put in the domain of discourse those entities that are responsible for the (alleged) truth of our propositions. These are called *truthmakers*. Re-visiting previous work, we propose in this paper a systematic analysis of *truthmaking patterns* for properties and relations based on the ontological nature of their truthmakers. Truthmaking patterns will be presented as generalization of reification patterns, accounting for the fact that, in some cases, we do not reify a property or a relationship directly, but we rather reify its truthmakers.

Keywords: Ontology-driven conceptual modeling · Reification

1 Ontological Analysis as a Search for TMs

Deciding what to put in the domain of discourse is a fundamental choice for conceptual modeling and knowledge representation. The things that are relevant for our conceptualization of reality—those that we implicitly assume to exist— are typically much more than those our language explicitly refers to. So, our *cognitive domain* is much bigger than our *domain of discourse* [12]. For example, when we say that John and Mary are married, our language only refers to them, although we *know* that there has been a wedding event and that there is an ongoing marriage relationship. It is up to us to introduce these further entities in our domain of discourse, should we need to represent and reason about them. Such process of making hidden entities explicit is called *reification*. Note that the new entities do not originate from a generic decision to expand the domain, but rather from a transformation of a language construct (typically, a predicate) into a domain element (a "first class citizen").

Reification is a standard technique in conceptual modeling and knowledge representation. Classic examples are the reification of relationships [3,23,24] and

© Springer Nature Switzerland AG 2018
J. C. Trujillo et al. (Eds.): ER 2018, LNCS 11157, pp. 151–165, 2018.
https://doi.org/10.1007/978-3-030-00847-5_13

events [4,6]. But how to decide what should be reified? Recent work on formal ontology offers us a simple answer: put in the domain of discourse those entities that are responsible for the (alleged) truth of our propositions. These are called *truthmakers* (TMs for short) [17].

Discovering TMs may be not always simple, of course, and requires some acquaintance with the basic tools of *formal ontological analysis*. Indeed, we can see conducting ontological analysis as a way of employing some special *detective magnifying lens*, which helps us in searching for TMs. Putting ourselves in this detective spirit, the basic questions we need to ask to analyze a proposition P are similar to the famous *Wh-questions*: *What* is responsible for making P true? *When* and *Where* will P be true? Of course, the answers to these questions depend on the kinds of properties and relations we use in our language. In this paper we shall adopt a systematic approach to account for the various *truthmaking patterns* (TMPs) associated to different kinds of properties and relations. A TMP is for us a generalization of a *reification pattern*, which accounts for the fact that, in some cases, when we want to 'talk' of a property or a relationship, we don't reify *it* directly, but we rather reify its TMs.

In the following, relying on earlier work [9–11] which will be revised and presented here in a systematic form, we first focus on properties and their TMs, distinguishing between strong and weak truthmaking, introducing qualities and descriptive properties, and presenting a number of TMPs at different levels of expressivity. Then we extend the analysis to relations, discussing the formal distinctions among them according to the ontological nature of their TMs, and presenting the corresponding TMPs. Finally, we conclude with some considerations on the practical implementation of TMPs in a conceptual modeling environment.

2   Properties and Their Truthmakers

We introduced the notion of truthmaking in a deliberately general way, saying that a TM is something that is *responsible* for the truth of a proposition. Strictly speaking, only propositions have TMs (they are the only *truthbearers*). However, in the case of atomic propositions, constituted of a property (or relation) plus its argument(s), we find it useful to see the TMs of such propositions as the TMs of the corresponding properties or relations, i.e., as something *in virtue of which* a property or a relation holds for certain entities; so, we shall talk interchangeably of TMs of properties or relations (holding for certain entities), and TMs of propositions. This move allows us to make distinctions among properties and relations according to the nature of their TMs.

2.1   Strong and Weak Truthmaking

Let us consider first the TMs of properties (we shall focus here on atomic properties, excluding logical combinations of them). What is it *in virtue of which* a property holds? Several attempts have been made by philosophers [17] to formally account for what *'in virtue of'* means. According to the mainstream doctrine, the TM of a property holding for a certain individual is something *whose very existence* entails that the property holds. The nature of such TM depends however on the kind of property, so that relevant distinctions may be drawn among properties based on the nature of their TMs.

Consider for instance two propositions such as (P1) *'a is a rose'* and (P2) *'a is red'*, where *a* denotes a particular rose. The very existence of *a* is enough for making P1 true, so *a* is a TM of P1. For P2, in contrast, the mere existence of *a* is not enough for P2's truth. What is its TM? A popular answer [17] is that it is a particular *occurrence of redness*, that is, a particular *event*[1] (intended in the most general sense that includes states).

So, *being a rose* and *being red* are properties whose TMs are of a very different nature. As we shall see, the latter is a *descriptive* property, while the former is a *non-descriptive* property. Intuitively, non-descriptive properties account for *what* something is, on the basis of its *nature and structure*; descriptive properties account for *how* something is, on the basis of its *qualities*. However, to better account for this and other intuitions concerning the different kinds of properties, we need to go deeper in the nature of their TMs.

There is indeed another notion called *weak truthmaking*, introduced by Josh Parsons [26], according to which a TM makes a proposition true not just because of its existence (i.e., because of its *essential* nature), but because of *the way it contingently is* (i.e., because of its *actual* nature). Differently from the *strong truthmaking* relation mentioned above, the weak truthmaking relation does therefore hold *contingently*.

Let us explain the difference between the two notions by considering again the example above. Suppose that *a* is red at time *t*1, and becomes brown at time *t*2. According to the mainstream TM theory, the *strong* TM of *a is brown at t*2 will be very different from that of *a is red at t*1, being a different event that is an occurrence of brownness and not an occurrence of redness[2]. According to Parsons' theory, however, the *weak* TM at both times is the rose itself: since the rose changes while keeping its identity, it is the very same rose, *in virtue of the way it (contingently) is* at *t*1 and at *t*2, which is a TM of the two propositions. A weak TM is something that, because of the way it intrinsically is, makes a proposition true; in Parsons' words, *the proposition cannot become false without an intrinsic change of its weak TM*.

So, whereas under the strong view there are two different TMs (namely two different events), under the weak view there is only one entity, namely the rose itself, responsible for the truth of the two propositions at different times.

2.2 Individual Qualities, Descriptive Properties, Intrinsic Properties

We have seen from the previous example that the rose undergoes a change, while keeping its identity, from *t*1 to *t*2. What kind of change? Of course, a change in

---

[1] Of course, there may be many of such events. Each of them would be a TM.

[2] Even if the color does not change, multiple strong TMs are necessary as time passes by, since each occurrence is different from the previous or future occurrences.

its *color*. So, as discussed in [10], there is something more specific than the whole rose that is responsible for the truth of the two propositions: their *minimal*[3] weak TM is the rose's *color*. Indeed, it is exactly *in virtue of* its color that the rose is red at $t1$ and brown at $t2$. This color is modeled as an *individual quality* in DOLCE [2] and in UFO [14].

Individual Qualities as Weak TMs. Individual qualities (qualities for short) may be seen as specific aspects of things we use to compare them. They *inhere* in things, where inherence is a special kind of *existential dependence* relation, which is irreflexive, asymmetric, anti-transitive and functional [14]. They are directly comparable, while objects and events can be compared only with respect to a certain *quality kind* (e.g., to compare physical objects, one resorts to the comparison of their shapes, sizes, weights, and so on). Qualities are distinct from their *values* (a.k.a. *qualia*), which are abstract entities representing what exactly resembling qualities have in common, and are organized in spaces called *quality spaces*; each quality kind has its own quality space. For instance, weight is a quality kind, whose qualia form a linear quality space. At different times, qualities can keep their identity while occupying different regions of their quality space; they are considered therefore *endurants* in UFO. Quality spaces may have a complex structure with multiple dimensions, each corresponding to a simple quality that inheres in a complex quality. Typical examples of complex qualities are colors and tastes, but we shall also consider mental entities such as attitudes, intentions and beliefs as complex qualities, collapsing, for the sake of simplicity, UFO's distinction between *qualities* and *modes* [14, p. 213].

An important class of qualities are *relational qualities*, which, besides being existentially dependent on the thing they inhere in, are also existentially dependent on something else. An example may be *John's love for Mary*, which inheres in John but is existentially dependent on Mary. Another example would be *Mary's commitment to marry John*, which inheres in Mary and is externally dependent on John. As we shall see, relational qualities typically come in bundles called *relators*.

Summing up, individual qualities, which were introduced for different reasons in DOLCE and in UFO, can be seen now under a new perspective, in their role of weak TMs for descriptive properties. From the above discussion, it is natural to say that an quality is what in virtue of which a descriptive property holds, legitimating therefore the interpretation of 'in virtue of' in terms of minimal weak truthmaking. Note also that, for the needs of ontological analysis, looking for qualities as minimal weak TMs has a clear advantage over looking for strong TMs: while negative truths are notoriously a problem for the strong truthmaking view, so that it is difficult to individuate the strong TM of *a is not red*, it is immediate to see that its minimal weak TM is its color, and not, say, its weight.

---

[3] Space does not allow to discuss the notion of minimality in detail. In short, we assume that an entity x is *internal* to y iff x inheres in, is a proper part of or participates to y, and *external* to y otherwise. Then, if t is a TM for a proposition P, it is a *minimal TM* for P iff no entity internal to t is itself a TM of P.

Descriptive Properties. Let us now define descriptive properties more carefully. As a first attempt, we may define them as properties holding in virtue of one or more individual qualities inhering in their argument, so that *being red* is descriptive, while *being an apple* is not. However, some observations are due. First, it seems plausible to assume that a descriptive property may hold for an object x in virtue of a quality inhering in a proper part of x, rather than in x itself. So, *having a big nose* counts as descriptive since it holds in virtue of the nose's size, while *having a nose* is non-descriptive since it holds in virtue of the object that has the nose, which is not a quality.

Second, we should account also for descriptive properties that hold in virtue of relational qualities. Considering a generic descriptive property holding for John, there are three possibilities: First, the weak TM consists of just one relational quality inhering in John, as in the case of *being in love with Mary*; Second, the truthmaking qualities are distributed between John and an external entity. This is the case of *being married with Mary*, which presupposes the existence of commitments and obligations (and possibly love) inhering in Mary and depending on John, as well as reciprocal ones inhering in John; Third, there is only one truthmaking quality inhering in something external to John, and existentially depending on it. This is the case of so-called Cambridge properties [5], like *being loved by Mary*. To include the last two cases, we refine our definition as follows: a property P is *descriptive* iff, for every x, $P(x)$ holds in virtue of (at least) a quality q being existentially dependent on x.

Intrinsic Properties. The notion of intrinsic property is well-established in philosophy, despite some debate on precise definitions holding for arbitrary properties [19]. We shall say that a property holding for x is *extrinsic* iff it requires the existence of something else *external* to x in order to hold (where *external* is defined as in footnote 3), and *intrinsic* otherwise. The *intrinsic/extrinsic* distinction turns out to be orthogonal to the *descriptive/non-descriptive* one, and each of the four combinations has its own peculiarities in terms of TMs.

Being red and *being married* are examples of, respectively, intrinsic descriptive and extrinsic descriptive properties. In the former case the minimal weak TM is a non-relational quality, in the latter it is a relational quality. *Being an apple* or *having a nose* are examples of intrinsic non-descriptive properties, whose argument coincides with the minimal weak TM. *Being proper part of a car* and *being Italian* are examples of extrinsic non-descriptive properties. The minimal weak TM for the former is a car, while the one for the latter (which is an *historical property*) is a birth event.

2.3   Truthmaking Patterns for Properties

Let us now discuss the practical impact of the above considerations on reification choices concerning properties, commenting the *truthmaking patterns* (TMPs for short) that emerge once we decide to put TMs in the domain of discourse. For the sake of space, we discuss here only two of the four combinations of descriptive/non-descriptive and intrinsic/extrinsic properties mentioned above, since no reification is necessary for intrinsic non-descriptive properties (their weak TM being already present in the domain of discourse), while the case of extrinsic descriptive properties is very similar to that of external descriptive relations which we shall discuss later. We shall introduce three broad classes of patterns: *partial* TMPs, in turn distinguished in *strong* or *weak*, depending on the kind of TMs reified, and *full* TMPs including both strong and weak TMs *as well as the relationship between them*. All the patterns will be discussed by means of examples.

Intrinsic Descriptive Properties. Note that these properties do rarely correspond to classes, because they do not carry a principle of identity [13,14]. So, the property of *being red* for a rose is typically expressed as an attribute-value pair within the class *Rose* (Fig. 1a)[4], where the attribute name implicitly denotes the color quality [8]. We have three reification options, corresponding to different TMPs. A *weak* TMP emerges when the quality is reified as a separate class (Fig. 1b). Note the 1-1 cardinality constraint, showing that a quality inheres in exactly one object, and an object has exactly one quality of a given kind. A *strong* TMP is exemplified in Fig. 1c, where an event of "color occurrence" is reified. The first option is generally more flexible, making it possible to describe the way the quality interacts with the world (*Mary likes the color of this rose*), or further information about the quality itself (*the color of a rose is located in its corolla*). The second option is however necessary when we need to account for temporal information (e.g., how long the redness lasted), or for the spatiotemporal context (what happened meanwhile and where...).

To achieve the maximum expressivity, a third option is that of a *full* TMP, including both strong and weak TMs plus the relationship among them (Fig. 1d). Concerning the latter, note that there is a formal ontological connection between qualities and events, discussed in [10]: events can be seen as *manifestations* of qualities, and qualities as the *focus* of events.

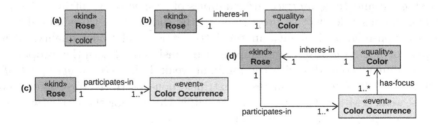

Fig. 1. Truthmaking patterns for an *intrinsic descriptive property*.

Extrinsic Non-descriptive Properties. For those of them that are anti-rigid, it certainly makes sense to reify the event during which they hold, i.e., their

---

[4] For clarity purposes, all models here are represented in OntoUML [14]. No commitment on OntoUML is however assumed.

strong TM. For example, a strong TMP applied to the class *InstalledCarPart* (Fig. 2a) would include the class *CarPartInstalled* (Fig. 2b), which denotes the state of having that part installed. A weak TMP that includes the car itself as a weak TM is exemplified in Fig. 2c[5]. The full TMP is shown in (Fig. 2d).

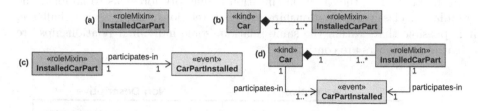

Fig. 2. Truthmaking patterns for a *non-descriptive property*.

3   Relations and Their TMs

In his early work, Guizzardi borrowed from [16] a crisp distinction between *formal* and *material* relations, describing the former as holding between two or more entities "directly without any further intervening individual" [14, p. 236], and the latter as requiring the existence of an intervening individual. His modeling proposal was to systematically introduce—for all material relations—a specific construct, called the *relator*, standing for such intervening individual.

In the philosophical literature, the formal/material distinction varies significantly among different authors both in content and terminology, and overlaps with other distinctions. In the following, building on recent work [9,10], we shall revise these distinctions in the light of TMs analysis, aiming at clarifying some conceptual and terminological problems which resulted in some confusions and inconsistencies in the way the relator construct was used in the past [11]. We shall discuss here three orthogonal distinctions, presented in compact form in Fig. 3: *internal/external, essential/contingent,* and *descriptive/non-descriptive*.

3.1   Kinds of Relationships

The formal/material distinction defined above overlaps with two main distinctions proposed in the literature: *internal/external* and *essential/contingent*. They both originate from a core idea of internal relations as holding only in virtue of the 'internal nature' of their relata. However, different definitions have been proposed depending on how such nature is understood [18]. A first definition, due to Moore [21], says that a relation is internal iff it necessarily holds

---

[5] The choice of reifying a weak TM only arises for those non-descriptive properties whose minimal weak TM does not coincide with their argument. In such cases, the weak TM is typically an argument's proper part (say, a nose for *having a nose*) or something that includes the argument as a proper part.

just in virtue of the *mere existence* of its relata, i.e., it is *essential* to its relata. A second definition, originally due to Russell [27], says that *a relation is internal iff it is definable in terms of the intrinsic properties of its relata, and external otherwise*. We shall adopt Russel's definition for *internal/external*, using *essential/contingent* for the distinction based on Moore's definition. Note that using both distinctions in the same classification framework forces us to adopt a fine granularity, classifying relationships and not relations, since, as we shall see, it is possible that, within the same relation, some individual relationships are essential and others are contingent.

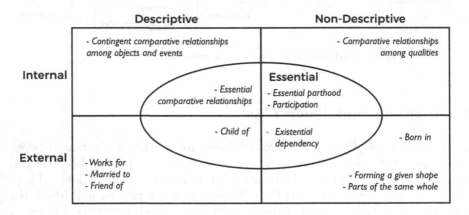

Fig. 3. Kinds of relationships (revised from [10]).

Let us see now how Guizzardi's formal/material distinction, which is crucial for his notion of relator, is mapped to the two definitions above. In his book [14, p. 236], he pointed explicitly to Moore essential/contingent distinction while talking of formal relations, but, in retrospective, what he actually had in mind was more in line with Russel's internal/external distinction, since he included within formal relations also comparative relations like *taller than*, which is not essential as it does not necessarilyt hold when the relata exist. For him, a material relation holds in virtue of the existence of a *relator* composed of particularized properties called *modes* (*qualities* according to our analysis) that inhere in the relata and are historically dependent on a common external *foundation* (an event). The typical example he makes is that of a marriage relationship, whose relator (the marriage itself) is composed of modes (mutual commitments and claims between the spouses) existentially dependent on a common foundation (a wedding event). So, Guizzardi's 'material' is narrower than 'external', and, since the formal/material distinction is exhaustive, his 'formal' turns out to be broader than 'internal'.

Let us now go back to the main reason for the formal/material distinction in conceptual modeling, which is deciding whether or not a relationship can be reified. In a recent paper [10], Guarino and Guizzardi showed that *none* of the distinctions considered so far (essential/contingent, internal/external[6], formal/material) can help in this decision. Their analysis was mainly motivated by the confusing behavior of *comparative relations*. They were considered as formal by Guizzardi, and therefore not deserving reification. However, there may be good reasons to *talk* about them [9]: for instance, one may want to keep track of the difference in height between a mother and her son, or of the temperature difference between two bodies. So, comparative relations seem to share something in common with other relations that deserve to be reified, although it is difficult to characterize them in terms of the distictions considered so far.

On one hand, as observed by Simons [28], within comparative relations some individual relationships are essential, but others are contingent. For instance, the mere existence of an electron e and a proton p is enough to conclude that $heavier(p, e)$ holds (since both of them have their mass essentially), but the mere existence of John and Mary is not enough to conclude that $taller(John, Mary)$ holds, since they do not have that particular height essentially. Moreover, even within the same relation, some individual relationships (like $heavier(p,e)$) may be essential, while others (like $heavier(John,Mary)$) may be just contingent.

On the other hand, although all comparative relations are internal (since they are definable in terms of the intrinsic properties of their relata), those among objects or events hold in virtue of the qualities of their relata, while those among qualities (e.g., perfect resemblance) hold in virtue of the relata themselves, so they don't deserve reification (otherwise we would have an infinite regress). In conclusion, the distinctions mentioned so far are not able to discriminate between reifiable and non-reifiable relations.

3.2   Descriptive Relations and Relators

Analogously to the case of descriptive properties, we define a descriptive relation as *a relation that holds in virtue of some qualities that are existentially dependent on one or both its relata*. The mereological sum of such qualities forms what we call a *relator*, which (recalling the discussion in Sect. 2) is therefore the *minimal weak TM* of the relation. As Fig. 3 shows, the descriptive/non-descriptive distinction is orthogonal to those discussed so far. So, there are descriptive relations that are internal (such as comparative relations among objects or events), and others that are external (those originally called *material relations* by Guizzardi). The relators in these two cases, however, are very different. Since internal relations are defined as derivable from *intrinsic* properties of their relata, the relators of internal descriptive relations are just formed of qualities depending only on the relatum they inhere in. On the contrary, relators of external descriptive relations include at least some qualities that, besides depending on the relatum they inhere in, are also depending on the other relatum. The latter is the notion of

---

[6] In the original paper [10], we labeled this distinction 'intrinsic/extrinsic', aiming at extending to relations the terminology adopted for properties. However, in the philosophical literature 'external relation' is not synonym of 'extrinsic relation', since the latter requires the existence of something completely external to the relata.

relator discussed in Guizzardi's thesis, which was generalized and simplified in a previous paper [10], in order to work for all kinds of descriptive relations, and here has been analysed in terms of truthmaking. The link between a relator and its relata was originally called 'mediation', but recently we decided to adopt a more neutral term, 'involvement'. An entity is involved in a descriptive relationship, reified as a relator, if one of its qualities is part of such relator.

Having relators (i.e., reified relationships) in the domain of discourse has been recognized as a solution to many practical problems in conceptual modeling, including disambiguation of cardinality constraints, transitivity of part-whole relations, and proper modeling of anadic relations, among many others [14]. Moreover, having relators as full-fledged endurants (being them bundles of qualities, which are endurants themselves) allows us to describe their behavior in time exactly like an object [9]. So, for example, relators can undergo different phases, have essential and accidental qualities of their own, and change in a qualitative way while remaining the same.

3.3   Truthmaking Patterns for Relations

Similarly to the case of properties, let us now discuss the reification options and the corresponding TMPs concerning relations. We shall only consider quadrants 2, 3 and 4 of Fig. 3, since the TMs of relationships in quadrant 1 are their own arguments, so that no further reification is necessary.

Internal Descriptive Relations. The main representatives of this class are comparative relations among objects or events. A first option is to reify their weak TM, i.e., their relator, composed of exactly two qualities of the same kind. For example, for the *heavier-than* relation, the relator is the class *WeightRelationship* shown in (Fig. 4b).

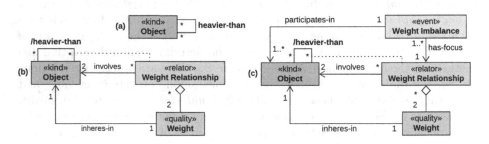

Fig. 4. Applying weak and strong truthmaking patterns to *comparative relations*.

Note that, since relators are weak TMs, multiple relations can be derived from the same relator, depending on the actual values of the qualities composing the relator. Consider for instance Fig. 4b. Since it is a weight comparison relationship, one could also derive relations such as *lighter-than, same-weight-as, twice-as-heavy,* and so on. In a sense, the relator generalizes the relation, maintaining the possibility to represent all the relevant cases by means of the qualities. Moreover, reifying the relator helps when one wants to make explicit which qualities ground the relation (e.g. *heavier-than* is derived from a comparison of weights). As discussed in depth in [14], explicitly acknowledging from which qualities a comparative relation is derived allows us to also account for the specific meta-properties on that relation. For example, *heavier-than* is a total order relation because it is founded on weight qualities, which take their values (qualia) in a linear (i.e., totally ordered) weight space.

Figure 4c shows the addition of a strong truhmaker to the previous case, which achieves the maximum expressivity. Adding the *WeightImbalance* event (which is actually a state, i.e., a static event) allows us to capture more explicitly the state of affairs corresponding to a *heavier-than* relationship, and specify as well details concerning its actual duration its spatiotemporal context.

External Descriptive Relations hold in virtue of at least one relational quality inhering in at least one relatum. We distinguish two main cases: *single-sided* relations holding in virtue of one or more qualities inhering in just one relatum, and *multi-sided* relations holding in virtue of at least two qualities each inhering in a different relatum. An example of the first kind is an attitudinal relation such as *desires*, represented in Fig. 5a. A weak TMP is shown in Fig. 5b, where a desire quality inhering in an agent and depending on some resources is reified. Note that we have represented it as a quality, but it could be seen as as well as a relator consisting of just one quality. The addition of a strong TM, resulting in a full TMP, is shown in Fig. 5c. The event labeled 'DesireEvolution' describes whatever happens in reality whose focus is that particular desire, such as the arising of the desire and its satisfaction.

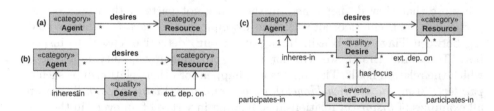

Fig. 5. Weak and full truthmaking patterns for a *single-sided relation*.

Multi-sided Relations are arguably the most frequent case of external descriptive relations. Reifying their TMs is often necessary to model social and legal relationships, such as marriages, economic contracts, employment relationships, and so on. An example of full TMP is presented in Fig. 6, which describes a *subscribes* relation holding between service providers and service customers. The relator is shown as a contractual relationship consisting of reciprocal commitments and claims inhering in the customer or the supplier (and externally dependent on each other).

Note that, just by explicitly representing the contract, this model clarifies the cardinalities of the *subscribes* relation. Here we assume that contracts are always bilateral, i.e., a contract involves exactly one customer and one provider. Thus, this pattern rules out the possibility of multi-party contracts that could be inferred otherwise. This is an example of how the explicit representation of relators can eliminate the aforementioned problem of cardinalities ambiguity.

The pattern described by Fig. 6 may also be applied to role-playing relations such as *president-of* between a person and an organization. In this case the relator accounts for the social commitments and obligations related to the particular role, while the event accounts for the period of time when the role is played. So, this pattern can be seen as the well-founded version of the *"time indexed person role"* defined in [7].

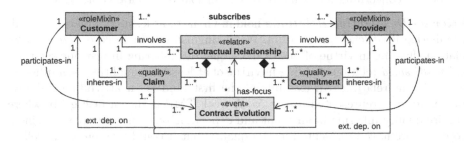

Fig. 6. The full truthmaking pattern for a service subscribing relation.

External Non-descriptive Relations. Unlike the cases we have just discussed, the TMs of these relations are completely external to their relata, in the sense that they do not inhere in them, are not parts of them and do not participate to them (footnote 3). One example is the *born-in* relation holding between Tiago and Brazil. If we put on our "detective lenses" to inspect both Tiago and Brazil, we will not be able to find any quality in virtue of which the relation holds. The same would happen for other relationships such as *painter(SistineChapel,Michelangelo)*, or *veteran(Jack,VietnamWar)*. These are all examples of *historical* relationships, holding in virtue of an event in the past in which at least one of the relata participated.

Figure 7 depicts a pattern in which the TM of the *born-in* relation, holding between a *Person* and a *Country*, is reified by means of a *Birth* event. The reification of TMs for historical relations is particularly useful when one needs to represent additional properties of these events, such as their duration or the presence of other participants (e.g. the doctor who assisted a birth). It may also be useful to do so for properly representing cardinality constraints, similarly to the case of relators. Finally, if one accepts the classical view that events are immutable entities [9], differentiating between TMs that are events and TMs that are endurants allows us to represent these entities' properties in a way that properly accounts for this constraint (e.g., in UML, representing all relevant properties of events as readOnly attributes).

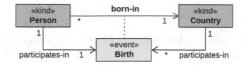

Fig. 7. An application of the *historical relation* pattern.

Complex Truthmaking Configurations. In certain cases, the TM is not just a single entity, like the birth event, but rather a *complex truthmaking configuration*. Take for instance the *colleague-of* relation, between people working in the same organization. For this relation, the weak TM is a larger entity of which the two relata are parts, namely the organization itself, while the strong TM is the actual event of working for the same organization at the same time, so that the relation is *derived* from a particular configuration of data that are external to the relata. More complex cases of truthmaking configurations have been discussed in [11] in the context of an ontological analysis of the REA accounting model [20].

Differently from the other cases we discussed in this section, *complex truthmaking configurations* do not follow a common structural pattern. Still, we presented them as a separate case to highlight that some relationships may not be simply reified by means of a single entity, but rather by a complex truthmaking configuration.

4   Conclusions

Differently from other approaches that look mainly at reification techniques from the modeling language point of view [3,15,24,25], we have focused in this paper on understanding the ontological nature of what should be reified, by systematically investigating why properties and relations hold, and providing guidelines for reification choices according to the nature of their TMs. A crucial contribution in this analysis was the recognition of qualities as minimal weak truthmakers, which turned out to be a very useful application of Parsons' quite original views.

We also clarified Guizzardi's distinction between formal and material relations, proposing a new classification of relations based on the orthogonal distinctions *descriptive/non-descriptive* and *internal/external* (a refinement of that presented in [10] that helped us to develop systematic TMPs aimed at facilitating the ontological analysis of actual modeling cases. These patterns may be easily incorporated in conceptual models based on foundational ontologies that support the notion of individual quality, such as DOLCE [2], UFO [14], or BFO [1]. In particular, we plan to implement them in the OntoUML [14] language and in its modelling evironments, such as the Menthor Editor [22], to support modelers in systematically investigating and representing the TMs of properties and relations.

References

1. Arp, R., Smith, B., Spear, A.D.: Building Ontologies with Basic Formal Ontology. MIT Press, Cambridge (2015)
2. Borgo, S., Masolo, C.: Foundational choices in DOLCE. In: Staab, S., Studer, R. (eds.) Handbook on Ontologies. IIIS, pp. 361–381. Springer, Heidelberg (2009). https://doi.org/10.1007/978-3-540-92673-3_16
3. Dahchour, M., Pirotte, A.: The semantics of reifying n-ary relationships as classes. In: 4th International Conference on Enterprise Information Systems (ICEIS), pp. 580–586 (2002)
4. Davidson, D.: The individuation of events. In: Rescher, N. (ed.) Essays in Honor of Carl G. Hempel. Synthese Library, vol. 24, pp. 216–234. Springer, Dordrecht (1969). https://doi.org/10.1007/978-94-017-1466-2_11
5. Francescotti, R.: Mere Cambridge properties. Am. Philos. Q. **36**(4), 295–308 (1999)
6. Galton, A.: Reified temporal theories and how to unreify them. In: IJCAI, pp. 1177–1183 (1991)
7. Gangemi, A., Presutti, V.: Ontology design patterns. In: Staab, S., Studer, R. (eds.) Handbook on Ontologies. IHIS, pp. 221–243. Springer, Heidelberg (2009). https://doi.org/10.1007/978-3-540-92673-3_10
8. Guarino, N.: The ontological level: revisiting 30 years of knowledge representation. In: Borgida, A.T., Chaudhri, V.K., Giorgini, P., Yu, E.S. (eds.) Conceptual Modeling: Foundations and Applications. LNCS, vol. 5600, pp. 52–67. Springer, Heidelberg (2009). https://doi.org/10.1007/978-3-642-02463-4_4
9. Guarino, N., Guizzardi, G.: "We Need to Discuss the *Relationship*": revisiting relationships as modeling constructs. In: Zdravkovic, J., Kirikova, M., Johannesson, P. (eds.) CAiSE 2015. LNCS, vol. 9097, pp. 279–294. Springer, Cham (2015). https://doi.org/10.1007/978-3-319-19069-3_18
10. Guarino, N., Guizzardi, G.: Relationships and events: towards a general theory of reification and truthmaking. In: Adorni, G., Cagnoni, S., Gori, M., Maratea, M. (eds.) AI*IA 2016. LNCS (LNAI), vol. 10037, pp. 237–249. Springer, Cham (2016). https://doi.org/10.1007/978-3-319-49130-1_18
11. Guarino, N., Guizzardi, G., Sales, T.P.: On the ontological nature of REA core relations. In: 12th International Workshop on Value Modeling and Business Ontologies (2018)
12. Guarino, N., Oberle, D., Staab, S.: What Is an ontology? In: Staab, S., Studer, R. (eds.) Handbook on Ontologies. IHIS, pp. 1–17. Springer, Heidelberg (2009). https://doi.org/10.1007/978-3-540-92673-3_0
13. Guarino, N., Welty, C.A.: An overview of OntoClean. In: Staab, S., Studer, R. (eds.) Handbook on Ontologies. IHIS, pp. 201–220. Springer, Heidelberg (2009). https://doi.org/10.1007/978-3-540-92673-3_9
14. Guizzardi, G.: Ontological foundations for structural conceptual models. CTIT, Centre for Telematics and Information Technology (2005). https://research.utwente.nl/en/publications/ontological-foundations-for-structural-conceptual-models
15. Halpin, T.: Objectification of relationships. In: Siau, K. (ed.) Advanced Topics in Database Research, vol. 5, p. 106. IGI global (2006)
16. Heller, B., Herre, H.: General ontological language (GOL): a formal framework for building and representing ontologies. Technical report 7/2004, Institute for Medical Informatics, Statistics and Epidemiology, University of Leipzig, Germany (2004)

17. MacBride, F.: Truthmakers. In: Zalta, E.N. (ed.) The Stanford Encyclopedia of Philosophy. Stanford University (2016). fall 2016 edn
18. Marmodoro, A., Yates, D. (eds.): The Metaphysics of Relations. Oxford University Press, New York (2017)
19. Marshall, D., Weatherson, B.: Intrinsic vs. extrinsic properties. In: Zalta, E.N. (ed.) The Stanford Encyclopedia of Philosophy. Stanford University (2018)
20. McCarthy, W.E.: ISO 15944-4 - REA Ontology. ISO, pp. 1–82, June 2007
21. Moore, G.E.: External and internal relations. In: Proceedings of the Aristotelian Society, vol. 20, pp. 40–62 (1919). JSTOR
22. Moreira, J.L.R., Sales, T.P., Guerson, J., Braga, B.F.B., Brasileiro, F., Sobral, V.: Menthor editor: an ontology-driven conceptual modeling platform. In: 2nd Joint Ontology Workshops (JOWO) (2016)
23. Noy, N., Rector, A.: Defining n-ary relations on the semantic web. Technical report, W3C (2006). https://www.w3.org/TR/swbp-n-aryRelations
24. Olivé, A.: Relationship reification: a temporal view. In: Jarke, M., Oberweis, A. (eds.) CAiSE 1999. LNCS, vol. 1626, pp. 396–410. Springer, Heidelberg (1999). https://doi.org/10.1007/3-540-48738-7_29
25. Olivé, A.: Conceptual Modeling of Information Systems. Springer Science & Business Media, Heidelberg (2007). https://doi.org/10.1007/978-3-540-39390-0
26. Parsons, J.: There is no 'truthmaker' argument against nominalism. Australas. J. Philos. **77**(3), 325–334 (1999)
27. Russell, B.: Philosophical Essays. Longmans, Green, and Co., New York (1910)
28. Simons, P.: Relations and truthmaking. Aristot. Soc. Suppl. **84**(1), 199–213 (2010)

Semi-structured Data Modeling

Conceptual Modeling of Legal Relations

Cristine Griffo[1(✉)], João Paulo A. Almeida[1],
and Giancarlo Guizzardi[1,2]

[1] Ontology and Conceptual Modeling Research Group (NEMO),
Federal University of Espírito Santo (UFES), Vitória, ES, Brazil
clbeccalli@tjes.jus.br, jpalmeida@ieee.org,
giancarlo.guizzardi@unibz.it
[2] Facoltà di Scienze e Tecnologie Informatiche,
Free University of Bozen-Bolzano, Bolzano, Italy

Abstract. Legal relations abound in conceptual modeling. Despite that, the representation of these relations in the area has not yet received sufficient theoretical support. We address this by establishing a basis for the well-founded representation of legal relations. We capture a comprehensive set of legal relations (and legal positions within these legal relations) that arise from widely accepted legal theories into a legal core ontology called UFO-L [14, 15], which is grounded on the Unified Foundational Ontology (UFO). We rely on the Theory of Constitutional Rights proposed by the philosopher of law Robert Alexy. This theory extends the system of legal positions proposed by the jurist Wesley Hohfeld, which has been used as a theoretical basis for several works in the conceptual modeling literature. The result is a modeling strategy for legal relations that in- corporates patterns from the legal core ontology. We present here a synthesis of empirical studies conducted to evaluate the aforementioned results. The studies show that the approach based on UFO-L produces models that are more comprehensible and clear for the representation of legal contracts.

Keywords: Legal relations · Conceptual modeling · Legal ontology
UFO-L

1 Introduction

The Law permeates our social lives. It is not surprising thus that legal phenomena have made their way into our information systems and have become a frequent subject in conceptual modeling [1, 2]. In the last decades, several approaches have been proposed to address the conceptualization and representation of legal phenomena [3, 4], addressing normative notions, e.g., rights, prohibitions, duties, claims, etc.

Despite these advances, we have observed that existing approaches fail to capture some important aspects of the legal phenomena. For example, many of the existing approaches lack support for legal relations between parties, given their root in monadic (standard) deontic logics. As a consequence, they are unable to make explicit the legal positions of the various parties in a legal relation and to capture the roles they play in the scope of a relation. This poses a challenge to the adequate representation of a number of real-life settings involving legal relations, such as legal contracts in general,

© Springer Nature Switzerland AG 2018
J. C. Trujillo et al. (Eds.): ER 2018, LNCS 11157, pp. 169–183, 2018.
https://doi.org/10.1007/978-3-030-00847-5_14

and service contracts in particular [5]. Moreover, much of the existing work in the conceptual modeling of legal aspects has emerged from the disciplines of computer science (and logics) and has failed to incorporate the state-of-the-art in legal theories. As a consequence, they often leave out key concepts underlying complex legal phenomena (such as *power* and *legal liberty*).

We address these gaps in this paper with a principled approach. We establish a basis for the well-founded representation of legal relations by capturing a comprehensive set of legal relations (and legal positions within these legal relations) into a legal core ontology called UFO-L. UFO-L is based on Hohfeld's seminal theory of fundamental legal concepts and Alexy's relational theory of constitutional rights [6]. As a result, UFO-L accounts for legal notions that include: *rights* and *duties*, *no-rights* and *permissions*, *powers* and *liabilities*, *disabilities* and *immunities*, as well as *liberties*.

A key aspect of the approach is the representation of legal relations based on the notion of *relator* from the UFO foundational ontology [7, 8]. At first sight, relators can be seen simply as a well-founded relationship objectification. However, as dis- cussed in depth in [8] they are genuine ontological entities that bundle relational properties and serve as truthmakers of material relations. As such, they can account for essential and accidental properties as well as change and modality in the scope of same relationship. Moreover, they can account for the roles that entities play in a relationship and capture the subtle ways in which they are related. The applicability of the approach is illustrated here with the representation of the legal aspects of a real-world cloud service contract (Amazon Web Service Contracts). Furthermore, we report on empirical studies that show that the approach based on UFO-L produces conceptual models that are more comprehensible and clear for the representation of legal contracts.

This paper is further structured as follows: Sect. 2 discusses the kinds of legal relations and legal positions that ought to be represented in conceptual models of legal aspects; Sect. 3 discusses an overall approach for the representation of relations based on the relator pattern [9]; Sect. 4 presents the UFO-L fragment employed here with a taxonomy of legal relator types; Sect. 5 shows how UFO-L and the relator pattern are applied to the modeling of contracts; Sect. 6 presents the results of the experiment that was conducted to assess models built following UFO-L, Sect. 7 discusses related work and Sect. 8 presents concluding remarks.

2 Legal Relations

In a seminal work in the legal literature, Hohfeld examined legal positions between subjects in legal relations [10]. He observed that key legal terms such as "right" were often misunderstood because of semantic overload. For instance, in the expression "right to smoke", the term "right" has the meaning of *permission*; in the expression "right to charge taxes" it takes on the meaning of *power*; in the expression "right to receives salary at the end of the month" it takes on the meaning of an *entitlement*.

After an analysis of legal concepts, he identified eight fundamental legal concepts (right, duty, no-right, privilege, power, liability, disability, and immunity), and established relations between them. Table 1 shows these concepts, grouping them in pairs of correlative legal positions. Correlative positions are those with a counterpart in the same legal relation. For instance, the correlative of "John's duty to pay his debt to Mary" is "Mary's right that John pay his debt to her". A right in this precise or 'narrow' sense is a legal position in which one may demand from another the performance of a certain conduct. Likewise, "John's permission to use Mary's car" correlates to "Mary's no-right that John refrain from using her car". Hohfeld observed further that the various legal positions are classified into two main categories: (i) those that arise from norms of conduct, namely: right, duty, permission, and no-right (at the left-hand side of Table 1); and (ii) those that arise from norms of power, namely: power, liability, disability, and immunity (at the right-hand side of Table 1). While norms of conduct have mainly a coordinative nature, norms of power presuppose a subordinate nature [6], and concern the creation and change of other legal positions. Norms of conduct regulate behavior which is possible independently of the law (e.g., smoking, exchanging goods, killing). Norms of power in their turn have a constitutive nature, in the sense that they create the possibility of institutional action (e.g., voting, marrying, hiring an employee).

Table 1. Fundamental legal concepts according to hohfeld

Correlatives			
Right	Privilege (Permission)	Power	Disability
Duty	No-Right	Liability	Immunity

Alexy [6] proposed a system of legal positions embedding Hohfeldian legal positions in *triadic legal relations* and considering also the possibility of denying the legal relation's object (thereby augmenting Hohfeld's theory). As a result, for each legal concept *right, duty, privilege,* and *no-right* to an action, there exists a concept of *right, duty, privilege,* and *no-right* to an *omission*. These legal positions are relevant because they define duties to negative actions (effectively prohibitions). For instance, in e- mail service contracts, the customer often has a duty to omit sending the same message indiscriminately to large numbers of recipients on the Internet (unsolicited e-mail or spam). The following categories are proposed by Alexy combining the legal positions of Hohfeld's theory with the new legal positions.

Right to Positive Action. Subject a has the right R, against subject s, to an act ϕ: Ras (ϕ). In this case, the addressee (s) has the *duty to perform* action ϕ. For instance, in a service contract with warranty, the *service customer* has the right that the *service provider* fixes the service in case of defect or failure.

Right to Negative Action. Subject a has the *right R,* against subject $s,$ to an omission ϕ: Ras ($\neg\phi$). In this case, the addressee (s) has the *duty to omit* to perform action ϕ. For instance, a *service provider* must not disclose a customer's private information.

Permission to Act. Subject a has permission P towards subject s to perform action ϕ: $Pas(\phi)$. In this case, the addressee (s) has *no-right* to demand that the permission holder (a) omit action ϕ. For instance, in a messaging service, a *service customer* has the permission to send messages using the provider's infrastructure.

Permission to Omit. Subject a has permission P to refrain from acting (abstain to perform action ϕ) towards subject s: $Pas(\neg\ \phi)$. In a relational sense, the addressee (s) has *no-right* to demand that the permission holder (a) perform action ϕ. For instance, a *service customer* has the permission to abstain from paying contractual interest established by a *service provider* if it exceeds permitted by law in delayed payments.

Unprotected liberty. Subject a has liberty L in face of subject s with respect to an action ϕ if a has the permission to perform ϕ and the permission to abstain from performing ϕ, i.e.: $Las(\phi) \equiv_{def} Pas(\phi) \wedge Pas(\neg\ \phi)$. The idea of liberty is related with an *alternative of action* and is defined in terms of the fundamental legal concept of *permission*. Because of the relational nature of permissions, if a has a liberty against s, then s has no-right to demand that a perform or abstain from performing ϕ. For instance, airline customers usually have the liberty to use in-seat entertainment.

Power. Subject a has the legal power K in face of subject b to create, change or extinguish a legal position X for subject b by means of institutional actions: $Kab(Xb)$. *Power* is created by a *competence norm*. The exercise of a legal power is an institutional action, which gives ability to act to a power holder. Since *power* has a converse position, it means that subject b is in a *subjection* position toward subject a (*subjection* is also called *liability*). For instance, often a *service provider* has the power to cancel the service agreement unilaterally in the case of contract violations.

Disability. A subject a has, in face of subject b, no power to create, change or extinguish a legal position X for subject b by means of institutional actions: $\neg Kab(Xb)$. The converse position of a disability is *immunity,* and the subject b is immune to changes in his/her legal position. For instance, often a *service provider* is immune to cancellation of a service agreement unilaterally in the cases of *force majeure*.

3 Ontology-Driven Conceptual Modeling of Relations

The Unified Foundational Ontology (UFO) [11], [12], [13] is an axiomatic theory that has been developed based on a number of contributions from Formal Ontology, Philosophical Logics, Philosophy of Language, Linguistics and Cognitive Psychology. For an in-depth discussion, empirical support and formalization see [11, 12].

At the core of UFO, we have four fundamental categories forming what is termed an *Aristotelean Square*, namely, *Substantials* (or Objects) and *Moments* (also termed tropes, aspects, particularized properties). Substantials are entities that exist on their own. Examples include a car, a dog, the moon and Mick Jagger; Moments, in contrast, are parasitic entities that can only exist by *inhering* in other individuals (inherence is a type of existential dependence relation). Example of moments include (a) John's headache or John's capacity of speaking Greek (which can only exist insofar John exists) as well as the (b) marriage between John and Mary (which can only exist insofar both John and Mary exist). Moments of type (a) are termed *intrinsic moments*, including *qualities* (e.g., the charge in this conductor) as well as *modes/dispositions* (e.g., the disposition of a magnet to attract metallic material; John's capacity of speaking Greek). Within the category of modes, we have *relational modes* (or externally dependent modes). These are entities that, while inhering in an individual A, are still existentially dependent on another individuals B disjoint from A. An example is John's commitment to meet Mary tomorrow (or Paul's love for Clara). Although this commitment inheres in John (it is a commitment *of* John), it is still existentially dependent on Mary (it is a commitment *towards* Mary). Moments of type (b) are termed *relators*. Relators and their constituting externally dependent modes are fundamental for the purposes of this article and are discussed in the sequel.

As discussed in depth in [8], relators are the real *truthmakers* of relations. At least for the so-called *material relations*, which are the vast majority of the relations we are interested in not only in conceptual modeling and information systems engineering but also in social and legal reality.

So, for instance, it is true that "John is married to Mary" because there is a relator (a particular *marriage*) binding them; likewise, it is true that "Paul works for the United Nations" because there is a relator of another type (an employment) connecting them. As a result, many of the fundamental tasks in enterprise and information systems management requires a proper understanding of the nature and lifecycle of relators such as employments, enrollments, marriages, contracts, presidential mandates [8]. In UFO, a relation of *mediation* is defined to connect relators to their relata. Mediation is, like inherence, a type of existential dependence.

Moreover, as also elaborated there, relators are full-fledged endurants, i.e., proper object-like entities as opposed to just n-uples of relata. They are entities that can bear their own individualized properties and can have their own complex mereological structures, i.e., they can have their own parts, some of which are essential (i.e., which they must have necessarily) and, conversely, they can be part of other complex relators. For example, a marriage has as an essential part several mutual commitments and claims and as an inseparable part a *conjugal society* [14]. Thus, if John and Mary are married, there is an individual relator *m* of type "Marriage" that mediates John and Mary. This relator consists of all properties that John and Mary have in the scope of the marriage. For instance, John has a fidelity commitment towards Mary and, *mutatis mutandis*, the same for Mary.

In Fig. 1, we illustrate the general *Relator Pattern* the marriage example discussed above. We use the OntoUML profile [11], which comprises modeling primitives that represent the ontological distinctions put forth by UFO (captured as stereotypes in the profile) and *ontology design patterns* that represent micro-theories in UFO [9]. In the relator pattern, entities in a relation instantiate types that are termed "roles" (here "Husband" and "Wife"). Roles classify, in a contingent and relationally dependent way, instances of the same kind, in this case persons classified as husbands and wives. The model reveals the roles played in a marriage and the relational modes that are part of it (e.g., mutual commitments and claims).

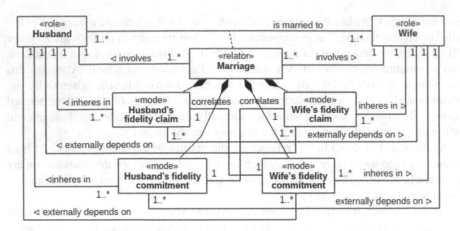

Fig. 1. Marriage modeled with the UFO relator pattern

4 Legal Relations in UFO-L

Based on Alexy's legal concepts, we have built a legal core ontology called UFO-L [14]. This core ontology uses the *Unified Foundational Ontology* (UFO) [11] as basis specializing ontological categories from UFO (both its UFO-A *upper* fragment and its UFO-C *social fragment*) [13]. A key notion in UFO-L is that of a legal relator, which is a relator that is composed of externally dependent legal moments, each of which represents a legal position following Alexy. A Legal Relator specializes *Social Relator* (UFO-C) which in turn specializes the basic notion of *Relator* (UFO-A).

Legal relators can be simple or complex. A Simple Legal Relator is composed simply by a pair of legal positions (categorized in UFO-L as legal moments), such as: Right/Duty, NoRight/Permission, Power/Subjection, and Disability/Immunity. In contrast, a Complex Legal Relator is composed of other legal relators (thus more than one pair of legal moments). For instance, a Liberty Relator is composed of NoRight to an Action–Permission to Omit Relator and NoRight to an Omission–Permission to Act Relator. Legal moments are related to each other by a correlation association and are essential and inseparable parts of the legal relator which they form. A fragment of UFO-L is shown in Fig. 2 with the existing legal positions, the taxonomy of legal relators and its connection with UFO. UFO-L has a catalog of modeling patterns composed of eight patterns based on the taxonomy of legal relators (Fig. 2) and the *relator* pattern [7, 9] (Fig. 1). A modeling pattern was created for each pair of legal moments and consists of the pattern name/code, the rationale, competence questions, restrictions, usage guidelines, questions to check the chosen pattern, and the model [14]. In [16], for instance, we applied two of these patterns in an ontological analysis of the universal notion of *right to life*. For the sake of space, we restrict our presentation to five of the UFO-L patterns in the next section and applied them to the modeling of service contracts.

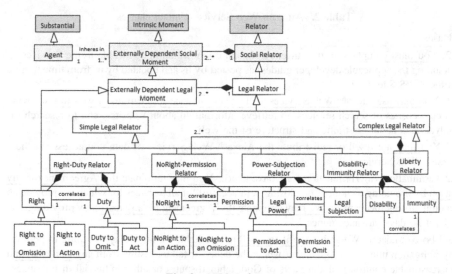

Fig. 2. UFO-L fragment

5 Modeling Amazon Web Service Contracts

The combination of the taxonomy of legal relations with the relator modeling pattern leads to a number of more specialized patterns for the modeling of legal relations. In this section, we discuss the application of some of these patterns in the modeling of Amazon Web Services (AWS) contract clauses (AWS Terms [17] and AWS Customer Agreement [18]). These legal documents were also used earlier in [5] in which we model service contracts based on the ArchiMate language. Here, we return to the study but focus only on some of the clauses for the sake of illustration. Table 2 shows the text of the clauses as obtained directly from the legal documents.

Figure 3 shows the model that was produced to account for the legal relations in clauses 1.2 and 5.3. It reveals the mereological structure of the Right-Duty to an Action relator ("Compliance with the technical documentation") with a pair of correlated modes, one of which is a Right to an Action ("Right that customer follows technical documentation") and the other of which is a Duty to Act ("Duty to follow technical documentation"). These modes inhere in the legal agents involved in this relation, each of which plays a particular role. Also, if a mode inheres in a Legal Role, then this mode is externally dependent on the counterpart Legal Role and *vice versa*. Also, Fig. 3 shows a legal relation that contains a pair of legal moments categorized as Right to an Omission and Duty to Omit (Clause 5.3). In this case, the Right Holder has a right (in a narrow technical sense) against to Duty Holder in such a way that the Duty Holder must abstain from performing a specific act. In the case of Clause 5.3, the duty is imposed on the customer not to resell or redistribute the AWS. Both legal relations in Fig. 3 are represented by *Simple Legal Relators*.

Table 2. Amazon Web Service contract clauses

Clauses
1.2. <u>You must comply</u> with the current technical documentation applicable to the Services (including the applicable developer guides) as posted by us and updated by us from time to time on the AWS Site. (…)
5.1. <u>You may use Alexa® Web Services to create or enhance applications or websites,</u> to create search websites or search services, to retrieve information about websites, and to research or analyze data about the traffic and structure of the web
5.3 <u>You may not resell or redistribute the Alexa® Web Services</u> or data you access via the Alexa® Web Services
7.2 Termination. (a) Termination for Convenience. <u>You may terminate this Agreement for any reason</u> by providing us notice and closing your account for all Services for which we provide an account closing mechanism. <u>We may terminate this Agreement for any reason</u> by providing you at least 30 days' advance notice
13.3 Force Majeure. <u>We and our affiliates will not be liable</u> for any delay or failure to perform any obligation under this Agreement where the delay or failure results from any <u>cause beyond our reasonable control</u>, including acts of God, labor disputes or other industrial disturbances, electrical or power outages, utilities or other telecommunications failures, earthquake, storms or other elements of nature, blockages, embargoes, riots, acts or orders of government, (…)

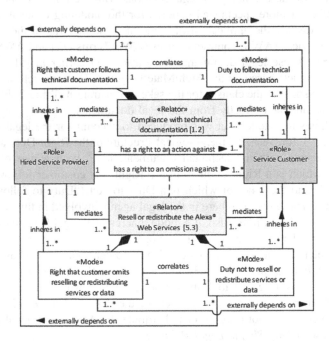

Fig. 3. Clauses 1.2 and 5.3 modeled on UFO-L

Figure 4 captures Clause 5.1 with a *Liberty Relator*. As discussed in the previous section, the liberty relator is a complex *legal relator*, composed of two NoRight-*Permission Relators*. The *Service Customer* and the *Hired Service Provider* have two legal positions at the same time in the scope of the relation: *Service Customer* holds both *Permission to use AWS* and *Permission not to use AWS*. In its turn, *Hired Service Provider* holds both *NoRight that customer omits using AWS* and *NoRight to demand customer use AWS*.

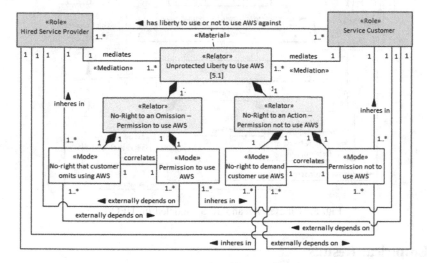

Fig. 4. Clause 5.1 modeled on UFO-L

Some legal positions empower the bearer to create, modify or extinguish other legal positions. This is the case of the legal position of *power*. For example, Clause 7.2(a) gives the *Service Customer* the power to terminate the contract when certain requirements are filled. This clause is modeled in the upper part of Fig. 5 with a Power-*Subjection* relation. The *Hired Service Provider* is in the legal position of *subjection*, and the *Service Customer* in the legal position of *power*.

It is also possible for contractual clauses to rule out the creation, modification or extinction of a legal relation. An agent affected by such a clause holds a *disability*. Consequently, the other player is in an immunity *legal position*, i.e., he/she is not submitted to an act (or an omission) from the *Disability Holder* that aims to create, modify or terminate a legal relation. The type of legal relation is called *disability-immunity*. Figure 5 shows (in the lower part) the modeling of Clause 13.3 that removes any responsibility from the Service *Provider in* case of service failure due to force majeure. In this case, the *Service Customer* has no power to hold *Service Provider* responsible (and thus the *Hired Service Provider* is immune in case of force majeure).

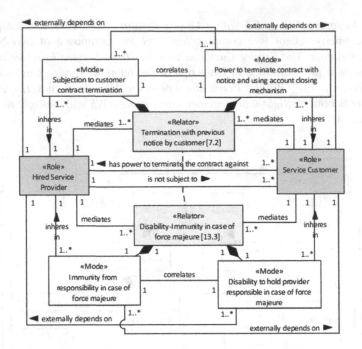

Fig. 5. Clauses 7.2 and 13.3 modeled on UFO-L

6 Empirical Results

We have performed an empirical study to investigate the quality of models built with the UFO-L patterns for legal relations. In particular, we have tested whether models built with UFO-L facilitate the correct interpretation of contract text, improving question answering performance and perceived clarity. In total, 37 subjects participated in the experiment (students and professionals in computer science and law, 92% of which indicated some experience in conceptual modeling, and 65% of which indicated no experience in legal aspects). First, they were given Amazon Web Service (AWS) contract clauses solely in text and were prompted to answer a number of questions concerning the content of these clauses. After answering these questions, 20 participants were given UFO-L-based models to represent the AWS contract relations. They were again prompted to answer questions on the legal aspects of the services.

When answering solely based on the textual representations, they were able to provide in total 215 correct answers (72.6% out of 296 total answers of all subjects). 36 answers were incorrect, and 45 answers were left blank (subjects indicated they were unable to answer). After participants were given access to UFO-L-based models, they managed to provide 233 correct answers (83.2% out of 280 total answers of all sub- jects), with 39 incorrect answers and 8 no answers. The McNemar test was applied to verify if the introduction of UFO-L diagrams resulted in improved correctness of the participants' responses. We selected five questions on the text treatment and related them to five questions on UFO-L treatment. Then, we related the answers given to these questions in order to form pairs of answers (200 pairs). A chi-squared ($\chi 2$) test was applied with the discordant pairs of the 200 responses. The McNemar test statistic was calculated and $\chi 2$ found to be 6.26 (degree of freedom 1). The p-value was found to be lower than 0.0125. Hence, we consider there is significant statistical difference between the number of correct answers in the text and the UFO-L treatment, with the UFO-L-based models improving average question answering performance.

In addition to performance in question answering, we have also considered the subjects' perception of clarity. They were asked whether specific legal relations and positions were clearly expressed in the text (or diagram). In case of a positive answer, the representation scheme was awarded one point. In total, the textual representation obtained 45 out of 111 points (40.5%) and the UFO-L-based diagrams, 97 out of 140 points (69.3%). A chi-squared ($\chi 2$) test was applied to the clarity points and $\chi 2$ found to be 12.19 (degree of freedom 1). The p-value was found to be lower than 0.0005. With this result, we concluded there is a relation between the increase of clarity and the introduction of UFO-L-based-models. In addition to the objective clarity point questions, we have also provided the opportunity for subjects to provide free feedback concerning clarity. After introducing the diagrams in the second part of the experiment, the participants were able to perceive concepts in the diagrams that went undetected in the textual representation, as well as relations (associations) that were not expressed in the text (there was an open question that allowed participants to provide any feedback they might see fit). To cite one example, three participants explicitly reported that they perceived, after the introduction of the diagrams, the concepts of *immunity, power*, and *liberty*.

An additional experiment was conducted with 10 participants that were exposed solely to the UFO-L based models (and not to the text). Our goal was to assess the threat that the first experiment was biased towards UFO-L because of the order of treatment (text first) and a possible learning effect. In this additional experiment, the percentage of correct answers was 79.3%, which, when contrasted to 83.2% in the first experiment, suggests that the learning effect, if any, had minor influence.

Considering the quantitative results of the experiment described above, we can conclude that the UFO-L-based models added greater clarity and comprehensibility to the representation of service contract clauses. The full dataset is available in [14] (published in Portuguese; a translation of the chapter concerning the experiment can be found at http://purl.org/nemo/griffoch7). Further, from a qualitative analysis of the questions that have had most incorrect answers, we are led to believe that UFO-L may benefit from more support for inferences concerning legal situations, especially in the case of violation of legal positions. This is a topic for further investigation.

7 Related Work

Several efforts that address the representation of legal aspects are reported in the literature, notably in the context of enterprise information systems.

A number of approaches, under the umbrella term "contract languages" [20, 21], are devoted to the formal representation of the contents of contracts, specifying rigorously the ways in which parties ought to act in the scope of (service) contracts. For example, in [20], a formal system for reasoning is proposed based on a Business Contract Language (BCL). The authors raised some issues for further investigation, such as an improved separation of subject and target roles in a policy expression and the expressiveness of BCL with respect to other legal concepts (right, authorization and delegation). About the first issue, we have suggested in our work that roles are explicitly represented and as well as their legal positions [16]. In this case, not only one party is modeled but two parties in the legal relation, each of which plays a different role in the scope of the legal relation.

In [22], a Contract Language (CL) was defined based on deontic logic. The authors stumble on a case of *semantic overload* when they do not distinguish right from permission. For instance, in the example cited to instantiate Postulate 3.8 ("Obligation to an action implies that the action is permitted") it is not correct to state that "the client has the right to pay". The correct assertion is that "the client has permission to pay". This is an instructive example of how the reduction of legal positions to a unique form of right-duty position results in loss of meaning and misunderstanding as discussed in [10]. In [23], the authors propose a Formal Language for Writing Contracts (FCL) that is based on monadic deontic logic operators of obligations and prohibitions. In the aforementioned languages, we observe the use of monadic operators: obligation, prohibition and permission as the unique way to represent legal positions. There is no representation of *power norms* and other relevant legal concepts (particularly missed is the notion of *right* in a narrow sense discussed here). Further, while these approaches have proven useful, e.g., in the analysis of business process compliance and in the verification of formal properties of contracts [4, 24], they do not aim to include legal aspects in the overall practice of structural conceptual modeling. Here, we have opted to instead, propose an approach that enables the integration of legal positions and legal relations into structural conceptual models.

In addition to contract languages, there has been some efforts on contract ontology in the literature, for example: the ontology for international contract law [25]; the Uniform Commercial Code Ontology, on legal contract formation [26]; the MPEG Media Contract Ontology (MCO) [27] to deal with rights concerning multimedia assets and intellectual property content; and, the contract ontology based on the SweetDeal rule-based approach [28]. All these approaches employ the monadic operators of deontic logics, hence, not being able to fully capture the relational aspect that is at the core of our service contract ontology. With the exception of [25], none of the approaches employ Hohfeld's legal concepts, failing thus to account for rights in a narrow sense. Additionally, none of these explicitly address powers. Also, we observed cases of semantic overload concerning the concept of right in some ontologies of contracts (e.g., [29]). Thus, we conclude that all these efforts could have benefited from an ontological foundation such as UFO-L.

Concerning legal core ontologies covering broad aspects of the legal domain such as UFO-L, we have observed in a systematic mapping of the literature [14, 19] that most extant legal core ontologies, including LKIF core, CLO and FOLaw, focus on the representation of norms or on the representation of isolated legal positions. Differently from these other core ontologies, UFO-L emphasizes the relational perspective, revealing explicitly the roles and positions in (simple and complex) legal relations.

Because of this, UFO-L has provided us with a basis for the conceptual modeling of legal relations which we could not obtain directly in the existing legal core ontologies.

In order to model legal aspects in the scope of enterprises and information systems, a number of dedicated representation schemes have been proposed including, e.g., RuleML [30], LegalRuleML [31], and Nòmos 3 [4]. LegalRuleML builds up on RuleML and uses notions of defeasible logics to treat violation of obligations; in the treatment of violations (something that we still need to address systematically in our approach). With respect to the legal positions that can be represented, it does not cover powers or rights in a narrow sense (capturing only the corresponding obligations). Note that the notion of "Right" adopted in LegalRuleML corresponds to the notion of protected liberty, which can be accounted for in our ontology with a complex relator composing an unprotected liberty with obligations, following Alexy [6]. In its turn, Nòmos 3 is a framework for representing laws and regulations that uses the conception of goals and Hohfeld's theory to reason about compliance of requirements. Consequently, its concept of liberty as synonym of privilege does not cover all the existing permissions (negative and positive permissions).

8 Conclusions and Future Work

This work presents an approach for the conceptual modeling of legal relations taking as theoretical basis Alexy's Theory of Constitutional Rights [6], the Unified Foundational Ontology (UFO) [11], and the legal core ontology termed UFO-L [14]. To illustrate this approach, the *relator* pattern proposed in [7, 9] and its specializations in terms of the UFO-L modeling patterns were employed here to model clauses of Amazon Web Service contracts. With UFO-L, both the model creator and its users are led to understand the differences between the notions of *right* in a narrow sense, *permission*, *liberty*, and *power*. Thus, the UFO-L-based-models expose nuances of each legal position and, therefore, of each legal relation. Understanding and representing the subtleties of these relations is of fundamental importance to the proper conceptual modeling and information systems engineering in critical domains such as finance, healthcare, and risk management, among others.

The modeling patterns proposed here were evaluated in an empirical experiment aimed to assessed whether a representation scheme including these patterns brings some benefit to stakeholders. Participants reported that, after being exposed to these diagrams, they were able to perceive concepts that they were not able to identify in contractual texts. We attribute this to the fact that all correlative legal positions are made explicit in the UFO-L based representation, which is not the case in legal texts. In fact, it is common for service contracts to be written from the provider's perspective. This means that the provider's advantageous positions are emphasized along with customer burden positions. On the other hand, customer vantage positions and provider burden positions are (sometimes deliberately) "hidden" [5].

Regarding future work, we intend to investigate the modeling of the lifecycle of legal relations, the detailed representation of power positions and the exercise of power, issues involving compliance and non-compliance, as well as specialized visual syntaxes for the legal domain (also experimenting with areas other than contracting).

Acknowledgments. This work is partially supported by CNPq (407235/2017-5, 312123/2017-5, and 312158/2015-7), CAPES (23038.028816/2016-41), FAPES (69382549) and FUB (OCEAN Project).

References

1. Breuker, J., et al.: OWL Ontology of Basic Legal Concepts (LKIF-Core), Deliverable 1.4. Information Society Technologies (2007)
2. Gangemi, A., Sagri, M.-T., Tiscornia, D.: A constructive framework for legal ontologies. In: Benjamins, V.R., Casanovas, P., Breuker, J., Gangemi, A. (eds.) Law and the Semantic Web. LNCS (LNAI), vol. 3369, pp. 97–124. Springer, Heidelberg (2005). https://doi.org/10.1007/978-3-540-32253-5_7
3. Palmirani, M., Governatori, G., Rotolo, A., Tabet, S., Boley, H., Paschke, A.: LegalRuleML: XML-based rules and norms. In: Olken, F., Palmirani, M., Sottara, D. (eds.) RuleML 2011. LNCS, vol. 7018, pp. 298–312. Springer, Heidelberg (2011). https://doi.org/10.1007/978-3-642-24908-2_30
4. Ingolfo, S., Siena, A., Mylopoulos, J.: Goals and compliance in nomos 3. In: CAiSE 2014 Proceedings, pp. 1–6 (2014)
5. Griffo, C., Paulo, J., Almeida, A., Guizzardi, G., Nardi, J.C.: From an ontology of service contracts to contract modeling in enterprise architecture. In: EDOC Proceedings (2017)
6. Alexy, R.: A Theory of Constitutional Rights. Oxford University Press, Oxford (2009)
7. Guizzardi, G., Wagner, G.: What's in a relationship: an ontological analysis. In: Li, Q., Spaccapietra, S., Yu, E., Olivé, A. (eds.) ER 2008. LNCS, vol. 5231, pp. 83–97. Springer, Heidelberg (2008). https://doi.org/10.1007/978-3-540-87877-3_8
8. Guarino, N., Guizzardi, G.: "We need to discuss the *relationship*": revisiting relationships as modeling constructs. In: Zdravkovic, J., Kirikova, M., Johannesson, P. (eds.) CAiSE 2015. LNCS, vol. 9097, pp. 279–294. Springer, Cham (2015). https://doi.org/10.1007/978-3-319-19069-3_18
9. Ruy, F.B., Guizzardi, G., Falbo, R.A., Reginato, C.C., Santos, V.A.: From reference ontologies to ontology patterns and back. Data Knowl. Eng. **109**, 41–69 (2017)
10. Hohfeld, W.N.: Some fundamental legal conceptions as applied in judicial reasoning. Yale Law J. **23**, 16–59 (1913)
11. Guizzardi, G.: Ontological Foundations for Structural Conceptual Model. Universal Press, Veenendaal (2005)
12. Guizzardi, G., Wagner, G., de Almeida Falbo, R., Guizzardi, R.S.S., Almeida, J.P.A.: Towards P_{{CS}}. In: Ng, W., Storey, V.C., Trujillo, J.C. (eds.) ER 2013. LNCS, vol. 8217, pp. 327–341. Springer, Heidelberg (2013). https://doi.org/10.1007/978-3-642-41924-9_27
13. Guizzardi, G., Wagner, G., Paulo, J., Almeida, A., Guizzardi, R.S.S.: Towards ontological foundations for conceptual modeling: the unified foundational ontology (UFO) story. Appl. Ontol. **10**, 259–271 (2015)
14. Griffo, C.: UFO-L: Uma Ontologia Núcleo de Aspectos Jurídicos construída sob a Perspectiva das Relações Jurídicas. Published PhD Dissertation, Departamento de Informática, Universidade Federal do Espírito Santo (Ufes), Vitória, ES, Brasil (2018)
15. Griffo, C., Almeida, J.P.A., Guizzardi, G.: Towards a legal core ontology based on Alexy's theory of fundamental rights. In: MWAIL-ICAIL 2015, San Diego, CA (2015)
16. Griffo, C., Almeida, J.P.A., Guizzardi, G.: A pattern for the representation of legal relations in a legal core ontology (2016)

17. Amazon: AWS Service Terms. https://aws.amazon.com/service-terms/
18. Amazon: AWS Customer Agreement. https://aws.amazon.com/agreement/
19. Griffo, C., Almeida, J.P.A., Guizzardi, G.: A systematic mapping of the literature on legal core ontologies. In: Brazilian Conference on Ontologies, Ontobras 2015., São Paulo (2015)
20. Governatori, G., Milosevic, Z.: A formal analysis of a business contract language. Int. J. Coop. Inf. Syst. **6**, 1–26 (2006)
21. Prisacariu, C., Schneider, G.: A formal language for electronic contracts. In: Bonsangue, M. M., Johnsen, E.B. (eds.) FMOODS 2007. LNCS, vol. 4468, pp. 174–189. Springer, Heidelberg (2007). https://doi.org/10.1007/978-3-540-72952-5_11
22. Prisacariu, C., Schneider, G.: A dynamic deontic logic for complex contracts. J. Log. Algebr. Program. **81**, 458–490 (2012)
23. Farmer, W.M., Hu, Q.: FCL : A formal language for writing contracts. In: IEEE 17th International Conference on Information Reuse and Integration (IRI), pp. 134–141 (2017)
24. Kabilan, V.: Contract workflow model patterns using BPMN. In: CEUR Workshop Proceedings, pp. 171–182 (2005)
25. Kabilan, V., Johannesson, P.: Semantic representation of contract knowledge using multi tier ontology. In: 1st International Conference on Semantic Web and Databases, pp. 395–414 (2003)
26. Bagby, J., Mullen, T.: Legal ontology of sales law application to eCommerce. Artif. Intell. Law **15**, 155–170 (2007)
27. Rodriguez-Doncel, V., Delgado, J., Llorente, S., Rodriguez, E., Boch, L.: Overview of the MPEG-21 media contract ontology. Semant. Web. **7**, 311–332 (2016)
28. Grosof, B.N., Poon, T.C.: SweetDeal: representing agent contracts with exceptions using XML rules, ontologies, and process descriptions. In: WWW 2003 Proceedings, pp. 340–349 (2003)
29. Pace, G.J., Schapachnik, F., Schneider, G.: Conditional permissions in contracts. Front. Artif. Intell. Appl. **279**, 61–70 (2015)
30. Governatori, G., Rotolo, A.: Modelling contracts using RuleML. In: T.G. (ed.) Jurix 2004: The 17th Annual Conference, pp. 141–150. IOS Press, Amsterdam (2004)
31. Palmirani, M., Governatori, G., Athan, T., Boley, H., Paschke, A., Wyner, A.: LegalRuleML core specification version 1.0. http://docs.oasis-open.org

Inferring Deterministic Regular Expression with Counting

Xiaofan Wang[1,2] and Haiming Chen[1(✉)]

[1] State Key Laboratory of Computer Science, Institute of Software,
Chinese Academy of Sciences, Beijing 100190, China
{wangxf,chm}@ios.ac.cn
[2] University of Chinese Academy of Sciences, Beijing, China

Abstract. Since many XML documents are not accompanied by a schema, it is essential to devise algorithms for schema inference. In this paper, we extend the single-occurrence regular expressions (SOREs) to single-occurrence regular expressions with counting (ECsores) and give an inference algorithm for ECsores. First, we present a *countable finite automaton* (CFA). Then, we construct a CFA by converting the *single-occurrence automaton* (SOA) built for the given finite sample. Next, after the CFA recognizes the given finite sample, we obtain the minimum and maximum number of repetitions of the subexpressions derivable from the CFA, possibly updating them by using an optimization method. Finally we transform the CFA to an ECsore. Moreover, our algorithm can ensure the result is a *minimal* generalization (such generalization is also called *descriptive generalization*) of the given finite sample.

Keywords: Schema inference · Regular expressions · Counting
Descriptive generalization

1 Introduction

As a major file format for data exchange, the eXtensible Markup Language (XML) has been widely used on the Web [1]. The presence of a schema (such as DTD (Document Type Definitions) and XSD (XML Schema Definitions), which are two popular schema languages recommended by W3C (World Wide Web Consortium) [26].) provides many conveniences and advantages for various applications such as data processing, automatic data integration, and static analysis of transformations [20,21,23]. However, in practice, many XML documents are not accompanied by a schema. For example, [2,22] showed that approximately half of the XML documents available on the Web do not have a corresponding schema. It has also been noted that approximately two-thirds of XSDs

Work supported by National Natural Science Foundation of China under Grant No. 61472405.

J. C. Trujillo et al. (Eds.): ER 2018, LNCS 11157, pp. 184–199, 2018.
https://doi.org/10.1007/978-3-030-00847-5_15

gathered from schema repositories and from the Web are not valid with respect to the W3C XML Schema specification [5,6]. Therefore, it is essential to devise algorithms for schema inference.

Deterministic regular expressions [10] are used in DTD and XML Schema. Intuitively, determinism requires that given one word in the language, each symbol in the word can directly match the symbol in the expression without knowing the next symbol. For example, $(a|b)a^*$ is deterministic. On the other hand, $a^*(a|b)$ is not deterministic. In practice, many subclasses of deterministic regular expressions are used, including single-occurrence regular expressions (SOREs). However, SOREs are defined on standard regular expressions, which do not support counting. Regular expressions with counting are extended from standard regular expressions with numerical occurrence constraints (i.e., expressions of the form $E^{[m,n]}$), and are used in XML Schema [9,14,16–19]. In this paper, we extend SOREs to single-occurrence regular expressions with counting (ECsores). Table 5 (see Sect. 5) shows that the proportion of ECsores is 96.53% for the 32,750 real-world XSD files grabbed from Google, Maven, and GitHub, where 378,558 regular expressions were extracted. This indicates the practicability of ECsores. Therefore, it is necessary to study the inference algorithm for ECsore.

Gold [15] showed the class of regular expressions cannot be learned only from positive data. Using techniques from Gold [15], Bex et al. proved in [4] that even the class of deterministic regular expressions cannot be learned from positive data. Therefore for practical purposes many researchers turned to focus on the study of inference algorithms for subclasses of deterministic regular expressions [3,4,7,8,11,12], including SOREs. Compared with Gold-style learning, the descriptive generalization can lead to a compact and powerful model [13]. Thus, instead of using the technique from Gold-style learning, our inference algorithm is based on the descriptive generalization [12,13].

For inference algorithms of SOREs, Bex et al. [8] proposed the algorithms RWR and RWR_ℓ^2 [8]. Freydenberger et al. [12] presented the algorithm $Soa2Sore$ [12]. Additionally, both [8] (Sect. 8. EXTENSIONS) and [12] (Sect. 8. Conclusions and Further Work) identified the problem that SOREs cannot count, they mentioned that it can be used to infer SOREs extended with counting by an additional post-processing step following the algorithm $Soa2Sore$ or RWR. However, this is simply mentioned as future work, without technical details. According to our detailed analysis, the simple post-processing may result in the problem of overgeneralization, as the following example illustrates. Using the sample $S = \{aaaa, aab, ba, b\}$, $Soa2Sore$ generates the expression $(a|b)^+$, and RWR generates the same result. The additional post-processing will convert $(a|b)^+$ to $E = (a|b)^{[1,4]}$. However, the expression $E' = (a^{[1,2]}|b)^{[1,2]}$ is a descriptive generalization (see Sect. 2) of the finite sample S (i.e., $\mathcal{L}(E) \supset \mathcal{L}(E') \supseteq S$). This implies that simple post-processing will produce an ECsore that is overly generalized for the finite sample. Therefore, in this paper, a new method for inferring ECsore is proposed, which infers an ECsore that is a descriptive generalization of any given finite sample. As the above example illustrates, the inference algorithm of ECsores is different from that of SOREs.

The main contributions of this paper are as follows. First, we define *countable finite automaton* (CFA), which is used for counting the minimum and maximum number of repetitions of the subexpressions derivable from the CFA by accepting the given finite sample. Then, we present an inference algorithm for ECsores, where the main steps are as follows: (1) Construct a CFA by converting the *single-occurrence automaton* (SOA) [8,12] built for the given finite sample; (2) The CFA recognizes the given finite sample to obtain the minimum and maximum number of repetitions of the subexpressions derivable from the CFA, possibly updating them by using an optimization method; (3) Transform the CFA to an ECsore, and prove that the ECsore is a descriptive generalization of any given finite language.

The paper is structured as follows. Section 2 gives the basic definitions. Section 3 describes the CFA and provides an example of such an automaton. Section 4 presents the inference algorithm of the ECsore. Section 5 presents experiments. Section 6 concludes the paper.

2   Preliminaries

2.1   Regular Expression with Counting

Let Σ be a finite alphabet of symbols. A standard regular expression over Σ is inductively defined as follows: ε, $a \in \Sigma$ are regular expressions, for any regular expressions r_1 and r_2, the disjunction $(r_1|r_2)$, the concatenate $(r_1 \cdot r_2)$, and the Kleene-star r_1^* are also regular expressions. Usually, we omit concatenation operators in examples. The regular expressions with counting are extended from standard regular expressions by adding the *numerical occurrence constraints* [14]: $r^{[m,n]}$ is a regular expression for regular expression r, where $m \in \mathbb{N}$, $n \in \mathbb{N}_{/1}$, $\mathbb{N} = \{1, 2, 3, \cdots\}$, $\mathbb{N}_{/1} = \{2, 3, 4, ...\} \cup \{+\infty\}$, and $m \leq n$. Note that r^+, $r?$, and r^* are used as abbreviations of $r^{[1,+\infty]}$, $r|\varepsilon$, and $r^{[1,+\infty]}|\varepsilon$, respectively. In [24], $[m, n]$ is used on symbols, this is the case that $r^{[m,n]} = a^{[m,n]}$ ($r = a \in \Sigma$). Note that, here $\mathcal{L}(r^{[m,n]}) = \{w_1 \cdots w_i | w_1, \cdots, w_i \in \mathcal{L}(r), m \leq i \leq n\}$, instead of $\mathcal{L}(r^{[m,n]}) = \{w_1 \uplus \cdots \uplus w_i | w_1, \cdots, w_i \in \mathcal{L}(r), m \leq i \leq n\}$ specified in [9]. $\uplus$ denotes multiset union for words. For instance, $aacc \uplus ab = aaabcc$. For a regular expression E, $|E|$ denotes the length of E, which is the number of symbols and operators occurring in E plus the size of the binary representations of the integers [14]. For a finite sample S, $|S|$ denotes the number of strings in S. And for a string $s \in S$, L_s is length of s. An n-tuple is a sequence of n elements. $\pi_i^n(t)$ denotes the i-th component of an n-tuple t. Σ_s denotes the set of all symbols from Σ that appear in a string s. $\varnothing$ denotes empty set. For space consideration, all omitted proofs can be found at http://github.com/GraceFun/Inf-ECsore.

2.2   SORE, ECsore and SOA

SORE is defined as follows.

Definition 1 (SORE [7,8]). *Let Σ be a finite alphabet. A single-occurrence regular expression (SORE) is a standard regular expression over Σ in which every terminal symbol occurs at most once.*

In this paper, a SORE uses the following operations: the disjunction ($|$), the concatenation ($\cdot$), the iteration ($^+$), and the optional (?). For a SORE E, since $\mathcal{L}(E^*) = \mathcal{L}((E^+)?)$, a SORE does not use the Kleene-star operation (*), and expressions of the forms $(E?)?$, $(E^+)^+$, and $(E?)^+$ are forbidden.

Example 1. $(ab)^+$ is a SORE, while $(ab)^+a$ is not. The expressions $(a?)?$, $(a^+)^+$, and $(a?)^+$ are forbidden.

ECsore extends SORE with counting and does not use the Kleene-star and the iteration operations, is defined as follows.

Definition 2 (ECsore). *Let Σ be a finite alphabet. A single-occurrence regular expression with counting (ECsore) is a regular expression with counting over Σ in which every terminal symbol occurs at most once.*

ECsores are deterministic by definition. In this paper, an ECsore forbids immediately nested counters, and expressions of the forms $E?$, $(E)?$, and $(E?)^{[m,n]}$.

Example 2. $a?b^{[1,2]}(c|d)^{[1,+\infty]}(e^{[1,2]})?$, $(c|d)^{[1,2]}$, and $a?b(c|d)e$ are ECsores, while $a(b|c)^+a$ is not a SORE, therefore not an ECsore. $(a^{[3,4]}|b)^{[1,2]}$ and $(a^{[3,4]}b)^{[1,2]}$ are ECsores. However, the expressions $(a^{[3,4]})^{[1,2]}$, $((a^{[3,4]})?)^{[1,2]}$, and $((a^{[3,4]})?)?$ are forbidden.

Definition 3 (SOA [8,12]). *Let Σ be a finite alphabet, and let q_0, q_f be distinct symbols that do not occur in Σ. A single-occurrence automaton (SOA) over Σ is a finite directed graph $\mathscr{A} = (V, E)$ such that (1) $\{q_0, q_f\} \in V$, and $V \subseteq \Sigma \cup \{q_0, q_f\}$. (2) q_0 has only outgoing edges, q_f has only incoming edges, and every $v \in V \setminus \{q_0, q_f\}$ is visited during a walk from q_0 to q_f.*

A string $a_1 \cdots a_n$ ($n \geq 0$) is accepted by an SOA $\mathscr{A}$, if and only if there is a path $q_0 \to a_1 \to \cdots \to a_n \to q_f$ in $\mathscr{A}$.

2.3   Descriptivity

We give the notion of descriptive expressions.

Definition 4 (Descriptivity [12]). *Let $\mathcal{D}$ be a class of regular expressions or finite automata over some alphabet Σ. A $\delta \in \mathcal{D}$ is called $\mathcal{D}$-descriptive of a non-empty language $S \subseteq \Sigma^*$, if $\mathcal{L}(\delta) \supseteq S$ and there is no $\gamma \in \mathcal{D}$ such that $\mathcal{L}(\delta) \supset \mathcal{L}(\gamma) \supseteq S$.*

If a class $\mathcal{D}$ is clear from the context, we simply write *descriptive* instead of $\mathcal{D}$-descriptive.

Proposition 1. *Let Σ be a finite alphabet. There exists an ECsore-descriptive ECsore E of every language $\mathcal{L} \subseteq \Sigma^*$.*

3   Countable Finite Automaton (CFA)

A CFA is defined to count the minimum and maximum number of repetitions of the subexpressions derivable from the CFA by accepting the given finite sample. For an automaton which has the functions of counting, there are two classic ones: counter automaton [14] and finite automaton with counters (FAC) [16]. Both of these automatons must be constructed from the known subclasses of regular expressions with counting, and each of these automatons does not have exactly the same counting functions as the CFA. Therefore, a new defined automaton is necessary. The structure of CFA is similar to SOA [8,12]. The state-transition diagram of a CFA is a finite directed graph where the symbols of the states are distinct. We employ the notion of update instruction from FAC, but identify a part of the states as the counter states, which are different from FAC.

3.1   Counter States and Update Instructions

Counter states, which transit to next states, will be associated with update instructions to compute the minimum and maximum number of repetitions of the subexpressions derivable from the CFA. Update instructions are as follows:

Let q be a counter state, and $\mathcal{C}$ be the set of counter variables. We define the mapping $\theta: \mathcal{C} \mapsto \mathbb{N}$ as a function assigning a value to each counter variable $c_q \in \mathcal{C}$. θ_1 denotes that all counter variables in $\mathcal{C}$ are initialized to 1. Let partial mapping $\beta: \mathcal{C} \mapsto \{\mathbf{res}, \mathbf{inc}\}$ (**res** for reset, **inc** for increment) represent an update instruction for each counter variable. β also defines mapping g_β between mappings θ. If $\beta(c_q) = \mathbf{res}$, then $g_\beta(\theta)(c_q) = 1$, if $\beta(c_q) = \mathbf{inc}$, then $g_\beta(\theta)(c_q) = \theta(c_q) + 1$.

For each counter state q, we associate a lower bound variable $l(q) \in \mathbb{N}$ and an upper bound variable $u(q) \in \mathbb{N}$. Let L and U be the set of lower bound variables and the set of upper bound variables, respectively. We also define partial mapping $\gamma: \mathsf{L} \times \mathsf{U} \mapsto \mathbb{N} \times \mathbb{N}$ as a function assigning values to lower bound and upper bound variables: $l(q) \in \mathsf{L}$ and $u(q) \in \mathsf{U}$. γ_∞ denotes all upper bound variables that are initialized to $-\infty$ and all lower bound variables that are initialized to $+\infty$. Let partial mapping $\alpha: \mathsf{L} \times \mathsf{U} \mapsto (\mathbf{Min}(\mathsf{L} \times \mathcal{C}), \mathbf{Max}(\mathsf{U} \times \mathcal{C}))$ be an update instruction for $(l(q), u(q))$. **Min** and **Max** are the functions that return the lowest value and the highest value for any pair of input values, respectively, such that $\alpha(l(q), u(q)) = (\mathbf{Min}(l(q), c_q), \mathbf{Max}(u(q), c_q))$. α also defines the partial mapping $f_\alpha: \gamma \times \theta \mapsto \gamma$, such that $f_\alpha(\gamma, \theta)((l(q), u(q)), c_q) = (\mathbf{Min}(\pi_1^2(\gamma(l(q), u(q))), \theta(c_q)), \mathbf{Max}(\pi_2^2(\gamma(l(q), u(q))), \theta(c_q)))$. Let $\emptyset$ denote the empty instruction. $g_\emptyset(\theta) = \theta$, $f_\emptyset(\gamma, \theta) = \gamma$.

3.2   Countable Finite Automaton

Definition 5 (Countable Finite Automaton).   *A Countable Finite Automaton (CFA) is a tuple* $(Q, Q_c, \Sigma, \mathcal{C}, q_0, q_f, \Phi, \mathsf{U}, \mathsf{L})$. *The members of the tuple are described as follows:*

- Σ *is a finite and non-empty alphabet.*
- q_0 *and* q_f : q_0 *is the initial state,* q_f *is the unique final state.*

- Q is a finite set of states. $Q = \Sigma \cup \{q_0, q_f\} \cup \{+_i\}_{i \in \mathbb{N}}$.
- $Q_c \subset Q$ is a finite set of counter states. Counter state is a state q $(q \in \Sigma)$ that can directly transit to itself, or a state $+_i$.
- C is finite set of counter variables that are used for counting the number of repetitions of the subexpressions derivable from the CFA. $C = \{c_q | q \in Q_c\}$, for each counter state q, we also associate a counter variable c_q.
- $U = \{u(q) | q \in Q_c\}$, $L = \{l(q) | q \in Q_c\}$. For each subexpression derivable from the CFA, we associate a unique counter state q such that $l(q)$ and $u(q)$ are the minimum and maximum number of repetitions of the subexpression, respectively.
- Φ maps each state $q \in Q$ to a set of triples consisting of a state $p \in Q$ and two update instructions.
  $\Phi: Q \mapsto \wp(Q \times ((L \times U \mapsto (\boldsymbol{Min}(L \times C), \boldsymbol{Max}(U \times C))) \cup \{\emptyset\}) \times ((C \mapsto \{\boldsymbol{res}, \boldsymbol{inc}\}) \cup \{\emptyset\}))$.

The configuration of a CFA is defined as follows:

Definition 6 (Configuration of a CFA). *A configuration of a CFA is a triple (q, γ, θ), where $q \in Q$ is the current state, $\gamma: L \times U \mapsto \mathbb{N} \times \mathbb{N}$, $\theta: C \mapsto \mathbb{N}$. The initial configuration is $(q_0, \gamma_\infty, \theta_1)$, and a configuration is final if and only if $q = q_f$.*

For string recognition, a CFA recognizes a string by treating letters in a string individually. Let the string to be recognized be denoted by s, and an end symbol $\dashv$ is added at the end of the string s. Suppose the current state is $q \in Q$, and the current letter y $(y \in \Sigma_s$ or $y = \dashv)$ is read. If there exists a transition from state q to state y in a CFA, then y is directly recognized at state y. Otherwise, according to the state-transition diagram of a CFA constructed in Sect. 4.1, there must exist a state $+_i$ $(i \in \mathbb{N})$ such that the current state q transits to state $+_i$, the letter y will not be recognized, and y will be still read as the current letter to be recognized, until CFA reaches state y (i.e., y is only

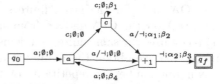

$\alpha_1 : (l(c), u(c)) \mapsto (\boldsymbol{Min}(l(c), c_c), \boldsymbol{Max}(u(c), c_c))$.
$\alpha_2 : (l(+_1), u(+_1)) \mapsto (\boldsymbol{Min}(l(+_1), c_{+_1}), \boldsymbol{Max}(u(+_1), c_{+_1}))$.
$\beta_1 : c_c \mapsto \boldsymbol{inc}; \beta_2 : c_c \mapsto \boldsymbol{res}; \beta_4 : c_{+_1} \mapsto \boldsymbol{inc}; \beta_3 : c_{+_1} \mapsto \boldsymbol{res}$.

Fig. 1. The CFA $\mathcal{A}$ for regular language $\mathcal{L}((a(c^+)?)^+)$. The label of the transition edge is $(y; \alpha; \beta)$, y $(y \in \Sigma \cup \{\dashv\})$ is a current letter, α is an update instruction for the lower bound and upper bound variables, and β is an update instruction for the counter variable.

recognized at state y). The next letter of y $(y \in \Sigma_s)$ is read if and only if y has been recognized at state y. When the last alphabet symbol in s has been recognized, $\dashv$ is not recognized until CFA reaches the final state q_f. The transition function of a CFA is defined as follows:

Definition 7 (Transition Function of a CFA). *The transition function δ of a CFA is defined for any configuration (q, γ, θ) and the letter $y \in \Sigma \cup \{\dashv\}$:*

(1) $y \in \Sigma$: $\delta((q, \gamma, \theta), y) = \{(z, f_\alpha(\gamma, \theta), g_\beta(\theta)) | (z, \alpha, \beta) \in \Phi(q) \wedge (z = y \vee ((y, \alpha, \beta) \notin \Phi(q) \wedge z \in \{+_i\}_{i \in \mathbb{N}}))\}$.

(2) $y = \dashv$: $\delta((q, \gamma, \theta), \dashv) = \{(z, f_\alpha(\gamma, \theta), g_\beta(\theta)) | (z, \alpha, \beta) \in \Phi(q) \wedge (z = q_f \vee z \in \{+_i\}_{i \in \mathbb{N}})\}$.

The construction of a CFA and $\Phi(q)$ will be given in Sect. 4.1.

Definition 8 (Deterministic CFA). *A CFA $(Q, Q_c, \Sigma, \mathcal{C}, q_0, q_f, \Phi, \mathsf{U}, \mathsf{L})$ is deterministic if and only if $|\delta((q, \gamma, \theta), y)| \leq 1$ for any $q \in Q$, $y \in \Sigma \cup \{\dashv\}$ and $\gamma : \mathsf{L} \times \mathsf{U} \mapsto \mathbb{N} \times \mathbb{N}$, $\theta : \mathcal{C} \mapsto \mathbb{N}$.*

Example 3. Let $\Sigma = \{a, c\}$, $Q = \{q_0, a, c, +_1, q_f\}$, $Q_c = \{c, +_1\}$, $\mathcal{C} = \{c_c, c_{+_1}\}$ and $\mathsf{U} = \{u(c), u(+_1)\}$, $\mathsf{L} = \{l(c), l(+_1)\}$. Figure 1 illustrates a deterministic CFA $\mathcal{A} = (Q, Q_c, \Sigma, \mathcal{C}, q_0, q_f, \Phi, \mathsf{U}, \mathsf{L})$ recognizing the language $\mathcal{L}((a(c^+)?)^+)$.

4   Inference of ECsores

Our inference algorithm works in the following steps.

(1) We construct a CFA by converting the SOA built for the given finite sample. (2) After the CFA derived from step (1) recognizes the same finite sample used in step (1), we obtain the minimum and maximum number of repetitions of the

---
Algorithm 1. *InfECsore*

Input: a finite sample S;
Output: an ECsore-descriptive ECsore;
1:   SOA $\mathscr{A}$ =2T-INF(S); CFA $\mathcal{A} = Soa2Cfa(\mathscr{A})$;
2:   **if** $Counting(\mathcal{A}, S)$ **then**
3:       $E = GenECsore(\mathcal{A})$;
4:       **return** E;

---

subexpressions derivable from the CFA, possibly updating them by using an optimization method. (3) We transform the CFA to an ECsore.

Algorithm 1 is the framework of our inference algorithm. Algorithm 2T-INF [8] constructs the SOA for the given finite sample S. Algorithm $Soa2Cfa$ is given in Sect. 4.1, algorithm $Counting$ is showed in Sect. 4.2, algorithm $GenECsore$ is presented in Sect. 4.3.

4.1   Constructing CFA

In this section, we present how to construct a CFA. A CFA should include the transitions between alphabet symbols. In addition, a CFA has counter states, also should have transitions between alphabet symbols and counter states and transitions between counter states. Thus, we can first construct the state-transition diagram of a CFA by modifying the SOA built by algorithm 2T-INF [8] for the given finite sample. We then give the detailed descriptions of the CFA.

Algorithm 2 constructs the state-transition diagram (a finite directed graph G) of a CFA by using Algorithm 3, followed by the detailed descriptions of the CFA. Initially, in Algorithm 3, graphs G_l and $\overline{G}$ are both graph $\mathscr{A}$ (SOA). G_t is used to search a strongly connected component (SCC) by using Tarjan's algorithm [25]. Note that, an SCC does not contain nodes q_0 and q_f, and excludes the singletons. The distinct nodes $+_i$ $(i \in \mathbb{N})$ are added into $\overline{G}$ by step. No new node $+_i$ is added in $\overline{G}$, if there is no specified SCC in G_t. Subroutines in Algorithm 3 are as follows.

Algorithm 2. $Soa2Cfa$

Input: SOA $\mathscr{A}(V, E)$;
Output: a CFA $\mathcal{A}$;
1: $G = Construct_G(\mathscr{A}, \mathscr{A})$;
2: CFA $\mathcal{A} = (Q, Q_c, \Sigma, C, G.q_0, G.q_f, \Phi, \cup, L)$;
3: return $\mathcal{A}$;

Algorithm 3. $Construct_G$

Input: graph $G_t(V, E)$, graph $\overline{G}(V, E)$;
Output: a new graph G, which possibly contains one or more distinct nodes $+_i$ $(i \in \mathbb{N})$.
1: **if** G_t has a cycle (excluding self loop) **then**
2:     Search an SCC (U) in G_t such that $|U| > 1$;
3:     Graph $G_p(V, E) = G_t.extract(U)$;
4:     $breakscc(G_p, G_t, \overline{G})$; $add(G_p, \overline{G}, +_i)$; inc i;
5:     **if** $|upre(G_p)| > 1$ and $\exists v \in upre(G_p) : \{q_0, q_f\} = G_p. \prec (v) \cup G_p. \succ (v)$ **then**
6:         add edge (v, v) in $\overline{G}$;
7:     **if** $|upre(G_p)| > 1$ and $\exists v \in upre(G_p) : |path(v)| > 1$ **then**
8:         $add(G_t.extract(path(v)), \overline{G}, +_i)$; inc i;
9:     **return** $Construct_G(G_t, \overline{G})$;
10: **else**
11:     **return** $\overline{G}$;

Subroutine $G_t.extract(U)$. It copies all nodes of U and all edges between nodes of U in graph G_t to return a new graph. New nodes q_0 and q_f, add edges $\{(q_0, v)|v \in U \wedge v \in G_t. \succ (v'), v' \in G_t.V \setminus U\}$, and add edges $\{(v, q_f)|v \in U \wedge v \in G_t. \prec (v'), v' \in G_t.V \setminus U\}$. $G_t. \prec (v')$ denotes the set of all direct predecessors of v' in G_t, $G_t. \succ (v')$ denotes the set of all direct successors of v' in G_t.

Subroutine $breakscc(G_p, G_t, \overline{G})$. W is the set of all nodes that can be reached from a node in $G_p. \prec (q_f)$ without crossing a node in $G_p. \succ (q_0)$. W contains the nodes in $G_p. \prec (q_f)$. Let $W' = \{v|v \in W \wedge \exists v' \in G_p. \succ (q_0) : (v, v') \in G_p.E, v \neq v'\}$. Remove edges $\{(v, v')|v \in W', v' \in G_p. \succ (q_0), v \neq v'\}$ in graph G_p, G_t, and $\overline{G}$, respectively. Add edges $\{(v, q_f)|v \in W'\}$ in graph G_p, and add edges $\{(v, v')|v \in W', v' \in G'. \succ (v_1) \setminus U, G' \in \{\overline{G}, G_t\}, v_1 \in U\}$. The SCC U ($U = G_p.V \setminus \{q_0, q_f\}$) is broken in graphs G_p, G_t, and $\overline{G}$, respectively.

Subroutine $add(G_p, \overline{G}, +_i)$. All nodes in G_p are also in graph $\overline{G}$. New a node $+_i$ $(i \in \mathbb{N}$, initially, $i = 1)$. Let $\mathcal{R}_{+_i} = \{v|v \in G_p. \succ (q_0)\}$. The set of $\mathcal{R}_{+_i}$ is established to specify the transition entrances for state $+_i$ to count the minimum and maximum number of repetitions of the corresponding subexpression derivable from the CFA. Each $\mathcal{R}_{+_i}$ is a global variable. Let $V' = G_p.V \setminus \{q_0, q_f\}$. In graph $\overline{G}$, add edges $\{(+_i, v)|v \in \mathcal{R}_{+_i}\}$ and add edges $\{(+_i, v)|v' \in G_p. \prec (q_f), v \in \overline{G}. \succ (v') \setminus V'\}$. Remove edges $\{(v', v)|v' \in G_p. \prec (q_f), v \in \overline{G}. \succ (v') \setminus V'\}$. Add edges $\{(v, +_i)|v \in G_p. \prec (q_f)\}$. $upre(G_p) = \{v|v \in G_p. \succ (q_0), \{q_0\} = G_p. \prec (v) \setminus \{v\}\}$. $path(v)$ $(v \in upre(G_p))$ denotes the set of the middle vertices (including v) passed through by all paths from v to q_f in G_p.

In Algorithm 2, after the state-transition diagram G of a CFA is constructed, the CFA $\mathcal{A}$ is then obtained. The detailed descriptions of the CFA $\mathcal{A}$ are as follows. $\mathcal{A} = (Q, Q_c, \Sigma, \mathcal{C}, G.q_0, G.q_f, \Phi, \mathsf{U}, \mathsf{L})$ where $\Sigma = G.V \setminus (\{q_0, q_f\} \cup \{+_i\}_{i \in \mathbb{N}})$, $Q = G.V$, $Q_c = \{q | q \in G. \succ (q) \wedge q \in \Sigma\} \cup \{G.+_i\}_{i \in \mathbb{N}}$, $\mathcal{C} = \{c_q | q \in Q_c\}$, $\mathsf{U} = \{u(q) | q \in Q_c\}$, $\mathsf{L} = \{l(q) | q \in Q_c\}$. Here we present $\Phi(q)$:

(1) $q = q_0$: $\Phi(q) = \{(p, \emptyset, \emptyset) | p \in G. \succ (q_0)\}$.

(2) $q \in \Sigma$: $\Phi(q) = \{(p, \{(l(q), u(q)) \mapsto (\mathbf{Min}(l(q), c_q), \mathbf{Max}(u(q), c_q)) | q \neq p \wedge q \in G. \succ (q)\} \cup \{\emptyset\}, \{c_q \mapsto \mathbf{res} | q \neq p \wedge q \in G. \succ (q)\} \cup \{c_q \mapsto \mathbf{inc} | q = p\} \cup \{\emptyset\}) | p \in G. \succ (q)\}$.

(3) $q \in \{+_i\}_{i \in \mathbb{N}}$: $\Phi(q) = \{(p, \{(l(q), u(q)) \mapsto (\mathbf{Min}(l(q), c_q), \mathbf{Max}(u(q), c_q))\}, \{c_q \mapsto \mathbf{res}\}) | p \in G. \succ (q) \wedge (p \in \{+_i\}_{i \in \mathbb{N}} \cup \{q_f\} \vee p \notin \mathcal{R}_q)\} \cup \{(p, \emptyset, \{c_q \mapsto \mathbf{inc}\}) | p \in G. \succ (q) \wedge p \in \Sigma \wedge p \in \mathcal{R}_q\}$.

Suppose the SOA $\mathscr{A}$ uses n_s alphabet symbols and contains t_s transitions ($t_s > n_s$). In Algorithm 3, it takes $\mathcal{O}(n_s + t_s) = \mathcal{O}(t_s)$ time to find an SCC each time in graph G_t. Subroutine *breakscc* breaks an SCC (U), in worst case, only one node in U is separated from U, and other nodes in U still form an SCC. Then it needs n_s times at most to break all SCCs. For subroutines *extract*, *breakscc*, and *add*, each takes constant time. It takes $\mathcal{O}(n_s)$ time to obtain $path(v)$. Thus, the time complexity of constructing a CFA is $\mathcal{O}(n_s t_s)$.

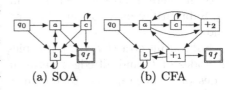

(a) SOA   (b) CFA

Fig. 2. The SOA (a) for the finite sample $S = \{a, acc, acbb, bab\}$. The state-transition diagram of the CFA (b) is constructed by modifying the SOA (a).

Example 4. For the given sample $S = \{a, acc, acbb, bab\}$, the SOA $\mathscr{A}$ is showed in Fig. 2(a). *Soa2Cfa* converts the SOA $\mathscr{A}$ into the CFA $\mathcal{A}$, the corresponding state-transition diagram of the CFA $\mathcal{A}$ is demonstrated in Fig. 2(b). For space consideration, we illustrate the CFA recognizing $\mathcal{L}((a(c^+)?)^+)$ in Example 3.

Proposition 2. *For any given finite sample S, if the CFA $\mathcal{A}$ is constructed from the SOA $\mathscr{A} = 2T\text{-}INF(S)$, then the CFA is deterministic and $\mathcal{L}(\mathcal{A}) \supseteq S$.*

4.2   Counting with CFA

Given a finite set of strings as input, the CFA counts the minimum and maximum number of repetitions of the subexpressions derivable from the CFA. The finite sample used to count with CFA is the same set of strings used to construct the SOA, which is the input of algorithm *Soa2Cfa*. Let S denote the set of strings. $\mathcal{A}$ is the CFA obtained from *Soa2Cfa*, the corresponding state-transition diagram is G. Counting rules are given by transition functions. Here, we present how to count the minimum and maximum number of repetitions of the subexpressions derivable from the CFA $\mathcal{A}$ (i.e., $\mathsf{C} = \{(l(q), u(q)) | q \in \mathcal{A}.Q_c\}$).

To obtain a descriptive ECsore, the optimal numerical predicates occurring in an ECsore should be obtained. Then, we first obtain C after the CFA $\mathcal{A}$ recognizes the given finite sample S, an optimal method is then presented to update C. Algorithm 4 presents the runs of the CFA. Before we give an optimal method in Algorithm 5 to update C, some parameters are described as follows.

Let $Q' = \{q|q, +_i \in G. \succ (q), q \in G. \prec (+_i) \cap \Sigma, i \in \mathbb{N}\} \cup \{|_j| |_j \in G. \prec (+_i), \mathcal{R}_{+_i} \cap \mathcal{R}_{+_j} \neq \varnothing, i, j \in \mathbb{N}, j \neq i\}$. Let $P_+ = \{+_i|| G. \prec (+_i) \cap Q'| \geq 1, i \in \mathbb{N}\}$. For any given $+_i \in P_+$, let $Q_{+_i} = \{q|q \in G. \prec (+_i) \cap Q'\}$. Note that $+_i \in P_+$ and $q \in Q_{+_i}$, which are mentioned below. Let S_{+_i} denote the set of strings recognised by the CFA $\mathcal{A}_s = \mathcal{A}.extract(U)$ (U is an SCC which contains $+_i$, but does not contain $+_j$ ($j \neq i$) being reachable from $+_i$). For any $s \in S_{+_i}$, s occurs in S. Let C' denote the updated C. Let S' (resp. S_1) denote the set of the strings in S_{+_i}, where for each string is accepted by CFA $\mathcal{A}_s$ such that $c_{+_i} = C.u(+_i)$ (resp. $c_{+_i} = 1$) before c_{+_i} is reset. Let $C_m(S', q)$ denote the maximum number of repetitions of the symbol $q \in \Sigma$ or the string $s' \in S_q$ ($q = +_l \neq +_i, l \in \mathbb{N}$), which occurs in S'. Let $n_s(+_i, q, j_q)$ denote the number of repetitions of the symbol $q \in \Sigma$ or the string $s' \in S_q$ ($q = +_l \neq +_i, l \in \mathbb{N}$), which is the j_qth time to repeatedly occur in $s \in S_{+_i}$. C is obtained before above parameters are derived.

For the given $+_i \in P_{+_i}$, $q \in Q_{+_i}$, the optimization is mainly to update $C.u(q)$ such that $C'.u(q)$ is as minimum as possible. Meanwhile, we ensure that the sample S can be generated by the ECsore derived from the CFA $\mathcal{A}$, when the optimal numerical predicates are introduced into the

$$\min_q \sum_q C'.u(q) \quad q \in Q_{+_i}, +_i \in P_+ \quad (1)$$

$$\text{s.t.} \quad C_m(S', q) \leq C'.u(q) \leq C.u(q) \quad (2)$$

$$\sum_q \sum_{j_q} \lceil \tfrac{n_s(+_i, q, j_q)}{C'.u(q)} \rceil \leq C.u(+_i) \quad (3)$$

$$\lceil \tfrac{n_s(+_i, q, j_q)}{C'.u(q)} \rceil \leq \lfloor \tfrac{n_s(+_i, q, j_q)}{C'.l(q)} \rfloor \quad (4)$$

ECsore. Let $C'.l(q) = C.l(q)$. If $C.l(+_i) = 1$, then let $C'.l(+_i) = \min\{\min_{s \in S_{+_i} \setminus S_1} \{\sum_q \sum_{j_q} \lceil \tfrac{n_s(+_i, q, j_q)}{C'.u(q)} \rceil\}, \min_{s \in S_1} \{\sum_q \sum_{j_q} \lfloor \tfrac{n_s(+_i, q, j_q)}{C'.l(q)} \rfloor\}\}$. Otherwise, let $C'.l(+_i) = \min_{s \in S_{+_i}} \{\sum_q \sum_{j_q} \lceil \tfrac{n_s(+_i, q, j_q)}{C'.u(q)} \rceil\}$. This problem is formatted as formulas (1)~(4), and was solved by using MIDACO-Solver[1]. Note that, formula (3) is satisfied for each $s \in S_{+_i}$, and formula (4) is satisfied for each $s \in S_{+_i}$, $q \in Q_{+_i}$, and $j_q \in \mathbb{N}$.

---

[1] http://www.midaco-solver.com/.

Given a finite set of strings as inputs, Algorithm 4 presents the runs of the CFA $\mathcal{A}$. Line 1 includes initializations for the lower bound variables, the upper bound variables and the counter variables. When the CFA has accepted a string, we do not reset the lower bound and upper bound variables (γ remains unchanged). According to the Proposition 2, there is unique configuration obtained from the transition function δ in line 7. Algorithm 5 is mainly to optimize the obtained C (in line 2) after the CFA $\mathcal{A}$ recognizes the sample S. If $P_+ = \varnothing$ in line 3, then C do not need any optimization. Otherwise, for each $+_i \in P_+$, we just optimize the $C.u(q)$ ($q \in Q_{+_i}$) by minimizing $\sum_q C'.u(q)$ in line 5, and set the appropriate value for $\mathcal{A}.L.l(+_i)$ in line 7. C is updated by this method, we prove the updated C (i.e., C') is optimal in Theorem 1.

For the given finite sample S, $N = |S|$, and a string $s \in S$. For every symbol a in s, it requires constant time to be recognized once the previous symbol of a has been recognized. Therefore, for a string s, it takes $\mathcal{O}(L_s C)$ (C is a constant.) time to be recognized. $\overline{L}$ is the aver-

---

Algorithm 4. *Running*

Input: CFA $\mathcal{A}$, and the finite sample S;
Output: Information about whether all strings in S are accepted by $\mathcal{A}$;
1: γ_∞; θ_1;
2: **for all** string $s \in S$ **do**
3:     $(q, \gamma, \theta) = (q_0, \gamma, \theta)$;
4:     Add $\dashv$ at the end of s; Read the first symbol y of s;
5:     **while** $q \neq q_f$ **do**
6:         **if** $\delta((q, \gamma, \theta), y) = \varnothing$ **then return** false;
7:         **else** $(z, f_\alpha(\gamma, \theta), g_\beta(\theta)) = \delta((q, \gamma, \theta), y)$;
8:         **if** $z = y$ **then**
9:             Read the next symbol x of symbol y; $y = x$;
10:            $(q, \gamma, \theta) = (z, f_\alpha(\gamma, \theta), g_\beta(\theta))$;
11: **return** true;

Algorithm 5. *Counting*

Input: CFA $\mathcal{A}$, and the finite sample S;
Output: Information about whether S is recognized by $\mathcal{A}$ and the optimal numerical predicates are obtained;
1: **if** $Running(\mathcal{A}, S)$ **then**
2:     Obtain C, Q', P_+, and Q_{+_i} ($+_i \in P_+$);
3:     **if** $P_+ = \varnothing$ **then return** true;
4:     **for all** $+_i \in P_+$ **do**
5:         Minimizing equation (1) s.t. equation (2)$\sim$(4);
6:         **for all** $q \in Q_{+_i}$ **do** $\mathcal{A}.U.u(q) = C'.u(q)$;
7:         Obtain $C'.l(+_i)$; $\mathcal{A}.L.l(+_i) = C'.l(+_i)$;
8:     **return** true;
9: **else**
10:    **return** false;

Table 1. The results of $(l(q), u(q))$ ($q \in \{c, b, +_2, +_1\}$) before and after the C is updated.

q	c	b	$+_2$	$+_1$
$(C.l(q), C.u(q))$	$(1,2)$	$(1,2)$	$(1,1)$	$(1,3)$
$(C'.l(q), C'.u(q))$	$(1,2)$	$(1,1)$	$(1,1)$	$(1,3)$

age length of the sample strings. Then, the runs of the CFA $\mathcal{A}$ takes $\mathcal{O}(N\overline{L})$ time. Suppose the state-transition diagram (G) of the CFA $\mathcal{A}$ contains n_g nodes (excluding q_0 and q_f) and t_g transitions. Q', P_+, and Q_{+_i} are derived from G in $\mathcal{O}(n_g + t_g)$ time. And the running time of MIDACO-Solver is in milliseconds. Note that, some statistics, such as $C_m(S', q)$, $n_s(+_i, q, j_q)$, and $C'.l(+_i)$ can be obtained in $\mathcal{O}(N\overline{L})$ time. We omit the processes to obtain them due to limited space.

Example 5. For the sample $S = \{a, acc, acbb, bab\}$, the CFA $\mathcal{A}$ is constructed in Sect. 4.1, *Counting* returns true. Table 1 lists the results of $(l(q), u(q))$ ($q \in \{c, b, +_2, +_1\}$) before and after the C is updated. Here, C' is the finally updated C. Note that the minimum numbers of repetitions of symbol c are both 0 in strings a and bab. In next section, we will convert expression $c^{[1,2]}$ to $(c^{[1,2]})$?

4.3   Generating ECsore

In this section, we transform the CFA constructed in Sect. 4.1 to an ECsore, where the values of the lower bound and upper bound of the counting operators are obtained in Sect. 4.2.

The state-transition diagram (G) of the CFA $\mathcal{A}$ is also an SOA if we also respect $+_i$ $(i \in \mathbb{N})$ as an alphabet symbol. A descriptive SORE of $\mathcal{L}(G)$ is generated by the algorithm $Soa2Sore$ [12], which is given G as input. The descriptive SORE contains symbols $+_i$. In order to obtain a descriptive ECsore, we first use the algorithm

---

Algorithm 6. $GenECsore$

Input: the CFA $\mathcal{A}$;
Output: an ECsore E;
1: Let G be the state-transition diagram of the CFA $\mathcal{A}$;
2: SORE $E_s = Soa2Sore(G)$;
3: Search all subexpressions E_b from E_s:
4:     **if** $E_b = a^+$ ($a \in \Sigma$ and $a \in \mathcal{A}.Q_c$) **then**
5:         Replace E_b by $a^{[\mathcal{A}.L.l(a), \mathcal{A}.U.u(a)]}$;
6:     **if** $E_b = (E_1+_i)^+$ (E_1 is an expression) **then**
7:         Replace E_b by $(E_1)^{[\mathcal{A}.L.l(+_i), \mathcal{A}.U.u(+_i)]}$;
8: **return** $E = E_s$;

$Soa2Sore$ which inputs the state-transition diagram (G) of the CFA $\mathcal{A}$ to generate a SORE. An ECsore is then obtained by introducing the counting operators according to the symbols $+_i$ $(i \in \mathbb{N})$ and the symbols $q \in \Sigma$ $(+_i$ and q are both counter states in the CFA $\mathcal{A})$ in the obtained SORE. The algorithm of generating ECsore is described by Algorithm 6, which inputs the CFA $\mathcal{A}$. Note that, for any expression E, $E^{[1,1]} = E$.

The time complexity of algorithm $Soa2Sore$ is $\mathcal{O}(n_g t_g)$ [12]. In line 3, subexpressions E_b can be searched by traversing the syntax tree of E_s, then this searching process takes $\mathcal{O}(|E_s|)$ time. Thus, the time complexity of algorithm $GenECsore$ is $\mathcal{O}(n_g t_g + |E_s|) = \mathcal{O}(n_g t_g)$ $(n_g t_g > |E_s|)$.

Example 6. For the given the state-transition diagram (G) of the CFA $\mathcal{A}$ in Fig. 2(b), and the pairs of numbers $\{(l(q), u(q)) | q \in \mathcal{A}.Q_c\}$ obtained from algorithm *Counting*, a SORE $((((a(c^+)?)+_2)^+|b^+)+_1)^+$ is generated by algorithm $Soa2Sore$. Then the ECsore is $((a(c^{[1,2]})?)|b)^{[1,3]}$.

Theorem 1. *For any given finite language S, let $E := InfECsore(S)$, then* C *updated by Counting is optimal, and E is a descriptive ECsore for S.*

5   Experiments

In this section, we evaluate our algorithm on real-world XML data and generated XML data. We also discuss the time performance of the algorithm. All experiments were conducted on a ThinkCentre M8600t-D065 with an Intel core i7-6700 CPU (3.4 GHz) and 8G memory. All codes were written in C++.

5.1   Data and Experiments

We analysed the practicability of ECsores in Table 5, then we evaluate our algorithm on XML data. In Table 2, we obtained DTD data from the DBLP Computer Science Bibliography corpus[2], from which we chose the

---

[2] http://dblp.org/xml/release/.

XML document *dblp-2018-02-01.xml*, and extracted the elements: `article`, `inproc(eedings)`, `phdth(esis)`, `procee(dings)`, and `incolle(ction)`. We obtained XSD data from Mondial corpus[3], from which the elements `count(ry)`, `city` and `provin(ce)` are extracted. We also obtained real-world XML data not accompanied by a schema from Nasa corpus, chosen from Miklau[4], from which the element `source` is extracted. Since $maxOccurs = unbounded$ and $minOccurs = 0$ or 1 in the above XSDs, in Table 3, a number of real-world XSDs are grabbed from Google in which $maxOccurs \neq unbounded$ or $minOccurs > 1$. However, we do not find the corresponding XML data, so we randomly generated them. All generated XML data in experiments are obtained by using ToXgene[5]. The samples employed in the experiments are available at http://github.com/GraceFun/Inf-ECsore.

In Tables 2 and 3, the left column gives element names, sample size for *Soa2Sore* and *InfECsore*, respectively. The right column lists original DTD/XSD, the results of *Soa2Sore* and the results of *InfECsore*, respectively. We replace all expressions of the form (E^+)? with E^* for better readability. *None* denotes there is no original schema for the corresponding elements.

Table 2 lists the results of the inference algorithms *Soa2Sore* and *InfECsore* on real-world XML data. The results indicate that the inferred results of our algorithm are more precise. For each of the elements `article`, `inproc(eedings)`, `phdth(esis)`, `procee(dings)`, and `incolle(ction)`, the corresponding expression produced by *InfECsore* is not only more precise than the corresponding one in the original DTD, but also more precise than the corresponding expression computed by *Soa2Sore*. For `count(ry)`, `city` and `provin(ce)`, the expressions produced by *InfECsore* are more precise than the corresponding one in the original XSD. For `source`, the sample size is smaller. It is easy to check whether the ECsore is descriptive of the corresponding sample or not. In addition, there is no original schema such that the inferred ECsore can be a reasonable schema.

Table 3 lists the results of the inference algorithms *Soa2Sore* and *InfECsore* on generated XML data. Compared with the original XSD, `ex.1` (`ex.` is an abbreviation of example) shows the inferred ECsore is more precise. For `ex.2`, the inferred ECsore indicates that more symbols or subexpressions can have numerical occurrence constraints, but are allowed to occur more times by the nested counters. For `ex.3`, the inferred ECsore is identical to the corresponding original XSD, so does even for `ex.4` with 64 symbols. It implies the original XSDs such as shown by `ex.3` and `ex.4` could be precisely learned by *InfECsore*. For `ex.5`, we specially generated XML data for *Soa2Sore* to learn the SORE $(a|(b(c|d)^+))^+$ such that the inferred ECsore has higher nesting depth of counting operators. Meanwhile, the sample size is smaller, it is easy to check whether the ECsore is descriptive of the corresponding generated sample or not.

---

[3] http://www.dbis.informatik.uni-goettingen.de/Mondial/#XML.

[4] http://www.cs.washington.edu/research/xmldatasets/.

[5] http://www.cs.toronto.edu/tox/toxgene/.

Table 2. Results of *Soa2Sore* and *InfECsore* on real-world XML data.

Element Sample Size	Original segment of DTD/XSD / Result of *Soa2Sore* / Result of *InfECsore*
article 1766146	$(a\|b\|\cdots\|v)^*$ $(b^*(((a^*(c\|e)?)\|m\|n\|q)$ $(((j\|((f\|r)d?)\|h\|i)k?)\|p\|l)^*o^*)^+)$
1766146	$((b^{[1,5]})?((((a^{[1,69]})?(c\|e)?)^{[1,2]}\|m\|n\|q)^{[1,3]}$ $((((j\|((f\|r)d?)\|h\|i)k?)^{[1,3]}\|p\|l)^{[1,3]})?$ $(o^{[1,116]})?)^{[1,3]}$
inproc. 2124105	$(a\|b\|\cdots\|v)^*$ $(b^*(ck?)?(r\|a\|m)?(o\|(dj?)\|f\|n\|q\|e\|l)^*)^+$
2124105	$(b?(ck?)?(r\|a^{[1,45]}\|m^{[1,3]}\|l^{[1,3]})?$ $((o^{[1,87]}\|(dj?)\|f\|n\|q\|e\|l)^{[1,6]})?)^{[1,5]}$
phdth. 64114	$(a\|b\|\cdots\|v)^*$ $(a^*c((((p\|(fk?)\|u)t?j?)\|e)(i\|l\|m\|s)^*)^+q?)$
64114	$((a^{[1,3]})?c((((p\|(fk?)\|u)t?j?)?\|e)$ $((i\|l\|m^{[1,5]}\|s^{[1,3]})^{[1,3]})?)^{[1,5]}q?)$
procee. 58257	$(a\|b\|\cdots\|v)^*$ $(((a?(b\|c))^+h?)?(i\|s\|d)?$ $(j\|q\|l\|(fr?)\|t\|e\|(pg?)\|m)^*)^*$
58257	$((((a?(b\|c))^{[1,32]}h?)?(i\|s\|d)?$ $((j\|q\|l\|(fr?)\|t\|e\|(pg?)\|m^{[1,3]})^{[1,5]})?)^{[1,4]})?$
incolle. 46338	$(a\|b\|\cdots\|v)^*$ $(a^*c((d(j\|p)?)\|f\|r\|(ev?)\|l\|m)^*(q\|n\|o^+)?)$
46338	$((a^{[1,49]})?c(((d(j\|p)?)\|f\|r\|(ev?)\|l\|m)^{[3,6]})?$ $(q\|n\|o^{[2,104]})?)$
count. 244	$(a^+b?c^*d?\cdots k?(l?\|m?)n?o^+p^*\cdots s^*(t^*\|u^*))$ $(ab?c^+(de?)?(f(g(hi)?)?j?k?)?(m?\|l)n?o^+$ $p^*\cdots t^*u^+)$
244	$(ab?c^{[1,25]}(de?)?(f(g(hi)?)?j?k?)?(l\|m?)n?$ $o^{[1,2]}(p^{[1,12]})?(q^{[1,8]})?(r^{[1,8]})?(s^{[1,16]})?$ $(t^{[1,2]})?u^{[1,306]})$
city 3383	$(a^+b?c?d?e?f^*g^*h^*)$ $(a^+b?(cde?)?f^*g^*h^*)$
3383	$(a^{[1,5]}b?(cde?)?(f^{[1,10]})?(g^{[1,4]})?(h^{[1,3]})?)$
provin. 1443	$(a^+b?c?d^*e^*)$ $(a^+b?c?d^*e^*)$
1443	$(a^{[1,4]}b?c?(d^{[1,6]})?(e^{[1,5]})?)$
source 26	*None* $(a\|b)^+$
26	$(a^{[1,2]}\|b^{[1,3]})^{[1,6]}$

Table 3. Results of *Soa2Sore* and *InfECsore* on generated XML data.

Element Sample Size	Original segment of XSD / Result of *Soa2Sore* / Result of *InfECsore*
ex.1 30	$(a^{[2,8]}(b^{[3,9]}\|c^{[4,10]})^+$ $d^{[5,11]}(o^{[6,12]}\|f^{[7,13]})^+)$
30	$(a^+(b\|c)^+d^+(e\|f)^+)$ $(a^{[2,7]}(b^{[4,9]}\|c^{[4,9]})^{[1,5]}d^{[5,7]}$ $(e^{[6,12]}\|f^{[7,9]})^{[1,6]})$
ex.2 941	$((a\|b\|c\|d\|e\|f)^{[1,10]})?$ $(a\|b\|c\|d\|e\|f)^+$
941	$(a^{[1,3]}\|b^{[1,4]}\|c^{[1,3]}\|d^{[1,4]}\|e^{[1,3]}\|f^{[1,4]})^{[2,6]}$
ex.3 188	$(a^{[10,20]}\|b^{[30,40]})^{[3,5]}$ $(a\|b)^+$
188	$(a^{[10,20]}\|b^{[30,40]})^{[3,5]}$
ex.4 192	$(a\|\cdots\|z\|A\|\cdots\|Z\|0\|\cdots\|9\|\kappa\|\lambda)^{[2,2]}$ $(a\|\cdots\|z\|A\|\cdots\|Z\|0\|\cdots\|9\|\kappa\|\lambda)^+$
192	$(a\|\cdots\|z\|A\|\cdots\|Z\|0\|\cdots\|9\|\kappa\|\lambda)^{[2,2]}$
ex.5 48	*None* $(a\|(b(c\|d)^+))^+$
48	$(a^{[1,3]}\|(b(c^{[1,2]}\|d^{[1,2]})^{[1,8]})^{[1,2]})^{[1,9]}$

Table 4. (a) and (b) are average running times in seconds for *InfECsore* as the functions of sample size and alphabet size, respectively.

(a)

sample size	time(s)
100	0.040
1000	0.059
10000	0.166
100000	1.256
1000000	11.183

(b)

alphabet size	time(s)
5	0.052
10	0.061
20	0.074
50	0.360
100	1.150

Table 5. Proportions of SOREs and ECsores.

Subclasses	% of XSDs
SOREs	93.74
ECsores	96.53

5.2 Performance

To illustrate the efficiency and scalability of algorithm *InfECsore*, we provide the statistics about running time in different size of samples and different size of alphabets. Table 4(a) shows the average running times in seconds for *InfECsore* as a function of sample size. We randomly extracted 1000 expressions of alphabet size 10 from DTDs and XSDs, which were grabbed from Google, Maven, and GitHub. To learn each expression, we randomly generated corresponding XML data, the samples are extracted from the XML data, each sample size is that listed in Table 4(a). The running times listed in Table 4(a) are averaged over 1000 expressions of that sample size. Table 4(b) shows the average running times in seconds for *InfECsore* as a function of alphabet size. For each alphabet size listed in Table 4(b), we also randomly extracted 1000 expressions of that alphabet size from the above DTDs and XSDs. To learn each expression, we also randomly generated corresponding XML data, but for each sample extracted from the XML data, the sample size is 1000. The running times listed in Table 4(b) are averaged over 1000 expressions of that alphabet size.

Table 4(a) presents that the running times do not over 2 s, when the sample size is less then 10^5. Table 4(b) illustrates that the running times do not over a second, when the alphabet size is less than 50. Thus, the performance of *InfECsore* demonstrates that the algorithm *InfECsore* is suitable for processing large data sets and generating the ECsores with more alphabet symbols.

6   Conclusion

This paper proposed a series of strategies for inferring a subclass of deterministic regular expressions: ECsores. The main strategies include: (1) Use *Soa2Cfa* to construct a CFA from the SOA built for the given finite sample; (2) Use *Counting* to run the CFA to obtain and possibly update the lower bound and upper bound values of counting operators; and (3) Use *GenECsore* to transform the CFA to the ECsore. The inference algorithm can infer a descriptive ECsore for any given finite language. The inferred ECsores, which represent the main information of the corresponding XML schemas, can be used to generate the corresponding database schema that are transformed from the created conceptual model. A future work is extending the SORE with counting and interleaving, and studying the corresponding inference algorithms. And then we can study the algorithm for generating database schema from the inferred XML schemas.

References

1. Abiteboul, S., Buneman, P., Suciu, D.: Data on the Web: From Relations to Semistructured Data and XML. Morgan Kaufmann, Burlington (2000)
2. Barbosa, D., Mignet, L., Veltri, P.: Studying the XML web: gathering statistics from an XML sample. World Wide Web 9(2), 187–212 (2006)
3. Bex, G.J., Gelade, W., Martens, W., Neven, F.: Simplifying XML schema: effortless handling of nondeterministic regular expressions. In: ACM SIGMOD International Conference on Management of Data, SIGMOD 2009, Providence, Rhode Island, USA, 29 June–July, pp. 731–744 (2009)
4. Bex, G.J., Gelade, W., Neven, F., Vansummeren, S.: Learning deterministic regular expressions for the inference of schemas from XML data. ACM Trans. Web 4(4), 1–32 (2010)
5. Bex, G.J., Martens, W., Neven, F., Schwentick, T.: Expressiveness of XSDs: from practice to theory, there and back again. In: Proceedings of the 14th International Conference on World Wide Web, pp. 712–721. ACM (2005)
6. Bex, G.J., Neven, F., Van den Bussche, J.: DTDs versus XML schema: a practical study. In: Proceedings of the 7th International Workshop on the Web and Databases: Colocated with ACM SIGMOD/PODS 2004, pp. 79–84. ACM (2004)
7. Bex, G.J., Neven, F., Schwentick, T., Tuyls, K.: Inference of concise DTDs from XML data. In: International Conference on Very Large Data Bases, Seoul, Korea, September, pp. 115–126 (2006)

8. Bex, G.J., Neven, F., Schwentick, T., Vansummeren, S.: Inference of concise regular expressions and DTDs. ACM Trans. Database Syst. **35**(2), 1–47 (2010)
9. Boneva, I., Ciucanu, R., Staworko, S.: Schemas for unordered XML on a DIME. Theor. Comput. Syst. **57**(2), 337–376 (2015)
10. Brüggemann-Klein, A., Wood, D.: One-unambiguous regular languages. Inf. Comput. **142**(2), 182–206 (1998)
11. Freydenberger, D.D., Kötzing, T.: Fast learning of restricted regular expressions and DTDs. In: Proceedings of the 16th International Conference on Database Theory, pp. 45–56. ACM (2013)
12. Freydenberger, D.D., Kötzing, T.: Fast learning of restricted regular expressions and DTDs. Theor. Comput. Syst. **57**(4), 1114–1158 (2015)
13. Freydenberger, D.D., Reidenbach, D.: Inferring descriptive generalisations of formal languages. J. Comput. Syst. Sci. **79**(5), 622–639 (2013)
14. Gelade, W., Gyssens, M., Martens, W.: Regular expressions with counting: weak versus strong determinism. SIAM J. Comput. **41**(1), 160–190 (2012)
15. Gold, E.M.: Language identification in the limit. Inf. Control **10**(5), 447–474 (1967)
16. Hovland, D.: Regular expressions with numerical constraints and automata with counters. In: Leucker, M., Morgan, C. (eds.) ICTAC 2009. LNCS, vol. 5684, pp. 231–245. Springer, Heidelberg (2009). https://doi.org/10.1007/978-3-642-03466-4_15
17. Kilpeläinen, P., Tuhkanen, R.: Towards efficient implementation of XML Schema content models. In: Proceedings of the 2004 ACM Symposium on Document Engineering, pp. 239–241. ACM (2004)
18. Kilpeläinen, P., Tuhkanen, R.: One-unambiguity of regular expressions with numeric occurrence indicators. Inf. Comput. **205**(6), 890–916 (2007)
19. Latte, M., Niewerth, M.: Definability by weakly deterministic regular expressions with counters is decidable. In: Italiano, G.F., Pighizzini, G., Sannella, D.T. (eds.) MFCS 2015. LNCS, vol. 9234, pp. 369–381. Springer, Heidelberg (2015). https://doi.org/10.1007/978-3-662-48057-1_29
20. Manolescu, I., Florescu, D., Kossmann, D.: Answering XML queries on heterogeneous data sources. In: International Conference on Very Large Data Bases, pp. 241–250 (2001)
21. Martens, W., Neven, F.: Typechecking top-down uniform unranked tree transducers. In: Calvanese, D., Lenzerini, M., Motwani, R. (eds.) ICDT 2003. LNCS, vol. 2572, pp. 64–78. Springer, Heidelberg (2003). https://doi.org/10.1007/3-540-36285-1_5
22. Mignet, L., Barbosa, D., Veltri, P.: The XML Web: a first study. In: Proceedings of the 12th International Conference on World Wide Web, pp. 500–510. ACM (2003)
23. Papakonstantinou, Y., Vianu, V.: DTD inference for views of XML data. In: Proceedings of the Nineteenth ACM SIGMOD-SIGACT-SIGART Symposium on Principles of Database Systems, pp. 35–46. ACM (2000)
24. Staworko, S., Boneva, I., Gayo, J.E.L., Hym, S., Prud'Hommeaux, E.G., Solbrig, H.: Complexity and expressiveness of ShEx for RDF. In: 18th International Conference on Database Theory (ICDT 2015) (2015)
25. Tarjan, R.: Depth-first search and linear graph algorithms. SIAM J. Comput. **1**(2), 146–160 (1972)
26. Thompson, H., Beech, D., Maloney, M., Mendelsohn, N.: XML Schema Part 1: Structures, 2nd Edn. W3C Recommendation (2004)

Towards Quality Analysis for Document Oriented Bases

Paola Gómez[1]([✉]), Claudia Roncancio[1]([✉]), and Rubby Casallas[2]([✉])

[1] Institute of Engineering Univ. Grenoble Alpes, LIG, CNRS, Grenoble INP,
38000 Grenoble, France
{paola.gomez-barreto,claudia.roncancio}@univ-grenoble-alpes.fr
[2] TICSw, Universidad de los Andes, Bogotá, Colombia
rcasalla@uniandes.edu.co

Abstract. Document-oriented bases allow high flexibility in data representation which facilitates a rapid development of applications and enables many possibilities for data structuring. Nevertheless, the structural choices remain crucial because of their impact on several aspects of the document base and application quality, e.g., memory print, data redundancy, readability and maintainability. Our research is motivated by quality issues of document-oriented bases. We aim at facilitating the study of the possibilities of data structuring and providing objective metrics to better reveal the advantages and disadvantages of each solution with respect to user needs. In this paper, we propose a set of structural metrics for a JSON compatible schema abstraction. These metrics reflect the complexity of the structure and are intended to be used in decision criteria for schema analysis and design process. This work capitalizes on experiences with MongoDB, XML and software complexity metrics. The paper presents the definition of the metrics together with a validation scenario where we discuss how to use the results in a schema recommendation perspective.

Keywords: NoSQL · Structural metrics
Document-oriented systems · MongoDB

1 Introduction

Nowadays, applications and information systems need to manage a large amount of heterogeneous data while meeting various requirements such as performance or scalability. NoSQL systems provide efficient data management solutions while offering flexibility in structuring data. Our work focuses on document-oriented systems, specifically those storing JSON documents, including MongoDB[1]. These systems are "schema-free". They support semi-structured data

---

[1] MongoDB is the top NoSQL database (https://www.mongodb.com, https://db-engines.com). It uses BSON, a binary-encoded serialization of JSON-like documents (http://bsonspec.org).

© Springer Nature Switzerland AG 2018
J. C. Trujillo et al. (Eds.): ER 2018, LNCS 11157, pp. 200–216, 2018.
https://doi.org/10.1007/978-3-030-00847-5_16

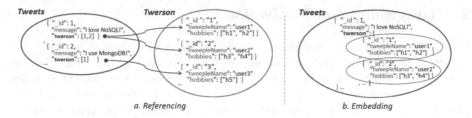

Fig. 1. Example of data in MongoDB using referencing and embedding documents

without a previous creation of a schema (unlike relational DBMS) [1]. Data can be stored in collections of document with atomic and complex attributes. This flexibility enables rapid initial development and permits many data structure possibilities for the same information. The choice is quite crucial for its potential impact on several aspects of application quality [2]. Indeed, each structure may have advantages and disadvantages regarding several aspects, such as the memory footprint of the document base, data redundancy, navigation cost, data access or program readability and maintainability.

It becomes interesting to consider several data structure candidates to retain a single choice, a temporal choice or several parallel alternatives. The analysis and comparison of several data structures is not easy because of the absence of common criteria for analysis purposes[2] and because there are potentially too many structuring possibilities.

Our research in the *SCORUS* project is a contribution in this direction. Even if document-oriented systems do not support a database schema, we propose to use a "schema" abstraction. The goal is to assist users in a data modeling process using a recommendation approach. We seek to abstract and to work with a "schema" to facilitate comprehension, assessment and comparison of document oriented data structures. The purpose is to clarify the possibilities and characteristics of each "schema" and to provide objective criteria for evaluating and assessing its advantages and disadvantages. The main contribution of this paper is the proposal of a structural metrics set for JSON compatible scheme abstraction. These metrics reflect the complexity of the data structures and can be used to establish quality criteria such as readability and maintainability. The definition of these metrics is based on experiments with MongoDB, XML-related work and metrics used in Software Engineering for code quality.

In Sect. 2, we provide background on MongoDB and the motivation of our proposal. Section 3 presents a brief overview of *SCORUS* and introduces the schema abstraction AJSchema. In Sect. 4, we propose structural metrics to measure AJSchemes. Section 5 is devoted to the validation. It presents a scenario for schema comparison using our metrics. Related work is discussed in Sect. 6. Conclusions and research perspectives are presented in Sect. 7.

---

[2] For document oriented data there are no design criteria analogous to normalization theory in the relational model.

2   Background and Motivation

We are interested in quality issues of document-oriented databases. We focus on JSON documents managed by systems like MongoDB [3]. Here, data is managed as collections of documents (see Fig. 1) where a document is a set of `attribute:value` pairs. The value type can be atomic or complex. Complex data type means either an array of values of any type or another *nesting* document. An attribute value can be the identifier of a document in another collection. This allows *referencing* one or more documents.

This simple type system provides a lot of flexibility in creating complex structures. Collections can be structured and connected in various forms considering or not data replication. e.g. completely nested collections or combination of nesting and referencing. Figure 1 depicts two ways of structuring information about *tweets* and *twerson*.

Figure 1a shows a `Tweets` collection and a `Twerson` collection. Documents in `Tweets` reference documents in `Twerson`. The choice in Fig. 1b is different, there is a single collection `Tweets` with nested documents for their *"twerson"*. In this example, there is no duplication of data.

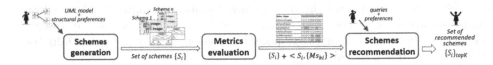

Fig. 2. *SCORUS* overview

There is not a definitive best structure because it depends on the current needs and priorities. However, the characteristics of the data structure have a strong impact on several aspects such as the size of the database, query performance and code readability of queries. The experiments presented in [2] confirm that influence. Our work is motivated by the analysis of how such aspects influence the maintainability and usability of the database as well as applications. In particular, it appears that collections with nested documents are favorable to queries following the nesting order. However, access to data in another order and queries requiring data embedded at different levels in the same collection will be penalized. The reason is that the complexity of manipulations required in such cases is similar to joining several collections. In addition, collections with nested documents have a larger footprint than the equivalent representation with references. When structuring data, priorities may lead to diverging choices, as replicating documents in multiple collections while reducing memory requirement and storage cost.

3   SCORUS Overview and Schema Abstraction

Our research focuses on helping users to understand, evaluate and make evolve semi-structured data in a more conscious way. The main contribution of this paper are the structural metrics presented in Sect. 4. Hereafter, we provide a brief overview of our larger project, *SCORUS*; we introduce the schema abstraction (Sects. 3.2 and 3.3) and tree representation (Sect. 3.4) we have defined to ease the evaluation of the metrics.

3.1   SCORUS Overview

The *SCORUS* project aims at facilitating the study of data structuring possibilities and providing objective metrics to better reveal the advantages and disadvantages of each solution with respect to the user needs. Figure 2 shows the general strategy of *SCORUS* which involves three steps: (1) from a UML data model, generating a set of schemes alternatives; (2) evaluating such schemes by using the metrics proposed in this paper; and (3) providing a top k of the most suitable schemes according to user preferences and application priorities. The sequence of the three steps can take place in a design process, but each step can be carried out independently for analysis and tuning purposes[3].

a. AJSchema abstraction of Figure1a   b. AJSchema abstraction of Figure1b

Fig. 3. AJSchemes for examples presented in Fig. 1

3.2   Schema Abstraction for Structural Analysis

To facilitate reasoning about the data structuring choices in document oriented systems, we define a schema abstraction, called here *AJSchema*. It is based on JSON schema approach [4] and on the data types supported by MongoDB.

AJSchema allows us to provide a concise representation of the collections and types of the documents to be stored in the base. Figure 3 shows the abstracted AJSchema for the examples in Fig. 1. For each collection the schema describes the type of its documents. A document type is a set of `attribute:type` pairs enclosed by { }. Types are those of MongoDB. Arrays are symbolized by [ ] and inside them, the type of its elements (atomic or document). In this paper, we use indistinctly the terms AJSchema or schema to refer to this abstraction. For schema-free systems like MongoDB, such abstractions can be considered as a data structure specification for construction and/or maintainability process.

---

[3] Schemes generation and recommendation are beyond the scope of this article.

3.3   An AJSchema from UML Model

Based on a UML model, $SCORUS$ provides several AJSchemes that are then compared to each other to improve the selection. Hereafter we illustrate the correspondence between the classes and relationships of the UML model and an AJSchema (Figs. 4 and 5) (see footnote 3).

Considering a UML model, $E = \{e_1, ..., e_n\}$ are the classes. The attributes of a class e_i are designated in the following by the type te_i and its relationships with the set $R(e_i) = \{r_1, ..., r_n\}$. Roles of relationships are known and noted by r_{irol}. Figure 4 shows a UML model with classes Agency, BusinessLines, Owner, Creative and Publicity. The properties of the class Agency are designated by the type tAgency[4].

An AJSchema describes the types of the collections that will be used. A collection type includes the attributes of a UML class (te_i) and extra attributes representing the relationships of this class $(R(e_i))$. The latter are named by the target role of r_i from the class e_i. Figure 5 presents a possible schema for the UML data model of Fig. 4. It has three collections Agencies, Owners and Creatives. The type of Agencies is formed by the attributes agencyName and id, corresponding to type tAgency, and attributes bLines and ows, corresponding to relationships r1 and r3 respectively.

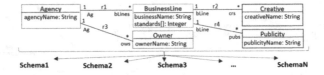

Fig. 4. From UML to several options of schemes

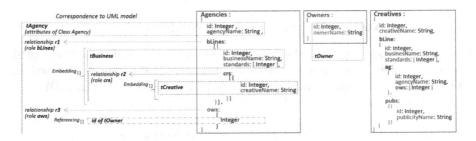

Fig. 5. Example of AJSchema option for the UML model of Fig. 4

---

[4] Type te_i has by default the attribute id corresponding to the MongoDB document identifier.

Relationships can be materialized by referencing or by embedding of documents. This choice, so as the cardinality, determine the type of the attribute. By referencing, the type is Integer, for the id of the referenced document. By embedding, the type is a document of type te_j. A cardinality "many" (in 1-many or many-many relationships) implies an array of types. Materialization by embedding induces a nested level. Referencing can occur at any level and forces the existence of a collection with type te_j.

In our example, the attribute bLines represents the 1-to-many relationship r1 by embedding of documents. Its type is therefore an array of documents of type tBusinessLines. Attribute ows corresponds to the role of relationship r3. The 1-to-many cardinality and a representation by referencing lead to an attribute of type array of Integers. Referencing forces the existence of a collection of Owners.

3.4   Tree Representation

To facilitate metrics evaluation, we use a tree representation of the AJSchema. The tree contains information about data types, nested levels, and embedded/referenced elements. The tree semantics is illustrated in Fig. 6 representing the AJSchema of Fig. 5.

The root node has a child per collection in the AJSchema. Agencies, Owners and Creatives in our example. The collection type is represented by the child sub-tree. Nodes of the form *typename@li* indicate that attributes of the *typename* appear in the documents at level *li* (starting with level 0). Attributes representing the relationships $R(e_i)$ appear as follows. A node with the name (and role used) of each relationship is created (e.g. $r1_{bline}$) with a child node, either *REF*, either *EMB* according to the choice of referencing or embedding documents. Arrays, denoted [], can be used for 1-to-many relationships. For example, the subtree on the left of Fig. 6, shows relationship $r1_{bline}$ materialized as an attribute (added to agency) of type array of business, $tAgencyEMB[]tBusiness$. Relation r2 of $tBusiness$ causes the embedding of an array of $tCreative$. The nodes indicating

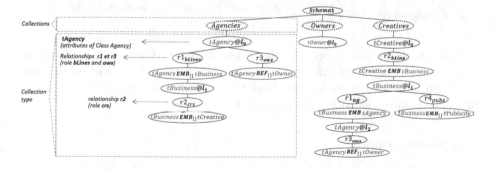

Fig. 6. Tree structure representing AJSchema of Fig. 4

a level (e.g. $tBusiness@l_1$) allow to easily identify the depth of a type and its extra attributes associated with the relationships.

4   Structural Metrics

In this section, we propose a set of metrics that reflects key aspects of the semi-structured schema complexity. The purpose is to facilitate schema analysis and comparison. We have defined a set of metrics grouped into 5 categories presented in the Sects. 4.1 to 4.5. A summary is presented in Sect. 4.6. In the following, φ denotes a collection, t a document type and x a schema.

4.1   Existence of Types and Collections

Having a collection can be mainly motivated by the access improvement to its document type at first level or its nested types. On the other hand, nesting a document into another one can be motivated by the fact that information is often accessed together. It may also be interesting to realize if a document type is nested in many places to help reducing the collection complexity.

In this section, we have defined metrics that allows us to identify the existence of a document type t in a schema. We consider two cases: (1) the existence of a collection whose type t is at the first level (l_o), and (2) the presence of such documents nested within other documents. These cases are covered, respectively, by metrics *colExistence* and *docExistence*.

Existence of a collection of documents of type t

$$colExistence(t) = \begin{cases} 1 & : \text{node } t@l_0 \text{ exists in schema } x \\ 0 & \end{cases} \tag{1}$$

Existence of embedded documents of type t: this is materialized in the graph by a node $*EMBt$.

$$docExistence(\varphi, t) = \begin{cases} 1 \ t \in \varphi \text{ node } *EMBt \text{ exists in the paths child of node } \varphi \text{ in } x \\ 0 \ t \notin \varphi \end{cases} \tag{2}$$

Figure 6 shows collections for the types *tAgency, towners and tCreative* (nodes $@l_0$) but not for *tPublicity*. Documents of type *tPublicity* exist exclusively embedded in *tCreative* documents. Note that documents of type *tBusiness* are embedded in two collections, *Agencies and Creatives*.

4.2   Nesting Depth

In general, the deeper the information is embedded, the higher is the cost to access it. This is true unless the intermediary information is also required. Knowing the nesting level of a document type facilitates the estimation of the cost of going down and back through the structure to access the data or to restructure the extracted data with the most suitable format. We propose a set of metrics to evaluate the complexity induced by embedded data. The following two metrics allow us to know the maximum depth levels of collections and schema.

Collection Depth: The *colDepth* (3) metric indicates the level of the more deeply embedded document in a collection. Embedded documents are represented by the EMB nodes in the graph.

$$colDepth(\varphi) = max(depth(p_i)) \quad : p_i \text{ is a valid child path of node } \varphi \quad (3)$$

$$depth(p) = n \qquad \text{number of nodes } EMB \text{ in path } p \quad (4)$$

Schema Depth: The *globalDepth* (5) metric indicates the deepest nesting level of a schema by considering all collections.

$$globalDepth(x) = max(colDepth(\varphi_i)) \qquad : \forall \text{ collection } \varphi_i \in x \quad (5)$$

Having recurring nested relations increases the schema complexity without necessarily improving query performance. A very nested collection can be advantageous if frequent queries require the joined information. Besides, having such a schema can be ideal if the access pattern matches with the schema. Otherwise, projections and other structuring operations will probably be required, introducing complexity in the data manipulation (see following metrics). This affects code readability and maintainability.

In Fig. 6, the depth of collection *Owners* is 0 and the depth of collections *Agencies* and *Creatives* is 2. The maximum depth of the schema is 2. Note that in the *Creatives* collection, the type *tAgency* adds no nesting level as it uses an array with references to *Owners*.

Depth of a Type of Document: The metric *docDepthInCol* (6) indicates the embedding level of a document of type t in a collection φ. If the items of the collection are of type t (node $t@l_0$), then the depth is zero. Otherwise, the metric is the level of the deepest embedded document of this type ($EMB\,t$ node) according to the root-leaf paths.

$$docDepthInCol(\varphi, t) = \begin{cases} 0 & : t \text{ corresponds to node } t@l_0 \text{ son of node } \varphi \\ max(docDepth(p_i, t) & : p_i \text{ is a valid root-leaf child path of node } \varphi \end{cases}$$
$$(6)$$
$$docDepth(p, t) = n \quad \text{number of nodes } EMB \text{ between } root \text{ and } *EMB\,t \quad (7)$$

In the *Creatives* collection of the example, the nesting level of *tPublicity* is 2, that of *tCreative* is 0. *tCreative* is also nested at level 2 in the *Agencies* collection.

We also introduce the *maxDocDepth* (8) and *minDocDepth* (9) metrics to measure the most and shallowest levels where a document type appears in a schema.

$$maxDocDepth(t) = max(DocDepthInCol(\varphi_i, t)) \quad : \varphi_i \in x \wedge t \in \varphi_i \quad (8)$$

$$minDocDepth(t) = min(DocDepthInCol(\varphi_i), t) \quad : \varphi_i \in x \wedge t \in \varphi_i \quad (9)$$

Knowing the minimum and maximum levels eases to estimate how many intermediate levels should be treated for the more direct or the less direct access to a document of a certain type. In the example, $minDocDepth(tBusiness) = 1$ as there is no collection of documents of that type.

4.3   Width of the Documents

Now we can look at the complexity of a document type in terms of its number of attributes and the complexity of their types. These metrics are motivated by the fact that documents with more complex attributes are more likely to require more complex access operations and projections. The reason is that to extract the attributes required by a query, it is necessary to "remove" the other attributes and data stored together. This operation is more expensive for documents with a larger number of attributes, i.e., with a high width. It may be interesting to choose a scheme by analyzing its "wide" and its nesting level.

The *docWidth*[5] (10) metric of a document type is based on the number of atomic attributes (coefficient a = 1), the number of attributes embedding a document (coefficient b = 2), the number of attributes of type array of atomic values (coefficient c = 1) and array of documents (coefficient d = 3). Arrays of documents have the highest weight as the experiments revealed them as the more complex to manage.

$$
\begin{aligned}
docWidth(t, \varphi) = \; & a * nbrAtomicAttributes(t, \varphi) + \\
& b * nbrDocAttributes(t, \varphi) + \\
& c * nbrArrayAtomicAttributes(t, \varphi) + \\
& d * nbrArrayDocAttributes(t, \varphi)
\end{aligned}
\quad (10)
$$

The metrics for each type of attributes can also be used separately. The size of the arrays is not considered here because it is not necessarily available in a design phase. If the size is available, it seems interesting to differentiate the orders of magnitude of the arrays, i.e. small ones vs very large ones (less than ten elements, around thousands elements, etc.).

In Fig. 4, *Agencies* and *Creatives* collections use documents of type *tBusiness* but do not have the same attributes. In *Creatives*, the type includes arrays of agencies and publicity, *docWidth (tBusiness, Creatives) = 8*, unlike *Agencies* where *docWidth(tBusiness, Agencies) = 4*.

---

[5] This metric is close to the fan-in metric for graphs.

4.4   Referencing Rate

Referential integrity becomes difficult to maintain for collections which documents are referenced by many other collections. For a collection with documents of a certain type t, the metric $refLoad$ (11) indicates the number of attributes (of other types) that are potential references to documents of type t.

$$refLoad(\varphi) = n \qquad n \text{ - number of nodes } *REF\,t \text{ where } t{=}t@l_0\,of\,node\varphi \qquad (11)$$

For the *Owners* collection in Fig. 4, $refLoad(tOwner) = 2$: collection *Agencies* references *towner* at level 0 while collection *Creatives* references it in a document embedded at level 2.

Table 1. Structural metrics

Category	Metric	Description	Schema	Collection	Type
Existence	colExistence	Existence of a collection		x	
	docExistence	Existence of a document type in a collection		x	x
Depth	colDepth	Maximal depth of a collection		x	
	globalDepth	Maximal depth of a schema	x		
	docDepthInCol	Level where a document type is in a collection		x	x
	maxDocDepth	The deepest level where a document type appears	x		
	minDocDepth	The least deep level where a document type appears	x		
Width	docWidth	"Width" of a document type		x	x
Referencing	refLoad	Number of times that a collection is referenced		x	
Redundancy	docCopiesInCol	Copies of a document type t in a collection		x	x
	docTypeCopies	Number of times a type is present in the scheme	x		

4.5   Redundancy

We are interested in estimating potential data redundancy during the schema design because it impacts several aspects. Data redundancy can speed-up access and avoid certain expensive operations (i.e. joins). However, it impacts negatively the memory footprint of the base and makes coherency enforcement more difficult. There is a cost and writing program complexity is increased and impacts the maintainability. As we are working on a structural basis, we do not use data replication information for the metric definition. The metric *docCopiesInCol* (12) is calculated by using the cardinality information of the relationships together with the representation choices in the semi-structured schema. Redundancy occurs for some cases of representation of the relationship by embedding documents.

$$docCopiesInCol(t,\varphi) = \begin{cases} 0 & : t \notin \varphi \; docExistence(\varphi,t) = 0 \\ 1 & : t \text{ corresponds to node } t@l_0 \text{ of } \varphi \\ \prod card(r_{rol}, t) & : \begin{array}{l} r_{rol} \text{ a valid node } r_{rol} \text{ father of a node} \\ EMB \text{ of the path } \varphi \text{ and } *EMBt \end{array} \end{cases}$$

$$(12)$$

$$card(r, \varepsilon) = n \qquad n - \text{ cardinality of } r \text{ on the } \varepsilon \text{ side in the UML model} \qquad (13)$$

In the *Creatives* collection of Fig. 2, the attribute for business, named *bline* , introduces redundancy for agencies. The relationship r1 may associate an agency A to n1 business instances. This leads to n1 copies of the document A. In the case where a business is referenced by n2 creatives (relationship r2), there will be n1 x n2 copies of the document A.

Moreover, we propose the metric *docTypeCopies(t)* indicating the number of times a document type is used in the schema. It reflects the number of structures that can potentially store documents of type t. This metric uses the metric of existence already introduced.

4.6 Summary

The proposed metrics are summarized in Table 1. These metrics are evaluated in the scope of the collections, types, or the whole schema, to reveal the data structure complexity.

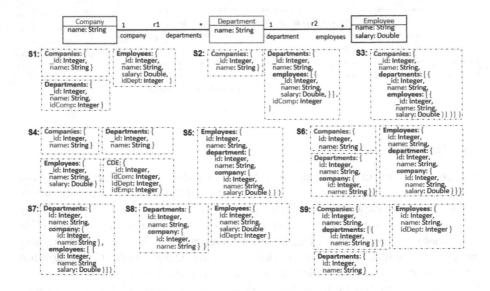

Fig. 7. Case study: UML data and AJSchema alternatives

5   Validation Scenario

As mentioned, our work aims at assisting users in the choice of document-oriented schema. The proposed metrics, together with application priorities, will be used to establish criteria for choosing and comparing schemes. This is primarily to bring out the most suitable schema according to certain criteria but also to exclude unsuitable choices or to consider alternative schemes that were not necessarily considered initially.

In the following we present a usage scenario for the proposed metrics. In Fig. 7, we introduce an example with nine structuring alternatives. This case study was already used in the work with MongoDB databases presented in [2], where we discussed the impact of the data-structures on the query evaluation. Here we are considering a similar application to analyze schema alternatives using the metrics. We evaluate the metrics for the nine schemes. Table 2 reports a subset of them.

Schema analysis will be based on user priorities such as efficient access, data consistency requirements and other user preferences. A criterion corresponds to a preference (maximization or minimization) over one or several metrics to privilege a set of schemes[6].

In our case study, the priorities of the application concern efficient access to companies information, including the names of their departments (high priority), and getting the employee with the highest salary in the company from its identifier or its name. Considering these priorities, the collection *Companies* plays an important role (criterion 1)[7] as well as the manipulation of its instances (criterion 5). The departments are accessed via the companies (criterion 6). Furthermore, it is known that the consistency of business data is important. It is therefore preferable to limit the copies to these data (criterion 2). Moreover, access to all employees (criterion 4) is not a priority.

In Table 3, each line represents one of the six criteria already mentioned. Each criteria has been evaluated for the nine alternative schemes. The values of the criteria evaluation were normalized (between 0 and 1). These values introduce a relative order between the schemes. For example, considering criteria 4, the schema S1, S4, S5, S6 and S9 are preferred over the others.

The analysis of schemes is multi-criteria (6 criteria in our case). Each criterion can have the same weight, or it can be some more important than others. The evaluation function of a schema, noted *schemaEvaluation* is the weighted sum of criteria.

$$schemaEvaluation(s) = \sum_{i=1}^{|Criteria|} weight_{criterion_i} * f_{criterion_i}(s) \qquad (14)$$

We evaluated three different weights: same weight for all criteria (case 1), priority criteria focused on companies (case 2), and priority addition on employees motivated by a new access pattern to its information (case 3). Figure 8 shows the result of the evaluation of 9 schemes for the three cases.

The estimates place schemes S5, S7 and S8 as the worst in the three cases. S5 and S7 are based on a single collection that is not a priority in the current

---

[6] A criterion is represented by a function in terms of maximization or minimization of metrics.

[7] The criterion number facilitates the presentation. It doesn't correspond to a priority.

Table 2. Case study: sub-set of metrics of schema S1 to S9

Metrics \ Schema	S1	S2	S3	S4	S5	S6	S7	S8	S9
colExistence(tCompany)	1	1	1	1	0	1	0	1	1
docCopies(tCompany)	1	1	1	1	1	3	1	1	1
refLoad(Employees)	0			1	0	0		0	0
colExistence(tEmployee)	1	0	0	1	1	1	0	0	1
docWidth(Companies,l1)	1	1	3	1		1			3
docExistence(tDepartment,Companies)	0	0	1	0		0			1

Table 3. Criteria evaluation on schema S1 until S9

Criteria \ Schema		S1	S2	S3	S4	S5	S6	S7	S8	S9
Criteria 1	$f_{c_1}(s) = colExistenceCompanies(s)$	1.00	1.00	1.00	1.00	0.00	1.00	0.00	0.00	1.00
Criteria 2	$f_{c_2}(s) = docCopiestCompany^{min}(s)$	1.00	1.00	1.00	1.00	1.00	0.33	1.00	1.00	1.00
Criteria 3	$f_{c_3}(s) = refLoadEmployees^{max}(s)$	1.00	1.00	1.00	0.00	1.00	1.00	1.00	1.00	1.00
Criteria 4	$f_{c_4}(s) = colExistenceEmployees(s)$	1.00	0.00	0.00	1.00	1.00	1.00	0.00	1.00	1.00
Criteria 5	$f_{c_5}(s) = levelWidthCompaniesL_1^{min}(s)$	1.00	1.00	0.33	1.00	0.00	1.00	0.00	0.00	0.33
Criteria 6	$f_{c_6}(s) = docDptInCompanies^{min}(s)$	0.00	0.00	1.00	0.00	0.00	0.00	0.00	0.00	1.00

criteria. On the other hand, S3 stands out in case 2 due to the high priority of its unique collection *Companies*. In addition, this collection embeds data in an appropriate order regarding the criterion 6.

Some scheme, as S9 and S6, are stable in their scores for all three cases. S9 is the best because it matches all criteria; its good results in the three cases denotes a form of "versatility" of the schema that can withstand changes in priorities. S6 introduces redundancy which is penalized by criterion 2. Meanwhile, criterion 6 penalizes it by not having embedded documents in collection *Companies*.

The criteria to be considered and their associated weight depends on the applications and the user. They may reflect good practices advocated for development or general priorities. For example, a very "compact" schema limiting memory footprint can be preferred for rarely used data. Knowing that the criteria may evolve and lead to divergent choices, the use of metrics and criteria for

Criteria	Factor		
	Case 1	Case 2	Case 3
Criterion 1	16.67	50	30
Criterion 2	16.67	10	10
Criterion 3	16.67	0	0
Criterion 4	16.67	0	20
Criterion 5	16.67	15	15
Criterion 6	16.67	25	25

a. Sets of weights

Schema \ Case	s1	s2	s3	s4	s5	s6	s7	s8	s9
Case 1	83.33	66.67	72.22	66.67	50.00	72.22	33.33	50.00	88.89
Case 2	75.00	75.00	90.00	75.00	10.00	68.33	10.00	10.00	90.00
Case 3	75.00	55.00	70.00	75.00	30.00	68.33	10.00	30.00	90.00

b. Schemes evaluation by case

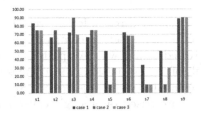

Fig. 8. Schema evaluation

a scheme analysis can help in a continuous process of "tuning" of the base. This can lead to schema evolution and data replication with heterogeneous structures. For a while, a document base may have, a copy (or partial copy) of the data with schema Sx and another copy with scheme Sy.

6   Related Work

We studied works concerning NoSQL systems [3,5–14], complex data [4,15–17], XML documents [18,19] and software metrics [20–25].

Concerning XML, Klettke et al. [18] propose 5 structural metrics based on the software quality model ISO 9126. They work on a graph representation of the DTD and metrics consider the number of references, nodes and make a link with the cyclomatic complexity [22]. In [19], Pušnik et al. propose six metrics each one associated with a quality issue such as the structure, clarity, optimality, minimalism, reuse and flexibility. These metrics use 25 variables that measure the number of elements, annotations, references and types, among others. We extended and adapted these proposals to take into account particularities of JSON as embedded documents and complex attribute types.

Our metrics are also influenced by software metrics [20–25]. Metrics proposed in [20–23] reflect, for example, the coupling levels between components, the size of the class hierarchies, the size of objects and the number of methods. [25] is an excellent survey of software metrics considering those based on the complexity of code and object oriented concepts.

Concerning NoSQL approaches, some works investigate about data modelling alternatives [5,6,26]. In [26], Abdelhedi et al. propose to translate an UML model into several alternatives of "schema" for Cassandra, Neo4J and MongoDB. For Cassandra in [5,6] the main concerns are the storage requirements and query performance. Queries are implemented with SET and GET primitives. Lombardo et al. [6] propose the creation of several versions of the data with different structures. Each version is best suited for a different query in the style of pre-calculated queries. Zhao et al. [7] propose a systematic schema conversion model of a relational database to NoSQL database. It creates a collection per entity type and the documents embed recursively the entities they reference. The structure is in the style of schema S6 in our validation scenario. Using the vocabulary of the relational model, this choice corresponds to a de-normalized structure with pre-calculation of natural joins (represented with embedded documents). The authors propose a metric for the data redundancy generated which uses the data volume. Among the existing tools working on operational bases, MongoDBCompass [8] allows to monitor the query execution time and data volume of a collection of documents. JSON schema [4] is the result of efforts to facilitate the validation of JSON documents. Tools as json-schema [17] analyze JSON documents in order to abstract a "scheme" with explicit collection and type definitions. Other researchers, as [9–12], work on ≪schemes≫ deduction for existing schema-free document-oriented bases. Their motivation is helping to understand data structuring and explaining its variants.

Guidelines to consider in modelling semi-structured data are discussed in [3,13–16]. Sadalage et al. [13] analyze various data models and NoSQL systems including MongoDB, Cassandra and Neo4j. Their main concerns are the issues in the migration of a relational database towards BigTables, documents and graphs. [3,14] propose guidelines for creating Mongo databases based on several use cases. These "best practices" can be formalized in our work as criteria to be taken into account in the schema analysis. To the best of our knowledge, no structural metrics are currently defined in the literature.

7   Conclusion and Perspectives

This work is motivated by quality issues in document-oriented bases. We focus on data structuring in JSON documents, supported by MongoDB. The flexibility offered by such systems is appreciated by developers as it is easy to represent semi-structured data. However, this flexibility comes at a cost in the performance, storage, readability and maintainability of the base and it's applications. Data structuring is a very important design decision and should not be overlooked. In this work, we briefly described *SCORUS*, our larger project, which aims at helping user to clarify the possibilities of data structuring and to provide metrics allowing to take decisions in a more conscious way. We defined a schema abstraction called AJSchema to reason about semi-structured data. We proposed a set of 11 structural metrics covering aspects as existential, nesting depth, nesting width, referencing, and redundancy. These metrics are evaluated automatically on a tree representation of the schema. The proposed metrics reflect the complexity of schema elements that play a role on quality aspects.

We presented a usage scenario of the metrics to analyze several schema variations and certain application criteria and priorities. The criteria analysis can rule out certain schema and highlight others. These findings on structural aspects were compared, and are well in line, with the results of performance evaluation experiments we conducted with databases containing data. It is interesting to note that when working on the structures, it is possible to consider more schema variants than when experimenting with the databases. This brought an unexpected result, that is the identification of a different schema with very good characteristics.

The proposed metrics form a set that is likely to evolve. Further work includes validation on a larger scale and the development of the *SCORUS* system to complete the automatic schema generation. We will also work in formalizing a recommendation system to facilitate the definition of criteria by using the metrics, important queries and other functional or non-functional preferences of potential users.

Acknowledgements. Many thanks to G. Vega, J. Chavarriaga, M. Cortés, C. Labbé, E. Perrier, P. Lago and the anonymous referees for their comments on this work.

References

1. Nayak, A., Poriya, A., Poojary, D.: Type of NOSQL databases and its comparison with relational databases. Int. J. Appl. Inf. Syst. **5**, 16–19 (2013)
2. Gómez, P., Casallas, R., Roncancio, C.: Data schema does matter, even in NOSQL systems! In: 2016 Tenth International Conference Research Challenges in Information Science (RCIS) (2016)
3. Copeland, R.: MongoDB Applied Design Patterns. O'Reilly, Sebastopol (2013)
4. jsonSchema: Json schema. http://json-schema.org/. Accessed 26 Mar 2018
5. Mior, M.J., Salem, K., Aboulnaga, A., Liu, R.: NoSE: schema design for NOSQL applications. IEEE Trans. Knowl. Data Eng. **29**(10), 2275–2289 (2017)
6. Lombardo, S., Nitto, E.D., Ardagna, D.: Issues in handling complex data structures with NOSQL databases. In: 14th International Symposium SYNASC, Romania, September 2012
7. Zhao, G., Lin, Q., Li, L., Li, Z.: Schema conversion model of SQL database to NOSQL. In: 2014 Ninth International Conference on P2P, Parallel, Grid, Cloud and Internet Computing (3PGCIC), pp. 355–362, November 2014
8. MongoDBCompass. https://docs.mongodb.com/compass/master/. Accessed 12 Feb 2018
9. Klettke, M., Störl, U., Scherzinger, S.: Schema extraction and structural outlier detection for JSON-based NOSQL data stores. Datenbanksysteme für Business, Technologie und Web (BTW) (2015)
10. Wang, L., et al.: Schema management for document stores. Proc. VLDB Endow. **8**(9), 922–933 (2015)
11. Sevilla Ruiz, D., Morales, S.F., García Molina, J.: Inferring versioned schemas from NoSQL databases and its applications. In: Johannesson, P., Lee, M.L., Liddle, S.W., Opdahl, A.L., López, Ó.P. (eds.) ER 2015. LNCS, vol. 9381, pp. 467–480. Springer, Cham (2015). https://doi.org/10.1007/978-3-319-25264-3_35
12. Gallinucci, E., Golfarelli, M., Rizzi, S.: Schema profiling of document-oriented databases. Inf. Syst. **75**, 13–25 (2018)
13. Sadalage, P.J., Fowler, M.: NoSQL Distilled: A Brief Guide to the Emerging World of Polyglot Persistence. Pearson Education, London (2012)
14. MongoDB: RDBMS to MongoDB migration guide. White Paper, November 2017
15. Abiteboul, S.: Querying semi-structured data. In: Afrati, F., Kolaitis, P. (eds.) ICDT 1997. LNCS, vol. 1186, pp. 1–18. Springer, Heidelberg (1997). https://doi.org/10.1007/3-540-62222-5_33
16. Herden, O.: Measuring quality of database schemas by reviewing-concept, criteria and tool. Oldenburg Res. Dev. Inst. Comput. Sci. Tools Syst. Escherweg **2**, 26121 (2001)
17. jsonschema.net. https://jackwootton.github.io/json-schema/. Accessed 26 Mar 2018
18. Klettke, M., Schneider, L., Heuer, A.: Metrics for XML document collections. In: Chaudhri, A.B., Unland, R., Djeraba, C., Lindner, W. (eds.) EDBT 2002. LNCS, vol. 2490, pp. 15–28. Springer, Heidelberg (2002). https://doi.org/10.1007/3-540-36128-6_2
19. Pušnik, M., Heričko, M., Budimac, Z., Šumak, B.: XML schema metrics for quality evaluation. Computer science and information systems **11**(4), 1271–1289 (2014)
20. Li, W., Henry, S.: Object-oriented metrics that predict maintainability. J. Syst. Softw. **23**(2), 111–122 (1993)
21. Chidamber, S.R., Kemerer, C.F.: Towards a metrics suite for object oriented design, vol. 26. ACM (1991)

22. McCabe, T.J.: A complexity measure. IEEE Trans. Softw. Eng. **SE–2**(4), 308–320 (1976)
23. Fenton, N.E., Neil, M.: Software metrics: roadmap. In: Proceedings of the Conference on the Future of Software Engineering, pp. 357–370. ACM (2000)
24. Fenton, N., Bieman, J.: Software Metrics: A Rigorous and Practical Approach. CRC Press, Boca Raton (2014)
25. Timóteo, A.L., Álvaro, A., De Almeida, E.S., de Lemos Meira, S.R.: Software metrics: a survey. Citeseer (2008)
26. Abdelhedi, F., Ait Brahim, A., Atigui, F., Zurfluh, G.: MDA-based approach for NoSQL databases modelling. In: Bellatreche, L., Chakravarthy, S. (eds.) DaWaK 2017. LNCS, vol. 10440, pp. 88–102. Springer, Cham (2017). https://doi.org/10.1007/978-3-319-64283-3_7

Process Modeling and Management

A Holistic Approach for Soundness Verification of Decision-Aware Process Models

Massimiliano de Leoni[1], Paolo Felli[2($\boxtimes$)], and Marco Montali[2]

[1] Eindhoven University of Technology, Eindhoven, The Netherlands
m.d.leoni@tue.nl
[2] Free University of Bozen-Bolzano, Bolzano, Italy
{pfelli,montali}@inf.unibz.it

Abstract. The last decade has witnessed an increasing transformation in the design, engineering, and mining of processes, moving from a pure control-flow perspective to more integrated models where also data and decisions are explicitly considered. This calls for methods and techniques able to ascertain the correctness of such integrated models. Differently from previous approaches, which mainly focused on the local interplay between decisions and their corresponding outgoing branches, we introduce a holistic approach to verify the end-to-end soundness of a Petri net-based process model, enriched with case data and decisions. In addition, we present an effective, implemented technique that verifies soundness by translating the input net into a colored Petri net with bounded color sets, on which standard state space analysis techniques are subsequently applied. Experiments on real life illustrate the relevance and applicability in real settings.

1 Introduction

The fundamental problem of verifying the correctness of business process models has been traditionally tackled by exclusively considering the control flow perspective, namely by only considering the ordering relations among activities present in the model. In this setting, one of the most investigated formal notions of correctness is that of *soundness*, originally introduced by van der Aalst in the context of workflow nets (a class of Petri nets that is suitable to capture the control flow of business processes) [2]. Intuitively, soundness guarantees the two good properties of "possibility of clean termination" and of "absence of deadlocks". This ensures that a process instance (*i*) always has the possibility of reaching its completion; (*ii*) when it does so, no running concurrent thread is still active; and (*iii*) all parts of the process can be executed, i.e., the process does not contain dead activities that are impossible to enact in some scenario.

The control-flow perspective is certainly of high importance as it can be considered the main process backbone; however, many other perspectives should also be taken into account. In fact, the last decade has witnessed an increasing transformation in the design, engineering, and mining of processes, moving

© Springer Nature Switzerland AG 2018
J. C. Trujillo et al. (Eds.): ER 2018, LNCS 11157, pp. 219–235, 2018.
https://doi.org/10.1007/978-3-030-00847-5_17

from a pure control-flow perspective to more integrated models where also data and decisions are explicitly considered. This trend is also testified by the recent introduction and development of the Decision Model and Notation (DMN), an OMG standard [1]. This calls for methods and techniques able to ascertain the correctness of such integrated models, which is important not only during the design phase of the business process lifecycle, but also when it comes to decision and guard mining [13,14], as well as compliance checking [12].

This work introduces a holistic, formal and operational approach to verify the end-to-end soundness of Data Petri nets (DPNs) [13], which models the order with which activities can occur (a.k.a. control flow) as well as the decision and the data of the process. Their solid formal foundation allows DPNs to unambiguously extend soundness to incorporate the decision perspective.

In the general case, verifying soundness of DPNs is undecidable due to the presence of case data and the possibility of manipulating them so as to reconstruct Turing-powerful computational devices. This applies, in particular, when case data can be updated using arithmetical operators. We isolate here a decidable class of DPNs that employs both non-numerical and numerical domains, and is expressive enough to capture data-aware process models equipped with S-FEEL DMN decisions [1], such as those recently proposed in [3,4]. Such DPNs cannot be directly analyzed algorithmically, since they induce an infinite state space even when employing bounded workflow nets. Hence, inspired to predicate abstraction [7], and in particular the approach of [12], we present an effective (i.e., sound, complete) approach for verifying soundness by translating the input net into a colored Petri net (CPN) with bounded color domains, which can then be analyzed with conventional state space exploration techniques. This has been implemented as a plug-in for the well-established ProM process mining framework.

The paper is organized as follows. In Sect. 2 we discuss related work. In Sect. 3 we recall DPNs, define their execution semantics and soundness. Section 4 illustrates our approach for translating a DPN into a corresponding CPN. Section 5 discusses the ProM implementation and reports on experiments on real-life processes, either designed by hand or obtained though manual-design and process discovery combined, showing that the technique is applicable to real-life case studies. Section 6 concludes the paper.

For space reasons, full proofs and additional technical details of this work are given in an extended technical report [8].

2   Related Work

Within the field of database theory, many approaches have been proposed to formalize and verify very sophisticated variants of data-aware processes [5], also considering data-aware extensions of soundness [15]. However, such works are mainly foundational, and do not currently come with effective verification algorithms and implementations. Within the field of business process management and information systems engineering, a plethora of techniques and tools exists for verifying soundness of process models that only capture the control-flow perspective, but not much research has been carried out to incorporate the data and decision perspective in the analysis. Sidorova et al. proposed a conceptual extension of workflow nets, equipped with an abstract, high-level data model [17]. Here, activities read and write entire guards instead of reading and writing data variables that affect the satisfaction of guards. Although simple (reading and writing guards is equivalent to reading and writing boolean values), this is not realistic: as testified by modern process modeling notations such as BPMN and DMN, the data perspective requires (at least) data variables and full-fledged guards and updates. [6] focuses on single DMN tables to verify whether they are correct or contain inconsistent, missing or overlapping rules. This certainly fits in the context of data-aware soundness, although the analysis is only conducted locally to decision points, and local forms of analysis do not suffice to guarantee good behavioral properties of the entire process [12]. A similar drawback is also present in [4], where decisions are considered either individually or in relation to their immediate outgoing sequence flows. Further, as mentioned in the introduction, soundness verification plays a key role in decision and guard mining [13,14]. Here, an initial process model is discovered by first considering only the control-flow perspective and then by enriching its decision points with decisions and conditions which are again inferred from the event data in the log. However, this "local enrichment" clearly cannot guarantee the soundness of the process, so soundness verification techniques must be employed to discard incorrect results.

The two closest works to our contribution are [3,12]. In [3], the authors consider the interplay between BPMN and DMN, providing different notions of data-aware soundness on top of such process models (once the BPMN component is encoded into a Petri net, which can be seamlessly tackled by known techniques [9,11]). Our representation of processes is expressive enough to capture the soundness properties in [3], therefore our verification technique based on an encoding into CPNs guarantees that the obtained CPN is behaviorally equivalent to the input DPN and, in turn, that the notion of soundness we introduce as well as all variants of soundness defined in [3] can be actually verified using this approach. Additional details can be found in [8]. In [12], the authors introduce an abstraction approach which shares the same spirit of our technique: it is faithful (i.e., it preserves properties), and it is based on the idea of considering only boundedly many representative values in place of the entire data domains. There are however four fundamental differences. First, in [12] abstractions are used to shrink the state space of the analysis, not to tame the infinity brought by the presence of data and the possibility of updating them. Second, [12] defines abstract process graphs that do not come with a formal execution semantics, hence not allowing to formally prove the correctness of the abstraction. Since our approach is expressive enough to capture the model of [12] (see [8]), our correctness result in Theorem 1 can be also lifted to [12]. Third, [12] focuses on compliance checking against LTL-based rules, which cannot capture soundness (in particular, the "possibility of termination", as this has an intrinsic branching nature); on the other hand, our encoding produces a CPN that is behaviorally equivalent to the original DPN, so it preserves all the runs and thus all LTL properties. Finally, while [12] translates the problem of compliance checking into a temporal model checking problem, we resort to Petri net-based techniques.

3   Syntax and Semantics of DPNs

We provide the necessary background on the DPN model [13], then providing for the first time a full account of their execution semantics, and lifting the standard notion of soundness to their richer, data-aware setting. We first define the notion of domain for case variables, assuming an infinite universe of possible values $\mathcal{U}$.

Definition 1 (Domain). *A domain is a couple $\mathcal{D} = \langle \Delta_\mathcal{D}, \Sigma_\mathcal{D} \rangle$ where $\Delta_\mathcal{D} \subseteq \mathcal{U}$ is a set of possible values and $\Sigma_\mathcal{D}$ is the set of binary predicates defined on $\Delta_\mathcal{D}$.*

Consider a set of domains $\mathfrak{D}$, and in particular the notable domains $\mathcal{D}_\mathbb{R} = \langle \mathbb{R}, \{<, >, =, \neq\} \rangle$, $\mathcal{D}_\mathbb{Z} = \langle \mathbb{Z}, \{<, >, =, \neq\} \rangle$, $\mathcal{D}_{bool} = \langle \{True, False\}, \{=, \neq\} \rangle$, $\mathcal{D}_\mathbb{S} = \langle \mathbb{S}, \{=\} \rangle$ for real numbers, integers, booleans, and strings ($\mathbb{S}$ denotes the set of all strings).

Given a set of variables V, for each $v \in V$ we write v^r or v^w to denote that the variable v is read or written, hence we consider two distinct sets $V^r = \{v^r \mid v \in V\}$ and $V^w = \{v^w \mid v \in V\}$. When we do not which to distinguish, we still use the symbol v to denote any member of $(V^r \cup V^w)$. Moreover, in order to talk about their possible values, we need to associate domains to variables. If a variable v is assigned a domain $\mathcal{D} = \langle \Delta_\mathcal{D}, \Sigma_\mathcal{D} \rangle$, for brevity we denote by $v_\mathcal{D}$ the corresponding *typed variable*, that is a shorthand to specify that v can only assume values in $\Delta_\mathcal{D}$.

Definition 2 (Guards). *Given a set of variables V with associated domains, the set of possible guards $\Phi(V)$ is the largest set containing the following:*

- *v iff $v \in (V^r \cup V^w)$;*
- *$v_\mathcal{D} \odot \Delta_\mathcal{D}$ iff $v \in (V^r \cup V^w)$ and $\odot \in \Sigma_\mathcal{D}$;*
- *$\phi_1 \wedge \phi_2$ and $\phi_1 \vee \phi_2$ iff ϕ_1 and ϕ_2 are guards in $\Phi(V)$.*

Hence, a guard can be any conjunction or disjunction of atoms of the form variable-operator-constant, whereas it is not possible to have guards of the form variable-operator-variable, which is left as future work. A *variable assignment* is a function $\beta : (V^r \cup V^w) \rightarrow \mathcal{U} \cup \{\bot\}$ assigning a value to read and written variables, with the restriction that $\beta(v)$ is a possible value for v: if $v_\mathcal{D}$ is the corresponding typed variable then $\beta(v) \in \Delta_\mathcal{D}$. The symbol $\bot$ is used to denote a null value, i.e., that the variable is not set. Given a variable assignment β and a guard ϕ, we say that ϕ is satisfied by the variable assignment β, written $\phi_{[\beta]} = \textit{true}$, iff:

- if $\phi = v$ then $\bot \neq \beta(v)$;
- if $\phi = v \odot k$, then $\odot(x, k)$ for $x = \beta(v)$;
- if $\phi = \phi_1 \wedge \phi_2$ then $\phi_{1[\beta]} = true$ and $\phi_{2[\beta]} = true$;
- if $\phi = \phi_1 \vee \phi_2$ then $\phi_{1[\beta]} = true$ or $\phi_{2[\beta]} = true$.

In words, a guard is satisfied by evaluating it after assigning values to read and written variables, as specified by β. A *state variable assignment*, abbreviated hereafter as SV assignment, is instead a function $\alpha : V \to \mathcal{U} \cup \{\bot\}$ which assigns values to each variable $v \in V$, with the restriction that $\alpha(v_\mathcal{D}) \in \Delta_\mathcal{D}$. Note that this is different from variable assignments, which are defined over $(V^r \cup V^w)$. We can now define our DPNs.

Definition 3 (Data Petri Net). *A DPN* $\mathcal{N} = (P, T, A, V, dom, \alpha_I, read, write, guard)$ *is a Petri net* (P, T, A) *with places* P*, transitions* T *and arcs* A*, where:*

- *V is a finite set of process variables;*
- *dom is a function assigning a domain $\mathcal{D}$ to each $v \in V$;*
- *α_I is the initial SV assignment;*
- *$read : T \to pwr(V)$ returns the set of variable read by a transition;*
- *$write : T \to pwr(V)$ returns the set of variable written by a transition;*
- *$guard : T \to \Phi(V)$ associates each transition t with a guard, so that v^r appears in $guard(t)$ only if $v \in read(t)$, and v^w appears in $guard(t)$ only if $v \in write(t)$.*

3.1   Execution Semantics

By considering the usual semantics for the underlying Petri net together with the guards associated to its transitions, we define the resulting data-aware execution semantics for DPNs. The set of possible states of any such DPN is formed by all pairs (M, α) where:

- $M \in \mathbb{B}(P)^1$, i.e., is the marking of the Petri net (P, T, A), and
- α is a SV assignment, defined as in the previous section.

In any state, zero or more transitions of a DPN may be able to fire. Firing a transition updates the marking, reads the variables specified in $read(t)$ and selects new values for those in $write(t)$. We model this through a variable assignment β which assigns a value to *all and only* those variables that are read or written. A couple (t, β) is called *transition firing*, said to be legal when consistent with the current state.

Definition 4 (Legal transition firing). *A DPN $\mathcal{N}$ as above evolves from state (M, α) to state (M', α') via the transition firing (t, β) with $guard(t) = \phi$ if and only if:*

---

[1] $\mathbb{B}(X)$ indicates the set of all multisets of elements of X.

- $\beta(v^r) = \alpha(v)$ *if* $v \in read(t)$: β *assigns values as* α *for read variables;*
- *the new SV assignment is s.t.* $\alpha'(v) = \begin{cases} \alpha(v) & \textit{if } v \notin write(t), \\ \beta(v^w) & \textit{otherwise;} \end{cases}$
- β *is* valid*, namely* $\phi_{[\beta]} = $ **true**: *the guard is satisfied under* β;
- $(M(p) > 0)$ *for any* $p \in P$ *such that* $(p, t) \in A$: *each input place of t has tokens;*
- *the new marking is calculated as usual, namely* $M \xrightarrow{t} M'$.

We denote this by writing $(M, \alpha) \xrightarrow{t,\beta} (M', \alpha')$, and extend the notation to sequences $\sigma = \langle (t^1, \beta^1), \ldots, (t^n, \beta^n) \rangle$ of n legal transition firings, called *traces*, and denote the corresponding *run* by $(M^0, \alpha^0) \xrightarrow{t^1,\beta^1} (M^1, \alpha^1) \xrightarrow{t^2,\beta^2} \ldots \xrightarrow{t^n,\beta^n} (M^n, \alpha^n)$ or equivalently by $(M^0, \alpha^0) \xrightarrow{\sigma} (M^n, \alpha^n)$. By restricting to the initial marking M_I and the initial variable assignment α_I, we define the traces of $\mathcal{N}$ as the set of sequences σ as above, of any length, such that $(M_I, \alpha_I) \xrightarrow{\sigma} (M, \alpha)$ for some $M \in \mathbb{B}(P)$ and α.

Example 1. Figure 1 shows a DPN modelling a process for managing credit requests and corresponding loans. There are two case variables, *amount* and *ok*, respectively representing the requested amount and whether the credit request is accepted or not. The process starts by acquiring the amount of the credit request (writing *amount*), which must be positive. Then a verification step is performed, determining whether to accept or reject the request (writing *ok*). If rejected, a new verification may be performed provided that *amount* exceeds 15000 euros. If accepted, depending on the amount, a simple or advanced assessment is performed (updating *ok*). The second phase of the process deals, concurrently, with the opening of a loan (for positive assessments) and with a communication sent to the customer, which depends again on the case variables.   □

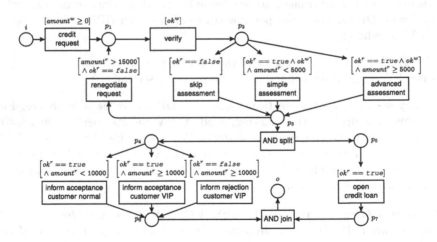

Fig. 1. Our working example of a DPN. Writing operations exist every time guards mention ok^w or $amount^w$. Terms ok^w in the guards of verify, simple assessment and advanced assessment explicitly indicate that the variable is written and can take on either **true** or **false**.

3.2   Data-Aware Soundness

We now lift the standard notion of soundness [2] to the case of DPNs. This notion requires not only to quantify over the markings of the net, but also on the assignments of its case variables, thus making soundness *data-aware* (we use 'data-aware' to distinguish our notion from the one of decision-aware soundness in the literature). In what follows, we write $(M, \alpha) \xrightarrow{*} (M', \alpha')$ to implicitly quantify *existentially* on traces σ, and denote by M_F the *final marking* of a DPN: it is the marking that, when reached, indicates the conclusion of the execution of the process instance.

Definition 5 (Data-aware soundness). *A DPN is* data-aware sound *iff:*

P1: $\forall (M, \alpha). ((M_I, \alpha_I) \xrightarrow{*} (M, \alpha) \Rightarrow \exists \alpha'. (M, \alpha) \xrightarrow{*} (M_F, \alpha'))$

P2: $\forall (M, \alpha). (M_I, \alpha_I) \xrightarrow{*} (M, \alpha) \wedge M \geq M_F \Rightarrow (M = M_F)$

P3: $\forall t \in T. \exists M_1, M_2, \alpha_1, \alpha_2, \beta. (M_I, \alpha_I) \xrightarrow{*} (M_1, \alpha_1) \xrightarrow{t, \beta} (M_2, \alpha_2)$

The first condition checks the reachability of the output state, i.e., that it is *always* possible to reach the final marking of $\mathcal{N}$ by suitably choosing a continuation of the current run (i.e., legal transitions firings). The second condition captures that the output state is always reached in a "clean" way, i.e., without having *in addition* other tokens in the net. The third condition verifies the absence of dead transitions, where a transition is considered dead if there is no way of assigning the case variables so as to enable it.

Example 2. The DPN in Fig. 1 is unsound. Suppose that the verification and assessment steps assign *ok* to `false`. Once the execution assigns a token to p_3, and the following AND-split transition is fired, two tokens are produced, in p_4 and p_5. Since the guard of open credit loan is false, token p_5 cannot be consumed and it is not possible to properly complete the execution. Also, if *amount* is less than 10000 an analogous situation occurs for the token in p_4.   □

4   Soundness Verification

In this section we propose an effective technique to check soundness of any DPN, by encoding it into a corresponding Colored Petri net (CPN) that enjoys two key properties: it employs finite color domains that suitably abstract away the source of infinity coming from the manipulation of data, but still preserves soundness, i.e., the original DPN is data-aware sound if and only if its corresponding CPN satisfies a corresponding, effectively checkable variant of soundness. This allows us to employ conventional Petri-net state space analysis techniques to ascertain data-aware soundness of DPNs. Indeed, the reachability graph may still be infinite, although the source of such infiniteness is handled by native Petri nets state-space analysis (cf. coverability graph or analogous well known techniques [2]). CPNs are an extension to DPNs that have a better support for time and resource [10], and can also be simulated through CPN Tools [16], making it possible to build on existing techniques to verify soundness. Differently from DPNs, where variables are global, CPNs encode the data aspects in the tokens, attaching them a data value: the *color*. Each place in a CPN can contain tokens of one type, called *color set* of the place.

Definition 6 provides a simplified definition of a CPN, which is enough to cover all the cases necessary in this paper. Note that tokens can also be associated with no values: in this case we introduce the color set $\bullet = \{\circ\}$, so that places with color set $\bullet$ can only contain tokens with value $\circ$, corresponding to black tokens in normal Petri nets. Similarly, variable $v_\bullet$ is a special variable that can only take on one value, i.e., $\circ$.

Definition 6 (CPN). *A CPN is a tuple* $(P, T, A, \Sigma, V, C, N, E, G, I)$ *where:*

- (P, T, A) *is a Petri net with places* P, *transitions* T *and direct arcs* A;
- Σ *is a set of color sets defined within the CPN model and* V *a set of variables;*
- $C : P \to \Sigma \cup \{\bullet\}$ *is a* color function *mapping each place to a color set in* $\Sigma \cup \{\bullet\}$;
- $N : A \to (P \times T) \cup (T \times P)$ *is a node function that maps each arc to either a pair* (p, t) *indicating that the arc is between a place* $p \in P$ *and transition* $t \in T$, *or* (t, p) *indicating that the arc connects* $t \in T$ *to* $p \in P$;
- $E : A \to V \cup \{v_\bullet\}$ *is an arc expression function, assigning variables to arcs;*
- $G : T \to \Phi(V)$ *is a guard function that maps each transition* $t \in T$ *to an expression* $G(t)$ *with the additional constraint that* $G(t)$ *can only employ variables used to annotate the arcs entering in* t: $G(t) \in \bigcup_{a \in A.N(a)=(p,t)} E(a)$;
- $I : P \to \mathbb{B}(\Sigma \cup \{\bullet\})$ *is an initialization function assigning color values to places.*
  For $p \in P$, $I(p)$ *returns the token colors initially present in* p, *with* $I(p) \in C(p)$.

In general, for any arc $a \in A$, the expression $E(a)$ can be more complex than just being a single variable. However, this simplification covers all the expressions we consider here. The concept of a marking M can be easily extended to CPNs as a function $M : P \to \mathbb{B}(\Sigma \cup \{\bullet\})$ such that $M(p)$ is a multiset of elements, each of which is the data (a.k.a. color in CPN) associated to a different token in p.

A CPN run is of the form $M^0 \xrightarrow{t^1, \gamma^1} M^1 \xrightarrow{t^2, \gamma^2} \dots \xrightarrow{t^n, \gamma^n} M^n$ where $M^0 = I$ and $\gamma^i : V \to (\Sigma \cup \{\circ\})$ is the so-called *binding* function, defined over the set of variables of the arcs entering transition t^i, for all $i \in [1, n]$. When firing each transition t^i from marking M^i, only legal bindings are possible. Binding γ^i is legal for t^i in M^i if:[2]

1. The binding considers exactly those variables that annotate the input arcs of t^i: $dom(\gamma^i) = \bigcup_{a \in A, p \in {}^\bullet t^i.N(a)=(p,t^i)} E(a)$.

---

[2] In the remainder, given a transition $t \in T$, we denote by ${}^\bullet t = \{p \in P.\exists a \in A.N(a) = (p, t)\}$ and $t^\bullet = \{p \in P.\exists a \in A.N(a) = (t, p)\}$ the usual notions of preset and postset of t.

2. If the same variable v annotates multiple input arcs of t^i, then $\gamma^i(v)$ consistently assigns v to a value that is carried by at least one token present in each of the source places of such arcs, according to M^i: for each $v \in dom(\gamma^i)$ and for each $p \in P$ s.t. there exists $a \in A$ with $N(a) = (p, t^i)$ and $E(a) = v$, we have $\gamma^i(v) \in M^i(p)$.
3. The guard of t is true when variables are substituted as per γ^i: $\phi_{[\gamma^i]} = \mathit{true}$.

Firing t^i with γ^i in marking M^i leads to a marking M^{i+1}, constructed as follows:[3]

$$M^{i+1}(p) = \begin{cases} M^i(p) & \text{if } p \notin (\bullet t^i \cup t^{i \bullet}) \\ M^i(p) \setminus [\gamma^i(E(a_{(p,t^i)}))] & \text{if } p \in \bullet t^i \\ M^i(p) \uplus [\gamma^i(E(a_{(t^i,p)}))] & \text{if } p \in t^{i \bullet} \end{cases}$$

We denote this by writing $M^i \xrightarrow{t^i, \gamma^i} M^{i+1}$, which is legal if γ^i is a valid binding of t^i. As for DPN runs, a CPN run is legal if it is a sequence of legal firings.

4.1 Translating DPNs into Colored Petri Nets

We now illustrate how a DPN $\mathcal{N} = (P, T, A, V, dom, \alpha_I, read, write, guard)$ can be converted into a CPN $\mathcal{N}^c = (P^c, T^c, A^c, \Sigma^c, V^c, C^c, N^c, E^c, G^c, I^c)$ that suitably mimics its execution semantics. Intuitively, as exemplified in Fig. 2, the transitions and places of the DPN become transitions and places of the CPN. Each variable v of the DPN becomes one *variable place* associated with the color set as the variable type of v, e.g., place var_x in Fig. 2 (right). These places always contain exactly one token, holding the variable's current value. Guards are as in the CPN, and if a transition writes a variable v, the token in the variable place for v is consumed and a new token is generated to model the update of v. For instance, the fact that transition A of the DPN Fig. 2 (left) writes a new value for variable x (denoted x^w) is modelled in Fig. 2 (right) through the two red arcs annotated with x_r and x_w, respectively entering and exiting transition A. This allows the token holding the value of v to change value when returned back to the place. A read operation is modelled as shown by the blue arcs in Fig. 2 (right), which have the same annotation, so that the token from the variable place is consumed and then put back with the same value. The initial marking of the DPN is part of the initial marking of the CPN, and each variable places is initialized with a token holding the initial value of the variable. In Fig. 2 (right), the place var_x contains a token with value 0, assuming $\alpha_I(x) = 0$. We now formalize this intuition.

---

[3] Notation $a_{(p,t^i)}$ denotes the arc $a \in A$ s.t. $N(a) = (p, t^i)$ and cannot be employed if such an arc does not exist. The set-difference operator $\setminus$ is overridden for multisets: given two multisets A and B, for each element $x \in B$ with cardinality $b_x > 0$ in B and cardinality $a_x \geq 0$ in A, the cardinality of x in $B \setminus A$ is $\max(0, b_x - a_x)$; moreover, if $x \notin B$ then $x \notin B \setminus A$.

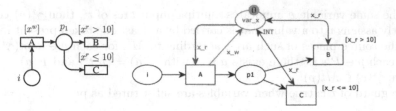

Fig. 2. Conversion of a simple DPN to CPN (left to right). The green token represents a token with value 0. Arcs without annotations are considered annotated by $v_\bullet$ and places with no color are associated with $\bullet$. Double-headed arcs stand for two arcs with same inscription in both directions. (Color figure online)

Places. The places of the CPN consist of all places of the DPN, plus one dedicated extra place $\xi(v)$, hereafter called *variable place*, for each DPN variable $v \in V$. Hence, $P^c = P \cup_{v \in V} \xi(v)$, where each variable place $\xi(v)$ always has one token, and precisely the one holding the current value of variable v at each step of the simulation of the CPN.

Transitions. The transitions of the CPN and DPN are the same: $T^c = T$.

Arcs. Each arc in A is preserved, and for each transition $t \in T$ and variable read and/or written in t, we add two extra arcs: $A^c = A \cup \{(t, \xi(v)), (\xi(v), t) \mid t \in T, v \in read(t) \cup write(t)\}$. The node function is defined as $N^c(a) = a$ for every $a \in A^c$.

Color Sets. The CPN supports the same variable types as the DPN, and we consider the color sets $\Sigma^c = \{\mathbb{Z}, \mathbb{R}, \mathsf{bool}, \mathsf{Strings}, \bullet\}$ corresponding to the domains defined at the beginning of Sect. 3 for integers, reals, booleans and strings, respectively (plus $\bullet$).

Variables. For each variable $v \in V$ the CPN considers the variables v^r and v^w, i.e., $V^c = \{v^r, v^w \mid v \in V\} \cup \{v_\bullet\}$, where $v_\bullet$ is the special variable defined earlier.

Color Functions. Recalling the shorthand notation $v_\mathcal{D}$ for typed variables in V, each place $p \in P_C$ is associated with a color set: if $p \in P$, i.e. it is also a place of the DPN, then $C^c(p) = \bullet$, otherwise the additional variable places are assigned the color set corresponding to the domain $\mathcal{D}$ of v. That is, $C^c(p) = \mathbb{Z}$ if there is $v_\mathbb{Z} \in V$ so that $p = \xi(v_\mathbb{Z})$, and the same for booleans, reals and strings (i.e. $\mathsf{bool}, \mathbb{R}, \mathbb{S}$, respectively).

Guards. Guards are not changed: $G^c(t) = guard(t)$ for each $t \in T^c$.

Arc Expressions. The expression associated with any arc between a source node $s \in P^c \cup T^c$ and a target $d \in P^c \cup T^c$ with $(s, d) \in A^c$ is as follows. If $(s, d) \in A$ then $E^c(s, d) = v_\bullet$, otherwise:

$$E^c(s, d) = \begin{cases} v^r & \text{if } d \in T \text{ and } s = \xi(v); \\ v^r & \text{if } s \in T \text{ and } d = \xi(v) \text{ and } v \notin write(s); \\ v^w & \text{if } s \in T \text{ and } d = \xi(v) \text{ and } v \in write(s); \end{cases}$$

The first case refers to arcs of the CPN that are also present in the original DPN (e.g. belonging to the set A of arcs in DPN); the places involved in these arcs contain tokens with no value associated, which we represent by $\circ$, and thus the arcs are annotated with the $v_{\bullet}$ variable. The remaining cases refer to arcs connecting the variable places for each $v \in V$ to a transition $t \in T^c$. If v is written by t then the incoming arc $(\xi(v), t)$ and the outgoing arc $(t, \xi(v))$ are annotated with v^r and v^w, respectively. This allows the token holding the value of v to change value when returned back to $\xi(v)$. If instead v is not written by t then both arcs are annotated with the same inscription v^r, guaranteeing that the value of the token does not change.

Initialization. Let M_I be the initial marking of the DPN. Places that are also in the DPN take on the same number of tokens as in the DPN, whereas each variable place $\xi(v)$ is initialized with a token holding the value specified by the initial SV assignment of the DPN. Namely, $I^c(p) = [\circ^{M_I(p)}]$ if $p \in P$, i.e., p is a place in the original net; otherwise $I^c(p) = [\alpha_I(v)]$ when $p = \xi(v)$.

This translation is correct (see Sect. 4.3): a DPN is sound if and only if its CPN translation is sound. As already discussed, this in turn allows one to exploit standard CPN analysis techniques [2] to ascertain the properties in Definition 5.

Example 3. Figure 3 shows how the working example is translated into a CPN. The red and the green elements implement the reading and the updating of the variables *ok* and *amount*. □

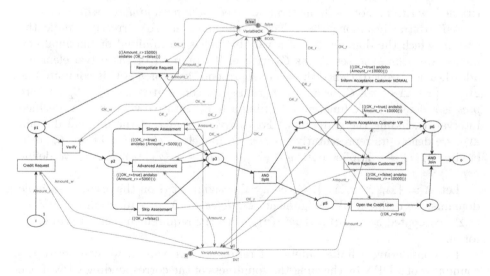

Fig. 3. Translation of DPN in Fig. 1. Double-headed arcs are a shortcut to indicate that there are two arcs with the same inscription in either of directions.

4.2 Taming Infinity via Representatives

Although the translation is correct, it is easy to see that the reachability graph can have infinitely many distinct states. The source of infiniteness is twofold: on the one hand, the original DPN and, hence, the resulting CPN is unbounded, namely the number of tokens in some places can grow indefinitely; on the other hand, the process variables of the original DPN determine color sets in the CPN over infinite domains, in turn requiring to consider potentially infinitely many different assignments for the tokens in the variable places. Standard techniques can be employed to check whether a DPN is unbounded and consequently unsound, as proven in [2]. If the DPN is bounded, the reachability graph of the CPN resulting from the translation needs to be built, which is however still infinite if the color sets of the CPN are defined over infinite domains. The remainder illustrates that soundness checking is still possible, by replacing the infinite domains of CPN color sets with a finite set of representative elements, chosen based on the set of constants appearing in the guards.

Definition 7 (Constants of the process). *The set of* constants $C_v \subset \Delta_D$ *related to a typed variable* $v_D \in V$ *of a DPN is defined as the set of all the values* k *such that either* $v^r \odot k$ *or* $v^w \odot k$ *appears in any guard of any* $t \in T$, *with* $\odot \in \Sigma_D$.

For each variable v, observing that the set C_v is finite and ordered, we partition the universe $\mathcal{U}$ into $|C_v| + 1$ intervals of values in which $\mathcal{U}$ can be partitioned w.r.t. v. For each interval we elect a *representative*, which can be chosen arbitrarily among the values in the interval. To correctly handle the case in which the domain Δ_D of a variable v has no minimal or maximal elements, we define the set C_v^+ as C_v with either or both of these two elements added, when needed. Hence the set of representatives for $v \in V$ is computed as $\bar{\Delta}_v := \{x \in \Delta_D \mid x \in C_v \text{ or } x = pick(x_1, x_2) \text{ for consecutive } x_1, x_2 \in C_v^+\}$ where $pick$ is a deterministic function returning a representative value in the specified interval, excluding the endpoints.[4] For a given value x in the original domain Δ_D, we denote its representative as $rep(x)$, namely $rep(x) := x$ iff $x \in C_v$, otherwise $rep(x) = y$ implies both $y \in (x_1, x_2)$ and $x \in (x_1, x_2)$. For $\bot$, we define $rep(\bot) := \bot$.

Let $\bar{\Delta} := \{\bar{\Delta}_{v_1}, \ldots, \bar{\Delta}_{v_q}\}$. Given a SV assignment α on the original variable domains, we define a SV assignment *restricted to* $\bar{\Delta}$ as a function $\alpha_{\bar{\Delta}} : V \to \cup_v \bar{\Delta}_v$ computed as $\alpha_{\bar{\Delta}}(v) := rep(\alpha(v))$, with the requirement that $\alpha_{\bar{\Delta}}(v) \in \bar{\Delta}_v$ for any v.

By considering a finite number of representative values, we can verify the soundness of a DPN by checking the soundness of the corresponding CPN if one restricts the values which can be assigned to each v to the finite set $\bar{\Delta}_v$, i.e., the set of representative values for v. As we are going to show, this suitably eliminates the infiniteness originating from the process data, leaving the unboundedness of

---

[4] For dense domains such as real numbers such intervals are always nonempty, whereas for non-dense domains they might be empty. In this case, we consider *pick* undefined.

the underlying net as the sole possible source of infiniteness in the reachability graph of the CPN. Such unboundedness can be detected and handled through standard Petri net analysis techniques.

To this end, we need to augment further the CPN $\mathcal{N}^c$, constructing a new CPN $\mathcal{N}_{\bar{\Delta}}^c$ which only makes use of representative values, as follows. For each variable $v \in V$ in the DPN, we add an additional place $\rho(v)$ to the set places P^c of the CPN, which is meant to represent the restricted set of possible values of v, namely $\bar{\Delta}_v$. Therefore $\rho(v)$ is assigned the same colorset as that of the variable place $\xi(v)$, and it holds one token for each possible representative value in $\bar{\Delta}_v$. This is achieved by extending the initialization function of $\mathcal{N}^c$, imposing $I(\rho(v)) = \uplus_{x \in \Delta_v}[x]$. Then, for any transition $t \in T^c$ and for each variable $v \in write^c(t)$, the representative value held by one token in $\rho(v)$ is used to update the value of the token in the variable place $\xi(v)$.

More formally, we add two arcs to A^c: an arc $(t, \rho(v))$ from t to the newly-introduced place $\rho(v)$ and a second arc $(\rho(v), t)$, and define the expression function E^c so that, for transition $t \in T$ and $v \in write(t)$, $E^c((\rho(v), t)) = E^c((\rho(v), t)) = v^w$.

Example 4. Consider, e.g., the transition credit request in the model in Fig. 3. This transition writes the integer variable *amount*. By inspecting all the guards, it is easy to see that the set of constants related to *amount* is $C_{amount} = \{5000, 10000, 15000\}$, from which we select the set of representatives $\bar{\Delta}_{amount} = \{4999, 5000, 5001, 10000, 10001, 15000, 15001\}$ by including an *arbitrary* value for each interval (e.g., in this case, 5001 was arbitrarily chosen to represent all the values in the interval $(5000, 10000)$). A token for each element of $\bar{\Delta}_{amount}$ is created in $\rho(amount)$. As it can be seen in Fig. 4(a), which depicts a small fragment of the resulting CPN $\mathcal{N}_{\bar{\Delta}}^c$, $\rho(amount)$ is called *Potential values for Amount* and its tokens can be used as possible values for the variable *Amount_w*, which in this figure stands for $amount^w$. □

4.3   Correctness of the Translation

We now discuss the correctness of the approach, summarized by Fig. 4(b). We need to show that the CPN $\mathcal{N}_{\bar{\Delta}}^c$ built as defined in the previous section (i.e. obtained by translating the original DPN $\mathcal{N}$ first into a CPN $\mathcal{N}^c$ and thus into $\mathcal{N}_{\bar{\Delta}}^c$) preserves soundness. This is based on the fact that the set of all legal runs of $\mathcal{N}$ and the set of all legal runs of $\mathcal{N}_{\bar{\Delta}}^c$ are in direct correspondence: for every run of $\mathcal{N}$ there exists a run of $\mathcal{N}_{\bar{\Delta}}^c$ that traverse the very same transitions and vice-versa. We call this property *trace-equivalence*.

Formally, we say that a DPN run $\tau = (M_I, \alpha_I) \xrightarrow{t^1, \beta^1} \cdots \xrightarrow{t^n, \beta^n} (M^n, \alpha^n)$ and a CPN run $\tau_c = M_{Ic} \xrightarrow{t_c^1, \gamma^1} \cdots \xrightarrow{t_c^n, \gamma^n} M_c^n$ are trace-equivalent iff $t^i = t_c^i$ for each $i \in [1, n]$, i.e. they have the same transitions (M_{Ic} is the initial marking of the CPN).

Theorem 1. *A DPN $\mathcal{N}$ and the CPN $\mathcal{N}_{\bar{\Delta}}^c$ defined as in the previous section are trace-equivalent. Moreover, the soundness properties in Definition 5 hold in $\mathcal{N}$ iff their translations hold in $\mathcal{N}_{\bar{\Delta}}^c$, hence $\mathcal{N}$ is sound iff $\mathcal{N}_{\bar{\Delta}}^c$ is sound.*

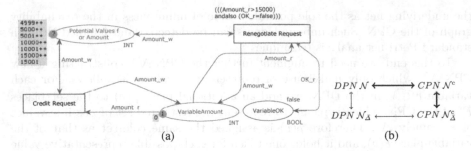

Fig. 4. (a) A fragment of $\mathcal{N}_{\bar{\Delta}}^{c}$, showing both additional places $\xi(amount)$ and $\rho(amount)$ for case variable $amount$ (see [8] for the complete figure). (b) An intuitive diagram of our approach.

The translation mentioned in the theorem is a simple rewriting of the properties in Definition 5 in terms of states and runs of CPNs instead of those of DPNs. Proving the result requires articulated intermediate steps, as translating the original DPN $\mathcal{N}$ first into CPN $\mathcal{N}^{c}$ and consequently into $\mathcal{N}_{\bar{\Delta}}^{c}$, as done in the previous section, implies not only comparing a DPN with a CPN that is syntactically very different but also handling infinite domains for variables. Due to lack of space, the proof is here omitted, but it is available in full in [8]. We give here the intuition: as illustrated in Fig. 4(a), we (*i*) first define a DPN $\mathcal{N}_{\bar{\Delta}}$ that employs only representative values in place of the possibly infinite domain values of variables, and prove that such $\mathcal{N}_{\bar{\Delta}}$ is a correct abstraction of $\mathcal{N}$; then (*ii*) show that for any run of $\mathcal{N}_{\bar{\Delta}}$ there exists a trace-equivalent run of $\mathcal{N}_{\bar{\Delta}}^{c}$. Since the properties in which we are interested only rely on the existence of legal traces, i.e. sequences of legal transition firings, and not on the values assigned to the case variables, it follows that we can analyse $\mathcal{N}_{\bar{\Delta}}^{c}$ to assess the soundness of $\mathcal{N}$.

Example 5. The CPN $\mathcal{N}_{\bar{\Delta}}^{c}$ obtained for Example 1 exhibits the same deadlocks of the original net, and it is indeed unsound. While in $\mathcal{N}$ there are infinite number of values of $amount$ for which a deadlock is reached, all these are represented by a finite number of runs in $\mathcal{N}_{\bar{\Delta}}^{c}$, one for each combination of representative values for $amount$ and ok.   □

Note that, since our encoding produces a CPN $\mathcal{N}_{\bar{\Delta}}^{c}$ that is behaviorally equivalent (i.e trace-equivalent) to the original DPN $\mathcal{N}$, the result of the previous theorem goes beyond soundness, and in particular to the different interesting properties discussed in [3] for decision-aware processes. In fact, our data-aware soundness coincides with the decision-aware soundness introduced in [3].

5   Implementation and Experiments

Our soundness-checking technique has been implemented as a Java plug-in for ProM, an established open-source framework for implementing process mining

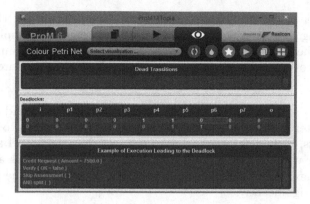

Fig. 5. Screenshot of the tool that implements the soundness-checking technique here described.

algorithms and tools (see http://www.promtools.org/). It supports both the PNML and the BPMN file formats for process models, and implements numerous algorithms for discovering process models that integrate the decision perspective (e.g. [13,14]). Thanks to this, we can employ our technique to validate the soundness of models where the decision perspective is mined from event data and models can be expressed in the two mentioned notations. In particular, the soundness-checking technique is available in the ProM *nightly* build after ensuring that the ProM package *DataPetriNets* is installed. The plug-in is named *Compute Soundness of a Data Petri Net* and takes a DPN as input. We performed a number of experiments with data-aware models of real-life processes in the literature. First, as shown in Fig. 5, our implementation correctly classify the process of Example 1 as unsound, showing a possible run that leads to a deadlock.

1. *Road-traffic fines example.* We used the model of the real-life process for the management of road-traffic fines, which is illustrated in Figs. 7 and 8 from [13]. By using our plug-in, we generated an execution that leads to a deadlock (i.e. with a token in place $pl10$), that is caused by the fact that transition *Appeal to Judge* can assign any value to variable *dismissal*. This transition is followed by a XOR split where two alternatives are possible, depending on the value of variable *dismissal*: NIL or #. However, the model does not impede *Appeal to Judge* to assign other values, e.g. G, causing a deadlock.
2. *Road-traffic fines with guard discovery.* We used the same model as at point 1 but, instead of keeping the pre-existing guards, we employed the guard discovery technique discussed in [14], which clearly does not guarantee, in general, the properties of Definition 5. The resulting model, not shown here for lack of space (see [8]), is data-aware sound. The analysis has not reported dead transitions nor deadlocks.
3. *Hospital example.* We checked the soundness of the data-aware models reported in Figs. 13.6 and 15.6 of the Ph.D. thesis by Mannhardt [13]. Both

models refer to processes that are executed within hospitals: the former is about curing patients with sepsis and the latter manages the hospital billing to patients. These models were partly hand designed and partly mined through process-discovery techniques and the analysis shows that they are data-aware sound.

The models at points 1 and 3 were analysed for deadlocks and dead transitions in a matter of seconds. The model at point 2 required 1.9 h to produce the analysis results. This difference is due to the fact that the model at point 2 is over-fitting for what concerns the decisions. In fact, the decisions are modelled through complex guards with several atoms; as a consequence, the search space to visit grows significantly.

6   Conclusions

In this paper we have introduced a holistic, formal and operational approach to verify the end-to-end soundness of Data Petri nets. Thanks to the solid formal foundation of DPNs, we defined a notion of soundness for these nets to incorporate the decision perspective, and developed an effective technique for assessing such property that can be directly implemented on existing tools. Since our DPNs are expressive enough to capture known data-aware process such as those equipped with S-FEEL DMN decisions (employing a translation similar to that in [6] – see [8] for details), this also allows us to consider various notions of soundness from the literature and in particular those in [3]. In future work, we plan to address more sophisticated guard languages, for instance by allowing to compare variables through guards such as $(v_1^w \geq v_2^r \wedge v_1^w \neq v_3^w)$. Note however that this goes beyond DMN S-FEEL and requires more sophisticated encoding techniques, although we believe this to be a decidable setting. Further, we aim at extending our results to other data domains. This is a quite delicate task, since even minimal extensions may lead to undecidability. For instance, by enriching integer domains by a successor predicate, we immediately get an undecidability result for soundness, even in the simple case of DPNs with two case variables. Finally, we plan to optimize the technique presented in this paper. In its current form, nondeterminism is managed *eagerly*, that is, by generated branches for possible values as soon as a variable is written. It appears instead promising to manage nondeterminism *lazily*, i.e., by postponing such choice to the moment where the variable actually appears in a guard, hence considering sets of possible representatives at the same time. This would not preserve trace-equivalence, but could still preserve data-aware soundness.

References

1. Decision model and notation (DMN) v1.1 (2016). http://www.omg.org/spec/DMN/1.1/
2. van der Aalst, W.M.P.: The application of petri nets to workflow management. J. Circ. Syst. Comput. **8**(1), 21–66 (1998)
3. Batoulis, K., Haarmann, S., Weske, M.: Various notions of soundness for decision-aware business processes. In: Mayr, H.C., Guizzardi, G., Ma, H., Pastor, O. (eds.) ER 2017. LNCS, vol. 10650, pp. 403–418. Springer, Cham (2017). https://doi.org/10.1007/978-3-319-69904-2_31
4. Batoulis, K., Weske, M.: Soundness of decision-aware business processes. In: Carmona, J., Engels, G., Kumar, A. (eds.) BPM 2017. LNBIP, vol. 297, pp. 106–124. Springer, Cham (2017). https://doi.org/10.1007/978-3-319-65015-9_7
5. Calvanese, D., De Giacomo, G., Montali, M.: Foundations of data aware process analysis: a database theory perspective. In: Proceedings of PODS 2013. ACM (2013)
6. Calvanese, D., Dumas, M., Laurson, Ü., Maggi, F.M., Montali, M., Teinemaa, I.: Semantics and analysis of DMN decision tables. In: La Rosa, M., Loos, P., Pastor, O. (eds.) BPM 2016. LNCS, vol. 9850, pp. 217–233. Springer, Cham (2016). https://doi.org/10.1007/978-3-319-45348-4_13
7. Clarke, E.M., Grumberg, O., Long, D.E.: Model checking and abstraction. ACM Trans. Program. Lang. Syst. **16**(5), 1512–1542 (1994)
8. de Leoni, M., Felli, P., Montali, M.: A holistic approach for soundness verification of decision-aware process models. CoRR Technical report arXiv:1804.02316, arXiv.org e-Print archive (2018). https://arxiv.org/abs/1804.02316
9. Dijkman, R.M., Dumas, M., Ouyang, C.: Semantics and analysis of business process models in BPMN. Inf. Softw. Technol. **50**(12), 1281–1294 (2008)
10. Jensen, K., Kristensen, L.M.: Coloured Petri Nets: Modelling and Validation of Concurrent Systems, 1st edn. Springer, Heidelberg (2009)
11. Kalenkova, A.A., van der Aalst, W.M.P., Lomazova, I.A., Rubin, V.A.: Process mining using BPMN: relating event logs and process models. Softw. Syst. Model. **16**(4), 1019–1048 (2017)
12. Knuplesch, D., Ly, L.T., Rinderle-Ma, S., Pfeifer, H., Dadam, P.: On enabling data-aware compliance checking of business process models. In: Parsons, J., Saeki, M., Shoval, P., Woo, C., Wand, Y. (eds.) ER 2010. LNCS, vol. 6412, pp. 332–346. Springer, Heidelberg (2010). https://doi.org/10.1007/978-3-642-16373-9_24
13. Mannhardt, F.: Multi-perspective process mining. Ph.D. thesis, Department of Mathematics and Computer Science (2018). https://pure.tue.nl/ws/portalfiles/portal/90463927
14. Mannhardt, F., de Leoni, M., Reijers, H.A., van der Aalst, W.M.P.: Decision mining revisited - discovering overlapping rules. In: Nurcan, S., Soffer, P., Bajec, M., Eder, J. (eds.) CAiSE 2016. LNCS, vol. 9694, pp. 377–392. Springer, Cham (2016). https://doi.org/10.1007/978-3-319-39696-5_23
15. Montali, M., Calvanese, D.: Soundness of data-aware, case-centric processes. Int. J. Softw. Tools Technol. Transf. **18**(5), 535–558 (2016)
16. Ratzer, A.V., et al.: CPN tools for editing, simulating, and analysing coloured petri nets. In: van der Aalst, W.M.P., Best, E. (eds.) ICATPN 2003. LNCS, vol. 2679, pp. 450–462. Springer, Heidelberg (2003). https://doi.org/10.1007/3-540-44919-1_28
17. Sidorova, N., Stahl, C., Trčka, N.: Soundness verification for conceptual workflow nets with data: early detection of errors with the most precision possible. Inf. Syst. **36**(7), 1026–1043 (2011)

Conceptual Modeling of Processes and Data: Connecting Different Perspectives

Carlo Combi[1], Barbara Oliboni[1], Mathias Weske[2], and Francesca Zerbato[1(✉)]

[1] Department of Computer Science, University of Verona, Verona, Italy
francesca.zerbato@univr.it
[2] Hasso Plattner Institute, University of Potsdam, Potsdam, Germany

Abstract. Business processes constantly generate, manipulate, and consume data that are managed by organizational databases. Despite being central to process modeling and execution, the link between processes and data is often handled by developers when the process is implemented, thus leaving the connection unexplored during the conceptual design. In this paper, we introduce, formalize, and evaluate a novel conceptual view that bridges the gap between process and data models, and show some kinds of interesting insights that can be derived from this novel proposal.

1 Introduction

The crucial role played by data in business process design, implementation, and execution is gaining considerable attention within the Business Process Management (BPM) and database communities [1,4].

Both the connection between processes and persistent data managed by organizational database systems and the notion of *data–aware* process modeling have been investigated by recent studies in BPM considering both data-centric [6,18] and activity-centric modeling paradigms [2,3,11,13].

However, despite being known that processes and data are "two sides of the same coin" [16], these two assets are still conceived separately in most organizational realities. On the one hand, activity-centric process modeling languages, such as the well-established Business Process Model and Notation (BPMN) [14], traditionally focus on modeling the control flow of a process by emphasizing the role of activities and their dynamic behavior. BPMN allows one to define process models at multiple levels of abstraction, starting from a high-level conceptual viewpoint to the specification of technical aspects needed for implementation. On the other hand, database design consists of three consolidated and distinct phases, namely conceptual, logical, and physical design [5]. At each design phase, modelers make use of different data models and related schemata to capture the aspects of interest at a particular level of abstraction. At the highest level, conceptual data models are used to create conceptual schemata that concisely represent how the information of interest for a specific domain is organized.

© Springer Nature Switzerland AG 2018
J. C. Trujillo et al. (Eds.): ER 2018, LNCS 11157, pp. 236–250, 2018.
https://doi.org/10.1007/978-3-030-00847-5_18

In this scenario, the connection between processes and data is often handled by developers who implement activities, thus creating a conceptual gap between processes and data [3,4]. However, being close to the human perception of the represented domain, conceptual approaches bring many advantages to process designers: they foster the representation of processes and related data, support conceptual reasoning, and improve system flexibility in terms of preventively detecting issues and data inconsistencies that may arise during implementation [2].

In this paper, we address the problem of connecting processes and data at the conceptual level by using BPMN to represent processes, and UML to model the conceptual database schema. BPMN is used at a level of abstraction tailored to meet the one of conceptual data modeling, that is, we consider dealing with detailed process models showing all the steps and complexities understandable by designers, but not having implementation details.

More specifically, we propose a novel *Activity View* aimed to capture the connection between a BPMN process model [14] and UML Class Diagram [15]. Activity Views are meant to support process designers in conceptually modeling the operations performed by process activities on persistent data stored in a database and to enable basic reasoning on the interplay between a process and a related database. Our approach is based on existing modeling standards to avoid defining yet another conceptual model and to ease the mapping of devised concepts to known (logical) frameworks [5]. Indeed, sitting in-between process models and data schemata, the Activity View provides a novel connected perspective, while leaving the original models untouched.

The remainder of the paper is as follows. Section 2 motivates our approach. Section 3 presents the Activity View, while Sect. 4 describes how our proposal fosters new conceptual insights. Section 5 shows the experimental evaluation of our approach. Section 6 discusses related work and Sect. 7 sketches some conclusions.

2   Motivating Example and Open Research Questions

In this section we introduce a sample scenario to motivate our approach, starting from the BPMN [14] and UML [15] standards.

Let us consider the procedure conducted by a professor to examine and grade students, as the one represented by the simple BPMN process of Fig. 1(a).

The process begins with a start event s, followed by activity Check attendance. To check student attendance, the professor must compare the matriculation numbers of the attending students with the data retrieved from the exam registration repository. The latter one is a source of persistent data and is represented by a *data store* named DB. Data stores are connected to one or more activities through directed *data associations*. The attendance of unregistered students is also recorded and this volatile information is passed along to activity Conduct exam as a *data object*. Then, each student is examined individually and, once the exam is concluded, the professor decides how to proceed: This decision is represented by exclusive gateway Exam passed? that splits the flow into

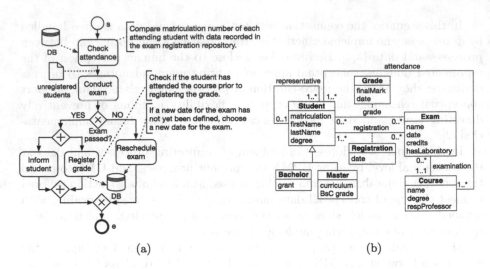

(a)   (b)

Fig. 1. (a) BPMN process diagram showing the actions performed by a professor to examine students. (b) UML class diagram of the exam registration repository. This figure shows that currently there is no formal relationship between a process model and a data model, hence providing a motivation for our work.

two branches. For those students who failed, the exam is rescheduled, whereas, only for those who passed the grade is registered in the repository. While registering the grade, the professor informs the student about the result. Hence, activities Inform student and Register grade are executed in parallel, as shown by the enclosing *parallel gateways*. Finally, end event e concludes the process.

The UML class diagram of the exam registration repository is shown in Fig. 1(b). Classes Student, Exam, and Course represent the main concepts of interest, related by associations grade, with multiplicity (0..*, 0..*), registration, examination, and attendance. Associations grade and registration have a related association class. Reflexive association representative links students with their student representative. Finally, classes Bachelor and Master specialize students.

Despite capturing informational aspects through data objects and data stores, BPMN process models provide little or no detail about the operation performed by process activities on a database. This lack of knowledge complicates the modeling of data in BPMN processes from different standpoints. (i) In BPMN data stores represent persistent data sources [14] and, most likely, conceptualize where in the process a database is accessed. However, BPMN defines them at a very high level, and there is currently no standard-compliant way of specifying the schema of the represented database. For example, the conceptual schema of the database represented by data store DB of Fig. 1(a) cannot be specified.

(ii) BPMN data objects can be used to represent volatile data at different granularities, spanning from single data entries to structured documents [12]. As a result, the correspondence between data objects or conceptual information entities related to a process and persistent data is not specified. For instance, it is not clear whether data object unregistered students is related to data class Student and, if so, how. More in general, let us suppose that a professor wants to read a student's grade transcript, containing the whole academic record. In the data schema of Fig. 1(b), there is no class "transcript" that captures such information directly, since the concept of grade transcript is realized by data classes Student and Exam, related through association grade, and by association class Grade. Therefore, the data object representing the transcript must correspond to a more complex conceptual object, which is identified by a view on the accessed database. (iii) Last but not least, too little detail about data operations is provided as these are often encoded within activities or their labels [4,12]. Despite directed data associations allow one to visualize when data are read or written by an activity, it is not possible to distinguish the granularity of the conceptual object(s) or of the sets of objects needed by a process.

As process models and data schemata are conceived separately, to foster data-aware process modeling it is necessary to support designers in understanding and capturing the connection between processes and databases [1,13].

3   Bridging the Gap Between Processes and Data

In this section, we propose a novel solution aimed to capture the connection between BPMN process models and UML class diagrams at a conceptual level. To this end, we devised the *Activity View*, a novel approach linking the conceptual representations of process models and data schemata by detailing which operations are performed by a process activity on a database and how.

We chose activities as a starting point, as data modeling in BPMN is often related to activities or whole processes. The final goal of the Activity View is to show which is the portion of a database schema (i.e., the view) that is accessed by a given process activity and to detail interesting aspects of the performed data operations.

Definition 1 (Activity View). Given an activity ac in a process model, its Activity View $av_{ac} = \{t_1, \ldots, t_n\}$ is a set of tuples, where each tuple t_i denotes a particular data access operation performed by activity ac on classes of a given data schema. The latter is composed of a set of classes Cl, a set of associations As, and a set of association classes AsC. Each tuple t_i is defined as follows:

$$t_i = \langle C_{set_i}, A_{set_i}, AccessType_i, AccessTime_i, NumInstances_i \rangle$$

where:

- $C_{set_i} = \{c_1, \ldots, c_j\} \subseteq (Cl \cup AsC)$, is the set of connected classes accessed by process activity ac. By "connected" we mean that each class $c_j \in C_{set_i}$ is reachable from at least another class $c_h \in C_{set_i}$ by navigating an association $a_f \in As$ that directly links c_h and c_j (i.e., c_h and c_j are at the ends of association a_f). Moreover, if a_f is associated to association class $c_f \in AsC$, then c_l must also belong to C_{set_i}. However, $c_f \in AsC$ may also be accessed individually, as other classes. If a class $c_j \in C_{set_i}$ specializes a more general class c_l, then it is sufficient that c_l is one end of association a_f for c_j to be considered connected to other classes of C_{set_i}. Instead, the opposite does not hold. Each class $c_j(attr_1, \ldots, attr_n) \in C_{set_i}$ is characterized by a unique name c_j and a set of attributes $\{attr_1, \ldots, attr_n\}$. If all the attributes of c_j are involved in the data operation, we write $c_j(*)$. Instead, if only a subset of attributes of c_j is accessed, we explicitly specify it by $c_j(attr_g, \ldots, attr_m)$ with $1 \leq g < m \leq n$.

- $A_{set_i} = \{a_1, \ldots, a_r\} \subseteq C_{set_i} \times C_{set_i} \subseteq As$ is a set of binary associations that directly link any two classes of C_{set_i} (i.e., the ends c_h and c_j of association a_i with $1 \leq i \leq r$ must belong to C_{set_i}). $A_{set_i} = \{*\}$ is the set of all associations that directly link any two classes of C_{set_i}. Moreover, the following consistency constraints hold: (i) If $|C_{set_i}| = 1$ and no (reflexive) association is accessed by the process, then $A_{set_i} = \varnothing$; (ii) If $|C_{set_i}| > 1$, then $A_{set_i} \neq \varnothing$ and all classes that are ends of $a_f \in A_{set_i}$ and $c_f \in AsC$ related to a_f must belong to C_{set_i};

- $AccessType_i \in \{R, I, D, U\}$ defines the type of access to the related information. R denotes a *read* of elements of C_{set}, whereas I, D, and U denote different kinds of write operations, namely I indicates an *insertion*, D stands for a *deletion*, while U denotes an *update*.

- $AccessTime_i \in \{start, end, during\}$ denotes when a data operation is performed with respect to activity execution. This qualitative information refines the description of data operations by specifying the moment they are performed. Access time defines a partial order among Activity View tuples.

- $NumInstances_i = (min, max)$, where $min \in \{0, 1, *\}$ and $max \in \{1, *\}$, denotes the number of objects involved in the considered operation. Indeed, activities must be able to access collections of objects and cardinalities should be properly managed [7]. Values $0, 1, *$ have the same meaning as in UML multiplicity specification, where $*$ means "multiple objects".   ◇

Sitting in-between two well-established standards, the Activity View blends the concepts of activity, borrowed from BPMN, with those of class, attribute, and association taken from UML. Moreover, being defined independently from data stores and data objects the Activity View provides a clear representation of the area of a data schema used by an activity as well as of the data operations performed on it, thus addressing the open issues discussed in Sect. 2.

As an example, consider the previously described process of Fig. 1(a). Given the data schema of the exam registration repository shown in Fig. 1(b), the described data access operations can be formalized as follows:

$$av_{CheckAttendance} = \{\langle\{Student(matriculation), Exam(name, date), Registration(*)\},$$
$$\{registration\}, R, during, (1, *)\rangle\}.$$

$$av_{RegisterGrade} = \{\langle\{Student(matriculation), Course(name)\}, \{attendance\}, R, start,$$
$$(1, 1)\rangle, \langle\{Grade(*)\}, \varnothing, I, during, (1, 1)\rangle\}.$$

$$av_{RescheduleExam} = \{\langle\{Exam(*)\}, \varnothing, I, during, (0, 1)\rangle\}.$$

$av_{RegisterGrade}$					
Tuple	C_{set}	A_{set}	AccessType	AccessTime	NumInstances
t_1	$\{Student(matriculation), Course(name)\}$	$\{attendance\}$	R	start	$(1,1)$
t_2	$\{Grade(*)\}$	$\varnothing$	I	during	$(1,1)$

Fig. 2. Tabular representation of the Activity View for activity Register Grade.

Activities Conduct Exam and Inform Student do not have a related Activity View as they do not require access to persistent data.

For better readability, the tuples of one Activity View can be represented in a tabular form, as exemplified in Fig. 2 for activity Register grade.

Graphically, the link captured by the three described Activity Views can be visualized over a process diagram and a data schema as shown in Fig. 3. Dashed arrows connect activities to the portion of the data schema specified in the Activity View. The same colors are used for the activity border, the dashed lines that frame the accessed data classes, and the full lines that highlight associations. Connecting arrows are labeled with the information related to access type, access time, and number of objects involved in the operation. Whenever multiple tuples of one Activity View represent different data operations on the same area of the data schema, the dashed arrow connecting the activity with that area of the data schema may be associated to multiple labels.

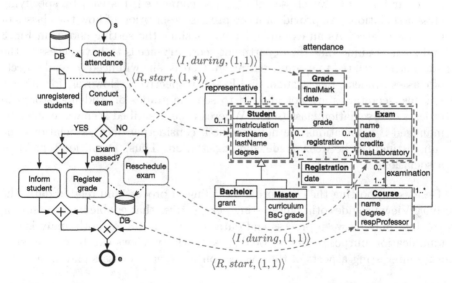

Fig. 3. Graphical representation of Activity Views for the introduced example. (Color figure online)

4   Novel Conceptual Insights

In this section we discuss some aspects of our research line that lead to the definition of the Activity View and show the novel perspectives that can be discovered by using the Activity View during process design and analysis.

To come up with the definition of Activity View presented in Sect. 3, we considered the following aspects related to linked processes and data.

> *Data classes and attributes.* Data classes define the conceptual objects of interest that are needed by a process activity. An Activity View allows designers to specify that only certain attributes of a class are read or written. This situation is quite common whenever the data schema represents a database that has not been specifically designed for process support. Moreover, the creation/update of a certain object may be realized by multiple activities, each one acting on a specific attribute [7]. Finally, when considering process roles and data access privileges, it is plausible that certain attributes may have restricted access and, thus, a data class may not necessarily be accessed as a whole.
>
> *Data associations and association classes.* Adding data associations to the specification of an Activity View changes the level of detail provided by the Activity View itself, especially for those data schemata having reflexive or multiple associations between any two classes. Specifically, if associations are not specified, the Activity View has a higher level of abstraction but it is not clear how any two classes of C_{set} are connected. Instead, by specifying class associations, we provide a more precise description of how the classes of C_{set} are related. As an example, let us consider the setting shown in Fig. 3 and let us assume that the department secretary needs to have access to the information related to exam registration, that is, she will need to read objects of classes Student, Registration, and Exam, respectively. However, for privacy reasons, secretaries are not allowed to see the grades of students, stored in objects of association class Grade. Whereas the described scenario cannot be managed if associations are not specified, considering data associations one can distinguish between grade and registration. Thus, Definition 1 includes associations.

Besides supporting the conceptual modeling of process and related data, the Activity View provides other interesting perspectives that enable basic reasoning on the interplay between process and data models, useful for both analysis and communication purposes. Below, we discuss Activity Views can be exploited to capture interesting aspects of the connection between processes and data.

- *Identifying the portion of a data schema accessed by a process activity.* As described in Sect. 3, the Activity View allows one to identify which are the classes and associations of a data schema that are accessed by a certain activity, thus providing a better specification than BPMN data stores. To identify the portion of the data schema accessed by an activity ac_k, all the

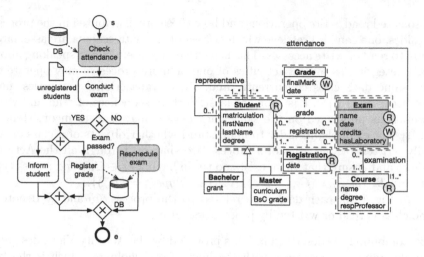

Fig. 4. Visualization of the conceptual insights provided by using Activity Views.

tuples $t_{1,k}, \ldots, t_{n,k}$ of av_{ac_k} must be properly combined: the comprehensive set of classes and association classes of a data schema accessed by ac_k is $\bigcup_{j=1}^{n} C_{set_{j,k}}$, where j denotes the tuple and k the activity, while the set of all associations accessed by ac_k is $\bigcup_{j=1}^{n} A_{set_{j,k}}$. In Figs. 3 and 4, the area of interest of the data schema is graphically rendered by framing classes with dashed lines colored as the border of the activity that access them.

- *Detecting which activities operate on a certain data class.* Under a different standpoint, Activity Views allow one to understand which process activities have access to objects of a certain data class. This is useful for several reasons, starting from easing the communication with domain experts during process modeling. Stakeholders are often interested in seeing where certain data are used in the process to understand which is the information that drives certain activities and used to make decisions. This holds also for data compliance. Indeed, in some circumstances, the quality of activity execution may drastically improve if proper information is available. Under an engineering perspective, understanding how data are used during process execution provides hints for data management support and re-engineering.

  For example, class Exam of Fig. 4 is accessed by tasks Check attendance and Reschedule exam, as highlighted by the filled background. By taking a look at the structure of the process, one can see that if the student succeeds, class Exam is only accessed at the beginning of the process.

  The set of process activities $ac_g, \ldots, ac_l$ that access a certain class c_i is $\{ac_k | \exists t_{j,k} (c_i \in C_{set_{j,k}})\}$ and it is obtained by going through all the Activity Views of a process and checking whether c_i belongs to at least one class set $C_{set_{j,k}}$ of a tuple $t_{j,k} \in av_{ac_k}$.

- *Understanding which classes are either read or written by a process.* The type of access to data allows designers to easily visualize whether data classes have associated read/write operations and how these are distributed in the process. Besides, one can also retrieve which classes of the data schema are associated only to read or write accesses. This is particularly useful when speaking about data integrity, as several activities of one or more processes may operate on the same data class concurrently and, thus, transactional properties must be discussed [13,18]. Last but not least, certain sequences of read and write operations performed on the same data classes may lead to inconsistencies in [2]. For example, in order to understand whether objects of a class c_k are only written by activities of a process, we shall go through all the Activity Views and ensure that there exists no tuple having $c_k \in C_{set}$ and access type of kind R: $\{c_k | \nexists j, i\left((C_{set_{j,i}} \ni c_k) \wedge AccessType_{j,i} = \text{"R"}\right)\}$.

In Fig. 4, for each data class related to the process, Ⓡ and Ⓦ denote if the class is read or written by process activities.

By combining the described insights provided by the Activity View, designers can understand and visualize, with the help of stakeholders, which is the key information needed to support process execution. This can be represented by one or more data classes, which we refer to as *core classes* for a given process. Informally, given a data schema, a core class is a class of the data schema that represents valuable process-related data and (i) it appears in a considerable number of Activity Views related to the process (i.e., it is shared by multiple process activities); (ii) its objects are frequently accessed by the process, that is, it appears in a considerable number of Activity View tuples; (iii) its objects are used by the most important activities of the process (i.e., activities that are crucial for the chosen application domain or are executed in (almost) all the instances of a process, if any); (iv) it is mostly subjected to mandatory access [7], that is, Activity View attribute $min \in NumInstances$ is never equal to 0.

With respect to the process of Fig. 4, classes Exam and Student are core classes, as they are the most accessed in the process. As for their use, they are accessed by exactly the same activities, but class Student is only read by process activities. Of course, to determine whether a read-only access is less important than a write access, domain experts should be consulted, as the idea is to exploit Activity Views in any way, to retrieve information useful for conceptual design. Indeed, whereas in such a simple example the identification of important data is quite straightforward, the concept of core classes becomes useful in complex and highly branched processes, where identifying the key information to support process execution is not straightforward. This latter issue is also open in the field of data-centric process modeling, since the same questions need to be answered to identify the data artifacts on which the processes are based [6,18].

5   Experimental Evaluation: Design and Results

This section describes the experimental evaluation of our approach and discusses the obtained results.[1]

---

[1] A complete description of the experiment can be found at http://hdl.handle.net/11562/976919.

Experiment Planning and Design. In order to analyze the usability of the Activity View, we conducted a human-oriented single factor experiment by following and combining the design principles described in [9, 10, 19].

The chosen factor is the *Activity View*, which represents our controlled variable, with factor levels *present* and *absent*. Specifically, we evaluated how the use of Activity Views can improve both the modeling and the understanding of the interplay between a process and a related database. In order to analyze such improvement quantitatively and qualitatively, we formulated two hypotheses.

1. *Perceived ease of understanding.* The Activity View improves the conceptual design of processes that operate on persistent data in terms of improved understanding of which data are needed by activities to be executed how they are used in the process. Improved understanding is quantified as better task performance in terms of increased speed and reduced error rate.
2. *Perceived ease of use.* It evaluates the ease of using the Activity View, that is, it assesses whether the Activity View can be easily read, understood, used, written, and adapted to different application domains.

Subjects are 21 students enrolled in the M.Sc. degree in Computer Science Engineering, 8 students enrolled in the M.Sc. degree in Medical Bioinformatics, and 4 researchers in the field of database design. All of the 33 subjects attended at least one information system course where BPMN is explained (about 8 teaching hours), and at least one complete database course (48 teaching hours). Among these, 8 subjects have working experience in the field of UML-based database design, whereas none of them has worked with BPMN at a professional level.

The experimental evaluation is organized as shown in Fig. 5. During *PHASE 1* the subjects attended a tutorial on the Activity View, where fundamental concepts and motivations were explained. Then, for both *PHASE 2* and *PHASE 3* subjects were asked to execute an experimental task on paper. *PHASE 2* was divided into two runs, where each run was based on a questionnaire containing 7 questions regarding diagrams insights (cf. Sect. 4). We used a within-groups approach, that is, we randomly divided the number of subjects into two groups,

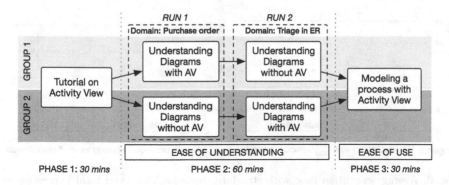

Fig. 5. Main steps of the experiment designed to evaluate the Activity View.

and each group performed the task with and without the Activity View. In detail, we provided all the subjects with a textual description of a process and its related data operations, and with the corresponding BPMN and UML diagrams. At each run, one group was also provided with the Activity Views related to the process and data diagrams. During the first run, "Group 1" was asked to execute the experimental task using also the provided Activity Views while "Group 2" was asked to execute the same task but relying only the textual description of the context and on the BPMN and UML diagrams. During the second run we switched groups and changed the application domain in order to avoid potential learning. The chosen domains were: Purchase order on a web-pharmacy for RUN 1 and triage in Emergency Room for RUN 2.

PHASE 3 was meant to test the actual Activity View ease of use during process modeling. All subjects were asked to model a BPMN process and to write the related Activity Views, given a textual description of the process and the UML class diagram of the referred domain database. We evaluated both the correctness of the designed processes and of the related Activity Views. Finally, we conducted a questionnaire-based interview to measure subjects perception the Activity View, both for process modeling and conceptual insight discovery.

Evaluation Results, Analysis, and Discussion. Overall, the obtained results confirmed that the Activity View improves the integrated design and understanding of processes and related data, both in terms of reduced task times and increased answers correctness.

The task executed during *PHASE 2* allowed us to quantitatively evaluate the use of the Activity View for diagram analysis. For each subject, we measured the task execution time and counted how many of the questions were answered correctly. We applied the most restrictive requirements for correctness, that is, answers were considered correct if answers were both right and complete. Finally, we analyzed the obtained results statistically with a paired t-test, where the execution times of the tasks carried out by one subject with Activity View were

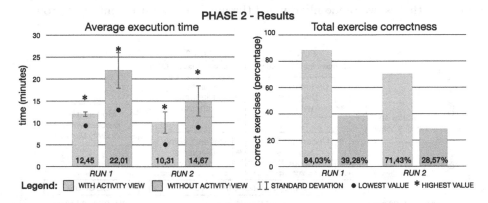

Fig. 6. Average execution time with standard deviation (left) and total percentage of correct answers (right) for the two runs of *PHASE 2*.

compared with those of the same subject without Activity View. In the first run, subjects provided with the Activity View, took an average of 12,45 min and the 84,03% of the answers was evaluated correct. Instead, the group without Activity View took an average of 22 min to complete the task, and only the 39,28% of the answers was correct. Results related to the second run showed a reduction in answering times for both groups, especially for the those not having the Activity View, as shown in Fig. 6. Accordingly, subjects claimed that they learned what the questions were asking for. However, the correctness of the answers also decreased. By combining the results of both runs, we see that by using the Activity View task times decrease by 37,94% on average, while the number of correct answers improves by 43,80%. The paired t-test applied to the difference of execution times and correctness obtained with the Activity View and without it revealed a p-value $< 0,001$. Thus, the obtained results, sketched in Fig. 6, are statistically significant and the first hypothesis is satisfied.

The experimental task of *PHASE 3* was reviewed by assigning one point for each correctly written Activity View tuple, thus considering each attribute of the tuple worth 0.20 points. Besides, we also considered the correctness of the BPMN process. Overall, the 58,89% of the written Activity Views was correct and the 83,94% of the BPMN process diagrams was designed properly. The percentage of correct Activity Views increases to 61,11% when excluding mistakes related to the access time, which was not easily determinable by reading the exercise.

The results of the modeling exercise are coherent with the outcome of the interviews conducted at the end of *PHASE 3*. Indeed, all the subjects declared that executing the first experimental task without the help of the Activity View was more difficult, and the 93% of them answered positively when asked whether the Activity View improves the modeling of the link between processes and related data. Then, we asked questions about to the use of the Activity View based on a rating scale from 1 to 5, with 1 meaning "strongly disagree" and 5 denoting "strongly agree". The average results of this questionnaire-based interview are reported in Fig. 7. Despite writing results harder, overall the Activity View was perceived as more than satisfactory, thus supporting our second hypothesis.

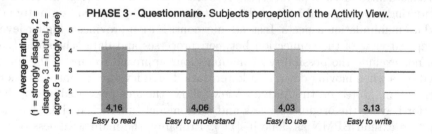

Fig. 7. Average rating of subjects perception of the Activity View.

Being most of the subjects computer science students, our evaluation app-roach is limited with respect to being generalizable to real organizational envi-ronments, where processes are more complex and people receive professional training. Indeed, exercises were designed in a didactic way to avoid task misun-derstanding. Moreover, results are limited to the kinds of asked questions and rely on the preparation of the students in the fields of process and data modeling.

6   Related Work

The relationship between data and processes has been tackled by several research efforts within the fields of high-level Petri nets [8,13], activity-centric process modeling [3,11,17], and data-centric process design [6,7,18].

In [13] db-nets are proposed as a novel three-layered approach to combine colored Petri Nets and relational databases, which communicate through an intermediate data logic layer. However, this approach goes beyond the conceptual modeling of data needed by a process, as it focuses on the modeling and formal verification [4] of a "connected system", where an instance of a database is subjected to changes imposed by the control layer.

The lack of well-founded conceptual modeling frameworks supporting process and data integration is motivated by recent proposals in the field of activity-centric modeling paradigms [2,3,11,17], which remain by all means the most used in practice. In [3], a BPMN process diagram is linked through OCL (Object Constraint Language) expressions to the information model of the process that incorporates a class "Artifact" containing all process variables. In [17] a logical framework is defined describe data object manipulation and to explicitly rep-resent the interaction of a process with an underlying database, while in [11] the authors address the automatic derivation of SQL queries from annotated data objects to check data requirements for activity execution. Finally, the con-nection between process and data diagrams is tackled by previous research [2], where a very simple form of Activity View is introduced to detect potential inconsistencies between data classes accessed by multiple activities.

Despite recognizing the need of linking processes and data conceptually, the approaches introduced in [3,11,17] address this issue at a lower (logical) level, by considering process variables and data instances, and by focusing on automat-ing data manipulation aspects. Instead, our contribution provides a higher-level, conceptual view of the connection between processes and data schemata, that does not exclude the possibility of mapping our approach to any of the intro-duced ones when moving down to a lower data design level that considers query specification. Moreover, the Activity View follows the idea presented in [13] that calls for leaving the original process and data models untouched (i.e., it is not meant to extend BPMN as done in [11]). Probably, the main weakness of the Activity View is its graphical representation, which suffers from the proliferation of connections. However, this issue seems to affect also the proposals in [11,13], as the number of data objects, respectively view places and labeled transitions, tends to increase with higher numbers of data instances and operations.

7   Conclusion

Bridging the gap between process and data models becomes necessary to support process designers in understanding the structure and semantics of conceptual data related to a process. In this paper, we introduced, formalized, and evaluated the Activity View, a novel approach aimed to realize the connection between a process model and a conceptual data schema, while allowing designers to also detail data operations. Then, we showed how using the Activity View allows one to obtain interesting insights related to the connected perspectives.

For future work, we aim to embody more abstraction levels within the Activity View to capture peculiar features of sub-processes and complete processes. Currently, we are working towards improving the graphical representation of the Activity View and its integration in existing process modeling tools.

References

1. Calvanese, D., De Giacomo, G., Montali, M.: Foundations of data-aware process analysis: a database theory perspective. In: 32nd ACM SIGMOD Symposium on Principles of Database Systems (PODS), pp. 1–12. ACM (2013)
2. Combi, C., Oliboni, B., Weske, M., Zerbato, F.: Conceptual modeling of inter-dependencies between processes and data. In: ACM Symposium on Applied Computing (SAC), pp. 110–119. ACM (2018). https://doi.org/10.1145/3167132.3167141
3. De Giacomo, G., Oriol, X., Estañol, M., Teniente, E.: Linking data and BPMN processes to achieve executable models. In: Dubois, E., Pohl, K. (eds.) CAiSE 2017. LNCS, vol. 10253, pp. 612–628. Springer, Cham (2017). https://doi.org/10.1007/978-3-319-59536-8_38
4. De Masellis, R., Francescomarino, C.D., Ghidini, C., Montali, M., Tessaris, S.: Add data into business process verification: bridging the gap between theory and practice. In: AAAI, pp. 1091–1099. AAAI Press (2017)
5. Elmasri, R., Navathe, S.B.: Fundamentals of Database Systems. Pearson, Boston (2015)
6. Hull, R.: Artifact-centric business process models: brief survey of research results and challenges. In: Meersman, R., Tari, Z. (eds.) OTM 2008. LNCS, vol. 5332, pp. 1152–1163. Springer, Heidelberg (2008). https://doi.org/10.1007/978-3-540-88873-4_17
7. Künzle, V., Reichert, M.: PHILharmonicFlows: towards a framework for object-aware process management. J. Softw. Maintenance Evol. Res. Pract. **23**(4), 205–244 (2011)
8. Lenz, K., Oberweis, A.: Inter-organizational business process management with XML Nets. In: Ehrig, H., Reisig, W., Rozenberg, G., Weber, H. (eds.) Petri Net Technology for Communication-Based Systems. LNCS, vol. 2472, pp. 243–263. Springer, Heidelberg (2003). https://doi.org/10.1007/978-3-540-40022-6_12
9. Maes, A., Poels, G.: Evaluating quality of conceptual models based on user perceptions. In: Embley, D.W., Olivé, A., Ram, S. (eds.) ER 2006. LNCS, vol. 4215, pp. 54–67. Springer, Heidelberg (2006). https://doi.org/10.1007/11901181_6
10. Mehmood, K., Cherfi, S.S.-S.: Evaluating the functionality of conceptual models. In: Heuser, C.A., Pernul, G. (eds.) ER 2009. LNCS, vol. 5833, pp. 222–231. Springer, Heidelberg (2009). https://doi.org/10.1007/978-3-642-04947-7_27

11. Meyer, A., Pufahl, L., Fahland, D., Weske, M.: Modeling and enacting complex data dependencies in business processes. In: Daniel, F., Wang, J., Weber, B. (eds.) BPM 2013. LNCS, vol. 8094, pp. 171–186. Springer, Heidelberg (2013). https://doi.org/10.1007/978-3-642-40176-3_14

12. Meyer, A., Weske, M.: Extracting data objects and their states from process models. In: 17th International Enterprise Distributed Object Computing Conference (EDOC), pp. 27–36. IEEE (2013)

13. Montali, M., Rivkin, A.: DB-Nets: on the marriage of colored petri nets and relational databases. In: Koutny, M., Kleijn, J., Penczek, W. (eds.) Transactions on Petri Nets and Other Models of Concurrency XII. LNCS, vol. 10470, pp. 91–118. Springer, Heidelberg (2017). https://doi.org/10.1007/978-3-662-55862-1_5

14. Object Management Group: Business Process Model and Notation (BPMN), v2.0.2. http://www.omg.org/spec/BPMN/2.0.2/

15. Object Management Group: Unified Modeling Language, v2.5. http://www.omg.org/spec/UML/2.5/

16. Reichert, M.: Process and data: two sides of the same coin? In: Meersman, R., et al. (eds.) OTM 2012. LNCS, vol. 7565, pp. 2–19. Springer, Heidelberg (2012). https://doi.org/10.1007/978-3-642-33606-5_2

17. Smith, F., Proietti, M.: Reasoning on data-aware business processes with constraint logic. In: 4th International Symposium on Data-driven Process Discovery and Analysis (SIMPDA), pp. 60–75 (2014)

18. Sun, Y., Su, J., Wu, B., Yang, J.: Modeling data for business processes. In: 30th International Conference on Data Engineering (ICDE), pp. 1048–1059. IEEE (2014)

19. Wang, W., Indulska, M., Sadiq, S., Weber, B.: Effect of linked rules on business process model understanding. In: Carmona, J., Engels, G., Kumar, A. (eds.) BPM 2017. LNCS, vol. 10445, pp. 200–215. Springer, Cham (2017). https://doi.org/10.1007/978-3-319-65000-5_12

Interactive Data-Driven Process Model Construction

P. M. Dixit[1]($\boxtimes$), H. M. W. Verbeek[1], J. C. A. M. Buijs[1],
and W. M. P. van der Aalst[2]

[1] Eindhoven University of Technology, Eindhoven, The Netherlands
{p.m.dixit,h.m.w.verbeek,j.c.a.m.buijs}@tue.nl
[2] Rheinisch-Westfälische Technische Hochschule (RWTH), Aachen, Germany
wvdaalst@pads.rwth-aachen.de

Abstract. Process discovery algorithms address the problem of learning process models from event logs. Typically, in such settings a user's activity is limited to configuring the parameters of the discovery algorithm, and hence the user expertise/domain knowledge can not be incorporated during traditional process discovery. In a setting where the event logs are noisy, incomplete and/or contain uninteresting activities, the process models discovered by discovery algorithms are often inaccurate and/or incomprehensible. Furthermore, many of these automated techniques can produce unsound models and/or cannot discover duplicate activities, silent activities etc. To overcome such shortcomings, we introduce a new concept to interactively discover a process model, by combining a user's domain knowledge with the information from the event log. The discovered models are always *sound* and can have duplicate activities, silent activities etc. An objective evaluation and a case study shows that the proposed approach can outperform traditional discovery techniques.

Keywords: HCI · Process discovery · Process mining

1 Introduction

Process discovery, a sub-field of process mining, aims at discovering process models from event logs. Most discovery algorithms aim to do so *automatically* by learning patterns from the event log. Automated process discovery algorithms work well in settings where the event log contains all the necessary (e.g. noise free, complete) information required by the algorithm, and the language of the underlying model is about the same as the language of the models discovered by the discovery algorithm. However, in many real world scenarios this is not the case.

First, the discovered process models might explain the event logs extremely well, but may still be completely incomprehensible to the end user. Therefore, it is imperative to enable the user to have control over the process model being

J. C. Trujillo et al. (Eds.): ER 2018, LNCS 11157, pp. 251–265, 2018.
https://doi.org/10.1007/978-3-030-00847-5_19

discovered, thereby also enabling incorporation of domain knowledge during process discovery. Second, the process models discovered by discovery algorithms are constrained by the vocabulary of the language used for representing the model, i.e., representational bias [1]. That is, some process discovery algorithms may not discover silent activities (i.e., skippable activities), duplicate activities (i.e., activities that occur more than once) etc. Third, many discovery algorithms may discover process models which are unsound. A sound process model guarantees an option to execute each activity in the process at least once and the ability to reach the final state (thereby terminating the process) from any valid reachable state. In practical settings, unsound process models are often not interesting to the user and hence are discarded. Ideally the discovery algorithm should limit the search space to only sound process models. However, for many discovery techniques this is not the case.

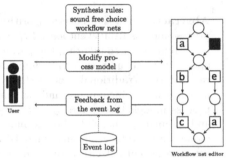

In the interactive process discovery approach presented in this paper and as shown in Fig. 1, the user discovers a process model incrementally. The user has total control over the discovery approach and can discover/model the process at a desired complexity level. Therefore, the user can balance simplicity, and to a certain extent, generalization of the process model (among other dimensions) in an efficient way. The interactive editor is further enhanced with process mining capabilities. Information from the event log is used to guide the user in modeling/discovering the process. The user can make informed decisions about where to place a particular activity, depending on the insights gained from the event log. The interactive approach gives the user total control over process discovery. Moreover, by default, the modeled/discovered processes are always in the realm of sound process models.

Fig. 1. Overview of interactive process discovery. The user can edit/discover a sound models guided by an event log.

The remainder of the paper is structured as follows. We review the related work in Sect. 2 and the preliminaries in Sect. 3 respectively. In Sect. 4 we discuss our approach followed by implementation overview in Sect. 5. In Sect. 6 we evaluate our approach and conclude in Sect. 7.

2   Related Work

In this section, we first review the state-of-the-art automated process discovery techniques followed by the user-guided process mining techniques.

Table 1. Representational bias of various process discovery and repair algorithms.

	α [2]	HM [3]	ILP [4]	SBR [5]	LBR [6]	ETM [7]	IM [8]	MBR [9]	IPD
Duplicate activities	−	−	−	+	−	+	−	−	+
Silent activities	−	+	−	+	−	+	+	+	+
Self loop activities	−	+	+	+	+	+	+	+	+
Non block structured	+	+	+	+	+	−	−	+	+
Classic soundness[a]	−	+	−	−	−	+	+	−	+

α: Alpha miner, HM: Heuristic miner, ILP: ILP miner, SBR: State based regions, LBR: Language based regions, ETM: Evolutionary tree miner, IM: Inductive miner, MBR: Model based repair, IPD: Interactive Process Discovery (this paper)
Classic soundness[a][10] is defined for the class of workflow nets.

2.1 Automated Process Discovery

Discovery algorithms which use the information from the event log about ordering relations in activities and their frequencies such as the α miner and the heuristic miner [2,3] are sensitive to noise. Alongside these, the techniques that use a semantic approach for discovery such as the state based region miner, numerical abstract domains, and the language based region miner [5,6,11] do not guarantee soundness. Furthermore, algorithms such as the ILP miner and the α miner [2,4] cannot discover silent activities. Genetic discovery techniques such as [7] have excessive run times and usually do not give any guarantees on any quality criterion in a short run. The Inductive Miner [8] can discover only block-structured process models and cannot discover duplicate activities. Techniques such as [12] focus on discovering complex process models represented as BPMN models containing sub-processes and events. Discovery techniques such as [13–15] discover either constraints based declarative models, or probabilistic models, and not an end-to-end process model.

Our approach differs from all these automated process discovery techniques in multiple ways (see Table 1 for an overview). The process models generated by our approach are always sound, since we use a synthesis rules kit (based on [16]) which guarantees soundness. In our approach, the user has control over the discovery (modeling) process, therefore addition of constructs such as duplicate activities, silent activities, (self-)loops etc. are all allowed as deemed appropriate by the user. Also, noisy (incomplete) information could be ignored (overcome) based on the extracted information from the event log presented to an informed user.

2.2 User-Guided Approaches

Of late, there has been an interest in using domain knowledge, along with the event logs, for process discovery [17–20]. However, the domain knowledge is usually represented by some pre-defined constructs or some sort of rules which are used as input during the process discovery. The language used to represent the domain knowledge severely limits the expressiveness of the domain expert. In our approach, the user has total control over the discovery phase and can intuitively use the domain knowledge, along with the event logs, to interactively visualize and discovery the process model. In [21], the authors provide a way for the user to include domain expertise in the α miner. However, this approach is tied to the underlying α algorithm, and thereby includes all the limitations of the α algorithm. AProMore [22] provides a host of techniques including interactive process model repair techniques [23]. However, the focus of these techniques is model repair using alignments, similar to [9]. These techniques have a different scope compared to our approach, whose focus is process model discovery.

3 Preliminaries

In this section, we introduce the so-called *activity logs* and *synthesized nets*.

3.1 Activity Logs

An *activity log* is a multi-set (or bag) of sequences of activities. Every sequence of activities in the activity log is called an activity trace. Table 2 shows an example activity log, which contains 20 activity traces, 7 different activity traces, 87 activities, and 6 different activities. In real life, we often see *event logs* instead of activity logs, where an event log is a (multi-)set of sequences of *events*. An event is typically a key-value pair containing values for several attributes that are related to the event, such as the time on which the event occurred and to which activity the event relates. Using the latter attribute, the mapping from such an event log to an activity log is then straightforward. As a result, although we use activity logs in this paper, the presented approach is also applicable on event logs.

Table 2. The example activity log

Activity trace	Freq
$\langle a, a, f, c \rangle$	8
$\langle a, f, a, e, c \rangle$	1
$\langle a, f, a, e, e, c \rangle$	1
$\langle b, d, e, c \rangle$	4
$\langle b, d, e, e, c \rangle$	1
$\langle b, e, d, c \rangle$	4
$\langle a, a, f, e, e, e, c \rangle$	1

3.2 Synthesized Nets

A *synthesized net* is a free-choice [16] workflow net [24] containing a source place i, a sink place o, a start transition $\top$, and an end transition $\bot$. Only the start transition may have an input arc from the source place, and only the end transition may have an arc to the sink place. The source and sink place are known from the workflow nets [24], but our approach also requires the additional start and end transition.

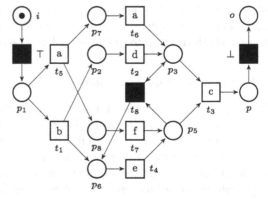

Fig. 2. The example synthesized net corresponding to Table 2.

We use the regular Petri net semantics [16] for the synthesized net. A transition is enabled if all its input places contain tokens. In the initial state of the example net, as shown in Fig. 2, only the $\top$ transition is enabled as only place i contains a token. An enabled transition may be fired, which removes a token from every input place and adds a token to every output place. Firing the enabled transition $\top$ in the example net results in the state where place p_1 contains one token and all other places are empty. This state enables the transitions t_1 and t_5, etc. Being a workflow net, the goal of the example net is to reach the state where only place o contains a token and all other places are empty, that is, the stop state of the net. The soundness property [10] guarantees that this stop state can always be reached.

Figure 2 also shows that we allow the transitions to be labeled (by activities, actually), but that we also allow a transition to be silent, that is, not labeled. As examples, the transition t_5 is labeled with activity a, and the transition t_8 is silent. The start and end transitions are required to be silent.

4 Approach

In this section, we discuss the approach used in order to enable interactive process discovery using the synthesis rules [16]. We begin with the initial net and the synthesis rules in the context of our approach.

4.1 Initial Net and Synthesis Rules

Figure 3 shows the initial synthesized net, which is the minimal synthesized net. As mentioned earlier, the places i and o, the transitions $\top$ and $\bot$, and the arcs connecting them are required, and we need the place p with its connecting arcs to make it a workflow net. We have three synthesis rules, where each of the rules is based on a synthesis rule from [16]: an abstraction rule, a place rule, and a transition rule.

The abstraction rule allows one to introduce a new place and a new transition, provided that we have a set of transitions R and a set of places S such that there is an arc from every transition in R to every place in S. Figure 4a shows a synthesized net that can be obtained from the initial net by applying this rule three times. First, place p_1 and transition t_1 are introduced using $R = \{\top\}$ and $S = \{p\}$. Second, place p_2 and transition t_2 are introduced using $R = \{t_1\}$ and $S = \{p\}$. Third, place p_3 and transition t_3 are introduced using $R = \{t_2\}$ and $S = \{p\}$.

Fig. 3. Initial net.

The other two synthesis rules use the fact whether a place or a transition is *linearly dependent*. For this, we need the *incidence matrix* of the net. A net can be short-circuited by removing the place o, and re-routing the outgoing arc from $\bot$ to i. Table 3 shows the incidence matrix of the short circuited version of Fig. 4b. In this matrix, every row vector corresponds to a place, and every column vector corresponds to a transition. The number at the intersection of a row vector and a column vector provides the net effect of firing that transition on that place. For e.g., the value for place p_1 and transition $\top$ is 1, as firing $\top$ adds a token to p_1.

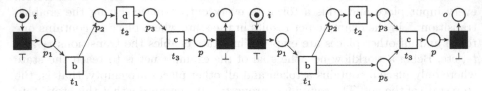

(a) A part of the example synthesized net.

(b) After having added the place p_5 using the place rule.

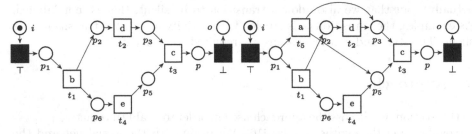

(c) After having applied the abstracton rule for the fifth time.

(d) After having added the transition t_5 using the transition rule.

Fig. 4. First steps in synthesis of the example synthesized net.

The place rule allows one to introduce a new place, provided that the new place is linearly dependent [16] on the existing places, from the incidence matrix. Figure 4b shows a synthesized net that can be obtained from the net in Fig. 4a by using a single application of this rule: the place p_5, which is linearly dependent on the places p_2 and p_3, has been introduced. In a similar way, the transition rule allows one to introduce a new transition, provided that the new transition is linearly dependent [16] on the existing transitions. Figure 4d shows a synthesized net obtained from Fig. 4c using a single application of this rule: the transition t_5, which is linearly dependent on the transitions t_1, t_2, and t_4, has been introduced. We can deduce the net of Fig. 2 from Fig. 4d by using two more abstraction rules and one more transition rule.

Table 3. Incidence matrix of 4b.

	$\top$	t_1	t_2	t_3	$\bot$
i	-1	0	0	0	1
p_1	1	-1	0	0	0
p_2	0	1	-1	0	0
p_3	0	0	1	-1	0
p	0	0	0	1	-1
p_5	0	1	0	-1	0

In [16] it has been proven that these rules preserve well-formedness [16], which is related to soundness but not exactly the same. The key difference is that for soundness we require a specific state (one token in the place i) whereas for well-formedness we do not require such a state. Hence, in our approach the place rule is restricted to forbid adding places which may preserve well-formedness but may not preserve soundness. The related intricacies are out of scope.

4.2   Using Activity Logs

In this section, we discuss how the information from the activity logs is linked to the synthesized nets. The activity logs are central for making decisions in all the automated process discovery techniques. Motivated by the usage of activity logs in state-of-the-art process discovery techniques, we derive three kinds of statistics, which guide the user in decision making. The synthesized net expands one transition and/or one place at a time. The user labels newly added transitions in the synthesized net with either an activity from the activity log, or the transition does not represent an activity, i.e. leaves the transition silent. The information from the activity log is aggregated in a pairwise manner between each activity from the synthesized net (labels of visible transitions) and the activity selected by the user to be added to the net, which assists the user in positioning the selected activity in the net. We first define the co-occurs value between two activities.

Definition 1 (Co-occurs ($C_{(a,b)}$)). *Let L be an activity log and let a and b be two activities that occur in L, the co-occurs value of (a, b), denoted $C_{(a,b)}$ is:*

$$C_{(a,b)} = \frac{|[\sigma \in L \mid a \in \sigma \wedge b \in \sigma]|}{|[\sigma \in L \mid a \in \sigma]|}$$

For a pair of activities a, b, if the co-occurs value is 0, then a and b do not occur together, whereas if the co-occurs value is 1 then b occurs if a occurs. It should be noted that, the co-occurs value is not commutative. This is because, the co-occurs value is calculated using the absolute occurrence of first activity in the denominator. Next, we define the eventually follows value, which indicates the number of times an activity is eventually followed by another activity.

Definition 2 (Eventually follows). *Let L be an activity log and let a and b be two activities that occur in L, and for a trace $\sigma \in L$, let $\#a >_\sigma b$ indicate number of occurrences of b after the first occurrence of a in σ. The eventually follows relationship between a and b for a trace σ, denoted $EF_\sigma(a, b)$, is:*

$$EF_\sigma(a, b) = \begin{cases} \dfrac{\#a >_\sigma b}{\#a >_\sigma b + \#b >_\sigma a} & , if \ \#a >_\sigma b \neq 0 \\[2mm] 0 & , otherwise. \end{cases}$$

The eventually follows value for a pair (a, b) w.r.t. the entire log L, denoted as $EF_L(a, b)$, is calculated as the average of all the traces which have a non-zero value for a pair, or is zero otherwise. If $EF_L(a, b) = 1$, then a and b co-occur, but a never occurs after a b, which hints that b should be following a. The *directly follows* relation w.r.t. activities a and b is calculated as follows:

Definition 3 (Directly follows ($DF_{(a,b)}$)). *Let L be an activity log and let a and b be two activities that occur in L, and for a trace $\sigma \in L$, let $\#a >_\sigma^d b$ indicate number of occurrences of b directly after the occurrence of activity a*

in σ. The directly follows relationship between a and b for a trace σ, denoted $DF_\sigma(a,b)$, *is:*

$$DF_\sigma(a,b) = \begin{cases} \dfrac{\#a >_\sigma^d b}{\#a >_\sigma^d b + \#b >_\sigma^d a} & , if\ \#a >_\sigma^d b \neq 0 \\[2ex] 0 & , otherwise. \end{cases}$$

The directly follows value for the entire log L, denoted as $DF_L(a,b)$, is calculated similar to $EF_L(a,b)$. Similarly, *eventually precedes* and *directly precedes* values are also calculated.

4.3   Activity Log Projection

In this section, we discuss the mechanism used for projecting information from the activity log on the synthesized net. The user selects an activity from the activity log, to be added to the synthesized net. Depending on the activity selected by the user, the coloring of the activities in the current synthesized net is updated. The colors indicate which activities (and to what extent) from the synthesized net occur before and/or after the selected activity. The projected information can be based either on the eventually follows (precedes) rela-

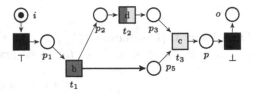

Fig. 5. Eventually follows/precedes projections on Fig. 4c, when the selected activity is e and the pairwise relations between activities are: $EF_L(b,e) = 1, EF_L(e,d) = 0.5, EF_L(e,c) = 1, C_{(e,d)} = 1, C_{(e,b)} = 1$ and $C_{(e,c)} = 1$. Purple (yellow) color indicates the degree to which selected activity occurs after (before) the activity represented by the transition. (Color figure online)

tion, or the directly follows (precedes) relation as desired by the user. The opacity of the colors indicate the co-occurrence of the two activities. All the information from the activity logs is also presented in a tabular format to the user. Thereby, in situations where the projected visualizations seem vague, the user can use the information from the tables for making an informed decision.

Figure 5 shows the projection on transitions when an activity e is selected by the user. The degree of *purple (yellow)* color in a transition indicates that the selected activity occurs *after (before)* the activity represented by the transition. As transition t_1 is completely colored purple (darker), we know that activity e occurs after activity b. Likewise, as transition t_3 is completely colored yellow (lighter), we know that activity e occurs before activity c. In contrast, as transition t_2 is fifty-fifty colored purple and yellow we know that activity e occurs about equally often before and after activity d. The opacity of the coloring indicates the co-occurrence values of the activity chosen and the activities represented by the transitions. Based on these insights, it is clear that activity e must be added in parallel to d, before c and after b, i.e., using the abstraction rule on the thicker edge in Fig. 5. Multiple transitions having the same label would also have the same coloring. If a transition is colored white, it implies that the activity selected and the activity represented by the transition never co-occur together. Furthermore, the user is also presented with raw information pertaining to the selected activity such as *% of the traces in which the selected activity occurred, average occurrence of the selected activity in a trace* etc.

5 Interactive Editing and Implementation

In this section, we discuss the implementation and user interaction details of our technique. The proposed technique is implemented and available in the "Interactive Process Mining" package in the process mining toolkit ProM[1]. The user interacts with a synthesized net (starting with the initial net), in order to deduce new nets, by applying one of the synthesis rules. In order to use the abstraction rule, the user clicks on a (set of) arc(s), and presses *enter*. The selected arcs are highlighted in green. The abstraction rule allows addition of a new place and a new transition in between a set of transitions and a set of places. The (optional) activity label of the new transition is pre-selected by the user, after which the rule is applied.

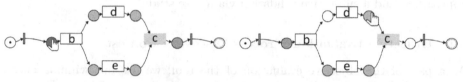

(a) Possible outputs when the place before *b* is selected. Dark grey colored place indicates possibility of a *self-loop*, i.e. a transition having same input and output. Blue colored places indicate candidate transition with multiple outputs, for the selected input place. Each green colored place indicate candidate with a single output place for the selected input place(s).

(b) After selecting the place after *d* as an output place. The input place (i.e. the place before *b*) is turned red, indicating it is no longer possible to use this place to add a *self-loop* transition. The green places are filtered out, as such candidate transitions are no longer valid for the selected input-output place combination. Moreover, the place before *d* is also colored white as there is no candidate transition which contains both the place after *d* and the place before *d* in its output places.

Fig. 6. User interaction for adding linearly dependent transitions in our tool. (Color figure online)

In the case of linear dependency rules, all possible applications of a rule are projected on the synthesized net based on the user interaction. We explain this with the help of an example from Fig. 6, which shows a screen shot of our tool (corresponding to Fig. 4c) and an application of the transition rule, when activity *a* is selected by the user. Note that activities *b, d* and *e* are colored white and activity *c* is colored yellow. Hence we would like to place activity *a* as an alternative to *b, d* and *e* and before *c*. Re-collect that the transition rule allows addition of a linearly dependent transition to a synthesized net. All the candidate linearly dependent transitions are pre-computed. This set of

[1] http://www.processmining.org/prom/start.

candidate transitions is finite, and independent of the activity label chosen by the user. Whenever a user navigates on a place, all the candidate transitions are explored and only those candidate transitions are presented which have the navigated place as an input place. For all such candidate transitions, the output places (and possible additional input places) are projected on the synthesized net, using color coding. The user first selects the desired input places of the candidate transition. The candidate transitions are filtered based on the selected input places, and the output places of the filtered candidate transitions are highlighted. Next, the user chooses output places. In case of multiple output places for a candidate transition, the user clicks on one of the desired multiple output place. The candidate transitions are further filtered based on the output places chosen. When a user has selected enough input and output places to pinpoint a single candidate transition, the selected candidate transition is added to the synthesized net. A similar approach is followed for the place rule.

6   Evaluation

In order to validate the approach presented in this paper, we presented two types of evaluations: an objective comparison with traditional process discovery approaches, and a subjective validation via a case study.

6.1   Objective Evaluation: Process Discovery Contest

As a part of the objective evaluation of the tool, we use our winning entry from the annual process discovery contest[2], organized at the BPI workshop of BPM 2017 conference. The aim of the process discovery contest is to evaluate tools and techniques which are able to discover process models from incomplete, infrequent event logs. In total, 10 process models had to be discovered from 10 corresponding event logs. Every model could contain sequences, choices and concurrent activities. Furthermore, the organizers provided participants with additional information about each process model, such as the presence of duplicate activities, loops, optional activities, inclusive choices and long-term dependencies. We consider this additional information as the *domain knowledge* in our approach. Every final process model was evaluated based on an unseen test log containing 20 traces, as well as by a jury of BPM practitioners to rate the simplicity and understandability of the discovered process models.

In order to keep the comparison across various discovery techniques fair, in this section we focus on 2 process models: model numbers 2 and 10. Other process models (1 and 3–9) contained artifacts such as inclusive choices. Even though our approach is able to model such artifacts, many state-of-the-art techniques can not. Both the training logs for models 2 and 10 are incomplete logs, and contain sequential, concurrent, exclusive choices and optional activities. Furthermore, model 2 contains loops and model 10 contains duplicate activities. We used

---

[2] https://www.win.tue.nl/ieeetfpm/doku.php?id=shared:process_discovery_contest.

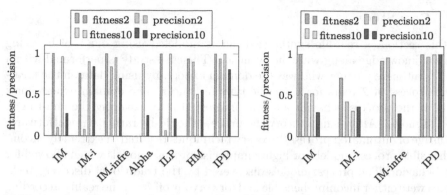

(a) Discovered models compared with training activity log using [25].

(b) Discovered models compared with the original models used to generate the activity logs of the process discovery contest, using [26].

Fig. 7. Discovery techniques - IM: Inductive Miner, IM-i: Inductive Miner incomplete, IM-infreq: Inductive Miner infrequent, Alpha miner, ILP Miner, HM: Heuristic Miner, IPD: Interactive Process Discovery - this paper.

this information in combination with simple log visualizers, in order to discover process models with our tool. Both the process models discovered were able to replicate all the behavior from the test activity logs. Furthermore, over all the 10 models, we received 98.5% accuracy w.r.t. the test activity logs: of the 200 test traces we had to classify, we classified 197 correctly.

In Fig. 7a we compare the fitness and precision scores of the training activity log used to discover the process model, with the discovered process model, using a prefix-based alignment technique [25]. For model 2, the α miner resulted in a model which could not be replayed using [25]. Furthermore, in Fig. 7b we compare the discovered process models with the original process models which are used to generate the activity logs (both training and test logs). This comparison is done using the technique from [26], which supports only the process models discovered by inductive miner variants and our approach. The fitness value indicates the part of the behavior of the activity log (or the original process model) that is captured by the discovered process model. The precision value on the other hand captures the part of behavior of the discovered model that is also present in the activity log (or the original process model). As evident from Fig. 7a and b, our approach typically outperforms all the automated discovery techniques. By using the properties of the net (such as possibility of loops, duplicates etc.) we were able to interactively discover process models which are strikingly similar to the original process models. For example, Fig. 8 shows the synthesized net we discovered from activity log 10. The only real difference when compared to the original process model is the incorrect presence of the silent transition just below the transition labeled f. Furthermore, Fig. 7 also demonstrates the ability of our approach to cope with complex constructs such as duplication of activities, silent activities etc., where many state-of-the-art techniques falter, as discussed in Table 1.

6.2   Case Study

In this section, a real-life synthesized net is modeled using our tool by using domain knowledge along with the event log. The case study is performed with a local healthcare partner who was the domain expert, by using data on the treatment process of 2 years for a specific type of cancer. The domain expert knew parts of the process behavior and was interested in discovering the end-to-end process model. Although in theory the process should be rather straightforward, the usage of automated process discovery techniques resulted in extremely incomprehensible process models, or highly imprecise process models which allowed for any behavior. The process models discovered by the traditional discovery techniques were either incomprehensible and/or very far off from the reality according to the domain expert. Therefore, the interactive process discovery approach was used to try to structure the process data by using domain knowledge.

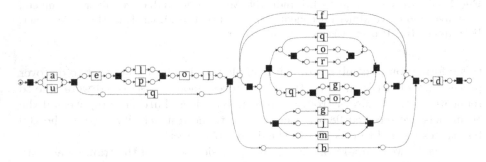

Fig. 8. Synthesized net 10 of the process discovery contest discovered using our approach.

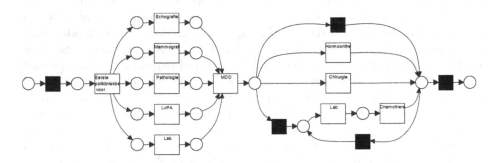

Fig. 9. Synthesized net for cancer patients in a Dutch hospital as discovered by the domain expert using our tool.

The synthesized net discovered using our approach is shown in Fig. 9. The domain expert had complete control about the modeling of the process. On several occasions, the domain expert took assistance from insights of the activity log gained via our technique. For e.g., the domain expert was not entirely sure if *LvPA* should be placed before or after *MDO*. However, after gaining insights from the data, the domain expert decided to add *LvPA* before *MDO*. On some other occasions, the domain expert chose to ignore the information from the data, deeming it inappropriate and/or inadequate. Finally, using the information from the event log and some background knowledge, the domain expert was able to discover, and was satisfied with, a very structured process model.

The interactively discovered process model had a low fitness with the activity log. Investigating the causes for the low fitness led to the conclusion that there were some serious data quality issues in the activity log, rather than non-compliance of protocols. Data quality problems are often the reason why all the automated discovery algorithms fail to discover a structured process model. However in our case, by not relying completely on an event log for process discovery, and using knowledge from a domain expert, a structured process model was discovered. Moreover, since the domain expert was directly involved in process modeling/discovery, the simplicity and generalization dimensions of the process model were implicitly taken into account.

7   Conclusion and Future Research

In this paper, we presented the concept of interactive discovery of a process model based on synthesis rules. The use of synthesis rules as an engine to expand the synthesized net guarantee soundness of the discovered process model. Furthermore, the information from the event log is extracted and presented to the user to assist the user in decision making. Giving users complete control over the discovery approach supported by the information from the event log enables critical decision making. This is true especially when all the information needed by the discovery algorithms is not present in the event log; which is often the case in many real-life event logs as apparent from the case study presented. The automated discovery algorithms fail to cope with insufficient information in the event log, and could produce process models which are incomprehensible and/or inaccurate. Moreover, our approach is able to discover constructs such as duplicate activities, inclusive choices and silent activities, that cannot be discovered by many state-of-the-art techniques. In the future, we aim to improve the assistance provided to the user in decision making. One future direction would be to provide online conformance results to the user during process discovery. Another future direction could be pre-populating a number of options to be presented to the user to add a particular activity.

References

1. van der Aalst, W.M.P.: On the representational bias in process mining. In: 2011 20th IEEE International Workshops on Enabling Technologies: Infrastructure for Collaborative Enterprises (WETICE), pp. 2–7, June 2011
2. van der Aalst, W.M.P., Weijters, A.J.M.M., Maruster, L.: Workflow mining: discovering process models from event logs. IEEE Trans. Knowl. Data Eng. **16**(9), 1128–1142 (2004)
3. Weijters, A.J.M.M., van der Aalst, W.M.P.: Rediscovering workflow models from event-based data using little thumb. Integr. Comput. Aided Eng. **10**(2), 151–162 (2003)
4. van der Werf, J.M.E.M., van Dongen, B.F., Hurkens, C.A.J., Serebrenik, A.: Process discovery using integer linear programming. In: van Hee, K.M., Valk, R. (eds.) PETRI NETS 2008. LNCS, vol. 5062, pp. 368–387. Springer, Heidelberg (2008). https://doi.org/10.1007/978-3-540-68746-7_24
5. Cortadella, J., Kishinevsky, M., Lavagno, L., Yakovlev, A.: Deriving Petri nets from finite transition systems. IEEE Trans. Comput. **47**(8), 859–882 (1998)
6. Bergenthum, R., Desel, J., Lorenz, R., Mauser, S.: Process mining based on regions of languages. In: Alonso, G., Dadam, P., Rosemann, M. (eds.) BPM 2007. LNCS, vol. 4714, pp. 375–383. Springer, Heidelberg (2007). https://doi.org/10.1007/978-3-540-75183-0_27
7. Buijs, J.C.A.M., van Dongen, B.F., van der Aalst, W.M.P.: A genetic algorithm for discovering process trees. In: 2012 IEEE Congress on Evolutionary Computation (CEC), pp. 1–8. IEEE (2012)
8. Leemans, S.J.J., Fahland, D., van der Aalst, W.M.P.: Discovering block-structured process models from event logs containing infrequent behaviour. In: Lohmann, N., Song, M., Wohed, P. (eds.) BPM 2013. LNBIP, vol. 171, pp. 66–78. Springer, Cham (2014). https://doi.org/10.1007/978-3-319-06257-0_6
9. Fahland, D., van der Aalst, W.M.P.: Repairing process models to reflect reality. In: Barros, A., Gal, A., Kindler, E. (eds.) BPM 2012. LNCS, vol. 7481, pp. 229–245. Springer, Heidelberg (2012). https://doi.org/10.1007/978-3-642-32885-5_19
10. van der Aalst, W.M.P., et al.: Soundness of workflow nets: classification, decidability, and analysis. Form. Asp. Comput. **23**(3), 333–363 (2011)
11. Carmona, J., Cortadella, J.: Process discovery algorithms using numerical abstract domains. IEEE Trans. Knowl. Data Eng. **26**(12), 3064–3076 (2014)
12. Conforti, R., Dumas, M., García-Bañuelos, L., La Rosa, M.: Beyond tasks and gateways: discovering BPMN models with subprocesses, boundary events and activity markers. In: Sadiq, S., Soffer, P., Völzer, H. (eds.) BPM 2014. LNCS, vol. 8659, pp. 101–117. Springer, Cham (2014). https://doi.org/10.1007/978-3-319-10172-9_7
13. Chesani, F., Lamma, E., Mello, P., Montali, M., Riguzzi, F., Storari, S.: Exploiting inductive logic programming techniques for declarative process mining. In: Jensen, K., van der Aalst, W.M.P. (eds.) Transactions on Petri Nets and Other Models of Concurrency II. LNCS, vol. 5460, pp. 278–295. Springer, Heidelberg (2009). https://doi.org/10.1007/978-3-642-00899-3_16
14. Bellodi, E., Riguzzi, F., Lamma, E.: Probabilistic declarative process mining. In: Bi, Y., Williams, M.-A. (eds.) KSEM 2010. LNCS (LNAI), vol. 6291, pp. 292–303. Springer, Heidelberg (2010). https://doi.org/10.1007/978-3-642-15280-1_28
15. Breuker, D., Matzner, M., Delfmann, P., Becker, J.: Comprehensible predictive models for business processes. MIS Q. **40**(4), 1009–1034 (2016)

16. Desel, J., Esparza, J.: Free Choice Petri Nets, vol. 40. Cambridge University Press, New York (2005)
17. Greco, G., Guzzo, A., Lupa, F., Luigi, P.: Process discovery under precedence constraints. ACM Trans. Knowl. Discov. Data **9**(4), 32:1–32:39 (2015)
18. Maggi, F.M., Mooij, A.J., van der Aalst, W.M.P.: User-guided discovery of declarative process models. In: 2011 IEEE Symposium on Computational Intelligence and Data Mining (CIDM), pp. 192–199. IEEE (2011)
19. Dixit, P.M., Buijs, J.C.A.M., van der Aalst, W.M.P., Hompes, B.F.A., Buurman, J.: Using domain knowledge to enhance process mining results. In: Ceravolo, P., Rinderle-Ma, S. (eds.) SIMPDA 2015. LNBIP, vol. 244, pp. 76–104. Springer, Cham (2017). https://doi.org/10.1007/978-3-319-53435-0_4
20. Rembert, A.J., Omokpo, A., Mazzoleni, P., Goodwin, R.T.: Process discovery using prior knowledge. In: Basu, S., Pautasso, C., Zhang, L., Fu, X. (eds.) ICSOC 2013. LNCS, vol. 8274, pp. 328–342. Springer, Heidelberg (2013). https://doi.org/10.1007/978-3-642-45005-1_23
21. Mathern, B., Mille, A., Bellet, T.: An Interactive Method to Discover a Petri Net Model of an Activity. working paper or preprint, April 2010
22. La Rosa, M., et al.: Apromore: an advanced process model repository. Expert. Syst. Appl. **38**(6), 7029–7040 (2011)
23. Armas-Cervantes, A., van Beest, N.R.T.P., La-Rosa, M., Dumas, M., García-Bañuelos, L.: Interactive and incremental business process model repair. In: Panetto, H. (ed.) On the Move to Meaningful Internet Systems. LNCS, vol. 10573, pp. 53–74. Springer, Cham (2017). https://doi.org/10.1007/978-3-319-69462-7_5
24. van der Aalst, W.M.P.: The application of Petri nets to workflow management. J. Circuits Syst. Comput. **8**(01), 21–66 (1998)
25. Adriansyah, A., van Dongen, B.F., van der Aalst, W.M.P.: Towards robust conformance checking. In: zur Muehlen, M., Su, J. (eds.) BPM 2010. LNBIP, vol. 66, pp. 122–133. Springer, Heidelberg (2011). https://doi.org/10.1007/978-3-642-20511-8_11
26. Leemans, S.J.J., Fahland, D., van der Aalst, W.M.P.: Scalable process discovery and conformance checking. Softw. Syst. Model. **08**, 1374–1619 (2016)

Spatio-Temporal Modeling

Efficient Multi-range Query Processing on Trajectories

Munkh-Erdene Yadamjav[1]([✉]), Farhana M. Choudhury[1], Zhifeng Bao[1],
and Hanan Samet[2]

[1] School of Science, RMIT University, Melbourne, Australia
munkh-erdene.yadamjav@rmit.edu.au
[2] Department of Computer Science, University of Maryland,
College Park, MD 20742, USA

Abstract. With the widespread use of devices with geo-positioning technologies, an unprecedented volume of trajectory data is becoming available. In this paper, we propose and study the problem of multi-range query processing over trajectories, that finds the trajectories that pass through a set of given spatio-temporal ranges. Such queries can facilitate urban planning applications by finding traffic movement flows between different parts of a city at different time intervals. To our best knowledge, this is the first work on answering multi-range queries on trajectories. In particular, we first propose a novel two-level index structure that preserves both the co-location of trajectories, and the co-location of points within trajectories. Next we present an efficient query processing algorithm that employs several pruning techniques at different levels of the index. The results of our extensive experimental studies on two real datasets demonstrate that our approach outperforms the baseline by 1 to 2 orders of magnitude.

Keywords: Spatio-temporal index · Multi-range query
Spatial database

1 Introduction

With the proliferation of GPS enabled devices, the spatio-temporal positions of the moving objects (e.g., cars, public transports, pedestrians, etc.), known as trajectory data, are being generated at an unprecedented rate.

Such information facilitates effective planning of urban area and intelligent infrastructure management. As an example, suppose that an urban planning authority wants to find the trajectories that involve travel from Richmond (a suburb in Melbourne) to the Melbourne city center via Kensington (another suburb in Melbourne) during office start hours, i.e., 7 am to 9.30 am. The result of such queries can facilitate the transportation authority to explore movement flows between different parts of a city at different time intervals and identify traffic congestion to improve road infrastructure and traffic throughput.

© Springer Nature Switzerland AG 2018
J. C. Trujillo et al. (Eds.): ER 2018, LNCS 11157, pp. 269–285, 2018.
https://doi.org/10.1007/978-3-030-00847-5_20

In this paper, given a set of spatio-temporal ranges, we address the problem of finding the trajectories that pass through each of the query ranges, where a query range consists of a spatial region (e.g., a rectangle or a circle) and a time interval. In this work, we consider both the spatial-only and spatio-temporal trajectories. We denote this problem as a *Trajectory multi-range query (MRQ)*. The size of the spatio-temporal query range can be arbitrary and does not have any pre-defined hierarchy [11]. Thus, it is difficult to precompute the intersections of all possible spatial range combinations.

There are several studies on trajectory data indexing [1,2,4,6,20] and processing different types of queries, such as finding regions of interest [18], popular routes [3,9,16], similarity etc. In the literature, the studies on answering *trajectory range queries* to find the trajectories that pass through a single spatial region during a specific time interval [2,10,12,19,21,22] are the closest to our work. However, the indexes and approaches for single range queries are not designed to capture the trajectory movements from one region to another, and thus they suffer from several drawbacks to answer a multi-range trajectory query (see Sect. 3.2 for details).

To overcome the drawbacks of existing studies, we propose a novel two-level index structure to maintain the trajectory movement relation between different regions and time intervals. The key idea of our index structure is to store the close-by trajectories together in the first level of the index, and maintain the close-by points of the trajectories together in the second level of the index. Such structure enables us to retrieve only the necessary trajectories that pass through all the query ranges in the given query time intervals efficiently. We denote this index as the *Location preserving two-level Trajectory Index (LoTI)*. We propose several pruning steps over this index as our query processing approach.

The contributions of the paper are as follows:

- We propose and study the problem of efficiently processing multi-range queries on trajectories that can be beneficial to many real life applications.
- We propose a novel two-level index structure, *LoTI* (Sect. 4.1) that supports storing the co-located trajectories together, and at the same time, stores the co-located points of trajectories together. We present several strategies to prune the unnecessary trajectories and points of the trajectories efficiently using this structure (Sect. 4.2).
- We evaluate our algorithms through an extensive experimental study on real datasets. The results demonstrate both the efficiency of our algorithm by 1 to 2 orders of magnitude over a baseline solution (Sect. 5).

2   Related Work

In this section, we review the closely related studies, specifically in the area of trajectory indexing and query processing. Since many of the trajectory indexes use variants of the traditional spatial indexes, and there is no index structure that can directly support *MRQ*, we start this section by briefly reviewing the spatial and spatio-temporal indexes that are used in the baseline and our proposed method.

2.1   Conventional Index Structures

The *R-Tree* [5] is a tree structure where the *Minimum Bounding Rectangle* (MBR) of the root covers the region containing all spatial data objects. Every node has a maximum capacity, where the leaf nodes contain pointers to the actual data objects and the non-leaf nodes cover the MBRs of all of its child nodes. Many researchers (e.g., [22]) propose a 3D R-tree where time is treated as the third dimension to index spatio-temporal data.

The *Bucket Quadtree* [17] is an adaptive form of the uniform grid data structure that handles spatial data of an arbitrary distribution. There are a number of quadtree variants, most of which repeatedly partition the underlying space into four equal-sized regions when (i) the number of objects in the space exceeds a threshold value (e.g., bucket quadtree for points [17] or PMR quadtree [7]), or (ii) when the underlying spanned space is not homogeneous (e.g., the region quadtree [8]).

The *Bucket Octree* [17] is a three-dimensional version of a bucket quadtree that handles three-dimensional data or spatio-temporal data by treating time as the third dimension where the space is partitioned into eight equal-sized cube regions.

2.2   Trajectory Index Structures

A number of spatial and spatio-temporal index structures have been proposed over the past decades to index trajectories. These indexes can be broadly categorized as (i) point-based indexes, and (ii) trajectory based indexes.

Point-Based Indexes. Several approaches propose to index trajectories, and spatio-temporal objects in general where the structure is constructed purely based on the locality of the points. As there are multiple traditional structures to index a set of points based on their locality (e.g., R-trees using MBRs or quadtrees using space partitioning), all the points of trajectories in a dataset are stored using such an index in point-based indexing techniques [6,10,19–21]. To maintain the relation between the points and trajectories (i.e., which point belongs to which trajectory), either an auxiliary lookup structure is maintained, or the trajectory identifiers are embedded into the point data.

The studies [6,20] both use variants of R-tree to store the points of trajectories. The *HR-tree* [10] stores a separate R-Tree for each timestamp to allow queries on the past states of a spatial point database. Spatial objects that do not move between consecutive timestamps can result in a substantial storage overhead. The HR-tree organizes consecutive trees to share the same branches to avoid indexing stationary objects multiple times. However, the entire node is replicated when only one object updates its location. The *Start/End timestamp B-tree (SEB-tree)* [19] partitions the search space into zones and the 'zoning information' of objects is stored in the database. An object requests an update when it enters into a new zone. A separate index is constructed for each zone based on the objects that moved in, moved out, or are currently in that zone.

All objects in one zone are indexed by start and end timestamps in the corresponding B-tree. This is similar to the active object data structure used in plane-sweep solutions [14]. The *MV3R-tree* [21] proposes to use a multi-version R-tree (MVR-tree) to index the data and all leaf nodes of the MVR-tree are used to build an auxiliary 3D R-tree. The idea of combining two different structures is to improve the query efficiency by choosing the appropriate structure. The MVR-tree is chosen for timestamp queries while the 3D R-tree is used to perform long time interval queries. A threshold parameter defines which structure is to be chosen to conduct short time interval queries.

In point based trajectory index structures, the points that are close in space are stored close-by, but points of the same trajectory may not be stored close-by. Thus the locality of trajectories are not preserved in such structures.

Trajectory-Based Indexes. In contrast to the point-based indexes, trajectory-based indexes [2,12] consider the relevance of points while indexing the data. Points from the same trajectory are more likely to be stored close-by, and in some structures, all the points of the same trajectory are stored together.

The *Scalable and Efficient Trajectory Index (SETI)* [2] assumes that a spatial dimension does not change as frequently as a temporal dimension. SETI partitions the search space into static and non-overlapping cells. An in-memory structure is used to store the last location of each object and to compute a trajectory segment when a new location update comes. Long segments that span more than one spatial cell are split at the cell boundaries. A time interval that covers all segments in a cell is computed and used to create a temporal R-tree of each cell. The *STR-tree* [12] extends R-tree to store line segments of trajectories by preserving trajectory locality partially, where the line segments of the same trajectory are more likely to be stored together. While the insertion criterion of the R-tree is based on the least enlargement of the bounding rectangles, the STR-tree introduces an additional insertion criterion based on the least number of nodes required to store the line segments of the same trajectory at different levels of the tree. The *TB-tree* [12] ensures that a leaf node contains only the line segments that belong to the same trajectory. All the leaf nodes that store a trajectory are linked by forward and backward pointers to reduce the time to retrieve a complete trajectory. In this structure, as the line segments of different trajectories that are close to each other are not stored together, their experimental results show that TB-tree has a higher cost than STR-tree for spatial range queries.

Other Index Structures. Other index structures, e.g., *TrajTree* [15], *SharkDB* [24] are also proposed to efficiently store trajectories. The idea of TrajTree [15] is to compute edit distances between trajectories by segmentation, which is not the focus of our problem. SharkDB [24] is an in-memory column oriented timestamped trajectory storage. This index can support k nearest neighbor queries and range queries in the spatio-temporal domain, but cannot be directly applied to solve multiple range queries.

2.3 Trajectory Query Processing

There are several studies that answer different types of queries on trajectory data [13,18,23,25]. The queries can be broadly categorized as (i) range queries, (ii) similarity based queries (k nearest neighbor, reverse k nearest neighbor queries, etc.). As the range queries are the closest to our problem, we focus on reviewing the studies on range queries in spatial and spatio-temporal databases, specifically on trajectory data.

In general, given a spatial range as a rectangle window or a circle, a range query on trajectories finds all the trajectories that pass through that range. For spatio-temporal range queries, a time interval is also given where a timestamp of the resulting trajectories in that region needs to fall in that time interval as well. As a range query is one of the most common search problems, all of the above mentioned point based indexing [6,10,19–21] and trajectory based indexing [2, 12] studies can answer this query. If the index is point based, the points that are inside the query range are retrieved, and then the auxiliary structure or the embedded trajectory identifiers are obtained to finally get the trajectories as the result of the range query. In trajectory-based indexes, the nodes that do not intersect with the query range are pruned, and then a further pruning is applied based on the intersection of the trajectory line segments with the query range. However, all of these approaches are designed to answer individual range queries. If these approaches are used to answer multiple range queries, many irrelevant trajectories and trajectory points are retrieved (details in Sect. 3.2).

3 Problem Formulation and the Baseline

3.1 Problem Formulation

As we consider both the spatial-only and spatio-temporal trajectories in this work, we present our problem formulation for both settings.

Let T be a set of spatio-temporal trajectories where each $t \in T$ is a sequence of tuples of the form $\langle loc, \tau \rangle$. Here, loc is a point location and τ is a timestamp associated with that location. For spatial-only trajectories, each $t \in T$ is a sequence of locations of the form $t = (loc_1, loc_2, \ldots, loc_m)$ where m is the number of points in the trajectory.

Definition 1 *Trajectory Range Query. Given a trajectory dataset T, a spatial range in space (e.g., a rectangle or a circle) and a time interval, a Trajectory Range Query returns all trajectories that intersect with both the spatial range and the time interval.*

Definition 2 *Trajectory Multi-range Query (MRQ). Given a trajectory dataset T, a sequence of spatio-temporal ranges $\{\langle R_1, I_1 \rangle, \langle R_2, I_2 \rangle, \ldots\}$, where R_i is a spatial region and I_i is a time interval, a Trajectory Multi-Range Query returns all trajectories from T that intersect with each region R_i in the corresponding time interval I_i.*

For the spatial-only case, only the spatial ranges are given. An MRQ query can be considered as a sequence of independent range queries. Based on this observation, we present a baseline approach in the following to answer the MRQ using existing methods.

3.2 Baseline Approach

As there are multiple studies that can find trajectories intersecting a single spatial region in a given time interval (details in Sect. 2), we can apply the following steps to find the solution of an MRQ in a straightforward way - (1) Find the answer of the single range queries individually using an existing approach. These are the set of candidate trajectories; (2) Compute the intersection of the individual responses to obtain resulting trajectories that intersect each of the ranges. We denote this straightforward method as our *Baseline Approach*.

Example 1. Let, Q_1 and Q_2 be two given query ranges as shown in the shaded region in Fig. 1a, where there are seven trajectories $\{T_1, \ldots, T_7\}$ in the dataset. The baseline approach finds $\{p_{41}, p_{71}\}$ for query range Q_1 and $\{p_{42}, p_{62}\}$ for query range Q_2. The candidate trajectories for Q_1 and Q_2 become $\{T_4, T_7\}$ and $\{T_4, T_6\}$, respectively, based on the retrieved points. The intersection of the candidates returns T_4 as the result for the baseline.

Limitations. Unfortunately, the baseline approach is computationally expensive for several reasons:

- Many irrelevant trajectories are retrieved in the first step that intersect at least one range, but not with all of the ranges.
- We need to compute the intersection of the set of candidate trajectories for all query ranges. As the intersection of just two sets takes $O(mn)$ where n and m are the number of trajectories in each set, computing the intersections of all these sets is computationally expensive.

4 Our Proposed Solution

To address the limitations of baseline, the key idea of our approach is to maintain both the co-location of trajectories and the points of the trajectories. We propose a two-level index to maintain such information to facilitate pruning at every level and check only the necessary trajectories. In Sect. 4.1 we present our index structure, *LoTI*, and in Sect. 4.2 we describe the multi-range query processing using that index.

4.1 Index Structure: LoTI

In our proposed index, *LoTI*, the first level is built over the trajectory MBRs to help pruning the trajectories that do not intersect with all query ranges. We use an R-tree as the first level index. The second level is created over the previous index for further pruning. We use a Bucket Quadtree [17] (details in Sect. 2) for each leaf node of the R-tree as the second level index. We present the details of our proposed index in the following. For ease of demonstration, we show the illustrations for the spatial-only query. For the spatio-temporal query, the R-tree and the Bucket Quadtree are replaced by 3D-Rtree and Bucket Octree, where time is the third dimension.

First level. We compute the MBR of each trajectory, enclosing all of its data points. The computed trajectory MBRs are indexed using an R-tree. Let the fanout of the R-tree be f, i.e., each leaf node contains at most f trajectories. This first-level index groups the trajectories into leaf nodes and helps to prune trajectories based on the MBRs rather than the trajectory points (explained in Sect. 4.2). Figure 1a shows seven trajectory MBRs indexed by an R-tree, where R_1 and R_2 are two leaf nodes of the R-tree.

Second Level. The MBR of each leaf node of the R-tree is indexed using a Bucket Quadtree by setting a *threshold* for the number of trajectories allowed in a leaf node of the Bucket Quadtree. If the number of trajectories exceeds the threshold, then the Bucket Quadtree node is split into four equal-sized sub-nodes. Each node of the Bucket Quadtree is assigned an identifier corresponding to the binary representation of the path from the root. The identifier known as the *location code* can be represented by 2 bits and 3 bits in a specific tree level for the Bucket Quadtree and Bucket Octree, respectively. Each time we split a region, the bottom-left, top-left, top-right, bottom right sub regions are assigned $00, 01, 10, 11$ binary values respectively as shown in Fig. 1b.

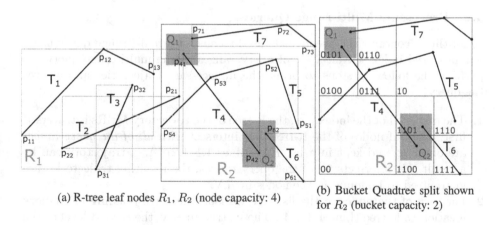

(a) R-tree leaf nodes R_1, R_2 (node capacity: 4)

(b) Bucket Quadtree split shown for R_2 (bucket capacity: 2)

Fig. 1. LoTI index components on the running example

As the size of the Bucket Quadtree increases, and there are multiple quadtrees to be stored, we need much more space to store the MBR information of each node. Thus, we use the *location code tree* based on the Bucket Quadtree to eliminate the use of MBR information. We can easily check whether a node in

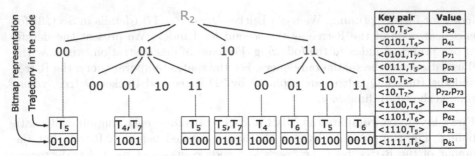

(a) The tree representation of the bucket decomposition given in Figure 1b   (b) Lookup table

Fig. 2. The location code tree for the Bucket Quadtree

the location code tree intersects with the query range using its code. Location code trees are much smaller than the original Bucket Quadtree and can be stored in main memory to increase query performance.

Moreover, as we know the number of trajectories in each Bucket Quadtree node, we assign a bit position for each trajectory as shown in Fig. 2a. Such a bit representation can facilitate the search of the trajectories intersecting with a query range (explained later in Sect. 4.2). A lookup table shown in Fig. 2b stores a pair (b, t) containing the location code b of the bucket and trajectory t of the trajectory that passes through b. These are the keys to the table, and the values stored in each table entry (b, t) are the points of trajectory t that occur in the bucket b.

4.2   Processing Multi-range Queries

Algorithm 1 contains the pseudocode used to answer a multi-range query using our proposed index. Given a set of query ranges as the multi-range query, we perform the following steps to prune the unnecessary trajectories and answer the query.

1. The first level of the index, i.e., the R-tree of the trajectory MBRs is traversed first. The leaf nodes of the R-tree that intersect with *all of the query ranges* are obtained and kept in a list RN (Lines 1.1–1.2). As a trajectory can be a result if it intersects all of the query ranges, the resulting trajectories are guaranteed to be found in the nodes in RN.
2. For each of the nodes in the list RN we load the corresponding in-memory location code tree (Line 1.3–1.4). These structures at the second level of the index are used to further prune the trajectories that cannot be in the result.
3. We find the nodes of the location code trees that intersect each of the query ranges. For each of these nodes, we obtain the trajectories that pass through the node and set corresponding bits in the $bitSet$ (Lines 1.6–1.8). $bitSet$ contains only trajectories where its contained nodes intersect with all query ranges. Thus, we minimize the number of trajectories that we need to check in a later step.

Algorithm 1. Multi-range query (MRQ) processing

Input: The *LoTI* index over a trajectory dataset D, a set of query ranges Q
Output: a list of trajectories R

1.1 Traverse the R-tree of *LoTI* from the root node for Q
1.2 $RN \leftarrow$ The leaf nodes L of the R-tree such that $\forall_{q \in Q} intersects(q, L)$
1.3 **foreach** $rn \in RN$ **do**
1.4 $lt \leftarrow$ get_LocationCodeTree(rn)
1.5 $bitSet \leftarrow$ assign a bit for each trajectory $t \in rn$
1.6   **foreach** $q \in Q$ **do**
1.7 $LN_q \leftarrow$ Get nodes from lt intersecting q
1.8   set bits of trajectories such that $\forall_{t \in rn} intersects(q, lt)$
1.9   **foreach** $t \in bitSet$ **do**
1.10   **foreach** $q \in Q$ **do**
1.11   **if** *contains(q, LN_q) is false* **then**
1.12   Get trajectory points from the lookup table of *LoTI*
1.13   **if** *no point of t in q* **then**
1.14   Eliminate trajectory t by setting $bitSet(t, false)$
1.15   **if** *True t $\in$ bitSet* **then**
1.16 $R.add(t)$
1.17 **return** R

---

4. After the above pruning steps, finally we need to perform a verification step to get the results. Using the trajectory identifier and its corresponding location code tree node identifier, we find the actual points that are stored in the lookup table (described in Sect. 4.1) and check whether there is a point of that trajectory that actually falls inside the query range. If yes, then that trajectory is a candidate result, obtained from the corresponding query range (Lines 1.11–1.14). If a node of the location code tree is completely contained inside a query range, then all trajectories that pass through that node are considered as candidate trajectories for that query range (Lines 1.11). Otherwise, we check actual points for each trajectory of that node and eliminate them from the candidate set without checking for remaining query ranges if no point intersects the query range. Finally, if a trajectory is a candidate for all of the query ranges, then that trajectory is a result of the multi-range query (Lines 1.15–1.16). We return all such result trajectories.

Example 2. We use the same example of query ranges as Example 1 to illustrate our algorithm's pruning power. Here, let all trajectories be indexed into two leaf nodes, R_1 and R_2. An MBR-based first level filter returns a candidate node R_2 as it intersects all query ranges. The corresponding location code tree for the candidate node R_2 is shown in Fig. 2a. The node 0101 intersects query range Q_1 while the nodes 1100, 1101 intersect query range Q_2. The bitset for the node 0101 is {1001}. The merged bitset of the nodes 1100, 1101 is 1010. The logical *AND* operation on the bitsets returns the value (1001 $\cap$ 1010 = 1000) where the position of the trajectory T_4 is set to *true*. In this example, no nodes are

Table 1. Dataset statistics

Statistics property	T-drive	Beijing
Total no. of points	12,806,605	176,470,199
Total no. of trajectories	933,518	608,603
Avg trajectory length	8.1 km	195.4 km
Avg points per trajectory	13	289

Table 2. Experimental parameters

Parameter description	Values
Number of ranges	2, 3, 4
Search region (km)	1, 2, 4, 8, 16
Search time interval (hour)	0.5, 1, 2, 4, 8
Quad(Oc)tree node capacity	5 trajectory

contained in query ranges. In the final step, we check points $\{p_{41}, p_{42}\}$ of the trajectory T_4 from the lookup table on the disk and return as a result.

5   Experimental Evaluation

In this section, we compare our proposed algorithm with the baseline approach (presented in Sect. 3.2) through an extensive experimental evaluation using real datasets.

5.1   Experimental Settings

All algorithms are implemented in JAVA. Experiments were run on a 24 core Intel Xeon E5-2630 2.3 GHz using 256 GB RAM, and 1TB 6G SAS 7.2K rpm SFF (2.5-in.) SC Midline disk drives running Red Hat Enterprise Linux Server release 7.2.

Datasets. All experiments were conducted using two real datasets, (i) T-drive dataset and (ii) Beijing dataset.

The T-drive dataset[1] contains 10,357 raw taxi trajectories in Beijing collected for a week in Feb 2008. The Beijing dataset contains 28,162 raw trajectories in Beijing collected for a month in March 2009. Each trajectory is a sequence of GPS locations (latitude and longitude) and the corresponding timestamp. As the locations recorded by a GPS device may contain some noisy data, and the start-end of a taxi trip is not explicitly specified in these datasets, we perform the following cleaning steps - (i) If the location of a taxi does not change for some specific time duration (i.e., the locations in a sequence of consecutive timestamps are very close to each other), it implies that the taxi is likely to be standby to pick up a passenger, and we split the raw trajectory into two separate trajectories at

---

[1] https://protect-au.mimecast.com/s/08IJCE8kAGuOQ1l9Twb4SR?
   domain=microsoft.com.

Table 3. 2D and 3D index sizes and construction times for T-drive and Beijing datasets

Dataset	2D				3D			
	Baseline		LoTI		Baseline		LoTI	
	Size (GB)	Time (min)	Size (GB)	Time (min)	Size (GB)	Time (min)	Size (GB)	Time (min)
T-drive	0.9	19.4	0.8	0.7	1.3	49.4	1.6	2.4
Beijing	12.4	424	14.2	32.4	17.2	605	12.6	16.2

such points. (ii) We have calculated the speed of a moving taxi at each location from its consecutive locations and their timestamps. In some cases we found the speed to be unrealistically high, which is likely to be caused by a wrong GPS location record or a time reset. We split the taxi trajectory into two at such points. We chose the suitable splitting parameters based on the properties of the dataset. (iii) As we want to explore the trajectory movements at different time intervals of the day, each trajectory also spans at most one day.

The cleaned T-drive dataset contains 933,518 trajectories with a total of 12,806,605 points, while the cleaned Beijing dataset contains 608,603 trajectories with a total of 176,470,199 points. Table 1 shows the detailed properties of the cleaned datasets.

Query Generation. We generate the query ranges based on the density distributions of the datasets. Specifically, we divide the dataset area into multiple grid cells, and find the cells with at least a threshold number of trajectories (1000, in our experiments). We choose grid cells randomly and use the center positions for our query ranges, where the number varies from 2–4. We take the timestamps of a trajectory in the dataset that passes through 2-dimensional query ranges to generate 3-dimensional query ranges. The ranges in a multi-range query do not overlap with each other. We generate such sets of 100 multi-range queries and report the performance. The experiments are conducted on both the spatial-only and spatio-temporal trajectories. We use the same set of queries for both setups, where the temporal information is removed in the spatial-only case.

Evaluation and Parameterization. We study the performance of both the baseline and our approach by varying several parameters as shown in Table 2, where the values in bold represent the default values. The fanout of the R-tree in baseline and the R-tree of *LoTI* are set to 100. For all experiments, a single parameter is varied while keeping the rest at the default values. We study the impact of each parameter by running 100 queries and report the values of: (i) the query execution time, (ii) the number of candidates considered, and (iii) the number of points checked. We also report the costs of constructing the indexes in Table 3.

5.2   Performance Evaluation on Spatial-Only Data

The performance for multiple runs is shown in boxplots, where the bounding box shows the first and third quartiles; the whiskers show the range, up to 1.5 times of the interquartile range; and the outliers beyond this value are shown as separate points. The average values are shown as connecting lines.

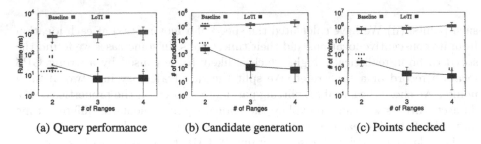

(a) Query performance   (b) Candidate generation   (c) Points checked

Fig. 3. Varying the number of query ranges in 2D space (T-drive dataset)

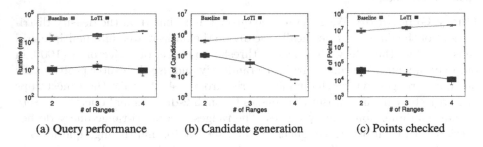

(a) Query performance   (b) Candidate generation   (c) Points checked

Fig. 4. Varying the number of query ranges in 2D space (Beijing dataset)

Effect of Varying the Number of Ranges. We conducted experiments by varying three different numbers of ranges and evaluated the performance as shown in Figs. 3 and 4 for the T-drive and Beijing datasets, respectively. Our algorithm, *LoTI* is up to 2 orders of magnitude faster than the baseline and checks up to 3 orders of magnitude fewer candidates and points in spatial-only space. As more points fall into a higher number of ranges, the costs of the baseline increases. In contrast, the benefit of our approach is more pronounced for a higher number of ranges as the trajectories that do not pass through all of the ranges can be quickly identified and pruned from further consideration with our two-level index. As shown with the ranges of the boxplots, our approach significantly outperforms the baseline for all of the 100 sets of queries. Table 4 shows the breakdown of runtime that different steps of our approach take. As shown in the table, the MBR filtering using the first level of *LoTI* is the quickest filter, the second level takes the most time as trajectories are pruned further, and the lookup table takes moderate time to finally obtain results.

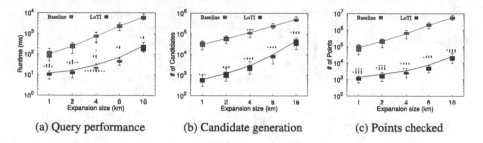

Fig. 5. Varying query expansion sizes in 2D space (T-drive dataset)

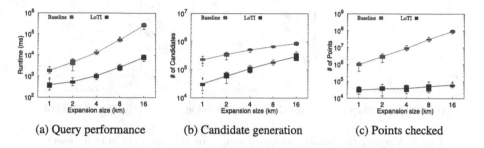

Fig. 6. Varying query expansion sizes in 2D space (Beijing dataset)

Table 4. Time to run each filtering on 2-dimensional LoTI index on Beijing dataset

Filter	# of ranges			Expansion (km)				
	2	3	4	1	2	4	8	16
	Runtime (ms)							
MBR	4	4	3	4	4	4	5	5
Quadtree	912	1177	928	272	414	912	2518	7696
Lookup table	143	148	41	112	130	143	176	261

Effect of Varying the Query Range Size. Figures 5 and 6 show the experimental result for different sizes of the query range expansions for the T-drive and Beijing datasets, respectively. As more trajectories are likely to fall inside larger query ranges, thus the cost of both methods increase with the increase of the query range size. In all cases, our method outperforms the baseline by 1 order of magnitude in runtime by efficient pruning, and the gaps in the performance do not vary much.

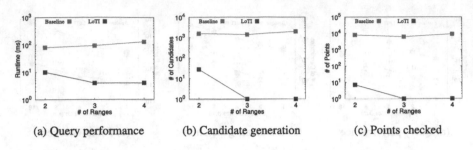

(a) Query performance   (b) Candidate generation   (c) Points checked

Fig. 7. Varying the number of query ranges in 3D space (T-drive dataset)

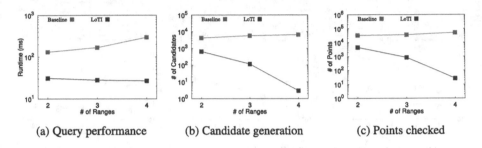

(a) Query performance   (b) Candidate generation   (c) Points checked

Fig. 8. Varying the number of query ranges in 3D space (Beijing dataset)

5.3   Performance Evaluation on Spatio-Temporal Data

As more trajectories are likely to be pruned by the temporal constraints, fewer trajectories and points are required to be checked for both methods than the spatial-only counterpart. As the number of candidates is very small for our approach in some experiments, a boxplot in log-scale is difficult to comprehend. Thus we present the average performance using line plots.

Effect of Varying the Number of Ranges. Figures 7 and 8 show the performances for varying the number of spatial-temporal query ranges, where the default time interval is set to one hour. Our approach, *LoTI* is up to 1 order of magnitude faster than the baseline and checks 2 to 3 orders of magnitude fewer candidates and points. The performance trends are similar to the spatial-only experiments. Trajectories in the Beijing dataset have more points in average than T-drive. Thus, we find more query results as the likelihood to find trajectories increase due to the point density.

Effect of Varying the Query Range Size. Figures 9 and 10 show the experimental result for different sizes of the query range expansions, where the time interval is set to its default value. As more points need to be checked for larger ranges, the number of candidates for both methods increases which lead to more runtime. As the additional dimension can prune some more trajectories, the costs of both methods are less than their spatial-only counterparts.

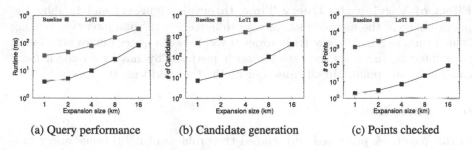

(a) Query performance   (b) Candidate generation   (c) Points checked

Fig. 9. Varying query expansion sizes in 3D space (T-drive dataset)

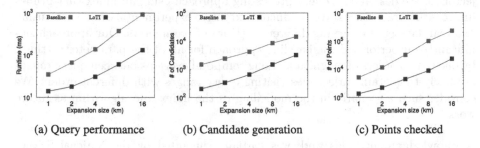

(a) Query performance   (b) Candidate generation   (c) Points checked

Fig. 10. Varying query expansion sizes in 3D space (Beijing dataset)

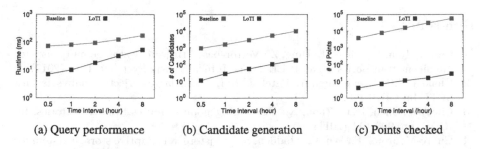

(a) Query performance   (b) Candidate generation   (c) Points checked

Fig. 11. Varying query time intervals in 3D space (T-drive dataset)

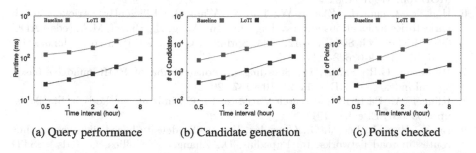

(a) Query performance   (b) Candidate generation   (c) Points checked

Fig. 12. Varying query time intervals in 3D space (Beijing dataset)

Effect of Varying the Query Time Interval. Figures 11 and 12 show the performance of the methods for varying different query time intervals, ranging from 30 min to 8 h. As more trajectories travel between query ranges for a larger time interval, the costs increase for both methods by having to check more candidates and points. In all cases, our method outperforms the baseline.

6　Conclusion

In this paper, we presented and studied the problem of multi-range query processing on trajectories. We proposed a novel two-level index, *LoTI* that preserves the co-location relationship between both trajectories and the points of the trajectories. We described a query processing approach using an index that prunes unnecessary trajectories. We conducted extensive experimental evaluations using two real datasets by varying different settings of parameters. Our approach significantly outperformed the baseline approach for all of the parameter settings, by one to two orders of magnitude. Long temporal overlaps between query ranges are likely to contain trajectories visiting query ranges with different orders. We consider such a scenario in the modelling of the index and algorithm as future work.

Acknowledgement. This work was partially supported by the National Science Foundation under Grant IIS-13-20791, ARC DP170102726, DP180102050, NSFC 61728204 and 91646204. Zhifeng Bao is a recipient of Google Faculty Award.

References

1. Cai, Z., Ren, F., Chen, J., Ding, Z.: Vector-based trajectory storage and query for intelligent transport system. IEEE Trans. Intell. Transp. Syst. **19**, 1–12 (2017)
2. Chakka, V.P., Everspaugh, A., Patel, J.M.: Indexing large trajectory data sets with SETI. In: CIDR (2003)
3. Chen, Z., Shen, H.T., Zhou, X.: Discovering popular routes from trajectories. In: ICDE, pp. 900–911 (2011)
4. Cudré-Mauroux, P., Wu, E., Madden, S.: Trajstore: an adaptive storage system for very large trajectory data sets. In: ICDE, pp. 109–120 (2010)
5. Guttman, A.: R-trees: a dynamic index structure for spatial searching. In: SIG-MOD, pp. 47–57 (1984)
6. Han, Y., Chang, L., Zhang, W., Lin, X., Wang, L.: Efficiently retrieving top-k trajectories by locations via traveling time. In: Wang, H., Sharaf, M.A. (eds.) ADC 2014. LNCS, vol. 8506, pp. 122–134. Springer, Cham (2014). https://doi.org/10.1007/978-3-319-08608-8_11
7. Hjaltason, G.R., Samet, H.: Speeding up construction of PMR quadtree-based spatial indexes. VLDB J. **11**(2), 109–137 (2002)
8. Klinger, A.: Patterns and search statistics. In: Optimizing Methods in Statistics, pp. 303–337. Elsevier (1971)
9. Li, X., Han, J., Lee, J.-G., Gonzalez, H.: Traffic density-based discovery of hot routes in road networks. In: Papadias, D., Zhang, D., Kollios, G. (eds.) SSTD 2007. LNCS, vol. 4605, pp. 441–459. Springer, Heidelberg (2007). https://doi.org/10.1007/978-3-540-73540-3_25

10. Nascimento, M.A., Silva, J.R.O.: Towards historical r-trees. In: Proceedings of the 1998 ACM Symposium on Applied Computing, pp. 235–240 (1998)
11. Papadias, D., Tao, Y., Kalnis, P., Zhang, J.: Indexing spatio-temporal data warehouses. In: ICDE, pp. 166–175 (2002)
12. Pfoser, D., Jensen, C.S., Theodoridis, Y.: Novel approaches to the indexing of moving object trajectories. In: VLDB, pp. 395–406 (2000)
13. Popa, I.S., Zeitouni, K., Oria, V., Barth, D., Vial, S.: Indexing in-network trajectory flows. VLDB J. **20**(5), 643–669 (2011)
14. Preparata, F.P., Shamos, M.I.: Computational Geometry - An Introduction. Springer, New York (1985). https://doi.org/10.1007/978-1-4612-1098-6
15. Ranu, S., Deepak, P., Telang, A.D., Deshpande, P., Raghavan, S.: Indexing and matching trajectories under inconsistent sampling rates. In: ICDE, pp. 999–1010 (2015)
16. Sacharidis, D., et al.: On-line discovery of hot motion paths. In: EDBT, pp. 392–403 (2008)
17. Samet, H.: Foundations of Multidimensional and Metric Data Structures. Morgan-Kaufmann, San Francisco (2006)
18. Shang, S., Chen, L., Jensen, C.S., Wen, J., Kalnis, P.: Searching trajectories by regions of interest. TKDE **29**(7), 1549–1562 (2017)
19. Song, Z., Roussopoulos, N.: Seb-tree: an approach to index continuously moving objects. In: MDM, pp. 340–344 (2003)
20. Tang, L.-A., Zheng, Y., Xie, X., Yuan, J., Yu, X., Han, J.: Retrieving k-nearest neighboring trajectories by a set of point locations. In: Pfoser, D. (ed.) SSTD 2011. LNCS, vol. 6849, pp. 223–241. Springer, Heidelberg (2011). https://doi.org/10.1007/978-3-642-22922-0_14
21. Tao, Y., Papadias, D.: Mv3r-tree: a spatio-temporal access method for timestamp and interval queries. In: VLDB, pp. 431–440 (2001)
22. Theodoridis, Y., Vazirgiannis, M., Sellis, T.K.: Spatio-temporal indexing for large multimedia applications. In: ICMCS, pp. 441–448 (1996)
23. Wang, H., Zimmermann, R.: Processing of continuous location-based range queries on moving objects in road networks. TKDE **23**(7), 1065–1078 (2011)
24. Wang, H., Zheng, K., Xu, J., Zheng, B., Zhou, X., Sadiq, S.W.: SharkDB: an in-memory column-oriented trajectory storage. In: CIKM, pp. 1409–1418 (2014)
25. Wang, S., Bao, Z., Culpepper, J.S., Sellis, T., Cong, G.: Reverse k nearest neighbor search over trajectories. TKDE **30**(4), 757–771 (2018)

A Volunteer Design Methodology
of Data Warehouses

Amir Sakka[1,2(✉)], Sandro Bimonte[1], Lucile Sautot[4], Guy Camilleri[2], Pascale Zaraté[2], and Aurelien Besnard[3]

[1] IRSTEA, UR TSCF, 9 Av. B. Pascal, 63178 Aubiere, France
{amir.sakka,sandro.bimonte}@irstea.fr
[2] IRIT, Toulouse University, Toulouse, France
{guy.camilleri,pascale.zarate}@irit.fr
[3] LPO Aquitaine, 433 Chemin de Leysotte, 33140 Villenave-d'Ornon, France
aurelien.besnard@lpo.fr
[4] AgroParistech, UMR TETIS, Maison de la télédétection, Montpellier, France
lucile.sautot@agroparistech.fr

Abstract. In the context of Volunteered Geographic Information (VGI), volunteers are not involved in the decisional processes. Moreover, VGI systems do not offer advanced historical analysis tools. Therefore, in this work, we propose to use Data Warehouse (DW) and OLAP systems to analyze VGI data, and we define a new DW design methodology that allows involving volunteers in the definition of analysis needs over VGI data. We validate it using a real biodiversity case study.

Keywords: OLAP · Data Warehouse · Volunteered Geographic Information
GDSS

1 Introduction

Crowd science (i.e., citizen science or volunteer science) has been defined as *"online, distributed problem-solving and production model"* [4]. Well-known examples of crowdsourcing systems are Wikipedia[1], forums, etc. In crowdsourcing systems the users of the community add, delete and modify contents (ex: forum answers, documents, etc.) until achieving an agreement. In the context of geographical data, crowdsourcing has been defined as VGI (Volunteered Geographic Information). VGI is *"the mobilization of tools to create, assemble and disseminate geographic data provided by volunteers"* [18]. VGI allows managing amounts of geo-localized data (e.g. Openstreetmap[2]), and it is widely used in different application domains i.e. urban, biodiversity, risks, etc. Usually, volunteers are data producers and passive consumers of VGI data analyses provided by organisms/enterprises. This *"bottom-up data supply and top-down data analysis"* paradigm represents an important barrier for the development of volunteers'

---

[1] https://www.wikipedia.org.
[2] https://www.openstreetmap.org.

© Springer Nature Switzerland AG 2018
J. C. Trujillo et al. (Eds.): ER 2018, LNCS 11157, pp. 286–300, 2018.
https://doi.org/10.1007/978-3-030-00847-5_21

observatories, since data producers feel excluded from the decision-making process [13]. Moreover, as highlighted in [2] VGI does not present analysis functionalities to scope huge volumes of geospatial data. Therefore, the analysis of VGI using Geo-Business Intelligence (GeoBI) has been proved as an effective solution [2]. In particular, VGI are designed for operational tasks and complex analysis on small spatial data, whereas Spatial On-Line Analytical Processing (SOLAP) systems are more relevant for analysis based on exploration of massive spatial datasets stored in Spatial Data Warehouse (SDW) [12, 17]. Since DWs are conceived according to data sources and users requirements, the more the DW model reflects stakeholders' needs, the more stakeholders will make use of their data [12, 15], implying social (e.g. welfare improvement) and economical (e.g. sustainable agriculture) benefits. Providing GeoBI applications fitting the VGI community's analysis needs, will represent important social and economic advances, since: (i) new required and effective analysis possibilities on numerous different crowdsourced data will be possible (urban, agricultural, risks, environmental data, etc.), and (ii) volunteers will be more and more motivated to collect data. Therefore, this work *aims at moving volunteers from data suppliers to volunteered data analysts by means of a new kind of OLAP systems, as described in the next.*

Our Vision: In the same way as methodologies of data validation adopted by existing crowdsourcing systems (OpenStreetMap, Wikipedia, etc.), in Fig. 1 we present our vision of a new OLAP system (OLAP2.0). The main idea is to allow volunteers to express separately their requirements for OLAP analysis, in a first step. These requirements will then be translated to multidimensional (i.e. DW) models. Next, these OLAP models are submitted to a set of particular volunteers called committers, who are fully involved in the project and highly experimented in the crowdsourced data. [12] emphasizes on the necessity of data stewardship (conducted by committers in our approach) to solve issues related to the lack of users' experience in queries specification and data ownership/sensitivity problems encountered by organizations during DWs implementation process. Hence thus, committers decide whether to implement crowdsourced requirements (i.e. multidimensional models) of volunteers or not, according to their expertise to judge the relevance of requirements. After that, the DW expert *designers* are in charge of implementing models agreed by committers (Fig. 1). Finally, the new OLAP models are implemented and made available to all *users* that can visualize, explore and analyze data (Fig. 1).

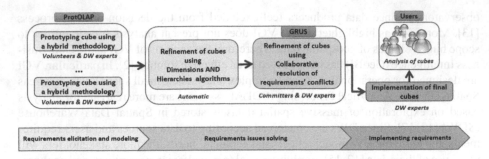

Fig. 1. OLAP2.0 methodology

Investigated Issues: In this work we focus on issues related to the design of multi-dimensional models from crowdsourced requirements. Let us note the fact that our use of VGI data does not cover the data quality validation/investigation, we apply our methodology on already cleansed VGI databases.

Several design methodologies for DW have been proposed [7, 15], however, when decision-makers are volunteers and they are different from those who decide the relevance of the requirements, they:

(i) Only represent few potential users of the OLAP system, so their specific analysis needs may be perhaps not those useful for most final users;

(ii) Can have different backgrounds (e.g. scientists, citizen, etc.), which can lead to multiple contradictious interpretations of the same requirement. When stakeholders have divergent goals, it becomes problematic to maintain an agreement between them from a requirement-engineering point of view [9, 21];

(iii) Are not skilled in DW, and sometimes, also in Information Technologies (IT), thus it remains possible that they do not correctly or clearly formalize most of their needs;

(iv) Can be numerous, making conflicts management an extremely complicated task;

(v) Are not "employed" by the project, their involvement time in the project is limited, and so they cannot exhaustively, accurately and correctly define their requirements;

(vi) They are geographically distributed over different locations;

Therefore, requirements elicited by volunteers can present [21]:

- *Similarities* i.e. the same multidimensional elements are separately defined,
- *Differences* i.e. Different definitions of the same multidimensional elements, and
- *Conflicts* i.e. irrelevant or erroneous multidimensional elements definition.

Hence, dealing with these particular stakeholders using the existing DW methodologies is not possible since the existing DW methodologies:

(a) Require advanced knowledge of OLAP main concepts (because of iii);

(b) Assume that users are effectively involved in the project, which makes all their needed requirements well and completely defined (because of i, ii and v);

(c) They do not handle the cases where large number of multidimensional models can be generated (because of ii, iv);

(d) They deal with domain experts only, so they have no need to manage Inconsistent definitions (because of ii, iv, v).

To address these issues, based on main principles of requirements engineering [21] and in particular using the Groupware tools approach [21], we propose an innovative collaborative DW design methodology using a Group Decision Support System (GDSS), to help committers to decide whether to implement or not the crowdsourced requirements of volunteers. Indeed, GDSSs are designed to support a group engaged in a collective and collaborative decision process with geographically distributed users, they are used in several domains e.g. workflows, user interfaces and databases design [19], but not for DW. Moreover, to allow volunteers to easily crowdsource their requirements (Fig. 1), we use the ProtOLAP methodology [3], a methodology for DW rapid prototyping when computer science inexpert users supply data.

We validate our proposal in the context of the French ANR project VGI4Bio, but several different VGI-based applications could be addressed in the same way.

The paper is organized as following: in Sect. 2 we describe our case of study; Sect. 3 illustrates the proposed methodology; Sect. 4 presents the implementation and validation of the methodology, and Sect. 5 overviews related works.

2 Case Study

In the context of project VGI4Bio[3], we mobilize two VGI databases (Visionature and Observatoire Agricole de la Biodiversité[4] - OAB) to build SOLAP applications to analyze farmland biodiversity indicators. Visionature and OAB have 7682 and 1500 volunteers that produce data, respectively. Among possible users interested in analyzing these data, we have identified a huge number of users belonging to diverse categories such as: volunteers that are interested in analyzing data to improve their data production quality, their related daily practices, etc.; public and private organisms (DREAL, Chambre d'Agriculture, etc.). At this phase of the project, we have identified some volunteers, and a set of committers. Figure 2 shows three multidimensional models defined by three different volunteers to analyze the abundance of animals, these models answers queries such as: "What is the total abundance of birds per altitude, species and week?" (Fig. 2a). On one hand, as for classical DW design methodologies, these requirements can present the following issues: *Similarities* such as "abundance + SUM", "day", etc., and *Differences* such as "Season_bio", "behavior", etc. On the other hand, since for different goals, different volunteers have defined these requirements, the multidimensional model can present some *Conflicts*. For example, for the "abundance" measure of some species, the data acquisition protocol requires that the observation last for a particular duration or distance (ex: 10 m for butterflies). Therefore, this measure makes no biological analysis sense, unless it is accompanied by the observation duration or

---

[3] www.vgi4bio.fr.

[4] Farmland biodiversity observatory www.observatoire-agricole-biodiversite.fr.

distance. These conflicts are not issued by the source data, but they are due to disparities of knowledge and expertise in the application domain. Thus, they cannot be solved using any automatic tool, but only by specialists. Moreover, due to the huge number of volunteers in any VGI project, providing an implementation for each proposed model is unrealistic because of its high human, temporal and economic costs. Therefore, we propose to design one or only few models that represent an agreement for all volunteers solving the Similarities, differences and Conflicts issues, instead of the classical DW implementation. In the rest of the paper, we use a simple graphical representation of multi-dimensional conceptual models for reasons of brevity.

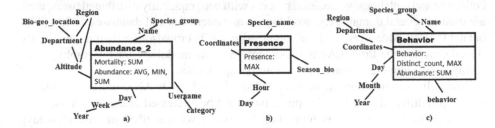

Fig. 2. SOLAP models of volunteers

3   OLAP2.0

In this section, we define the main steps of our volunteer design methodology for DW design (Fig. 1). In the rest of the paper we use the terms 'requirement' and 'model' for defining multidimensional requirement and multidimensional model, respectively. The methodology is composed of the following steps:

1. *Requirements elicitation, modelling and validation on data.* It aims at collecting requirements of each volunteer, translates them into models and validates them on data source (Sect. 3.1).
   The following two steps aim at solving issues of requirements as previously described. Since requirements are translated into validated models, these following two steps provide a refinement of the models from the previous step:
2. *Solving Differences and Similarities of requirements.* This step merges the different volunteers' models in order to solve Similarities and Differences issues, and generates refined models. (Sect. 3.2), it is based on existing works that integrates data marts.
3. *Collaborative resolution of requirements' conflicts.* This step allows committers to solve conflicts (Sect. 3.3).
4. The models that meet the committers' agreement are then implemented.

It is important to underline that the collaborative design step has not been added from the beginning of the design process for two important reasons:

(a) The lack of collaborative tools and methodologies for DW design,

(b) Since the impossibility of achieving an agreement among committers a priori, moving the collaborative task after the models' definition will grant us at least a set of possible models that can be implemented.

Let us provide some notations used in the next: (i) An Indicator is the measure + aggregation function; (ii) A cube is a model (dimensions and fact) (iii) A dimension d is a directed acyclic graph (iv) A hierarchy is a path from root to leaf of d, e.g. the Location dimension in Fig. 2a have 3 hierarchies: {*Altitude –> region, Altitude –> Department –> Region* and Altitude *–> Department –> Bio-geo_Location*}.

3.1 Requirements Elicitation, Modeling and Validation on Data

This step is composed of two phases: the first is the *requirements elicitation*, and the second is their *translation into valid multidimensional models*.

We use the ProtOLAP methodology and tool [3] for the elicitation of volunteers' requirements. According to the elicitation of requirements practices [2], ProtOLAP provides interviews, workshops, and prototyping. In particular with ProtOLAP, volunteers explain their analysis requirements during meetings in natural language and using word/excel documents [14]. Then, the DW experts transform them into a UML model, defined using the UML profile ICSOLAP implemented in the commercial CASE tool MagicDraw. Finally, the ProtOLAP tool generates a prototype cube from the expressed requirements. This prototyped cube is used in an iterative process to support volunteers eliciting their requirements e.g. models in Fig. 2.

After this elicitation phase, these cubes are validated by DW experts on data sources using an existing hybrid DW design methodology [7, 15], and the DW experts associate to each model a goal specification given by its definer to be used later at the third step. For example, for the model of Fig. 2b, the owner volunteer announce, "This model is for analyzing spatial and temporal coverages of VGI data".

Let us note that, by using ProtOLAP, multidimensional requirements are simply represented with pivot tables of prototyped cubes, which, as shown in [11], can be automatically translated into well-formed models. This allows us to avoid the usage of multidimensional requirements formalisms, which can be very complicated for our decision-makers (i.e. volunteers). Moreover, the volunteers know very well the dataset since they have already used and/or alimented it. Therefore, they can easily define some indicators over the source dataset, which eases the validation of the requirements on the data sources (such as in an on-demand data supply approach [3]). Finally, to avoid vocabulary alignment issues during the elicitation phase with ProtOLAP, DW experts check and oblige volunteers to use the same vocabulary when possible using a MagicDraw repository to keep a track of previously used terms for every specified requirement (such as [1]). For example, for the temporal dimension, the "Time" dimension name is imposed.

To conclude, *this step takes as input the requirements of each volunteer, and outputs a set of multidimensional models that are validated on data sources*.

3.2 Solving Differences and Similarities of Requirements

This step aims to solve Differences and Similarities among similar requirements that were differently defined in step 3.1. This step is based on previous existing methods of data marts design and integration [11, 15, 23]. In this paper, we provide our own methodology only for comprehensibility purposes. To this goal, it refines these models by merging them, which is achieved using the *Dimensions algorithm* of Fig. 3. For each common measure, the algorithm fuses all dimensions of different models in one model. In this way, when a volunteer expresses the same analysis subject i.e. the measure of other volunteers, but using different dimensions/hierarchies, the *Dimensions* algorithm returns the same analysis subject but enriched with dimensions of the other volunteers. For example "F1" model in Fig. 5 is the fusion of the two cubes "abundance_2" and "behaviour" based on their common measure "Abundance" with their common dimension "Species", their non-common dimensions "Behaviour" and "Users", and their non-conformed dimensions "Time" and "Location".

```
Input all cubes C
Output FinalCubes a set of cubes
1:M = measures of C;
2:For each m in M do
3:  Generate  new cube FusionCube;
4:  Add m to FusionCube;
5:Let  CommonDims  =  Common  dimensions of
cubes with m;
6: Let NonCommonDims = Non Common dimensions
of cubes with m;
7:   Add CommonDims and NonCommonDims to
FusionCube;
8:  Let  NonConformedDims  =    Non  Conformed
dimensions of
cubes with m;
9:   For each d of NonConformedDims do
10:   Let H = hierarchies of d;
11:   d = FusionHierarchies(H);
12:   Add d to FusionCube;
13: Endfor
14: Add FusionCube to FinalCubes;
15: Endfor
16: For each cubes set Cs having Common
dimensions do
17:Generate FinalCube with Common dimensions and
All
measures of Cs
18:  add FinalCube to FinalCubes
19:Endfor
20:return Cf
```

```
Input: hierarchies h1, …, hn
Output: Dimension d
1: G = Union(h1, …, hn);
2 V = all Bottoms of G  ;
4:   If  size(V) > 1 do
5:      choose vBottom among
V to
6:         forEach node    in
V_bottoms  do
7:          CreateEdge  (G,
vBottom, node);
8:            Endfor
9: Endif
10: d = Ø;
11: ForEach path P in G do
12:            add P  to  d
13: Endfor
14: return d
```

Fig. 3. Dimensions algorithm   **Fig. 4.** Hierarchies algorithm

Likewise, the *Hierarchies algorithm* Fig. 4 aims to return all possible hierarchies defined for a commonly, but differently defined, dimension.

The *Hierarchies' algorithm* merges all hierarchies in one graph, and then finds all possible paths from the leaf to the root nodes. When the graph has multiple bottom leaves, the DW expert must choose one, e.g. in "F1" model of Fig. 5b, the level "coordinates" was considered as the lowest level (lower than the level "Altitude") of the enriched hierarchy of the dimension "Location".

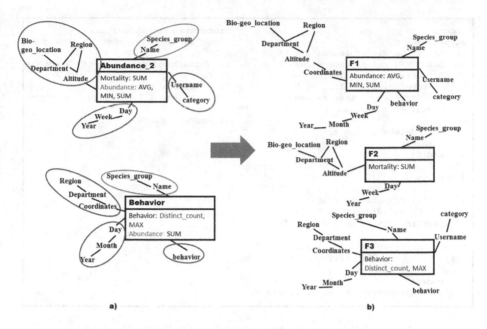

Fig. 5. Solving differences and similarities of requirements step example

To conclude, this step allows proposing useful dimensions and hierarchies that volunteers have unintentionally forgotten or consciously ignored, i.e. *Differences* requirement issue, and use, when possible, the same multidimensional elements, i.e. *Similarities* requirement issue.

3.3   Collaborative Resolution of Requirements Conflicts

The aim of this step is to solve the conflicts engendered at the previous step by means of another refinement of the previously obtained models (Sect. 3.2): *Are the multidimensional elements added by the Solving Differences and Similarities of requirements step needed by all volunteers?* The refinement is provided by the *Collaborative design algorithm* (Fig. 6), where the committers express their recommendations for each multidimensional element according to some criteria concerning their utility and usability. The algorithm finds a consensus among committers, and returns the agreed models. In the following, we firstly describe the algorithm, and then we explain the objective of each used method.

```
input: Cube c                          Procedure CleanIndicators
output Cube c                           Input set of Dimension D' subset of
1:  Let I the set of indicators of c;   D
2:  Let D the set of dimensions of c;   1:  ForEach dimension d of D not in
3:  SetConfidenceLevel(c);              D' do
4:  I'=VoteIndicators(I);               2:     delete d from c;
5:  if I' is empty then return;         3:     delete all Holistic indicators
6:    ForEach indicator i of I not in I' from c;
do                                      4: endFor
7:       delete i from c;
8:    endFor
9:  endif
10:      D'=RankDimensions(D);    //rank
dimensions
11   CleanIndicators(D',D) ;
12:  D''=VoteHierarchies(D');
13:    CleanIndicators(D'',D') ;
14:  ForEach dimension d of D'' do
15:
flag=VoteCubeDimensionUsability(d);
16:            if flag is false then
CleanIndicators(d,D'')
17: endFor
18: VoteImplementationCube(c);Return c;
```

Fig. 6. Collaborative resolution of requirements conflicts algorithm

Collaborative resolution of requirements conflicts algorithm: using the method 'SetConfidenceLevel' of Table 1 (Fig. 6-Line 3), committers define a confidence level for the cube according to their skills in the cube's application domain. This confidence level prioritizes the choices of committers with most appropriate skills regarding the under evaluation cube e.g. a committer specialized in ecology, sets his/her confidence level for the cube 'Behaviour' in Fig. 7a to "High".

Table 1. Resolution of requirements conflicts methods

For each committer	Input	Output	Method	Criteria
SetConfidenceLevel	Cube	Conf- level	auto-evaluation	Application skilled
VoteIndicators	Indicators	Indicators	Vote (Borda)	Indicator is useful
RankDimensions	Dimensions	Dims rank	Vote (Borda)	Dimension is useful
VoteHierarchies	Dimension hierarchies	Dims rank	Multicriteria (Weighted avg)	- Hierarchies richness - Fact-dim is accurate
VoteCubeDimensionUsability	Dimension	Dimension	Vote (Majority)	Cube with dimension is usable
VoteImplementationCube	Cube	Finale cube	Vote (Majority)	Cube must be implemented

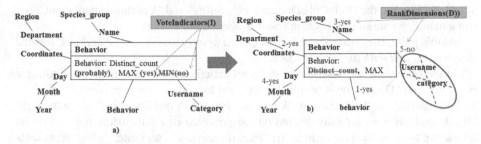

Fig. 7. Collaborative resolution of requirements conflicts example

Then, the committers evaluate the analysis relevance of each indicator in order to remove useless indicators from the final cube by the method 'VoteIndicators' of Table 1 (Fig. 6-Line 4). As an example, all committers estimate that "behaviour + Min" is not a relevant indicator, therefore, it is removed (Fig. 7a).

Afterwards, if at least one indicator is kept after the previous vote procedure, the committers evaluate the analysis relevance of each dimension in order to remove the useless dimensions, using the method 'RankDimensions' of the Table 1 (Fig. 6-Line 10), e.g. *committer1* considers relevant all but "Users" dimension, then it's removed (Fig. 7b). Note that holistic indicators [12] are removed when a dimension is not kept after the vote procedure (Fig. 6-Line 11) since this type of indicators becomes erroneous when it haven't access to the finest level of granularity after a dimension's removal. For other indicators (i.e. distributive and algebraic) the dimension elimination does not pose problems since measures can be aggregated on its 'All' member, and then reused for other aggregations (such as materialized views [12]).

Once all useless dimensions are gotten rid of, the committers must evaluate each retained dimension according to its hierarchies' richness and the accuracy of its lowest level of granularity, that is done by the method 'VoteHierarchies' of Table 1 (Fig. 6-Line 12). In our case study for example, the committers agreed that all the dimensions' hierarchies are well defined. Let us note that, this method eliminates the dimension if all its hierarchies are eliminated. Indeed, with the 'VoteCubeDimensionUsability' method of Table 1 (Fig. 6-Line 15) the committers must evaluate the usability of the cube with every dimensions [20], since it is well known that the number of used dimensions affects the usability of the cube, and so the decision-making process. For that goal, the algorithm, starting from the most important dimension, adds dimensions consecutively to the cube showing each time the resulting cube to committers. In this way, committers, exploring the cube with the new added dimension, decide of its usability, and thus to keep it or not.

Finally, the committers vote the implementation of the resulting cube made with 'VoteImplementationCube' method of Table 1 (Fig. 6-Line 18).

Methods description: The Table 1 illustrates the methods used by the *Collaborative resolution of requirements conflicts* algorithm.

'VoteIndicators' and 'RankDimensions', use a vote procedure with the Borda calculation method [16], since they have only one criterion. The 'VoteHierarchies' uses a weighted sum aggregation operator as well as a weighted sum, since it a multicriteria approach. Finally, the 'VoteCubeDimensionUsability' and 'VoteImplementationCube' use a majority vote since a boolean result is needed.

At this point, the obtained cube *is composed of only usable, useful and well-formed dimensions, and with useful indicators.*

In the following, we describe the different criteria used by the methods. "Indicator is useful" and "Dimension is useful" [5] are used to evaluate the necessity of indicators and dimensions for the decision-making goal. For the 'VoteHirarchies', since the right OLAP analysis does not only depend on the presence of a dimension, but also on its levels, we have defined the criteria: (i) "Fact-dimension is accurate", which represents whether or not the factual data are stored at the convenient dimension's granularity, and (ii) "The hierarchies of the dimension are rich enough", which means that sufficient aggregation possibilities exists over the dimension. "Cube with dimension is usable" [5, 20] is used to check the degree of usability of the cube using each dimension, and finally "Cube must be implemented" corresponds to the evaluation of the users' satisfaction about the obtained cube [5]. Let us note that we have used a scoring scale of [1–5] for all our GDSS evaluations.

4   Implementation and Validation

4.1   Implementation

The methodology has been implemented in a Relational OLAP architecture composed of Postgres as DBMS, Mondrian as OLAP server, and JRubik as OLAP client.

We use the ProtOLAP system for the first step of our methodology. ProtOLAP takes as input an UML model defined using ICSOLAP UML profile for SOLAP [3], which is implemented in the CASE tool MagicDraw. It automatically creates the SQL scripts for Postgres (tables creation and data insertion) and XML configuration Mondrian files. The collaborative design has been carried out by the GRUS system [19]. With a voting-oriented approach as well as a Multi-Criteria approach, we defined a specific group decision-making process. During the voting-oriented approach, users participated to GRUS system and sorted the alternative elements in order of their preferences. For the Multi-Criteria approach, participants gave to every element a mark based on each criteria. The system then, returned a report of results. Finally, it is important to underline that GRUS is a web-based system that allows asynchronous processes. Therefore, it is well adapted to our committers that are geographically located in different places, and work at different time.

4.2   Experiments and Validation

For the validation of our proposal, we engaged four volunteers with different skills, and we have identified four committers.

For the validation of the first step (Sect. 3.1) using ProtOLAP, we have counted the number of meetings between volunteers and DW designers and their duration. The time of implementing a DW prototype with ProtOLAP is negligible, since it is only a few minutes task. In average, there are three meetings by volunteer and each is one hour long. Therefore, we can conclude that only when the number of volunteers is small, the usage of the ProtOLAP methodology is possible. When the number of volunteers becomes significant, a new methodology must be provided to allow volunteers defining themselves their OLAP models without the intervention of DW designers.

To validate the proposed *collaborative resolution of requirements conflicts* methodology (Sect. 3.3), we considered one cube defined with one ornithology decision maker, which corresponds exactly to the experts' needs. Then, we have modified it by adding some dimensions and indicators that the ornithologist considers useless. In this way, we obtained a degraded cube. In particular, we have added a dimension "Users", and the indicator "Max behaviour". Finally, we submitted this cube to committers, and we tested whether or not using our design methodology, committers will be able to obtain the original 'good' cube. The experiments validated our methodology, since 'VoteIndicators' effectively classified "Max behaviour" as the last important indicator and the 'RankDimensions' function eliminated the "Users" dimension (with only 7.7% of votes). In GRUS, the Borda calculation method does not eliminate alternatives, and then we have chosen for each vote method a threshold for eliminating the multidimensional elements. For example, for 'RankDimensions' 10% or under would be eliminated. All other methods kept the other multidimensional elements. In this way, the exact original cube was returned by the end of the collaborative step.

Finally, since committers are not employed by the project and then they cannot spend too much time, it is important to note that the complete collaborative process has taken less than one hour, and it has been done during one meeting.

5   Related Work

(S) DWs design has been investigated in several works [7, 15]. Three types of approaches have been defined: (i) methods based on user specification (user-driven approach), which define the DW schema using users requirements only (i.e. analysis needs); (ii) methods based on data sources (data-driven approach), where the multidimensional schema is automatically derived from the data sources; (iii) mixed methods (hybrid approach), which merges data-driven and user-driven methodologies. it has been widely recognized that mixed approaches are the most effective for the design of successful DW projects. They provide mechanisms to map and validate users' requirements on data sources, and output a model [7]. However, as previously described in Sect. 1, they are not appropriate for our vision since they do not provide collaborative support needed to solve *conflicts* of requirements. Indeed, although conflicts management during the requirement elicitation phase has been explored in several domains [9], this software engineering theory has not yet been applied to the DW design. To the best of our knowledge, only [6] provides an agile questionnaire-based methodology to help decision-makers to work together in the conception of the DWs, but this approach is not supported by a computer tool. Contrary to DWs, users participation to design and collaborative design methodologies has been adopted in other fields (such as in GIS [8], to aid in mitigating semantic coherence, in socio-material design [10], e-learning, etc.). For the *similarities and differences* among users' requirements, several approaches for DW schema mapping and similarities have been proposed in literature (such as [1]), but they are too much complex to be used in our proposal, since, contrary to existing approaches, in our context all models are defined from the same source of data.

Existing computer tools for collecting analysis needs within user-driven approaches are formalized using complex formalisms [7, 15], or query languages (i.e. SQL, MDX, etc.). However, in our approach with ProtOLAP we use the same approach of [11] that formalizes requirements as pivot tables. This allows us to use pivot tables (i.e. cubes prototypes) to represent requirements but also to elicit them. Indeed, about the elicitation of multidimensional requirements, apart from manual approaches (meetings, reports, etc.), some existing works provide automatic tools for translating requirements defined in natural language into models [22]. Nevertheless, they require that decision-makers are OLAP skilled users, which is obviously not true for volunteers.

6   Conclusion and Future Work

In this work, we propose new collaborative DW design methodology that allows involving volunteers in the definition of analysis needs over VGI data. Our methodology allows DW and OLAP unskilled volunteers to participate to the design process. We implement the methodology and validate it using a real farmland biodiversity case study, thus, a better assessment would be by applying it on a case study in which, different volunteers with conflicting models can attend the collaborative evaluation step to validate the effectiveness of the conflicts resolution.

Our current work is dedicated to apply the collaborative methodology also on hierarchies' definition and to test other group decision methods. Moreover, with ProtOLAP, DW experts must assist volunteers in the elicitation process, which becomes impossible in a large-scale requirements crowdsourcing scenario. Then, our future work is to provide a user-friendly visual language based on the pivot table metaphor for the multidimensional requirements elicitation step. Finally, we will extend criteria used by our collaborative approach according to qualitative metrics defined for DW user satisfaction, as in [5], as well as integrating some quantitative ones such as what [20] highlighted.

Acknowledgment. This work is supported by the project ANR-17-CE04-0012. We thank Pr. Omar Boussaid and Stefano Rizzi for their precious advices.

References

1. Bakillah, M., Mostafavi, M.A., Bédard, Y.: A semantic similarity model for mapping between evolving geospatial data cubes. In: Meersman, R., Tari, Z., Herrero, P. (eds.) OTM 2006. LNCS, vol. 4278, pp. 1658–1669. Springer, Heidelberg (2006). https://doi.org/10.1007/11915072_72
2. Bimonte, S., Boucelma, O., Machabert, O., Sellami, S.: A new Spatial OLAP approach for the analysis of volunteered geographic information. Comput. Environ. Urban Syst. **48**, 111–123 (2014). https://doi.org/10.1016/j.compenvurbsys.2014.07.006

3. Bimonte, S., Edoh-alove, E., Nazih, H., Kang, M.-A., Rizzi, S.: ProtOLAP: rapid OLAP prototyping with on-demand data supply. In: DOLAP 2013, pp. 61–66. ACM, New York (2013). https://doi.org/10.1145/2513190.2513199
4. Brabham, D.C.: Crowdsourcing as a model for problem solving: an introduction and cases. Convergence. **14**, 75–90 (2008). https://doi.org/10.1177/1354856507084420
5. Chen, L., Soliman, K.S., Mao, E., Frolick, M.N.: Measuring user satisfaction with data warehouses: an exploratory study. Inf. Manage. **37**, 103–110 (2000). https://doi.org/10.1016/S0378-7206(99)00042-7
6. Corr, L., Stagnitto, J.: Agile Data Warehouse Design: Collaborative Dimensional Modeling, from Whiteboard to Star Schema. DecisionOne Consulting (2011)
7. Cravero, A., Sepúlveda, S.: Multidimensional design paradigms for data warehouses: a systematic mapping study. J. Softw. Eng. Applications. **07**, 53 (2013). https://doi.org/10.4236/jsea.2014.71006
8. Driedger, S.M., Kothari, A., Morrison, J., Sawada, M., Crighton, E.J., Graham, I.D.: Correction: using participatory design to develop (public) health decision support systems through GIS. Int J Health Geogr. **6**, 53 (2007). https://doi.org/10.1186/1476-072X-6-53
9. Egyed, A., Grunbacher, P.: Identifying requirements conflicts and cooperation: how quality attributes and automated traceability can help. IEEE Softw. **21**, 50–58 (2004). https://doi.org/10.1109/MS.2004.40
10. Ehn, P.: Participation in design things. In: Proceedings of the Tenth Conference on Participatory Design 2008, Indianapolis, IN, USA, pp. 92–101 (2008)
11. Nabli, A., Feki, J., Gargouri, F.: Automatic construction of multidimensional schema from OLAP requirements. In: AICCSA (2005). https://doi.org/10.1109/aiccsa.2005.1387025
12. Kimball, R., Ross, M.: The Kimball Group Reader: Relentlessly Practical Tools for Data Warehousing and Business Intelligence Remastered Collection. Wiley (2016)
13. Levrel, H., et al.: Balancing state and volunteer investment in biodiversity monitoring for the implementation of CBD indicators: a French example. Ecol. Econ. **69**, 1580–1586 (2010). https://doi.org/10.1016/j.ecolecon.2010.03.001
14. Nuseibeh, B., Easterbrook, S.: Requirements engineering: a roadmap. In: Conference on the Future of Software Engineering, pp. 35–46. ACM, New York (2000). https://doi.org/10.1145/336512.336523
15. Romero, O., Abelló, A.: A survey of multidimensional modeling methodologies. IJDWM **5**, 1–23 (2009). https://doi.org/10.4018/jdwm.2009040101
16. Gavish, B., Gerdes, J.H.: Voting mechanisms and their implications in a GDSS environment. Ann. Oper. Res. **71**, 41–74 (1997). https://doi.org/10.1023/A:1018931801461
17. Stefanovic, N., Han, J., Koperski, K.: Object-based selective materialization for efficient implementation of spatial data cubes. IEEE Trans. Knowl. Data Eng. **12**, 938–958 (2000). https://doi.org/10.1109/69.895803
18. Sui, D.Z., Elwood, S., Goodchild, M. (eds.): Crowdsourcing Geographic Knowledge: Volunteered Geographic Information (VGI) in Theory and Practice. Springer, Dordrecht (2013). https://doi.org/10.1007/978-94-007-4587-2
19. Zaraté, P.: Tools for Collaborative Decision-Making: Zaraté/Tools for Collaborative Decision-Making. Wiley, London (2013)
20. Golfarelli, M., Rizzi, S.: Data warehouse testing: a prototype-based methodology. Inf. Softw. Technol. **53**(11), 1183–1198 (2011). https://doi.org/10.1016/j.infsof.2011.04.002
21. Pohl, K.: Requirements Engineering: Fundamentals, Principles, and Techniques. Springer, Heidelberg (2010)

22. Naeem, M.Asif, Ullah, S., Bajwa, I.S.: Interacting with data warehouse by using a natural language interface. In: Bouma, G., Ittoo, A., Métais, E., Wortmann, H. (eds.) NLDB 2012. LNCS, vol. 7337, pp. 372–377. Springer, Heidelberg (2012). https://doi.org/10.1007/978-3-642-31178-9_50
23. Torlone, R.: Two approaches to the integration of heterogeneous data warehouses. Distrib. Parallel Databases **23**(1), 69–97 (2008). https://doi.org/10.1007/s10619-007-7022-z

Using a Conceptual Model to Transform Road Networks from OpenStreetMap to a Graph Database

Dietrich Steinmetz[1], Daniel Dyballa[1], Hui Ma[2], and Sven Hartmann[1(✉)]

[1] Clausthal University of Technology, Clausthal-Zellerfeld, Germany
{dietrich.steinmetz,daniel.dyballa,sven.hartmann}@tu-clausthal.de
[2] Victoria University of Wellington, Wellington, New Zealand
hui.ma@vuw.ac.nz

Abstract. We present a method for extracting road network data that has been crowdsourced in the OpenStreetMap project and transform it into a road network that is stored as a graph database. We propose an algorithm for the transformation, discuss opportunities for semantic enrichment of the road network, and for restricting the transformation to geographic regions of interest. Our approach is guided by a conceptual schema. To evaluate the practicability of our approach we have implemented it in a prototype tool, and conducted experiments that demonstrate the scalability.

Keywords: Geographic information · Data modelling
Graph database

1 Introduction

Graph databases represent data by graph structures with vertices, edges and properties. Recently a new class of DBMS has emerged that provide effective support for native storage, retrieval and management of relationship-centric data, such as Neo4j [14]. These systems are based on a simple property-graph model that permits efficient, highly scalable data traversals and avoids time-consuming joins. Meanwhile these systems also offer dedicated SQL-like query languages, such as Cypher in Neo4j. This makes them an attractive technology for applications that want to persistently store and process relationship-centric data.

In this paper, we focus on road networks that are relation-centric by nature. Road network data is an essential asset in many popular applications, such as route planning, navigation, ride sharing, traffic simulations, or urban planning. The database community has discovered road networks as a worthwhile study object, too [1,5,6,9–11,18–21,23,24]. While these works address a variety of different research problems, all of them conduct experiments to empirically explore the performance of proposed solutions. There is demand, both from industry and research, to make road network data readily available.

J. C. Trujillo et al. (Eds.): ER 2018, LNCS 11157, pp. 301–315, 2018.
https://doi.org/10.1007/978-3-030-00847-5_22

OpenStreetMap (OSM) [16] is a community project with the goal to generate a map of the world. The amount of road network data collected in OSM is comparable to commercial providers such as Google Maps. The advantage of OSM data is that it is free for personal and commercial use under the Open Data Commons Open Database License. OSM data can be downloaded in XML format. Unfortunately the way how OSM represents road network data is not relationship-centric. The conceptual data model of OSM can only represent geometric objects such as points, polylines or polygons. For example, road intersections and road segments are represented as points and polylines in OSM. However, the meta information that would allow the efficient traversal through these geometric objects is lacking. It is known that computing a route from a start point to an end point is not efficient when directly using OSM data [8].

We want to make road network data available in a graph DBMS so that it is directly accessible by applications that process such data. Our goal is to develop a method for extracting road network data from an OSM data set, and to transform it into a graph database which can be persistently stored and managed in a graph DBMS such as Neo4j. The input of the transformation will be an OSM data set in XML format. The elements of the conceptual data model of OSM will be mapped to the entities of a property-graph model.

We propose an algorithm for the transformation that defines when and how an element is transformed, and how its components are handled. Our main focus is on (1) single point objects that represent points of interest (POI) such as intersections, and (2) polylines that represent road segments, and determine course of the road. Obviously, not everything in an OSM data set is road network data. The input file has to be filtered to extract those elements that belong to the road network. Furthermore, it is important to control road types and the set of road properties that are considered during the transformation. The user should be able to select prior to the transformation which road properties are of interest for an application. While graph databases do not require a conceptual schema, we found it extremely useful to specify one to guide the transformation and show it is correct. It also provides valuable assistance in understanding the data, communicating requirements, and formulating semantic queries. The instance of the schema, i.e., the road network produced by the transformation, is stored in Neo4j where it is ready-to-use for applications, i.e., no further adaptation or conversion of the data is needed.

In many applications, it is essential to restrict the road network data to certain regions of interest. In order to support users in comfortably specifying regions of interest the course of borders such as district borders or national borders will be extracted from the OSM data and stored in the graph database, too. The stored border data will then be available for topological or analytical queries to be evaluated using a dedicated geographic library, such as Neo4j Spatial [15].

Organisation. This paper is organised as follows. In Sect. 2 we recall the conceptual data model of OSM. In Sect. 3 we give a formal definition of the problem and present the transformation for our proposed approach. In Sect. 4 we describe a prototypical implementation of our proposed approach. In Sect. 5

we report on experiments that we conducted to demonstrate the applicability of our proposed approach. In Sect. 6 we discuss related work. Finally, we conclude this paper in Sect. 7 and discuss suggestions for further research.

2 Background

The *conceptual data model of OSM* [16] provides three element type, see Fig. 1. *Node-elements* represent primitive objects located in a specific point on the map. A node-element may have attributes, such as *latitude* and *longitude* to specify the position on the map. *Way-elements* represent linear objects (called polylines) such as walkways, roads or boundaries. A polyline is a sequence of points. The points of a polyline will not be stored directly in the way-element, but rather references to the respective node-elements in an ordered list (called nd-list). If the last reference in the list is identical to the first one, then the polyline is actually a polygon. *Relation-elements* represent complex objects that often aggregate multiple node- and/or way-elements. A relation-element has a list of references to its member-elements (called member list). The attribute *type* of a member indicates whether it is a node- or a way-element. Each node-, way- and relation-element can have a set of *tags* associated to it. A tag provides a key-value pair and gives some property of the respective object.

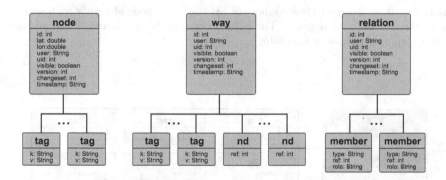

Fig. 1. Examples for the element types of the OSM data model.

Example 1. Figure 2 (left) shows a section of the OSM map for Bremen. In OSM, the road Frederikshavner Straße is represented by a set of polylines. The polyline that corresponds to the last road segment leading to the road intersection is marked in red.

The marked road segment is represented by the way-element with *id* "472359", see Fig. 3. Its nd-list contains references to two node-elements. It has 13 associated tags, each of them carrying a key-value pair that specifies one road property, such as *surface* or *maxspeed*.

This polyline references two node-elements. See Fig. 2 (right) for its nd-list. Both node-elements have a set of attributes, including *latitude* and *longitude* that give their position on the map. The road intersection is represented by the node-element with *id* "349072703". It has an associated tag carrying the key-value pair (*highway*, "traffic_signals"). Hence, there are traffic signals are located in this point, which will be visualised accordingly on the map, see Fig. 2 (left).

OSM data can be exported to XML format (called OSM/XML in this paper) and downloaded from the website of the OSM project. For our example above, the respective XML is shown in Fig. 4.

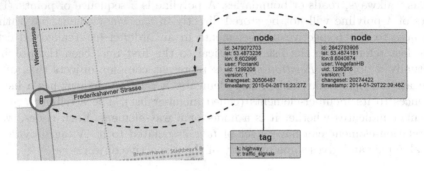

Fig. 2. A part of the road Frederikshavner Strasse. Two node-elements represent the ends of the marked road segment. The left-side one has traffic signals which are visualised accordingly. (Color figure online)

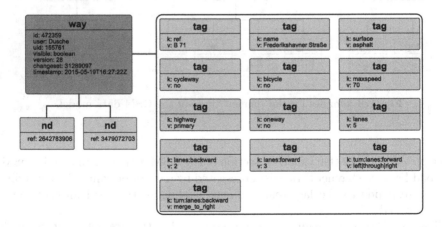

Fig. 3. The way-element that represents the marked road segment in Fig. 2.

```
<!--left node-element, traffic signal on intersection -->
<node id="3479072703" lat="53.4873236" lon="8.602996" ... >
    <tag k="highway" v="traffic_signals"/>
</node>
<!-- right node-element -->
<node id="2642783906" lat="53.4874181" lon="8.6040874" ... />
<!-- way-element -->
<way id="4723597">
    <!-- references to node-elements -->
        <nd ref="2642783906" />
        <nd ref="3479072703" />
    <!-- features and restrictions -->
    <tag k="name" v="Frederikshavner Strasse"/>
    <tag k="highway" v="primary"/>
    <tag k="surface" v="asphalt"/>
    ...
    <tag k="lanes:forward" v="3"/>
    <tag k="maxspeed" v="70"/>
</way>
```

Fig. 4. The OSM/XML fragment that corresponds to our example.

3   Method Conceptualization

In this section we will present our approach for extracting road network data from OSM and transforming them into a road network stored in a graph database.

3.1   Problem Definition

We introduce some terminology and formal notations. A *road network* is a directed graph $G = (V, E)$ where V is the set of vertices representing road nodes, and E is the set of edges representing road segments. Road nodes can be intersections, terminal points, bends, or similar. Each edge $e = (u, v) \in E$ connects two vertices $u, v \in V$.

Suppose we are given an *OSM data set* in the form of a triple $O = (N, W, R)$ where N is a set of node-elements, W is a set of way-elements, and R is a set of relation elements. Let $\mathcal{A}$ and K denote the sets of all attributes and keys, respectively. Every node-element $n \in N$ has a set of attributes $A_n \subseteq \mathcal{A}$ and a set of associated tags T_n, thus defining a partial function $tag_n : K \to dom$ such that $tag_n(k) \in dom(k)$ for all keys $k \in K$. Every way-element $w \in W$ has a set of attributes $A_w \subseteq \mathcal{A}$ and a set of associated tags T_w, thus defining a partial function $tag_w : K \to dom$ such that $tag_w(k) \in dom(k)$ for all keys $k \in K$. Moreover, it has a list $nd - list_w$ of references to nodes in N, and defines a partial function $next_w : N \to N$ such that $next_w(n)$ is the successor of n in the list $nd - list_w$.

Our goal is to transform an OSM data set into a road network and then store it in a graph database. The target of our transformation is a graph database that is based on the property-graph model, as for used e.g. in Neo4j [14]. We have developed the conceptual schema in Fig. 5 that will guide the transformation.

In our target schema, road nodes in V are captured by *node* vertices. They have at least the properties *latitude* and *longitude*. Further properties can be added if the user considers them relevant. The road type is captured by a label

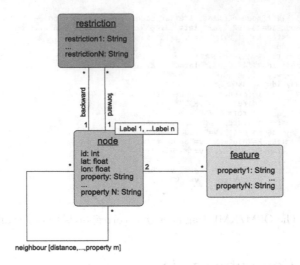

Fig. 5. Target schema of the transformation into a graph database

such as "motorway". In some applications it is important to distinguish between different road types. In taxi-ride sharing [6], for example, the pick-up and drop-off points of passengers cannot be on interstate highways. Labels can be easily accessed by taxi routing queries.

Road segments in E are captured by *neighbour* edges, each connecting two *node* vertices. They have at least the property *distance*. This enables shortest-path algorithms like Dijkstra or A^* to traverse the road network efficiently. Further properties can be added if the user considers them relevant.

Additional characteristics of road segments are captured by *restriction* and *feature* vertices. Experience shows that in Neo4j reading properties of an edge is much slower than reading properties of a vertex. To overcome this technical difficulty we propose a different solution: we simply add new vertices to the graph database to store the relevant properties of a road segment. In practical applications, it turned out to be advantageous to distinguish two kinds of tag keys: *features* that are independent of the direction in which the road is used, and *restrictions* that are dependent on the direction. For example, *surface* is a feature, while *maxspeed* is a restriction, see also Fig. 8.

3.2   Transformation of OSM Elements

To meet the requirements, we propose a transformation in two phases: first we focus on the type and course of the roads, and then we semantically enrich the road network. Our approach is outlined in Alg. 1. In principle, we map a node-element in the OSM data set O to a vertex in the road network G, and a way-element in O to one or more edges in G.

Note that not all information in an OSM data set is relevant for road networks. We are only interested in way-elements that represent road segments. These have an associated tag that carries a key-value pair (*highway*, *road_type*). The value herein specifies the road type. Possible values are "motorway", "trunk", "primary", "secondary", "tertiary", "unclassified", "residential", "service", etc. Further values are used to represent connections between roads of different types. For example, fedder roads that connect a normal highway to a motorway have the road type "motorway_link". We use the road type value to define a label for the vertices in G, see line 8 in Algorithm 1.

Example 2. Figure 6 shows two road segments of road type "motorway" as represented in OSM. Note that the two road segments are adjacent: both have a reference to the node-element with *id* "16" in their nd-list. The result of the transformation is shown in Fig. 7.

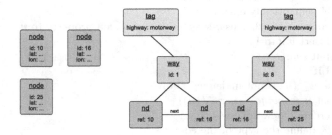

Fig. 6. Example of two way-elements that represent adjacent road segments

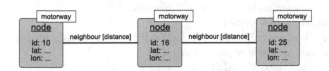

Fig. 7. The result of transforming the two way-elements in Fig. 6

The transformation of a node-element n is described in lines 5–15 in Algorithm 1. In this paper, we use the convention that only those node-elements will be mapped that belong to some way-element representing road segments. This ensures, e.g., that traffic signals will only be considered during the transformation if they are located on some road. Otherwise there could be traffic signals that are not related to any road, which is likely an error in the OSM data.

A node-element n may be referenced by multiple way-element, as in Fig. 6. We will map n to a single unique vertex in G, denoted by *image*(n). In G one would need to search for a vertex the *id* of n. If no such vertex exists, a new vertex v must be generated in G. Afterwards the key-value pair of the linked tag will be copied to the new vertex.

Algorithm 1 Transformation

Input: An OSM/XML file representing an OSM data set $O = (N, W, R)$
Output: A Neo4j graph database representing a road network $G = (V, E)$

1: **for all** way-elements $w \in W$ **do**
2:     **for all** $ref \in nd - list_w$ **do**
3:         let $n = id^{-1}(ref)$
4:         **if** $image(n)$ is undefined **then**
5:             create a new vertex v
6:             put $id(v) := ref$
7:             % define a label for v based on the roadtype
8:             add $tag_n(highway)$ as a label to v
9:             copy every other user-selected attribute of n as a property to v
10:           % check the semantic integrity of the tag set of n
11:           **if** T_n is not valid **then** clean T_n
12:           % semantically enrich v
13:           copy every other user-selected tag of n as a property to v
14:           insert v into V
15:           put $image(n) := v$
16:         **end if**
17:         **if** $next_w^{-1}(n)$ is not undefined **then**
18:             let $u = image^{-1}(next_w^{-1}(n))$
19:             create a new edge $e = (u, v)$
20:           % define the type of the edge e
21:           put $type(e) := neighbour$
22:           copy every user-selected attribute of w as a property to e
23:           insert e into E
24:           % check the semantic integrity of the tag set of w
25:           **if** T_w is not valid **then** clean T_w
26:           % semantically enrich e with features
27:           create a new vertex y
28:           copy every valid user-selected feature tag of w as a property to y
29:           create new edges (u, y) and (y, v)
30:           insert y into V and (u, y), y, v to E
31:           % semantically enrich e with restrictions
32:           create a new vertex z
33:           copy every valid user-selected feature tag of w as a property to z
34:           create new edges (u, z) and (z, v)
35:           put $type(u, z) := backward$ and $type(z, v) := forward$
36:           insert z into V and (u, z), (z, v) to E
37:         **end if**
38:     **end for**
39: **end for**

Note that not all attributes and tag keys stored in a node-elements will be of interest. It is up to the user to decide which ones she/he requires. We assume that the user selects those that she/he regards *relevant* prior to the transformation.

Then these ones will be copied during the transformation, see lines 9 and 13. In Fig. 7, e.g., the attributes *latitute* and *longitude* have been considered relevant.

The road course is determined by the nd-list of the way-element. It provides references to the node-elements that constitute the polyline, which are handled in lines 17–19. Often the tags associated with a way-element carry valuable information that should be kept. We copy relevant tags of w as properties to a new *feature* vertex and/or *restriction* vertex e, see lines 24–36. Note, however, that OSM itself does not provide such a classification of key tags. Hence, input from the user or a domain expert is needed.

Note that there are dependencies between the tags that have to respected for *semantic integrity*. We have indicated this in lines 11 and 25. In the tags associated with a node- or way-element, certain keys may only occur if others occur, too. For example, speed limits can only exist on road types. In Germany a speed limit is optional on motorways, while it is mandatory in New Zealand.

Example 3. To illustrate our approach, see Fig. 8. Tags that are relevant for the user can be stored in the feature node or in the restriction node. The results of the transformation are shown in Fig. 9.

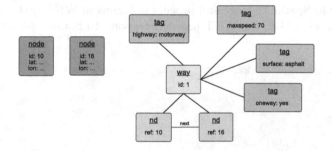

Fig. 8. An example of a way-element with tags carrying features and restrictions

For our transformation we can demonstrate that adjacency is preserved:

Proposition 1. *Two vertices u, v in G are adjacent (i.e., connected by a road segment $e = (u, v)$) if and only if there is a way-element w in O that has a tag with key* highway *and that references two node-elements m, n such that $image(m) = u$ and $image(n) = v$ and $next(m) = n$.*

3.3   Restriction to Regions

Finally, we extend our approach so that user queries against the road network can be restricted to specific geographic regions specified by the user. Actually, OSM provides a wealth of useful information on boundaries like national or regional borders. Neo4j Spatial [15] is a library of utilities for Neo4j that facilitates spatial

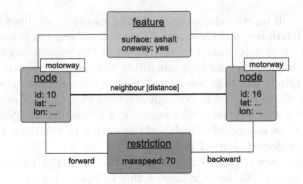

Fig. 9. The result of transforming the way-element in Fig. 8

operations on data. In particular, spatial indexes can be added to already located data, and spatial operations can be performed, e.g., searching for data within specified regions or within a specified distance of a point of interest.

To unlock these capabilities we have extended our conceptual data model to provide support for Neo4j Spatial. To achieve this we create a point layer in Spatial and add all vertices in our road network to this layer. Moreover, we create a WKT layer in Spatial to store and process polygons in WKT text format. The polygons that we add to the WKT layer correspond to boundaries on the map.

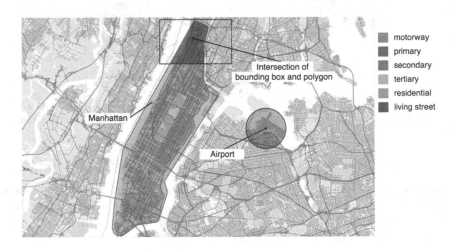

Fig. 10. Extension of our approach to support spatial queries

Therefore, we enhance our transformation to also extract and transform boundaries from OSM. They can be found using the key-value pair (boundary, "administrative"), and are usually represented by relation-elements in OSM. The points for the polygon are obtained from the member-list of the relation-element.

The polygons which we store in the WKT layer can be readily used as parameters for spatial queries against the road network in the graph database, see Fig. 10. Similarly, spatial queries that restrict query results to circles or bounding boxes are supported, too.

4   Proof of Concept

To demonstrate the practicability of our approach we have implemented a prototype system (called *OSM2RN transformer*) using Java. An OSM/XML file is used as input. It is imported, the OSM elements are filtered and transformed, and the results are stored into a graph database. For storing and managing the graph database we use the graph database management system Neo4j.

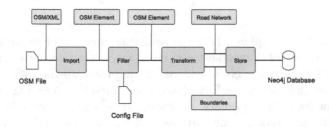

Fig. 11. Outline of the architecture of our OSM2RN transformer

The architecture of our tool, see Fig. 11, is based on the Pipes & Filter architectural pattern. It consists of four modules. The module *Import* takes an OSM/XML file as input and processes it using the SAX parser. Unfortunately, the ordering of the OSM elements in the OSM/XML file is not ideal for our purposes, as it starts with all node-elements followed by all way-elements followed by all relation-elements. For our transformation, the elements are needed in reverse order. Therefore we have to pass through the input file three times. The module *Filter* inspects each OSM element and its components whether they are needed, considering the general requirements and the user-specified rules. The module *Transform* executes the transformation. We obtain an instance of the target schema shown in Fig. 5. The module *Store* assembles the results of the transformation in Cypher operations and writes them into the Neo4j. When using Neo4j Spatial, then the extracted boundaries will be stored, too.

5   Case Study and Evaluation

In this section we will describe the test cases that we have used to evaluate our approach. In our experiments we have measured the execution time needed to transform an OSM data set and to store the results in the graph database. Every experiment has been repeated 30 times to eliminate measuring errors.

Testcases. We conducted experiments with two test cases to investigate the scalability of our approach. To explore the impact of using Neo4j Spatial, we conducted all our experiments with the library enabled and disabled. In *Testcase A*, we investigated how the size of the resulting road network impacts the run time. We used the OSM data set for Lower Saxony, see Fig. 12. The size of the resulting road network varied when different road types were considered during the transformation: in Test A1 we only considered motorway, motorway_link, in Test A2 also primary, primary_link, and in Test A3 also secondary, secondary_link.

	Testcase A: Lower Saxony	Testcase B: Berlin
#node-elements	26,897,284	4,622,951
#way-elements	31,676,139	4,957,448

Fig. 12. OSM data sets used for Testcases A and B

In *Testcase B*, we explored how the semantic enrichment impacts the run time. We kept the size of the road network constant, but varied the amount of features and restrictions that were considered during the transformation. We used an OSM data set for Berlin, see Fig. 12 and considered the following road types: motorway, motorway_link, primary, primary_link,secondary, secondary_link. In Test B1 we did not consider any features and restrictions, and then increased the amount stepwise in Tests B2 and B3 (in B1 features *tunnel, bridge, lanes, name* and restrictions *maxpseed*, in B2 additionally features *crossing, surface* and restrictions *motorcar, bicycle*).

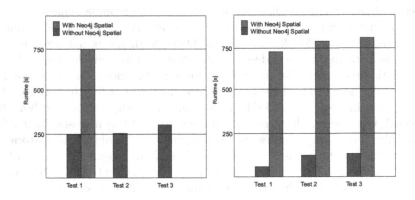

Fig. 13. Run times observed for Testcases A and B

Results. The run times are shown in Fig. 13, and the sizes of the resulting road networks in Fig. 14. In *Testcase A*, the size of the road network has at least doubled in each experiment. Without Neo4j Spatial, the run time increased from 248 s to 308 s for Test A1 to A3. When using Neo4j Spatial, Test A1 needed 748 s,

	Test A1	Test A2	Test A3
#road vertices	53,335	141,469	347,120
#road edges	53,896	143,519	353,214

	Test B1	Test B2	Test B3
#feature vertices	0	62,387	155,645
#restriction vertices	0	58,615	102,431

Fig. 14. Sizes of the resulting road networks for Testcases A and B

while Test A3 had to be aborted after 2 h. In *Testcase B*, the amount of semantic enrichment increased in each experiment. Without Neo4j Spatial, the run time increased from 58 s to 133 s for Test B1 to B3. When using Neo4j Spatial the run time increased from 731 s to 804 s for Test B1 to B3.

Our tests show that the use of Neo4j Spatial causes a performance loss. The insertion of geometry nodes into a layer in Spatial seems to be a time-consuming database operation. A possible explanation for this might be the build-in search index of the layer, which might require expensive reorganization after each insertion.

6 Related Work

The Neo4j Spatial library [15] also offers an importer for OSM data into Neo4j that can read OSM/XML files. In principle, it mimics the hierarchical organization of OSM/XML using node-elements, way-elements, and relation-elements. Different from our approach, it does not reflect the topology of the road network. Rather, the importer organizes the imported data such that it supports the spatial capabilities of Neo4j Spatial well. In particular, the importer does not generate a road network in form of a relationship-centric graph as defined in Sect. 3.1 which would simplify data traversal and speed-up computationally intensive applications such as navigation, ride sharing, traffic simulations or nearest neighbour queries. For that, further expensive transformations would be necessary to derive a suitable representation of the road network.

[22] studies the density and diversity of road networks in China based on OSM data for China, and finds that different parts of China observe different spatial patterns, and that road density reflects the intensity of traffic in the real world. This work could directly benefit from our tool as it unlocks the capabilities of Neo4j Spatial for analysing road network data and exploring spatial patterns.

[4, 17] present frameworks of methods for assessing OSM data quality solely based on the data's history. They propose quality indicators such as attribute completeness and average number of tags for points-of-interest (POI), or overall road length for road network completeness. When applying our approach to transform road networks into a graph database, such indicators can be efficiently evaluated by means of the data analytics capabilities of the Neo4j DBMS. Furthermore, the feature and restriction vertices that we have introduced in our conceptual database schema make the handling of historic data comfortable, as they can be versioned when data changes over time.

[3] explores the retrieval of geographic objects that are explicitly stored in OSM data, such as hospitals or lakes. [13] observes that there is a lack of methods for retrieving POI that are not explicitly stored, as this requires more detail

on geometries, topology and semantics. They propose rule-based spatial reasoning on OSM data using OWL and SQWRL, and discuss two applications for detecting entry points of footways and entrances of buildings. Similarly to our approach, it would be possible to design a conceptual model to extract the respective data for these applications from OSM, and store them in a graph database where they can be further analysed.

Conceptual modelling support for modelling and processing spatial data has been studied, e.g., in [7,12], but without focus on OSM, graph databases nor road networks. [2] proposes a rule-guided approach to resolve conceptual overlapping classes due to non-precise definition of geographic objects in OSM. [8] uses OSM data for personalized route planning, but without considering conceptual models nor graph databases. None of these works uses conceptual modelling to inform the transformation of road network data from OSM to graph databases.

7 Conclusions and Outlook

In this paper we have proposed a novel method for extracting road network data from OSM and transforming it into a road network in a graph database. To guide the transformation we have proposed a conceptual database schema for road networks that is based on the graph-property model and reflects the topology of the road network. We have discussed possible extentions and alternatives for the schema, described opportunities for semantic enrichment by user-selected relevant features and restrictions, and addressed the correctness of the transformation. We have further extended our approach by extracting and transforming boundaries from OSM that can then be used in the graph database to restrict user queries against the road network to geographic regions. We have implemented our approach in a prototype tool. Using this tool we have conducted experiments with several OSM data sets of different size. The results are very promising, and indicate that our approach is practical. Our tests have shown that our approach is more efficient without integrating the Neo4j Spatial library, but then the support for spatial queries is missing. It is up to the user to decide whether this extention is required for a particular application.

In future we plan to extend our approach to other information collected in the OSM project, such as railway network data.

References

1. Abeywickrama, T., Cheema, M.A., Taniar, D.: k-nearest neighbors on road networks: a journey in experimentation and in-memory implementation. PVLDB 9(6), 492–503 (2016)
2. Ali, A.L., Sirilertworakul, N., Zipf, A., Mobasheri, A.: Guided classification system for conceptual overlapping classes in OSM. Int. J. GeoInf. 5(6), 87 (2016)
3. Ballatore, A., Bertolotto, M., Wilson, D.C.: Geographic knowledge extraction and semantic similarity in OpenStreetMap. Knowl. Inf. Syst. 37(1), 61–81 (2013)
4. Barron, C., Neis, P., Zipf, A.: A comprehensive framework for intrinsic OpenStreetMap quality analysis. Trans. GIS 18(6), 877–895 (2014)

5. Chen, Z., Liu, Y., Wong, R.C., Xiong, J., Long, C.: Efficient algorithms for optimal location queries in road networks. In: ACM SIGMOD, pp. 123–134 (2014)
6. Cheng, P., Xin, H., Chen, L.: Utility-aware ridesharing on road networks. In: ACM SIGMOD, pp. 1197–1210 (2017)
7. Currim, F., Ram, S.: Modeling spatial and temporal set-based constraints during conceptual database design. Inf. Syst. Res. **23**(1), 109–128 (2012)
8. Graf, F., Kriegel, H.-P., Renz, M., Schubert, M.: MARiO: multi-attribute routing in open street map. In: Pfoser, D. (ed.) SSTD 2011. LNCS, vol. 6849, pp. 486–490. Springer, Heidelberg (2011). https://doi.org/10.1007/978-3-642-22922-0_36
9. Han, B., Liu, L., Omiecinski, E.: A systematic approach to clustering whole trajectories of mobile objects in road networks. IEEE TKDE **29**(5), 936–949 (2017)
10. Han, Y., Sun, W., Zheng, B.: COMPRESS: a comprehensive framework of trajectory compression in road networks. ACM ToDS **42**(2), 11:1–11:49 (2017)
11. Luo, S., Kao, B., Li, G., Hu, J., Cheng, R., Zheng, Y.: TOAIN: a throughput optimizing adaptive index for answering dynamic kNN queries on road networks. PVLDB **11**(5), 594–606 (2018)
12. Ma, H., Schewe, K.-D., Thalheim, B.: Geometrically enhanced conceptual modelling. In: Laender, A.H.F., Castano, S., Dayal, U., Casati, F., de Oliveira, J.P.M. (eds.) ER 2009. LNCS, vol. 5829, pp. 219–233. Springer, Heidelberg (2009). https://doi.org/10.1007/978-3-642-04840-1_18
13. Mobasheri, A.: A rule-based spatial reasoning approach for OpenStreetMap data quality enrichment; case study of routing and navigation. Sensors **17**(11), 2498 (2017)
14. Neo4j: Neo4j. https://neo4j.com/
15. Neo4j: Neo4j Spatial. http://neo4j-contrib.github.io/spatial/
16. OpenStreetMap: OpenStreetMap. https://wiki.openstreetmap.org/wiki
17. Singh Sehra, S., Singh, J., Singh Rai, H.: Assessing OpenStreetMap data using intrinsic quality indicators. Future Internet **9**(15), 1–22 (2017)
18. Steinmetz, D., Burmester, G., Hartmann, S.: A fast heuristic for finding near-optimal groups for vehicle platooning in road networks. In: Benslimane, D., Damiani, E., Grosky, W.I., Hameurlain, A., Sheth, A., Wagner, R.R. (eds.) DEXA 2017. LNCS, vol. 10439, pp. 395–405. Springer, Cham (2017). https://doi.org/10.1007/978-3-319-64471-4_32
19. Wang, S., Xiao, X., Yang, Y., Lin, W.: Effective indexing for approximate constrained shortest path queries on large road networks. PVLDB **10**(2), 61–72 (2016)
20. Yan, D., Zhao, Z., Ng, W.: Efficient algorithms for finding optimal meeting point on road networks. PVLDB **4**(11), 968–979 (2011)
21. Zhang, D., Yang, D., Wang, Y., Tan, K., Cao, J., Shen, H.T.: Distributed shortest path query processing on dynamic road networks. VLDB J. **26**(3), 399–419 (2017)
22. Zhang, Y., Li, X., Wang, A., Bao, T., Tian, S.: Density and diversity of OpenStreetMap road networks in China. J. Urban Manag. **4**(2), 135–146 (2015)
23. Zhao, J., Gao, Y., Chen, G., Jensen, C.S., Chen, R., Cai, D.: Reverse top-k geo-social keyword queries in road networks. In: IEEE ICDE, pp. 387–398 (2017)
24. Zheng, B., Su, H., Hua, W., Zheng, K., Zhou, X., Li, G.: Efficient clue-based route search on road networks. IEEE TKDE **29**(9), 1846–1859 (2017)

Cloud-Based Modeling

Modeling Conceptual Characteristics of Virtual Machines for CPU Utilization Prediction

Shengwei Chen, Yanyan Shen[✉], and Yanmin Zhu[✉]

Department of Computer Science and Engineering,
Shanghai Jiao Tong University, Shanghai, China
{shineway_chan,shenyy,yzhu}@sjtu.edu.cn

Abstract. Cloud services have grown rapidly in recent years, which provide high flexibility for cloud users to fulfill their computing requirements on demand. To wisely allocate computing resources in the cloud, it is inevitably important for cloud service providers to be aware of the potential utilization of various resources in the future. This paper focuses on predicting CPU utilization of virtual machines (VMs) in the cloud. We conduct empirical analysis on Microsoft Azure's VM workloads and identify important conceptual characteristics of CPU utilization among VMs, including locality, periodicity and tendency. We propose a neural network method, named Time-aware Residual Networks (T-ResNet), to model the observed conceptual characteristics with expanded network depth for CPU utilization prediction. We conduct extensive experiments to evaluate the effectiveness of our proposed method and the results show that T-ResNet consistently outperforms baseline approaches in various metrics including RMSE, MAE and MAPE.

Keywords: Cloud computing · CPU utilization · Residual network

1 Introduction

Recent years have witnessed the rapid growth of cloud computing technology. Many companies have migrated their workloads to cloud service platforms such as Microsoft Azure, Alibaba Cloud Compute Services, and Amazon Web Services. Under the pressure of market competition, cloud service suppliers have to provide attractive features to the customers while saving their platform costs, which, however, can be extremely hard to achieve without effective resource management. From the perspective of cloud resource management, understanding future demands of VM resources can help system administrators reallocate resources wisely in a dynamic manner. When it is foreseen that the demands for resources will increase, cloud providers could prepare more physical hosts to meet the growth of future demands in time. Similarly, when the demands of VM resources are predicted to experience a declining trend, cloud managers could stop allocating new resources and migrate the underloaded VMs properly

© Springer Nature Switzerland AG 2018
J. C. Trujillo et al. (Eds.): ER 2018, LNCS 11157, pp. 319–333, 2018.
https://doi.org/10.1007/978-3-030-00847-5_23

so that the idle physical hosts can be shut down to avoid waste of resources and improve the lifetime of equipments. Among various resources for VM workloads, CPU utilization is one of the most important indicators since it has a great impact on the total cost of the cloud service [1]. And it is more useful to understand the behaviors of maximum CPU utilization than average one, since the former keeps performance at high percentiles of the response time [2, 3].

In this paper, we focus on predicting maximum CPU utilization for long-running VMs based on historical utilization series. The key challenge of this problem lies in the instability of utilization series for each VM. Specifically, the series itself involves nonlinear short-term trends which increase the prediction difficulty. Furthermore, the volatility of the series varies with time, and it is hard to capture long-term temporal dependencies without delicately designed prediction methods. Most existing works on CPU utilization prediction leverage the conventional machine learning methods such as ARIMA [4], linear regression [5], hidden Markov model (HMM) [6, 7], and multilayer perceptron model (MLP) [8]. However, the performances of these methods are far from satisfactory. In particular, linear regression and ARIMA can only capture linear relationships for time series due to the restricted model complexity, and HMM-based methods can only predict the change of patterns according to pre-defined finite states. MLP can forecast nonlinear relationships but the model itself is too simple to catch long-term temporal dependencies.

To provide an effective method for CPU utilization prediction, we first conduct empirical analysis on real CPU utilization dataset from Microsoft Azure [9], from which we obtain two important observations and extract key concepts as follows. First, CPU utilizations for VMs within one deployment often fluctuate together over time due to the fact that multiple VMs in the same deployment are typically created to execute tasks collaboratively. Second, VM CPU utilization as a time series presents three conceptual characteristics: (1) locality: VM CPU utilization at present will impact the value in the near future; (2) periodicity: utilization series shows cyclical changes over time; (3) tendency: utilization will continue to increase or decrease in the long run with the change of work intensity. These refined characteristics as part of CPU utilization reflect the user's real-time requirements from different aspects.

Based on the above key observations, we develop a neural network method to solve the maximum VM CPU utilization prediction problem in consideration of VM behaviors. Given a target VM, we first use the Pearson correlation coefficient to select the most relevant VMs' utilization from a deployment, which are used as additional inputs to expand the target utilization series. We then divide the expanded utilization series into three parts at different time frequencies to represent locality, periodicity, tendency properties, respectively. Finally, we propose Time-aware Residual Networks (T-ResNet) based on residual networks [10]. Our T-ResNet takes three divided utilization series as inputs, and models each component by using an individual residual networks. The residual structure is able to capture both nonlinear short-term volatility and the long-term temporal dependencies simultaneously. The outputs of sub-networks are concatenated and fed to a fully-connected layer to re-weight the importance of latent features from different residual networks for final maximum CPU utilization prediction. The experimental results on Microsoft Azure data show that our model outperforms five benchmark methods in various metrics.

The contributions of this paper can be summarized as follows:

- We analyze the real Microsoft Azure dataset to disclose important conceptual characteristics of VM CPU utilization. We exploit both stochastic behaviors of VM CPU utilization as well as CPU utilization similarity among VMs.
- We propose Time-aware Residual Networks (T-ResNet) for CPU utilization prediction. T-ResNet dynamically aggregates the outputs of residual networks which model locality, periodicity, and tendency respectively, assigning different weights to different frequency patterns. The aggregation is further activated to generate utilization prediction.
- We conduct extensive experiments to evaluate the performance of our proposed method. The results show that our model outperforms five baseline approaches by reducing 19%–64% in RMSE, 38%–106% in MAE, 43%–132% in MAPE.

Organization. Section 2 provides definitions and problem statement. Section 3 presents the data set and empirical analysis for VM CPU utilization. Section 4 introduces the proposed model. Section 5 gives experimental results. We review related works in Sect. 6 and conclude this paper in Sect. 7.

2   Preliminaries

Definition 1 (VM CPU Utilization). *The virtual machine is the smallest executable unit hosted on physical servers. Let $V = \{v_t | t = 1, \cdots, T\}$ be a series of CPU utilization for one VM, where v_t is a 3-dimensional vector containing minimum, average, and maximum CPU utilization of the VM at time period t, denoted by $v_{min,t}$, $v_{avg,t}$ and $v_{max,t}$, where the time interval is fixed, e.g., 5 min.*

Definition 2 (Deployment). *A deployment contains a set of VMs that are managed together and allocated in the same cluster of servers. Let $D = \{V^i | i = 1, 2, \cdots, N\}$ be a deployment with N VMs, where the CPU utilization of the i-th VM is represented by V^i. Note that the VMs in the same deployment typically collaborate with each other for executing the same task.*

Since users prefer to execute tasks in batches by increasing (or shrinking) the number of VMs in a deployment, we focus on the deployments which contain VMs with long-running workloads in this paper. The prediction of CPU utilization of any newly created VMs is beyond the scope of this paper. Then we give the problem definition we would like to solve.

Problem Statement. Given a deployment $D = \{v_t^i | i \in [1, N] \wedge t \in [1, T]\}$ where v_t^i denotes the CPU utilization of the i-th VM during the t-th time interval, we aim to predict the maximum CPU utilization of all the VMs in D at time period $T + 1$, i.e., $v_{max,T+1}^i$ for all $i \in [1, N]$.

3   Empirical Analysis

We start with the description of the Microsoft Azure dataset[1] offered by Cortez et al. [9]. We then perform an empirical analysis on the dataset and present several key observations on CPU utilization behaviors of the VMs within a deployment.

3.1   Microsoft Azure Dataset

Data Description. Microsoft Azure is one of the largest and most influential cloud providers in the world, and contains both first-party and third-party VM workloads. The dataset from [9] only includes first-party workloads of more than two million VMs running on Azure spanning cross 30 consecutive days from November 16 to December 15, 2016. The first-party workloads mainly combine Microsoft development, test, and internal services.

Data Preprocessing. As we aim at predicting CPU utilization of long-running VMs, we first filter VMs with short lifetime from the original dataset. In this paper, we consider deployments where all VMs run throughout the entire life cycle of the dataset, i.e., 30 consecutive days. After filtering deployments that do not satisfy our requirement, we obtain 3,005 deployments with 16,065 VMs. And the following presented observations are based on these remaining deployments.

3.2   Stochastic Behaviors of VM CPU Utilization

Since VM may execute tasks for a long time, it is meaningful to analyze the CPU utilization patterns belonging to each VM itself. Seasonal decomposition is a statistical method used for time series decomposition, which decomposes a time series into several components: trend, seasonality, and residual [11]. The trend is defined as the increasing or decreasing value in the series; seasonality is defined as the repeating short-term period in the series; residual is defined as the random variation in the series. Based on seasonal decomposition, we are able to transform the CPU utilization series into several components as follows:

$$v_{max,t} = vt_t + vs_t + vr_t \tag{1}$$

where $v_{max,t}$ is the maximum CPU utilization during the t-th time interval, and vt_t, vs_s, vr_t are the respective decomposed components. Seasonal decomposition adopts smoothing technique to calculate the trend. Next, we estimate seasonality by averaging the de-trended values for a specific season. Finally, we get residual component by removing the trend and seasonality from the original series.

---

[1] https://github.com/Azure/AzurePublicDataset.

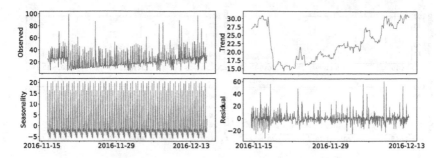

Fig. 1. Decomposed maximum CPU utilization by seasonal decomposition

Figure 1 illustrates the original CPU utilization series and its decompositions for a sampled VM. From the figure, we observe the following conceptual characteristics in terms of behaviors inside VM:

- *Locality*: CPU utilization is continuously changing over time and hence we should consider continuous utilization data in 5-min granularity, which reflects the short-term dependencies that tend to be similar.
- *Periodicity*: The de-trended utilization series shows stable periodicity every day, and this reveals some workloads consume CPU resource in a diurnal cycle.
- *Tendency*: There exists an increasing trend in CPU utilization after a sudden drop.

Inspired by the above observations, we explicitly derive three key fragments from the original 5-min CPU utilization series, to capture *locality*, *periodicity* and *tendency* behaviors, respectively. This is achieved by sampling data at 5-min, one-hour and one-day granularities.

3.3   CPU Utilization Similarity Among VMs

As we mentioned above, some VMs deployed in the same deployment tend to execute the same type of tasks and exhibit some common patterns. To better visualize the relationships between different VMs in the same deployment, we downsample the origin time series from five minutes granularity into one hour granularity by selecting a maximum point at one-hour interval. Figure 2 shows the sample maximum CPU utilization time series of 4 VMs collected in the same deployment covering a consecutive month. From Fig. 2, we can see that: VM1 and VM2 appear to fluctuate around an average line (marked in pink) and there is a trigger that makes the average CPU utilization smaller; VM3 and VM4 fluctuate steadily before an abrupt decline, followed by a continuous upward trend (marked in purple). These facts suggest that *some VMs in the same deployment should be relevant, i.e., they work in a parallel and collaborative manner*.

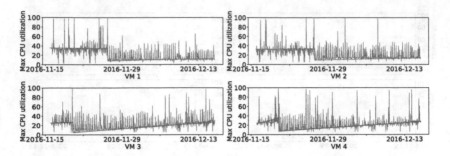

Fig. 2. Maximum CPU utilization on four VMs in the same deployment

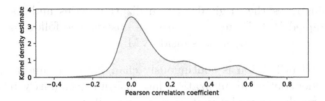

Fig. 3. Kernel density estimate of Pearson correlation

To measure the correlation between different VMs, we first randomly sample 320 VM CPU utilization series from the filtered deployments, and then calculate the Pearson correlation coefficient between different VMs. Figure 3 shows the Gaussian Kernel Density Estimate (KDE) [12] plot of pairwise VM correlation with setting bins as 100. The $x-$axis is the Pearson correlation, and the $y-$axis is the Gaussian weighted sum of nearby densities. We can see that there exists a positive correlation for pairwise VMs. To leverage this observation for improving prediction accuracy, we pick the maximum CPU utilization series of the K VMs that are most relevant to the target VM as external inputs. Note that these CPU utilization series are aligned over the timeline. Formally, we define the new CPU utilization series of target VM, as follows:

Definition 3 (Expanded VM CPU utilization). *Let $\mathcal{T} = \{x_t | t = 1, 2, \cdots, T\}$ be a series of expanded CPU utilization for one VM, where x_t is a (3 + K)-dimensional vector containing minimum, average and maximum CPU utilization of the target VM and maximum CPU utilization of other K most relevant VMs.*

In what follows, we introduce our neural method that incorporates all the above observations for predicting CPU utilization in next time period.

4 Time-Aware Residual Networks for CPU Utilization Prediction

Inspired by the model in [13], we propose our Time-aware Residual Networks (T-ResNet) model. Figure 4 depicts the structure of our T-ResNet, which contains three major components modeling temporal *locality*, *periodicity*, and *tendency*, respectively. As shown in top part of Fig. 4, we first choose K VMs that are most relevant to the target VM based on Pearson correlation coefficient and expand the target VM CPU utilization according to Definition 3. We then split the expanded utilization series from time axis into three fragments by sampling at 5-min, one-hour, and one-day frequencies. After that, the sampled 2D tensors of each fragment are fed into three residual networks accordingly to model *locality*, *periodicity*, and *tendency*, respectively. Finally, the flattened outputs of three residual networks are concatenated and further put into the fully-connected neural network to generate the predicted maximum CPU utilization value for the target VM in the next time period. Since the CPU utilization ranges from 0 to 1 after Min-Max normalization, we use the sigmoid function to activate final output and try to minimize the loss between true and predicted values through backward propagation [14] during model training.

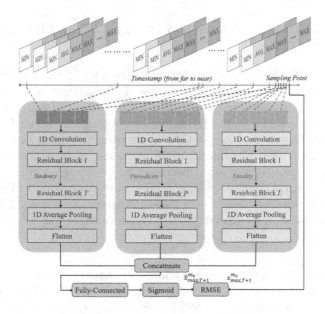

Fig. 4. Time-aware Residual Networks

The structure of each residual network in Fig. 4 is composed of convolution and the residual block. Convolutional neural network (CNN) is first proposed to handle image recognition problem by Yann et al. [15], and it can also be used for applications other than image recognition such as signal processing [16] and time series classification [17]. From Sect. 3.2, we observe repeatable local patterns in utilization series which could be detected via the convolutional operation. As single convolution captures local patterns, stacking multiple convolutions together can identify much more complex patterns and capture global temporal dependencies of VM utilization. However, deeper networks were difficult to train due to the notorious problem of gradient vanishing/exploding which blocks convergence [18]. Fortunately, the residual network solves this problem by adding shortcut connection to residual block (stacking of two layers of CNN, as shown in Fig. 5). Formally, the connection is defined as: $\mathbf{y} = \mathbf{W}\mathbf{x} + \mathcal{F}(\mathbf{x})$, where $\mathbf{x}$, $\mathbf{y}$ are inputs and outputs, respectively. $\mathbf{W}$ represents linear transformation and $\mathcal{F}$ is residual function [10]. The key idea of the residual network is to learn residual function $\mathcal{F}$. Therefore, we employ residual networks with deep structures to capture global temporal dependencies.

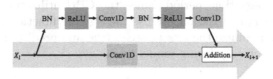

Fig. 5. Residual block. BN: Batch Normalization; Conv1D: 1-D Convolution

We develop T-ResNet to predict CPU utilization for all the VMs in a deployment. For illustration purpose, we describe the procedure of predicting one VM CPU utilization called target series. The prediction of other VMs in the same deployment is performed in the same way. Assume we select K most relevant VMs, and the extended target CPU utilization series can be expressed as $\mathcal{T} = \{x_t | t = 1, 2, \cdots, T\}$, according to Definition 3. For the residual network of *locality* in Fig. 4, we have the input fragment to be $[x_{T-l_l+1}, x_{T-l_l+2}, \cdots, x_T]$, where l_l is the length of intervals of *locality* component to look back. Then we concatenate them along time axis to be a two-dimensional tensor $\mathbf{X}_0^l \in \mathbf{R}^{l_l \times (3+K)}$. We use 1-D convolution to extract shallow characters (see Fig. 4):

$$\mathbf{X}_1^l = W_1^l * \mathbf{X}_0^l \tag{2}$$

where $*$ denotes the convolution operation, W_1^l denotes the filters of the first layer, and biases are omitted for simplifying notations. After that, we use multiple stacking residual blocks to model global temporal relationship. Figure 5 shows the residual blocks used in this paper, which can handle the dimension inequality problem in shortcut [19]. Each residual block (i.e., the upper branch in Fig. 5) involves two sets of "Batch Normalization + ReLU + Convolution". Batch Normalization accelerates deep network training by reducing internal covariate shift [20] and ReLU is a nonlinear activation function which helps network capture more complex patterns [21]. In order to extract features as much as possible and reduce information loss, we convolve to halve the dimension of time step while doubling the dimension of features by adding filters. To add the origin inputs of the residual network to transformed outputs, we perform a linear projection on the shortcut connection to match the dimension. In short, the layer k of the residual block can be formulated as:

$$\mathbf{X}_k^l = \mathcal{F}\left(\mathbf{X}_{k-1}^l, \{W_{k,1}^l, W_{k,2}^l\}\right) + W_{k,3}^l * \mathbf{X}_{k-1}^l \tag{3}$$

where $\mathcal{F}$ represents the residual mapping, i.e., $\mathcal{F} = W_{k,2}^l * \sigma\left(W_{k,1}^l * \sigma\left(\mathbf{X}_{k-1}^l\right)\right)$; σ denotes ReLU; $W_{k,1}^l$ and $W_{k,2}^l$ denote filters; $W_{k,3}^l$ are filters used for dimension transform. We omit batch normalization for simplicity. Noticeably, these time steps of inputs of our three residual networks may be different and will shrink as the depth deepen, hence the depth of networks required to reduce the dimension of time steps varies. For the residual network of *locality*, we stack L residual blocks upon the first layer to reduce the length of time steps to 2, and then get outputs $\mathbf{X}_{L+1}^l$. In addition, we use average pooling which can detect more background information and reduce noise followed by flattening to further transform features into one-dimensional vector denoted by x^l.

Similarly, based on above operation, we can construct the residual networks of *periodicity* and *tendency* in Fig. 4. We represent the *periodicity* fragment as $\left[x_{T-(l_p-1)*T_p}, x_{T-(l_p-2)*T_p}, \cdots, x_T\right]$, where l_p denotes the length of intervals of period, and T_p is the period span. After stacking P residual blocks, the output of *periodicity* is x^p. Meanwhile, the output of *tendency* is x^t, given the inputs $\left[x_{T-(l_t-1)*T_t}, x_{T-(l_t-2)*T_t}, \cdots, x_T\right]$. l_t is the length of *tendency* fragment and T_t denotes trend frequency, by stacking T residual blocks.

Finally, we concatenate x^l, x^p and x^t together as a new vector x_{res}, and feed it into a fully-connected layer to dynamically adjust weights of different features extracted from residual networks. Formally, we have:

$$\widehat{x}_{max,T+1}^{m_0} = \delta\left(W_{res}x_{res}\right) \tag{4}$$

where m_0 denotes target VM, W_{res} is weights of the fully-connected neural network, and δ denotes the sigmoid activation function. We train our model by minimizing the mean square error (MSE), between true and predicted maximum utilization values, as described below:

$$L = \frac{1}{N}\sum_{t=1}^{N}\left(\widehat{x}_{max,t}^{m_0} - x_{max,t}^{m_0}\right)^2 \tag{5}$$

where N is the number of training samples.

Algorithm 1 describes the preprocessing and training process of our T-ResNet method. We first pick K most relevant VMs for each target VM to expand the origin series (line 1). We then construct the training samples via the whole deployment (lines 2–11). Finally, we train the T-ResNet based on Adam optimizer [22], a variant of Stochastic Gradient Descent (SGD) (lines 12–16). Due to the fact that objective function is non-convex, the gradient-based optimization methods are usually trapped into the local optimum. Fortunately, Adam fuses the

---

Algorithm 1. Time-aware Residual Networks Training Algorithm

Input:

    VM CPU utilizations in a deployment: $\{\mathcal{V}^1, \cdots, \mathcal{V}^N\}$, where $\mathcal{V}^i = \{v_1^i, \cdots, v_T^i\}$;

    Lengths of *Locality, Periodicity, Tendency* fragments: l_l, l_p, l_t ;

    Periodicity span: T_p; Tendency frequency: T_t; Number of relevant VM: K;

Output:

    Learned Time-aware Residual Networks model

    //generate new CPU utilization series

1: Select K most relevant VMs in the deployment and expand the origin series for
    each VM to get: $\{\mathcal{T}^1, \cdots, \mathcal{T}^N\}$, where $\mathcal{T}^i = \{x_1^i, \cdots, x_T^i\}$

2: $\mathcal{U} \leftarrow \emptyset$

3: **for** *all expanded VM CPU utilization* $\mathcal{T}^i$ ($i \in [1, N]$) **do**

4:    **for** *all available training timestamps* ts ($ts \in [1, T]$) **do**

5:       $\mathcal{F}_l = \left[x_{ts-l_l+1}^i, x_{ts-l_l+2}^i, \cdots, x_{ts}^i \right]$

6:       $\mathcal{F}_p = \left[x_{ts-(l_p-1)*T_p}^i, x_{ts-(l_p-2)*T_p}^i, \cdots, x_{ts}^i \right]$

7:       $\mathcal{F}_t = \left[x_{ts-(l_t-1)*T_t}^i, x_{ts-(l_t-2)*T_t}^i, \cdots, x_{ts}^i \right]$

8:       $x_{max,ts+1}^i$ is the true maximum CPU utilization at next duration $ts + 1$

9:       add an training sample $\left(\{\mathcal{F}_l, \mathcal{F}_p, \mathcal{F}_t\}, x_{max,ts+1}^i \right)$ into $\mathcal{U}$

10:    **end for**

11: **end for**

12: Initialize all the network parameters θ in T-ResNet,

13: **while** θ *not converged* **do**

14:    Select a batch of samples $\mathcal{U}_b$ randomly from $\mathcal{U}$

15:    Find parameters θ by minimizing the loss defined in Eq. 5 with $\mathcal{U}_b$

16: **end while**

---

advantages of Momentum method [23] and RMSprop method [24] to overcome this problem. Specifically, Momentum considers the direction of last gradients, while RMSprop adopts the exponential moving average method to filter historical gradients. Such a combination effectively speeds up network learning process and helps the training process escape from local optimum.

5 Experiments

5.1 Experimental Setup

Datasets. We use the Microsoft Azure dataset for performance evaluation. As discussed in Sect. 3.1, we filter many deployments that do not sustain the entire lifetime of our dataset and leave 3005 deployments. After considering the rationality of experiments and the limitations of resources, we prepare two training sets, **Dep1** and **Dep2**, each of which includes 32 VMs in a deployment.

Baselines. To demonstrate the effectiveness of T-ResNet, we compare it against 5 baseline methods.

– **NAÏVE**: We predict the maximum VM CPU utilization by the maximum CPU utilization of that VM at last time interval.

- **ARIMA** [25]: Autoregressive Integrated Moving Average (ARIMA) model is a generalization of an autoregressive moving average (ARMA) model, which is widely used in time series analysis.
- **SARIMA**: Seasonal ARIMA which incorporates both non-seasonal and seasonal factors in ARIMA.
- **XGBoost** [26]: XGBoost is a scalable machine learning system for tree boosting.
- **LSTM** [27]: Long Short-Term Memory (LSTM) network are a special kind of recurrent neural network (RNN), which can learn long-term dependencies.

Experimental Settings. We train each model using the whole training set, i.e., one deployment. Since the value of maximum CPU utilization varies greatly, we first transform origin series into log-scale to balance the order of magnitude. We then use Min-Max normalization to further scale data into the range $[0, 1]$. For all residual blocks after the first convolutional layer, we halve the length of time step by setting filter stride as 2 while doubling the feature dimension by doubling the filter numbers. All the filter size in convolution is set to 3. In our T-ResNet, T_p and T_t are empirically fixed to one-hour and one-day. Based on our observations, we set l_l, l_p, and l_t to 12, 24, and 7, respectively. Once the hyperparameter K is determined, all the VMs in a deployment will select the same number of relevant VMs as additional inputs and we set $K \in \{0, 2, 4\}$. For each deployment, we select the first two-weeks data as the training set, the following week's data as the validation set, and the last week's data as the test set. The validation set is used for early stopping and selection of best network parameters.

Metrics. To measure the effectiveness of various methods, we introduce three different evaluation measures. Among them, root mean square error (RMSE) and mean absolute error (MAE) are scale-dependent metrics, while mean absolute percentage error (MAPE) is scale-independent metrics. Specifically, the RMSE is defined as $\textbf{RMSE} = \sqrt{\frac{1}{N} \sum_{t=1}^{N} (\hat{y}_t - y_t)^2}$, MAE is defined as $\textbf{MAE} = \frac{1}{N} \sum_{t=1}^{N} |\hat{y}_t - y_t|$, and MAPE is defined as $\textbf{MAPE} = \frac{1}{N} \sum_{t=1}^{N} |\frac{\hat{y}_t - y_t}{y_t}|$, where $\hat{y}$ is the predicted value and y is the true value.

5.2 Main Results

The maximum CPU utilization prediction results of T-ResNet and other baseline methods over the two selected deployments are shown in Table 1.

Results of Compared Methods In Table 1, we observe that the RMSE of NAÏVE method is generally worse than other methods. This result shows that not all VMs tend to be stable in short term. ARIMA and SARIMA are all linear regression with differential operation in nature [25]. Both of them have slightly better performance compared with NAÏVE method, however, the ability of them is limited which fails to capture nonlinear relationship. XGBoost is a tree-based model which combines classification and regression tree (CART) algorithm with gradient boosting method [26], hence it can capture more complicated relationship and perform better than ARIMA in RMSE. As shown, XGBoost achieves

Table 1. Comparison results over the Dep1 and Dep2 dataset (best performance displayed in **blodface**). Our original model does not consider relevant VMs, and kREL means adding extra most relevant k VMs into the model.

Models	Dep1 Dataset			Dep2 Dataset		
	RMSE $(\times 10^{-2})$	MAE $(\times 10^{-2})$	MAPE $(\times 10^{-2})$	RMSE $(\times 10^{-2})$	MAE $(\times 10^{-2})$	MAPE $(\times 10^{-2})$
NAÏVE	11.87	6.95	16.75	11.96	7.03	18.10
ARIMA	9.43	6.89	19.97	10.34	8.81	25.99
SARIMA	10.16	6.63	16.08	10.77	7.05	18.06
XGBoost	8.99	7.21	21.07	9.50	7.70	24.83
LSTM	8.78	5.68	14.20	8.83	6.43	17.60
T-ResNet	7.68	4.66	11.37	7.81	4.78	12.32
T-ResNet-2REL	**7.35**	**4.11**	**9.95**	**7.28**	**4.27**	**11.19**
T-ResNet-4REL	7.62	4.71	12.41	7.68	4.47	11.76

little improvement since it can only consider a few steps of inputs. LSTM is particularly designed to remember information for long periods of time due to the existence of memory cell, but it does not show significant performance improvement according to the results. The reason may be that the above-mentioned key drivers of our dataset may not be captured by LSTM.

Effect of Number of Related VMs. With the integration of extra inputs and multiple frequencies feature extraction, our T-ResNet models achieve the best RMSE, MAE, and MAPE across two datasets. Specifically, we attempt three variants of our model with setting different numbers of the relevant VMs. The results show that introducing extra inputs indeed help the model improve prediction accuracy, which confirms our observations. However, this effect is not continuously increasing. When the extra input dimension reaches the threshold, the earnings of improvement will decrease. The reason is that introducing more extraneous inputs will also introduce noise.

6   Related Work

VM Workload Prediction. With the development of cloud technology, many researchers have focused on predicting the workloads of VMs. Calheiros et al. [4] paid attention to predicting the requests from web servers, but the data does not really reflect the consumption of workload. Farahnakian et al. [5] used linear regression method to predict the CPU utilization based on past one-hour data. Such a method can only approximate short-term CPU utilization. Khan et al. [7] discovered repeatable workload patterns within groups of VMs that belong to a cloud customer and further designed an HMM-based method to predict the changes of these patterns. However, the method has restricted the states of workload levels and cannot model new states. Some works [8,28] focused on using neural network based model to predict the VM workload. Islam et al. [8] employed the extra sliding window technique to evaluate the impact of different windows on the prediction accuracy. Xue et al. [28] trained a group of networks models and further generate the prediction based on ensemble of these pre-trained networks. These works are different from ours where the proposed methods cannot consider both characteristics of VMs and effective algorithms that can capture specified temporal relations.

Time Series Prediction. Time series forecasting is the most common problem and we have many general methods to handle it. ARIMA is popular and widely used statistical method for time series forecasting, and it combines autoregression and differencing together to model linear relationship [25]. Hidden Markov model describes the process that the observations are produced by corresponding hidden states which randomly generated by hidden Markov chain before [7]. The Markov process is based on finite pre-defined states and can capture limited nonlinear patterns. Recurrent neural network (RNN) [29] is well designed neural network for sequence tasks, but vanilla RNNs suffer from gradient vanish problem [30]. Long short-term memory units (LSTM) [27] and gated recurrent units (GRU) [31] are proposed to handle this problem and keep long-term dependencies by memory cell. Moreover, convolutional neural network (CNN) has exhibited strong ability in image recognition which benefits from recent AlexNet [32]. Prior works [16,17] also showed the effectiveness of one-dimensional convolution in solving sequence problems. Neural networks with stacked convolutions can learn more complex patterns, and residual networks [10] make the learning of deep models possible. While RNN and CNN have the ability to capture nonlinear relations, our proposed model and experiments validate the view that networks are still hard to learn rules without specified structures.

7   Conclusion

In this paper, we studied the problem of VM CPU utilization prediction, which is an important task for cloud resource managers. We conduct an empirical analysis over real-world VM CPU utilization data, showing that deployment-based VMs have internal and external CPU utilization features. The internal features imply that the CPU utilization series has periodicity, tendency, and locality, while the external features reflect that VMs in a deployment tend to work in parallel and their CPU utilization behaviors are similar. Based on the observations, we propose Time-aware Residual Networks model named T-ResNet for prediction. T-ResNet consists of three residual networks to capture features at different frequencies and uses fully-connected layers to model deep feature interactions. The experimental results verify the effectiveness of our proposed method in terms of prediction accuracy, compared with various baseline approaches.

Acknowledgements. This research is supported in part by 973 Program (no. 2014CB340303), NSFC (No. 61772341, 61472254, 61572324, 61170238, 61602297 and 61472241) and the Shanghai Municipal Commission of Economy and Informatization (No. 201701052). This work was also supported by the Program for Changjiang Young Scholars in University of China, the Program for China Top Young Talents, and the Program for Shanghai Top Young Talents.

References

1. Gusev, M., Ristov, S., Simjanoska, M., Velkoski, G.: CPU utilization while scaling resources in the cloud. Cloud Comput. (2013)
2. Herodotou, H., et al.: Starfish: a self-tuning system for big data analytics. In: CIDR (2011)
3. Clark, C., et al.: Live migration of virtual machines. In: NSDI (2005)
4. Calheiros, R.N., Masoumi, E., Ranjan, R., Buyya, R.: Workload prediction using arima model and its impact on cloud applications' QoS. TCC **3**(4) (2015)
5. Farahnakian, F., Liljeberg, P., Plosila, J.: LiRCUP: linear regression based CPU usage prediction algorithm for live migration of virtual machines in data centers. In: SEAA (2013)
6. Gong, Z., Gu, X., Wilkes, J.: PRESS: predictive elastic resource scaling for cloud systems. In: CNSM (2010)
7. Khan, A., Yan, X., Tao, S., Anerousis, N.: Workload characterization and prediction in the cloud: a multiple time series approach. In: NOMS (2012)
8. Islam, S., Keung, J., Lee, K., Liu, A.: Empirical prediction models for adaptive resource provisioning in the cloud. FGCS **28**(1) (2012)
9. Cortez, E., Bonde, A., Muzio, A., Russinovich, M., Fontoura, M., Bianchini, R.: Resource central: understanding and predicting workloads for improved resource management in large cloud platforms. In: SOSP (2017)
10. He, K., Zhang, X., Ren, S., Sun, J.: Deep residual learning for image recognition. In: CVPR (2016)
11. Cleveland, R.B., Cleveland, W.S., Terpenning, I.: STL: a seasonal-trend decomposition procedure based on loess. J. Official Stat. **6**(1) (1990)
12. Silverman, B.W.: Density Estimation for Statistics and Data Analysis. Routledge (2018)
13. Zhang, J., Zheng, Y., Qi, D.: Deep spatio-temporal residual networks for citywide crowd flows prediction. In: AAAI (2017)
14. Rumelhart, D.E., Hinton, G.E., Williams, R.J.: Learning internal representations by error propagation. Technical report, California Univ San Diego La Jolla Inst for Cognitive Science (1985)
15. LeCun, Y., Bottou, L., Bengio, Y., Haffner, P.: Gradient-based learning applied to document recognition. Proc. IEEE **86**(11) (1998)
16. Van Den Oord, A., et al.: WaveNet: a generative model for raw audio. arXiv preprint arXiv:1609.03499 (2016)
17. Cui, Z., Chen, W., Chen, Y.: Multi-scale convolutional neural networks for time series classification. arXiv preprint arXiv:1603.06995 (2016)
18. Glorot, X., Bengio, Y.: Understanding the difficulty of training deep feedforward neural networks. In: AISTATS (2010)
19. He, K., Zhang, X., Ren, S., Sun, J.: Identity mappings in deep residual networks. In: Leibe, B., Matas, J., Sebe, N., Welling, M. (eds.) ECCV 2016. LNCS, vol. 9908, pp. 630–645. Springer, Cham (2016). https://doi.org/10.1007/978-3-319-46493-0_38

20. Ioffe, S., Szegedy, C.: Batch normalization: accelerating deep network training by reducing internal covariate shift. In: ICML (2015)
21. Nair, V., Hinton, G.E.: Rectified linear units improve restricted boltzmann machines. In: ICML (2010)
22. Kingma, D.P., Ba, J.: Adam: a method for stochastic optimization. arXiv preprint arXiv:1412.6980 (2014)
23. Polyak, B.T.: Some methods of speeding up the convergence of iteration methods. USSR Comput. Mathe. Mathe. Phys. 4(5) (1964)
24. Tieleman, T., Hinton, G.: Lecture 6.5–RmsProp: Divide the gradient by a running average of its recent magnitude. COURSERA: Neural Networks for Machine Learning (2012)
25. Makridakis, S., Hibon, M.: ARMA models and the box-jenkins methodology. J. Forecast. (1997)
26. Chen, T., Guestrin, C.: XGBoost: a scalable tree boosting system. In: SIGKDD (2016)
27. Hochreiter, S., Schmidhuber, J.: Long short-term memory. Neural Comput. 9(8) (1997)
28. Xue, J., Yan, F., Birke, R., Chen, L.Y., Scherer, T., Smirni, E.: Practise: robust prediction of data center time series. In: CNSM (2015)
29. Elman, J.L.: Finding structure in time. Cogn. Sci. 14(2) (1990)
30. Bengio, Y., Simard, P., Frasconi, P.: Learning long-term dependencies with gradient descent is difficult. IEEE Trans. Neural Netw. 5(2) (1994)
31. Cho, K., et al.: Learning phrase representations using rnn encoder-decoder for statistical machine translation. arXiv preprint arXiv:1406.1078 (2014)
32. Krizhevsky, A., Sutskever, I., Hinton, G.E.: Imagenet classification with deep convolutional neural networks. In: NIPS (2012)

Towards the Design of a Scalable Business Process Management System Architecture in the Cloud

Chun Ouyang[1], Michael Adams[1], Arthur H. M. ter Hofstede[1], and Yang Yu[2]([✉])

[1] Queensland University of Technology, Brisbane, Australia
{c.ouyang,mj.adams,a.terhofstede}@qut.edu.au
[2] Sun Yat-sen University, Guangzhou, China
yuy@mail.sysu.edu.cn

Abstract. The ubiquity of cloud computing is shifting the deployment of Business Process Management Systems (BPMS) from traditional on-premise models to the Software-as-a-Service (SaaS) paradigm, thus aiming to deliver *Business Process Automation as a Service* to multiple tenants in the cloud. However, scaling up a traditional BPMS to cope with simultaneous demand from multiple organisations in the cloud is challenging, since its underlying system architecture has been designed to serve a single organisation with a single workflow engine. A typical SaaS often deploys multiple instances of its core applications and distributes workload to these application instances via load balancing. But, for stateful and often long-running process instances, standard stateless load balancing strategies are inadequate. In this paper, we propose a conceptual design of a scalable system architecture for deploying BPMS in the cloud. In our design, Object Role Modeling (ORM) is used to conceptualise the data requirements of the system and UML sequence diagrams are used to capture the interactions between system components. A prototypical implementation using an open-source traditional BPMS offers focused load balancing strategies and demonstrates improved capabilities for supporting large volumes of work in a multi-tenanted cloud environment.

Keywords: Business Process Management System
Software-as-a-Service · System architecture · Scalability
Load balancing · Workflow engine

1 Introduction

The ubiquity of cloud computing is reshaping application architectures from locally-based deployments towards web service-based distributions. As a primary representative of process-aware information systems, Business Process Management Systems (BPMS) are dedicated to provide automated support for the execution of business processes in modern organisations. The advantages offered by

© Springer Nature Switzerland AG 2018
J. C. Trujillo et al. (Eds.): ER 2018, LNCS 11157, pp. 334–348, 2018.
https://doi.org/10.1007/978-3-030-00847-5_24

cloud computing have triggered increased demand for the deployment of BPMS to shift from traditional on-premise models to the Software-as-a-Service (SaaS) paradigm with the aim to deliver *Business Process Automation as a Service* to multiple organisations in a multi-tenant cloud.

However, a traditional on-premise BPMS architecture is designed to serve a single organisation and often relies on a single process engine (a.k.a. workflow engine). In a multi-tenant cloud, scaling up such a BPMS to address simultaneous demand from multiple organisations is challenging given the limited capacity of a single engine. Thus it has become necessary to reshape the architectural frameworks of BPMS for multi-tenancy. A review of the existing efforts towards a generic architecture for a scalable BPMS in the cloud reveals that this aim has not yet been achieved.

In cloud computing, to cope with the increasing load due to multi-tenancy, a SaaS approach often deploys multiple instances of its core applications and distributes workload (of tenant requests) to these application instances via load balancing [3]. The assumption in this approach is that the tenant requests are stateless, whereas a business process execution is *stateful* and often long-running, in which case existing mechanisms for handling stateless requests (e.g. load balancing strategies) become invalidated.

In this paper, we propose a conceptual design of a system architecture for deploying BPMS in a multi-tenant cloud with a focus on supporting scalability. In our design, Object Role Modeling (ORM) is used to conceptualise the data requirements of the system and UML sequence diagrams are used to capture the interactions between system components. To transform our conceptual design proposal into an effective implementation, a prototype has been developed using an open-source traditional BPMS with a focus on the implementation of certain load balancing strategies. A simulation of multiple concurrent case execution instances within the prototype demonstrates improved capabilities for supporting large volumes of work in a multi-tenanted cloud environment.

The contribution of our work is twofold. First, as guided by a conceptual modelling-based approach, the proposed design of a system architecture for scalable BPMS is *generic*, i.e., it is independent of any specific BPMS and any specific cloud platform in use. Such a system architecture can serve as an important reference for designing a specific BPMS in the cloud. Second, this research work in its entirety demonstrates the importance and impact of applying conceptual modelling methods and techniques to information systems development.

The rest of the paper is organised as follows. Section 2 provides the background and sets out the research problem to address. Section 3 presents the approach and underlying principles to guide our design. Section 4 proposes the design of an overall architecture of a scalable BPMS and elaborates the design of a scalable engine in BPMS. Section 5 discusses a prototypical implementation of a scalable BPMS. Section 6 reviews related work. Finally, Sect. 7 concludes the paper.

2 Background and Problem Description

2.1 Business Process

A business process consists of a number of tasks that need to be carried out to achieve a business goal and a set of conditions that determine the possible ordering of the tasks. A task is a logical unit of work that is performed by a human actor (or software application, machine, etc). When performing a business process, each execution of the process results in a process instance (i.e. a case), and each case can be distinguished from every other case. A case comprises instances of tasks in the corresponding process, and an instance of a task in a case is referred to as a work item.

Consider a simplified example of a travel booking process (taken from [2]). The process may consist of tasks such as registering travel itinerary, make bookings of flights, hotels and/or rental cars for each itinerary segment, and handling payment. A case of travel booking is launched whenever a traveller submits his/her itinerary, and each case can be uniquely identified by e.g. a travel booking reference number.

2.2 Traditional BPMS

Business Process Management Systems (BPMS), previously known as workflow management systems, are a class of software that aims to provide automated support to execution of business processes. The focus of BPMS is to support process execution by coordinating the right work (i.e. in terms of work items) to the right person (or application) at the right time, but not to actually perform the piece of work [1]. The *workflow reference model* [7] proposed by the Workflow Management Coalition (WfMC)[1] serves as a generic architecture of a BPMS and forms the basis of most of BPMS in use to date.

Figure 1(a) depicts the architecture of a BPMS according to this reference model. The core component is the *workflow engine*. Its main functionalities include creating, running, and completing cases for execution of a business process and, for each active case, generating work items to coordinate which tasks to work on next by which resource. The work items are managed by the *worklist handler* and made available to individual resources (e.g. staff in an organisation) responsible for performing the work. The resources (i.e. end users) interact with their own lists of work items, e.g. to check out a work item to start the work, or to check in a work item when the work is completed. Such interactions are conducted via the user interfaces of the worklist handler. In addition to the main functionalities, there are other operational matters in a BPMS including resource administration (e.g. addition or removal of staff, managing the access control of individual users) and process monitoring (e.g. tracking the progress of each running case, monitoring the performance of a business process). These are taken care of by the *administration and monitoring tools*.

---

[1] An organization formed to standardise workflow management terminology and define standards for workflow management systems (see http://www.wfmc.org).

A BPMS has two important data repositories (see Fig. 1(a)). One is the *process model repository*, while the other stores the *execution logs* of business processes. Before a process can be executed in the workflow engine, it needs to be defined using a proper modelling language, and the resulting process definition is called a process model. A process modelling tool is used to create process models. It is also part of a BPMS but is not further considered in this paper, as we are mainly interested in business process automation (rather than process modelling). Hence, we assume that a process model has always already been deployed to the workflow engine from the process model depository. Next, the workflow engine records the (step-by-step) execution of a process and exports the relevant data in the form of execution logs. Such data are valuable for process analysis and monitoring.

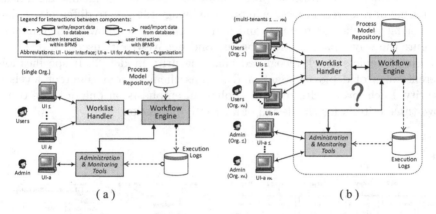

Fig. 1. A traditional BPMS (a) deployed on-premise for a single organisation *vs.* (b) deployed in a multi-tenanted cloud environment for multiple organisations

2.3   Problem Statement

The BPMS schematic shown in Fig. 1(a) is a typical example of a traditional BPMS deployed on-premise to serve the needs of a single organisation. When deploying BPMS in the cloud environment, it is expected that the system will provide process automation service to multi-tenants. As Fig. 1(b) shows, a number of (e.g. m) organisations are engaged as (m) multi-tenants where each tenant represents a single organisation. Each organisation interacts with the BPMS in the cloud via a worklist handler *just like* how it would interact with a traditional BPMS. However, in this setting, a BPMS faces a potentially greatly increased workload imposed by simultaneous requests from multiple organisations, making it challenging for the system to maintain the expected performance (e.g. response time) under such workload with a single workflow engine. Hence, the problem addressed in this paper is: *how to reshape the supporting architecture of a traditional BPMS so that the resulting system is capable of scaling up to cope with large volumes of work in a multi-tenanted cloud environment?*

3   Approach and Design Principles

We apply a conceptual modelling-based approach to the design of a scalable system architecture for BPMS in the cloud. As a key design principle, the system architecture proposed should be *generic*, that is, it should be independent of any specific BPMS and any specific cloud platform in use. Hence, our design is to be guided by established standards, reference frameworks, and/or maturity models that are relevant in the domain.

Deploying a BPMS in the cloud to provide service to multiple organisations is an offering of Software-as-a-Service (SaaS). A well-designed SaaS application should satisfy three key criteria, namely *scalable, configurable*, and *multi-tenant-efficient* [3]. Scaling an application means maximising concurrency to address the increasing workload; being multi-tenant-efficient means maximising the sharing of resources across tenants while being still able to differentiate data that belong to different tenants; and being configurable allows each tenant to configure the way an application appears and behaves to its users.

Figure 2 depicts a well-recognised SaaS maturity model [3]. It specifies four different ways of deploying SaaS and assigns them to four different levels of maturity. Each level addresses and applies an emphasis on one or more of the above three criteria for the design of an SaaS application.

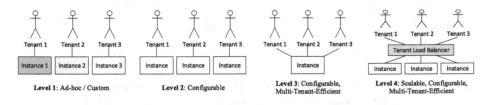

Fig. 2. Four-level SaaS maturity model [3]

At *Level 1*, each tenant has its own customised instance of an application on the host's servers. This approach is easy to deploy but offers very few of the benefits of an SaaS solution. At *Level 2*, a separate instance of an application is hosted for each tenant, while all instances of the application use the same code implementation. Each tenant is provided with options to configure how the application looks and behaves to its users. At *Level 3*, a single instance is hosted that serves every tenant with possible configuration of certain features and allows resource sharing between tenants. The scalability of the application is however limited. Finally, at *Level 4*, multiple identical instances of an application made available via a load balancer are hosted for multi-tenants. It allows resource sharing between tenants and also provides configuration for each tenant. The resulting SaaS application is scalable to a large number of tenants. It is worth noting that the appropriate maturity level to choose for the design of a specific SaaS application is determined by the nature and needs of the application.

4   Conceptual Design

This section presents a conceptual design of a scalable system architecture for BPMS in the cloud. The WfMC's workflow reference model (see Sect. 2.2) is used to form the basis of a traditional BPMS architecture. The SaaS maturity model (see Sect. 3) serves as the underlying principles to guide the design decisions.

4.1   Scalable BPMS Architecture

Figure 3 depicts an overall system architecture of a scalable BPMS that works for multiple organisations in a multi-tenanted cloud environment. From the user's endpoint, there are multiple worklist handlers and each of them is configured for an individual organisation to best capture their specific requirements regarding the management of the work items, e.g. how a list of work items is presented to the users, how a user interacts with his/her work list, etc. This is designed according to Level 2 of the SaaS maturity model, since configurability by each organisation is the key concern to address.

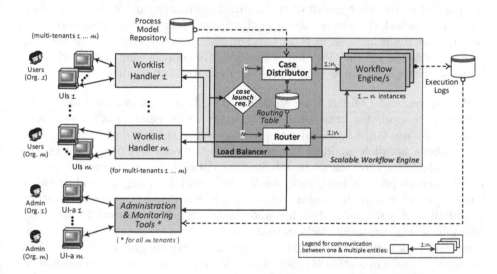

Fig. 3. A system architecture (block diagram) of a scalable BPMS

Secondly, there is a single instance of administration and monitoring tools shared by multiple organisations. This design is based on Level 3 of the SaaS maturity model, for the reason that sharing the functions to support operational matters, such as process progress tracking and monitoring, between the organisations' administrators is the main focus.

Finally, the *scalable workflow engine* consists of *multiple instances* of workflow engines coordinated via a *load balancer*. It is designed according to Level 4 of the SaaS maturity model so that it can be scaled to cater to a large number of organisations and at the same time it allows sharing of the resources of multiple workflow engines between the organisations. As the core element of a scalable BPMS, a scalable workflow engine is the focus of our design, and is therefore discussed in detail in the rest of this section.

4.2   Scalable Workflow Engine

Load Balancer. This is the key component of a scalable workflow engine. It is responsible for receiving requests from multiple tenants and distributing or directing the requests to the available workflow engines, and upon receiving responses from the workflow engines it directs them back to the corresponding tenants. The load balancer further comprises a *case distributor*, a *router*, and a *routing table* that work together to support the above functionalities.

Whenever a request for launching a case is received from a worklist handler, the case distributor allocates the case to one of the workflow engines based on the computation of a load balancing algorithm (e.g. to identify the least busy engine). Once a case is distributed to a workflow engine, the case will be executed in a *stateful* manner, meaning that all the work items belonging to the case will be handled by the same workflow engine until the completion of the case. After a case is launched, the router takes over the responsibility of communication and directs work items between worklist handlers and workflow engines. The routing table records the information that is necessary in coordinating each work item between the right workflow engine and the right worklist handler.

Data Requirements. Object-Role Modeling (ORM) is a fact-oriented modelling approach for specifying, transforming, and querying information at a conceptual level [4]. We use ORM (version) 2 to precisely define the key data requirements of a scalable workflow engine[2]. The resulting data model is depicted in Fig. 4.

A scalable workflow engine mainly handles data related to process execution. In addition to process, case, task, work item, (workflow) engine, and worklist handler, there is also the concept of a *session*. Each case is performed within the lifetime of a session and each session must be held by a worklist handler. The join subset constraint declares that if a work item is executed on an engine then that work item must belong to a case that is allocated to that engine, thus capturing stateful execution of cases. A so-called *busy-indicator* is a performance measure of an engine that quantifies how busy the engine is. At run-time, the busyness of an engine changes over time. This dynamic is captured by the fact that an engine's busy-indicator has a specific value during a certain time period uniquely identified by its start and end timestamps.

Component Interactions. UML sequence diagrams [9] is used to specify how the components of a scalable workflow engine collaborate to support case executions for multiple organisations. Figure 5 depicts five sequence diagrams that capture the component interactions in the five phases of a case execution, respectively.

---

[2] A summary of the ORM 2 Graphical Notation is available from www.orm.net/pdf/ORM2GraphicalNotation.pdf.

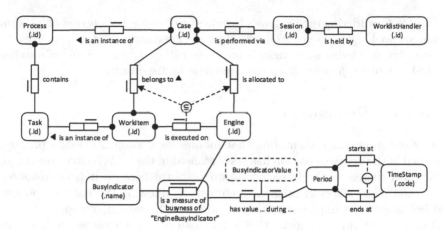

Fig. 4. A data model (ORM diagram) conceptualising the data requirements of a scalable workflow engine.

In (a), upon receiving a request to launch a new case, the case distributor will ask each engine for its current busyness in a certain time window, select the appropriate engine, and communicate with that engine to launch the case. Note that the decision about which engine to select is based on a specific load distribution strategy used by the case distributor (and this will be discussed next when different load distribution capability levels are defined).

After the case is launched on the engine, the case distributor will record into the routing table the information necessary for correctly routing the work items belonging to the case. During the case execution, the engine will compute the next enabled work item and the router will pass it to the worklist handler as in (b). The work item will be checked out (see (c)), performed by end users, and then checked in (see (d)) upon its completion. The phases (b) to (d) will be repeated till there are no more enabled work items, and the case will be completed as in (e).

Load Distribution Capability Levels. This is to set specific requirements for a load balancer to reach different levels of capability in distributing the load of cases to the appropriate workflow engine at run-time. Assume that the number of engines (i.e. the value of n in Fig. 3) is fixed. The load distribution can be worked out using different strategies which lead to the following four capability levels.

- *Level 0*: Applying a general work scheduling or resource allocation mechanism (e.g. random choice) *without* considering how busy each engine is.
- *Level 1*: Considering each engine's busyness status *at the point of time* when a case is needed to be distributed (to an engine).

– *Level 2*: Considering each engine's busyness *within the time period* of a sliding window looking *backward* from the time of case distribution.
– *Level 3*: Considering each engine's busyness *within the time period* of a sliding window looking *forward* from the time of case distribution.

5   System Realisation

A proof-of-concept implementation that follows the conceptual design proposal described in the previous section has been realised in the YAWL environment [6]. YAWL was chosen as the implementation platform because it is open-source, stable, and offers a service-oriented architecture, allowing the new component, load balancer, to be implemented independent to the existing components.

The focus of our implementation is the load balancer as it is the key component designed to achieve scalability. It manages a set of available engines, redirecting calls from worklist handlers to the appropriate engine for handling. Here we specify an engine's busyness indicator which is calculated with reference to the following four key parameters:

– R the number of requests processed per second.
– P the average processing time per processed request (in milliseconds).
– W the number of work item starts and completions per second.
– T the number of worker threads currently executing in the engine's container.

An upper limit for each parameter is configurable, to account for different hardware infrastructures, and weightings can be set for each so that an accurate factor for any platform can be established. The final busyness factor calculation (B) for an engine is given as:

$$B = (((((R_{count} / R_{limit}) \cdot R_{weight}) + \\ ((P_{time} / P_{limit}) \cdot P_{weight}) + \\ ((W_{count} / W_{limit}) \cdot W_{weight}) + \\ ((T_{count} / T_{limit}) \cdot T_{weight})) / 4) \cdot 100$$

The implemented load balancer in our prototype currently supports load distribution capability levels 0, 1 and 2 as described in the previous section.

Experiment Setup. We set up a test bed on the QUT High Performance Computing platform, using a discrete virtual machine for each engine, and for the load balancing component. Each VM ran on a single core of an Intel Xeon CPU E5-2680 @ 2.70 GHz, and 4 Gb of RAM was available to each.

For testing, a simulation tool was used to launch new cases given a process specification, with a new instance is launched every 1.5 seconds for a total of 300 instances. The tool allowed for the processing of work items created by resource 'robots', i.e. automated agents that mimic the role of human resources. These agents were configured to simulate processing of each work item for a randomised period within an upper limit of 100 to 3000 ms.

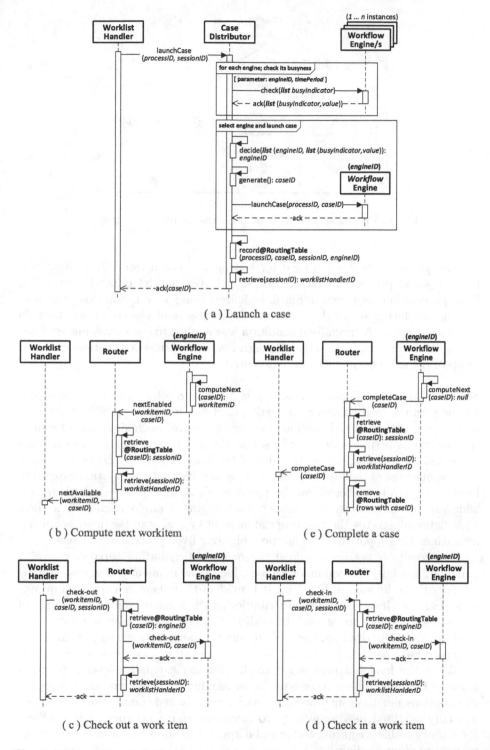

Fig. 5. UML sequence diagrams capturing the component interactions carried out to support case executions by a scalable workflow engine in five phases (a)–(e)

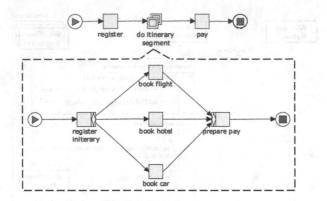

Fig. 6. Process specification used for the simulation testing [2]

The process specification used for testing was taken from [2]. It models a travel booking process of which a process narrative was introduced in Sect. 2.1. The specification, as shown in Fig. 6, is defined using YAWL notation. The portion in the dotted rectangle represents a subprocess of the composite task *do itinerary segment*. A specialised condition was applied to each outgoing arc from the *register itinerary* task so that for each instance of the specification, one, two or all of the subsequent tasks were executed.

Simulation Results. Figure 7 shows the simulation results of deploying our prototype with the load balancer supporting load distribution capability levels 0, 1 and 2 given the simulated work load set in the above. Each simulation result is represented as a time series graph, where the x- axis specifies a series of time slices and the y-axis specifies the value of the busyness factor (B) of an engine.

The outcome of the load balancer assigns new cases at distribution capability level 0 (i.e. random choice) can be found in Fig. 7(a). As expected, there is a wide variance in the busyness values of each engine at any particular time slice. This figure illustrates the inappropriateness of typical stateless load balancing algorithms for supporting stateful, possibly long lived process executions.

The results of setting the load balancer to distribution capability level 1 (calculating a busyness value for each engine at the moment a new case start is requested) can be seen in Fig. 7(b). It is evident that there is a strong improvement over the level 0 random distribution, with a narrowing of the range of engine busyness factors at each time slice. However, since snapshot values are used for level 1 distribution, there is still some variance between engines at times, as should be expected.

When the load balancer is set to distribution capability level 2 (using a backward sliding window to smooth factor readings), the narrowing of busyness distributions per time unit becomes even more marked. Level 2 uses an exponentially weighted moving average to calculate engine busyness, which allows for a configurable weighting factor to be applied in an exponentially decreasing manner to older readings.

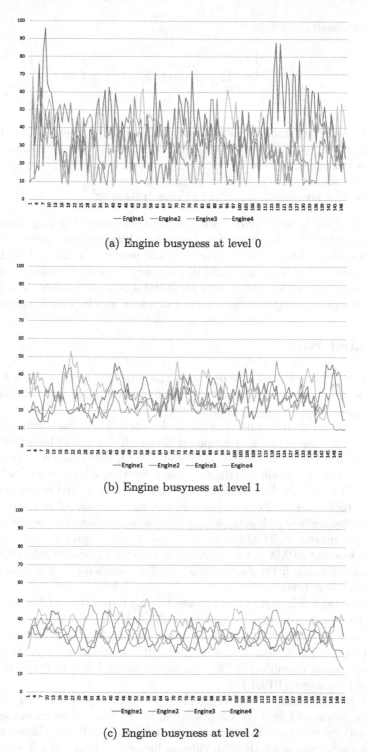

(a) Engine busyness at level 0

(b) Engine busyness at level 1

(c) Engine busyness at level 2

Fig. 7. Simulation results of engine busyness given load distribution capability levels 0, 1 and 2

It is calculated recursively:

$$S_t = \begin{cases} Y_1, & t = 1 \\ \alpha \cdot Y_t + (1 - \alpha) \cdot S_{t-1}, & t > 1 \end{cases}$$

where:

- α is the weighting factor (a.k.a. the forget factor), between 0 and 1, representing the degree of decrease for older readings, which diminish more quickly as values of alpha approach 1.
- Y_t is the engine busyness reading at time instance t.
- S_t is the weighted moving average at time instance t.

Figure 7(c) would indicate that the weighted exponential sliding average used for level 2 distribution has a marked impact on keeping load quite evenly balanced across time slices when compared to the previous levels, particularly the stateless random distribution of level 0. From our testing, it is clear that implementing an intelligent load algorithm for stateful process management provides improved load balancing outcomes over stateless or random-choice approaches.

6   Related Work

Generally, the related literature discusses the provision of cloud-based services (SaaS) that can be executed for multiple tenants on demand, but with the overall stateful process execution operated locally.

In [5] a hybrid architecture that offers distributed process executions across both client and server sides is proposed although the cloud-based executions are limited to invoked web services. Sun et al. [12] focus on an approach that separates process execution and data management through the use of 'self-guided artifacts', thereby providing data security in a multi-tenanted environment. On a relevant topic, the authors in [11] present an overview of the security and privacy challenges when dealing with multi-tenanted data in the cloud. These approaches require new BPMS to be developed from the ground up, rather than leveraging existing BPMS *in their entirety* in the cloud as discussed in this paper. Furthermore, there is little discussion about load balancing in a multi-tenanted, multi-engined environment.

A similar approach to that implemented in this paper is described in [10], The authors show an architecture for hosting multiple YAWL engines in the cloud for multiple tenants but only a basic load balancing strategy is offered. In [14] the authors propose a multi-tenant extension to jBPM, which is a *specific* workflow system supporting execution of business processes modelled in Business Process Execution Languages (BPEL).

Other related research efforts include a generic Quality-of-Service (QoS) framework for cloud-based BPMS presented in [8], which comprises QoS specification, service selection, monitoring and violation handling. The authors also present a QoS framework that provides a hierarchical scheduling strategy for costing purposes [13]. However, such frameworks guide the design of new support systems based in the cloud, rather than primarily informing the cloud-based hosting of existing BPMS.

7 Conclusions

We have proposed a conceptual design of a scalable system architecture for deploying BPMS in a multi-tenanted cloud environment. Our design proposal is generic and can be used to serve as a reference for designing of specific BPMS in the cloud. We have implemented our conceptual design using an open-source traditional BPMS with a focus on load balancing strategies.

The conceptual design proposal in this paper helps set out clear directions for future work. So far the underlying architecture focuses on scaling up the system capacity and hence the next step is to advance the design to support both scaling up and down of the system capacity depending on the varying demand of workload. Further extensions to the architecture to support data management set another direction for future work. In regard to load balancing, the next phase of our work is to develop more precise load balancing strategies, e.g. to implement load distribution capability at a higher level by applying predictive analysis techniques, and furthermore to allow dynamically increasing and/or reducing the number of workflow engines at run-time to achieve better scalability.

Acknowledgments. This work is supported by the Research Foundation of Science and Technology Plan Project in Guangdong Province (2016B050502006) and the Research Foundation of Science and Technology Plan Project in Guang-zhou City (2016201604030001).

References

1. van der Aalst, W., van Hee, K.: Workflow Management: Models, Methods, and Systems. MIT Press, Cambridge (2004)
2. van der Aalst, W.M.P., ter Hofstede, A.H.M.: YAWL: yet another workflow language. Inf. Syst. **30**(4), 245–275 (2005)
3. Frederick Chong and Gianpaolo Carraro. Architecture Strategies for Catching the Long Tail. MSDN Library, April 2006
4. Halpin, T.: Object-Role Modeling Fundamentals: A Practical Guide to Data Modeling with ORM. Technics Publications, LLC (2015)
5. Han, Y.-B., Sun, J.-Y., Wang, G.-L., Li, H.-F.: A cloud-based BPM architecture with user-end distribution of non-compute-intensive activities and sensitive data. J. Comput. Sci. Technol. **25**(6), 1157–1167 (2010)
6. ter Hofstede, A.H.M., van der Aalst, W.M.P., Adams, M., Russell, N.: Modern Business Process Automation: YAWL and Its Support Environment, 1st edn. Springer, Heidelberg (2009)
7. David Hollingsworth. Workflow Management Coalition: The Workflow Reference Model. Technical report, TC00-1003, January 1995

8. Liu, X., Yang, Y., Yuan, D., Zhang, G., Li, W., Cao, D.: A generic QoS framework for cloud workflow systems. In: 2011 IEEE Ninth International Conference on Dependable, Autonomic and Secure Computing, pp. 713–720. IEEE (2011)
9. OMG. OMG Unified Modeling Language (OMG UML), Version 2.5.1. Technical report, December 2017. https://www.omg.org/spec/UML/2.5.1/PDF
10. Schunselaar, D.M.M., Verbeek, H.M.W., Reijers, H.A., van der Aalst, W.M.P.: YAWL in the cloud: supporting process sharing and variability. In: Fournier, F., Mendling, J. (eds.) BPM 2014. LNBIP, vol. 202, pp. 367–379. Springer, Cham (2015). https://doi.org/10.1007/978-3-319-15895-2_31
11. Skouradaki, M., Ferme, V., Leymann, F., Pautasso, C., Roller, D.H.: BPELanon: protect business processes on the cloud. In: Proceedings of the 5th International Conference on Cloud Computing and Service Science, 20–22 May 2015. SciTePress (2015)
12. Sun, Y., Su, J., Yang, J.: Universal artifacts: a new approach to business process management (BPM) systems. ACM Trans. Manage. Inf. Syst. $7(1)$, 3:1–3:26 (2016)
13. Xu, R., Wang, Y., Huang, W., Yuan, D., Xie, Y., Yang, Y.: Near-optimal dynamic priority scheduling strategy for instance-intensive business workflows in cloud computing. Concurrency Comput. Pract. Experience $29(18)$ (2017)
14. Yu, D., Zhu, Q., Guo, D., Huang, B., Su, J.: jBPM4S: a multi-tenant extension of jBPM to support BPaaS. In: Proceedings of the 3rd Asia Pacific Business Process Management Conference (AP-BPM 2015), Busan, South Korea, pp. 43–56 (2015)

Towards an Ontology of Software Defects, Errors and Failures

Bruno Borlini Duarte[1], Ricardo A. Falbo[1], Giancarlo Guizzardi[1,2],
Renata S. S. Guizzardi[1], and Vítor E. S. Souza[1(✉)]

[1] Ontology and Conceptual Modeling Research Group (NEMO),
Federal University of Espírito Santo, Vitória, Brazil
bruno.b.duarte@ufes.br, {falbo,rguizzardi,vitorsouza}@inf.ufes.br
[2] Conceptual and Cognitive Modeling Research Group (CORE),
Free University of Bolzano-Bozen, Bolzano, Italy
gguizzardi@unibz.it

Abstract. The rational management of software defects is a fundamental requirement for a mature software industry. Standards, guides and capability models directly emphasize how important it is for an organization to know and to have a well-established history of failures, errors and defects as they occur in software activities. The problem is that each of these reference models employs its own vocabulary to deal with these phenomena, which can lead to a deficiency in the understanding of these notions by software engineers, potential interoperability problems between supporting tools, and, consequently, to a poorer adoption of these standards and tools in practice. We address this problem of the lack of a consensual conceptualization in this area by proposing a reference conceptual model (domain ontology) of Software Defects, Errors and Failures, which takes into account an ecosystem of software artifacts. The ontology is grounded on the Unified Foundational Ontology (UFO) and is based on well-known standards, guides and capability models. We demonstrate how this approach can suitably promote conceptual clarification and terminological harmonization in this area.

Keywords: Ontologies in software engineering · UFO
Software anomaly

1 Introduction

In software and systems engineering, the term *anomaly* denotes a condition that deviates from expectations, based on requirements specifications, design documents, user documents, standards as well as user's and/or modeler's perceptions and experiences [1]. A software anomaly (usually loosely referred by terms such as *bug*, *glitch*, *error* or *defect*) is a situation that suggests a potential problem in a software artifact [2]. In other words, these concepts are used to denote that an artifact is not behaving as expected or is not producing the desired results.

© Springer Nature Switzerland AG 2018
J. C. Trujillo et al. (Eds.): ER 2018, LNCS 11157, pp. 349–362, 2018.
https://doi.org/10.1007/978-3-030-00847-5_25

This informal use, as common and practical as it is used in our daily conversations, may be the source of ambiguity and false-agreement problems, since the concept anomaly is constantly overloaded, referring to many entities with distinct nature. In a more formal environment, this construct overload may lead to communication problems and material losses. Because of that, having a way to properly classify software anomalies is important. Proper classification enables the development of different types of anomaly profiles that could be produced as one indicator of product quality. Also, the information that is generated when an organization understands and systematically classifies software anomalies that may occur at design-time or runtime is a rich source of data that can be used to improve processes and avoid the occurrence of anomalies in future projects [3].

Defects, errors and failures have a negative impact on important aspects of software, such as reliability, efficiency, overall cost and even lifespan. A software with a fairly large "defect density" may go through heavy reconstruction, since it does not meet a minimal acceptance criteria [4]. In more extreme cases, it may be abandoned/discontinued in its early years of usage.

The Guide to Software Engineering Body of Knowledge (SWEBOK) [5] emphasizes the need of a consensus about anomaly characterization and how a well-founded classification could be used in audits and product reviews. A proper defect characterization policy can also provide a better understanding and facilitate corrections in the product or in the process. For an example of this necessity, CMMI [6] defines that organizations should create or reuse some form of defect classification method. It also suggests the use of a defect density index for many work products that are part of the software development process.

The most recent version of the Standard Classification of Software Anomalies [3] provides a classification for different types of anomalies, including information about how they are related. Other standards are more concerned about how to deal with anomalies in different perspectives. For instance, IEEE 1012 [7] is focused on the Verification & Validation phase of the software life-cycle. On the other hand, IEEE 1028 [4] focuses on anomalies in a software audit context, which affects clients and users of a (software) product.

Although there are some proposals for classifying different *terms* for software anomalies, there is no reference model or theory that elaborates on the *nature* of different software anomalies. In other words, to the best of our knowledge, there is no proper *reference ontology* of software anomalies. In order to address this gap, we propose a Reference Ontology of Software Defects, Errors and Failures (OSDEF). This ontology takes into account different types of anomalies that may exist in software-related artifacts and that are recurrently mentioned in the set of the most relevant standards in the area. OSDEF was developed following the process defined by the Systematic Approach for Building Ontologies (SABiO) [8] and grounded in the Unified Foundational Ontology (UFO) [9], including UFO's Ontology of Events [10]. In order to extract consensual information about the domain, we analyzed relevant standards, guides and capability models such as CMMI [6], SWEBOK [5], IEEE Standard Classification for Software Anomalies [3], IEEE Standard for System, Software, and Hardware Verification and Validation [7] as well as complementary current Software Engineering literature. Finally, the ontology was evaluated by verification and validation techniques.

The remainder of this paper is structured as follows. Section 2 introduces the ontological foundations used for developing OSDEF. Section 3 presents the Ontology of Software Defects, Errors and Failures. Section 4 evaluates the proposed ontology. Section 5 discusses related work. Finally, Sect. 6 concludes the paper.

2 Foundations: UFO and the Software Process Ontology

We ground the Ontology of Software Defects, Errors and Failures (OSDEF) in UFO [9]. This choice is motivated by the following: (i) UFO's foundational categories address many essential aspects for the conceptual modeling of the intended domain, including concepts like events, dispositions and situations; (ii) UFO has a positive track record in being able to successfully address different phenomena in Software Engineering [11–13]; (iii) A recent study shows that UFO is among the most used Foundational Ontologies in Conceptual Modeling and the one with a fastest growing rate of adoption [14]. By using an ontology that is frequently used, we increase the reusability of this work, also facilitating its future integration in *ontology networks* in software engineering [15].

UFO is originally composed of three main parts: UFO-A, an ontology of endurants [9]; UFO-B, an ontology of perdurants/events [10]; and UFO-C, an ontology of social entities (both endurants and perdurants) built on top of UFO-A and UFO-B [12]. However, for brevity, Fig. 1 presents only a fragment of UFO that contains the categories that are essential for the purpose of this article. Moreover, we illustrate these categories and their relations using UML diagrams that express typed relations connecting categories, cardinality constraints for these relations, subsumption constraints, as well as disjointness constraints relating sub-categories with the same super-category. UFO has been formally characterized in [9,10,16]. Thus, it is important to emphasize that the following UML diagrams are used here for illustration purposes only.

Endurants and Perdurants are Concrete Particulars (also called *concrete individuals*), i.e., entities that exist in time and space possessing a unique identity. Endurants do not have temporal parts, but are able to change in a qualitative manner while keeping their identity (e.g., a person). Perdurants (or Events, *occurrences, processes*), are composed by temporal parts (e.g., a trip): they exist in time, accumulating temporal parts and, unlike Endurants, they are immutable, i.e., cannot change any of their properties; cannot be different from what they are [10,17]. Moreover, Events are transformations from a portion of reality to another, which means that when a Situation S triggers an Event E, E can *bring about* another Situation S'. Finally, Events can *cause* other Events. This causality relation is a strict partial order (irreflexive, asymmetric and transitive) relation.

Actions are Events that are *performed* by Agents (persons, organizations or teams) with the specific purpose of satisfying *intentions* of that Agent. However, it is important to realize that although all Actions are based on intentions of Agents, if those intentions are based on the wrong assumptions, they can lead to problems, i.e., they can bring about situations that do not satisfy (or that even dent) the goals (propositional content of the intention) that motivated

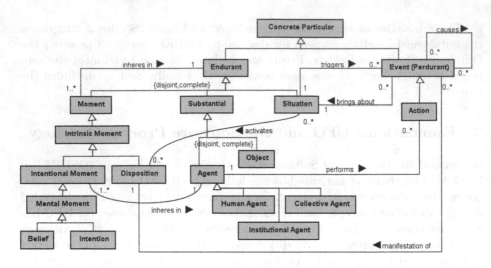

Fig. 1. Fragment of UFO showing events, agents and objects.

that action. Moments (also called *aspects*, *particularized properties* or *tropes*) are existentially dependent entities. This means that they need to *inhere in* other *Concrete Particulars* in order to exist. For example, if a person (as an Agent) or a chair (Object) cease to exist, their Moments (e.g., the Beliefs and Intentions of that person, the texture, a bump or a scratch on that chair) will also disappear. Dispositions are a special type of Moment that are only manifested in certain Situations and that can fail to be manifested, but when they are *manifested*, it is by the occurrence of an Event [10]. Examples of Disposition are the capacity of a magnet to attract metallic material, or John's competence for playing guitar. Situations are complex Endurants that are constituted by possibly many Endurants (including other situations). Situations are portions of reality that can be comprehended as a whole. See [10] for a deeper discussion about Events, Situations and Moments (including Dispositions).

For OSDEF, we reuse the concept of Software Artifact that is presented in the Software Process Ontology (SPO) [18]: objects intentionally made to serve a given purpose in the context of a software project or organization. Stakeholders, which are Agents (a single person, a group or an organization) interested or affected by the software process activities, may be responsible for them (e.g., a user or a development team). We also reuse the concept of Hardware Equipment, which are physical objects used for running software programs or to support some related action (e.g., a computer or a tablet). Moreover, Hardware Equipment are not considered Software Artifacts because they are not created in the context of a software project, although they can be considered resources in a software project. We also reused the Program concept that is present in the Software Ontology (SwO) [19]. According to SwO, a Program is an Artifact that is constituted by code but which is not identical to a code. In contrast, a Program owes its identity principle to a Program Specification, which the Program intends to implement. A complex aggregation of Programs can constitute a software system.

3   An Ontology of Software Defects, Errors and Failures

To build the Ontology of Software Defects, Errors and Failures (OSDEF), we apply SABiO [8], a method for building domain ontologies [20] that incorporates best practices from Software Engineering and Ontology Engineering. We chose SABiO because it is focused on the development of domain ontologies. Moreover, it has been successfully used on the development of several domain ontologies in Software Engineering, such as the Software Process Ontology (SPO) [11] and the Software Ontology [21] and other ontologies developed in the context of SEON, a Software Engineering Ontology Network [15]. Moreover, SABiO explicitly recognizes the importance of using foundational ontologies in the ontology development process to improve the ontology quality, representativity and formality.

SABiO's development process is composed of five phases: (1) purpose identification and requirements elicitation; (2) ontology capture and formalization; (3) operational ontology design; (4) operational ontology implementation; and (5) testing. These phases are supported by well-known activities in the Requirements Engineering life-cycle, such as knowledge acquisition, reuse, documentation, etc. Here, since our main goal is to produce a domain ontology as a *reference conceptual model*, we focus on the first two phases of SABiO, executed in an iterative way, refining the ontology at each iteration. As discussed in Sect. 4, we also conducted verification and validation of the proposed reference conceptual model. Phases (3) to (5), i.e., the design, implementation and testing of the reference ontology proposed here in a computational language (e.g., Common Logic, OWL, HOL-Isabelle, Alloy) are left for future work.

As previously mentioned, the term *anomaly* is commonly used to refer to a variety of notions of distinct ontological nature. Because of that, OSDEF was developed to provide an ontological conceptualization of the different types of software anomalies that exist throughout the software life-cycle. To elaborate on these different types of entities, we raised a set of Competency Questions (CQ), which are questions that the ontology should be able to answer [22]. In a Requirements Engineering perspective, CQs are analogous to the functional requirements of the ontology [8]. Moreover, CQs help to refine the scope of the ontology and can also be used in the ontology verification process. For OSDEF, CQs were raised and refined in a highly-interactive way, through analysis of the international standards mentioned in Sect. 1 and through several meetings with ontology experts. The CQs raised for OSDEF are listed below:

- **CQ1:** What is a failure?
- **CQ2:** What is a defect?
- **CQ3:** What is a fault?
- **CQ4:** What is an error?
- **CQ5:** What is a usage limit?
- **CQ6:** In which type of situation can a failure occur?
- **CQ7:** What are the situations that result from failure?
- **CQ8:** What are the cases of failures?

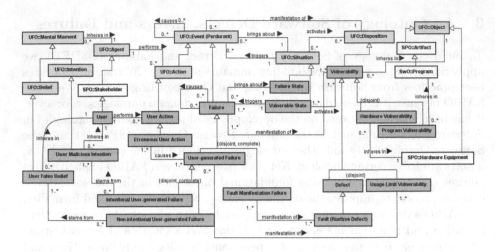

Fig. 2. Conceptual model of the ontology of software defects, errors and failures.

Figure 2 shows the conceptual model of OSDEF. The central concept of our ontology is Failure, since it is the occurrence of a failure that is usually perceived by an agent operating the software system. As defined in standards [1,3,7] and as employed in scientific literature [23], Failures are Perdurants (Events). In that respect, the conceptual basis provided by UFO can help us to understand how failures occur as events during the execution of software. In a software context, a Failure is defined as an event in which a program does not perform as it is intended to, i.e., an event that hurts the goals of stakeholders [24], which motivated the creation of that software. As Events, Failures can cause other Failures, in a chain of Events (e.g., a severe failure in a web server such as *Apache httpd* can make all of its hosted applications experience failures as well). As defined in UFO [10], *causation* is a relation of *strict partial order* and, hence, failures cannot be their own causes or causes or their causes but failures can (perhaps indirectly) trigger other failures in a chain of causation.

As Events, Failures are directly related with two distinct Situations, the first one is the Situation that exists prior to the occurrence of that Failure and that *triggers* the Failure. This Situation is represented in the ontology as a Vulnerable State and denotes the situation that *activates* the Disposition that will be manifested in that Failure. In other words, the state of being exposed to the possibility of a harm or an attack. The second one is the situation that is *brought about* by the occurrence of the Failure, which is defined in the ontology as the Failure State. The occurrence of the failure transforms a portion of reality to another: in its pre-situation, the software is executing, it has the disposition to manifest the failure (i.e., a Vulnerability) but the failure has not occurred yet, since the disposition was not yet activated; in its post-situation, the (failure) event was triggered and reality was "transformed" to a situation in which the software is not executing its functions (at least not as intended by stakeholders).

Although it is out of the scope of this ontology to provide vocabulary for the classification of post-failure situations, we note that Failure States can be: transient — when a failure happens but the software system is capable of recovering itself; continued — when after the occurrence of the failure the Failure State becomes permanent until some action is taken in order to bring the software system back to a execution state in which it is capable to properly execute its functions. Failures can also be classified by other properties, such as severity or effect. We are also not addressing these finer-grained classification here because, once again, we are more interested in providing an ontological analysis of the nature of different software anomalies than in providing a terminological systematization. In our view, the former is a prerequisite for the latter.

Failures are classified in two distinct subtypes: Fault Manifestation Failures and User-Generated Failures. The former are Failures that are manifestations of Faults and are not *caused by* User Actions; the latter are Failures that are directly *caused by* User Actions. These two subtypes have sub-distinctions of their own.

A Vulnerability[1] represents the Dispositions that can exist in software artifacts or in hardware equipments. We thus refined this concept in two distinct generalization sets. The first represents the types of the Dispositions that can be activated and manifest Failures and is composed by Defects and Usage Limit Vulnerabilities. The second represents the types of entities in which those Dispositions inhere: a Hardware Vulnerability inheres in a Hardware Equipment, while a Program Vulnerability inheres in a Program. With that said, we have that both dispositions, Defect and Usage Limit Vulnerability can *inhere in* both types of Objects and thus, we have the following definitions: defects that inhere in Programs are Program Defects; defects that inhere in Hardware Equipment are Hardware Defects. Moreover, we have that a Program Usage Limit Vulnerability is a Usage Limit Vulnerability that inheres in a Program, and that a Hardware Usage Limit Vulnerability is a Usage Limit Vulnerability that inheres in a Hardware Equipment.

A Defect is a type of Vulnerability that can exist in Programs or Hardware Equipments. It is defined by the Standard Classification for Software Anomalies [3] as *an imperfection in a work product (WP) where that WP does not meet its specification and needs to be repaired or replaced.* What this and other definitions in the literature [5, 27] have in common is that Defects are understood as properties of Objects. However, differently from moments that are manifested all the time (e.g., the color of a wall), Defects may never be activated and, consequently, never be manifested into Failures. For example, suppose that a program has a bad implementation of the method **retrieveUsersByLastName**, that can cause a Failure in the software system, which will not be able to execute the functionalities that are associated to that Defect. If that method is never invoked during a program execution, the system may never experience the Failure that is a manifestation of that particular Defect. Given this characteristic, we take Defects (Vulnerabilities in general) to be Dispositions that inhere in Objects.

---

[1] The notion of vulnerability is frequently used in a way that is restricted to defects that can be exploited by attacks. We take a more general *Risk Management* view [25, 26] of vulnerabilities as dispositions that can be manifested by events that can hurt stakeholder's goals [24] or diminish something's *perceived value* [26].

Defects can exist throughout the entire life-cycle of a software [28]. As previously mentioned, some Defects can (accidentally) refrain from being manifested across software executions. When a Defect is manifested in a Failure, we term that Defect a Fault (Runtime Defect). A Fault, hence, can be seen as a role played by a Defect in relation to a Failure. Furthermore, we countenance the occurrence of Failures that are directly caused by User actions. In this scenario, a User *performs* an Erroneous User Action that *causes* a User-Generated Failure. In other words, we name an Erroneous User Action a User action that *causes* such a Failure. As discussed in [29], software artifacts are designed taking into consideration *Domain Assumptions*. When a software artifact makes incorrect assumptions about the environment in which it will execute, we consider this a Program Defect. However, there are cases in which the software makes explicitly defined assumptions (disclaimers, usage guidelines), which are neglected by users in their actions. In this case, it is the Erroneous User Action itself that is the cause of the Failure.

As discussed in [30], events (including Failures) are *polygenic* entities that can result from the interaction of multiple dispositions. For instance, we take that a User-Generated Failure can be caused by a combination of certain dispositions of a software system combined with certain Mental Moments of Agents. These mental moments include Beliefs (including User False Beliefs about domain assumptions) as well as Intentions (including User Malicious Intentions). A particular case of a User-Generated Failure, is one in which this Usage Limit Vulnerability is exploited in an intentional malicious manner, in what is termed an *attack*, e.g., a User, with Malicious Intentions, can make a Web server fail with a DDoS (Distributed Denial of Service) attack. In this case, the server that is being attacked has no Defect (and, hence, no Fault), it just has a limited number of requests that it can answer in a period of time (a *capacity*, which is a type of disposition). If this number is exceeded for a long period, all system resources will be consumed and the server will experience an Intentional User-generated Failure. This failure can be as simple as a denial of service due to lack of resources, or as critical as a full system crash. In a different scenario, a Non-intentional User-generated Failure can stem from the User False Belief of a collective of users simultaneously accessing the system (e.g., as witnessed on Nike's website during the 2017 Black Friday).

4   Evaluation

In order to evaluate the Ontology of Software Defects, Errors and Failures (OSDEF), we applied verification and validation techniques, as prescribed by SABiO. Regarding ontology verification, SABiO suggests a table that shows the ontology elements that are able to answer the competency questions (CQs) that were raised. For validation, the reference ontology should be instantiated to check if it is able to represent real-world situations.

Table 1 illustrates the results of verification of OSDEF regarding the predefined CQs. Moreover, the table can also be used as a traceability tool, supporting ontology change management. The table shows that the ontology answers all of the appropriate CQs.

Table 1. Results of ontology verification.

CQ	Concepts and relations
CQ1	Failure is a *subtype of* Event that *brings about* a Failure State.
	A User-generated Failure is a *subtype of* Failure that is *manifestation of* a Usage Limit Vulnerability and is *caused by* an Erroneous User Action *stemming from* a User False Belief (Non-intentional) or a User Malicious Intention (Intentional).
	A Fault Manifestation Failure is a *subtype of* Failure that is *manifestation of* a Fault (a Runtime Defect).
CQ2	Defect is a *subtype of* Vulnerability (which is a subtype of Disposition) that *inheres in* an Object. A Defect that *inheres in* a Program (i.e., a Program Vulnerability) is called a Program Defect; A Defect that *inheres in* a Hardware Equipment (i.e., a Hardware Vulnerability), is called a Hardware Defect.
CQ3	Fault is a *subtype of* Defect which is manifested at runtime via a Fault Manifestation Failure.
CQ4	An error or, more precisely, an Erroneous User Action is a *subtype of* User Action (Action) that is *performed by* a User, which is a *subtype of* Stakeholder (Agent).
CQ5	Usage Limit Vulnerability is a *subtype of* Vulnerability that *inheres in* an Object. Analogous to Defect, it can *inhere in* a Program (Program Usage Limit Vulnerability) or a Hardware Equipment (Hardware Usage Limit Vulnerability).
CQ6	Vulnerable State is a *subtype of* Situation that *activates* a Fault (which, in turn, is manifested by a Failure) and *triggers* a Failure.
CQ7	Failure State is a *subtype of* Situation that is *brought by* a Failure.
CQ8	A Failure can be *caused by* another Failure, in a chain of Events.
	A Vulnerable State can activate a Fault that is manifested by a Fault Manifestation Failure.
	An Erroneous User Action can *cause* a User-generated Failure, which is a *manifestation of* a Usage Limit Vulnerability.

For a brief validation, we took real-world scenarios of famous cases of software failures and used the ontology to analyze them, showing that OSDEF is capable of representing and analyzing these situations.

Case 1: the Therac-25 disaster [31]. Therac-25 is a medical equipment that handled two types of therapy: a low-powered direct electron beam and a megavolt X-ray mode. The issue was that the software that was responsible for controlling the equipment was reused from a previous model, missing important upgrades and adequate testing. The Fault was manifested into a critical Failure when an operator changed the therapy mode of the equipment too quickly, causing, instructions for both treatments to be simultaneously sent to the machine. The first instruction to arrive would set the mode for the treatment to be applied (a kind of fault known as *race condition*). The consequences were devastating, as patients expecting an electro beam ended up receiving the X-ray and, because of that, died from radiation poisoning. This was an example of a Fault caused by a Program Vulnerability. Although the Fault Manifestation Failure was brought about by a User Action, however, this action cannot be considered an Erroneous User Action (since this cannot be considered a user's negligence of stated assumptions).

Case 2: in 1994, an entire line of Pentium processors could not calculate floating point operations precisely after the eighth decimal case [32]. No matter what software was executing the calculations, the Failure could be manifested since the Defect was intrinsic to the CPU of the computer. We can analyze this case based on OSDEF and on the reports that the Failures happened independently of which software was being executed. We can start our analysis by assuming that the whole Failure State started with a Hardware Vulnerability. The Vulnerability in this case was a Defect inhering in the chip that would prevent it from correctly process arithmetic operation with more than eight decimal cases. As a result, whenever a software execution would trigger the manifestation of that Vulnerability, aFault Manifestation Failure would occur.

Case 3: in 2013, Spamhaus, a nonprofit professional protection service, was the target of what might have been the largest DDoS attack in history. Hackers redirected hundreds of controlled DNS servers to send up to 300 gigabits of flood data to each server of the network, in order to stop them. For this type of situation, when the occurrence of a Failure is directly related with deliberate Actions of an Agent, OSDEF proposes the representation of the event as an Intentional User-generated Failure, since in this particular case the Agents that were responsible for the Failure were basing their Actions in a set of User Malicious Intentions.

5   Related Work

Del Frate [23] provides an ontological analysis of the notion of failure in engineering artifacts. A theory that distinguishes between three types of failures is built: *function-based failures*, *specification-based failure* and *material-based failure*. Del Frate also discusses the relation between a failure — an event that happens to an artifact — and a fault — a state of the artifact after the failure, for each of the three types of failures that are proposed. The ontological analysis provided by Del Frate shares with the work presented here the interpretation of failures as events. However, honoring the terminology employed in software engineering standards, we conceive faults as processual roles of defects in an existing (occurred) failure. In contrast, Del Frate considers faults as states (situations, in the sense of UFO) in a way that is similar to what we call a Failure State. Moreover, another important difference is that we take into account other types of anomalies, such as defects and errors (even taking in consideration the direct participation of human agents in the occurrence of failures). Other distinctions worth mentioning is that our work is focused on software and grounded on a foundational ontology, whereas Del Frate's work is more generic (covering all engineering artifacts) and does not reuse any particular foundational ontology.

Kitamura and Mizoguchi [33] propose an ontological analysis of the fault process and an ontology of faults that provides a categorization of different types of Faults considering different facts, providing a vocabulary for specifying the scope of a diagnostic activity. Characteristics, ontological aspects (e.g., causality and parthood relations) and constraints of different types of faults are presented, e.g., faults are differentiated between: externally or internally caused; structural or property-related; or depending on their ontological nature. The ontology is intended to be used as a tool for characterization of model-based diagnostic systems and as a formal vocabulary, for human use, during the diagnostic activity. It is also used in a diagnostic system that aims to enumerate deeper causes of Failures, providing "depth analysis" to diagnostic systems. In comparison with our ontology, this work has a different focus, which is centered in the fault process and in specifying different characteristics and constraints of Faults and Failures.

Avizienis et al. [34] proposes a taxonomy of faults, failures and errors in a context of dependability, reliability and security. In comparison with OSDEF, the taxonomy proposed there also understands Failures as Events and Faults and Vulnerabilities as properties of a system, composed of software, hardware and people. However, the concept of Error used by the taxonomy is different from the one that we used in OSDEF. Our notion of Error is the one of an Erroneous User Action, being based on the IEEE 1044 standard. This notion is similar to what is termed by Avizienis and colleagues as a Human Fault. Moreover the taxonomy presented by Avizienis et al. has a broader scope than OSDEF, presenting a larger vocabulary focused on properties such as criticality and consistency. On the other hand, OSDEF is more focused on defining the ontological nature of these concepts and the relations between then, using UFO as foundation.

Finally, we should emphasize that, unlike these efforts, OSDEF has been conceived in connection with other UFO-based Software Engineering domain ontologies [11,12] and with the purpose of contributing to a Software Engineering Ontology Network (SEON) [19]. Although these previous works do not address aspects related to software anomalies, they provide context to our work.

6   Conclusions

The main contribution of this paper is proposing an Ontology of Software Defects, Errors and Failures (OSDEF), developed using the SABiO approach, based on a series of standards and capability models, and grounded in UFO. This ontology contributes to the conceptual modeling and management of software anomalies in a number of ways that are summarized as follows.

Firstly, by making use of UFO's foundational categories, OSDEF provides a conceptual analysis of the *nature* of different types of anomalies, systematizing the overloaded use of the term *anomaly* in the Software Engineering literature. Furthermore, this ontology can serve as a reference model for supporting the ontological analysis and conceptual clarification of real-world failure cases. For instance, although sometimes used almost interchangeably, we manage to show that notions such as Failure, Fault, Defect and (User) Error (Erroneous User Action) refer to different types of phenomena. In a nutshell, a Failure is an Event caused by a Vulnerability (a Disposition). A Defect is a Vulnerability inhering in the Program itself or in a Hardware Equipment that is manifested at runtime, in which manifestation this Defect plays the role of a Fault. An Error (Error Erroneous Action) is an Action (an Event brought about by an Agent) that neglects the assumptions under which a Program was designed.

Secondly, as a domain reference model, OSDEF can be used for the development of issue trackers or other types of configuration management-related tools, since it is based on widely accepted standards. Moreover, it can also be used for enabling interoperability between existing tools developed for these purposes.

Thirdly, the ontology establishes a common vocabulary for improving communication among software engineers and stakeholders, avoiding construct overloads and other types of communication problems.

Fourth, in addition to these uses as a reference model, an operational version of OSDEF (for instance, implemented in a logical language such as Common Logic or OWL) can be used to semantically annotate configuration management and software testing data that are directly related to the occurrence of software anomalies. In fact, as future work, we intend to connect OSDEF to our Software Engineering Ontology Network (SEON) [15]. In particular, we intend to develop an ontology of configuration management artifacts and combined it with OSDEF and related ontologies. This, in turn, will enable the development of a traceability tool to relate requirements and stakeholders goals with change requests and issue reports that are tracked during configuration management.

We also intend to strengthen the connection between the work developed here and a common ontology of Value and Risk [26]. After all, the management of anomalies in software artifacts is a special case of Risk Management applied to software. Also, we pretend to investigate further properties of Event types, such as regularity and consistency failures in the Failure context, (e.g., the case of the Therac-25). Finally, we intend to improve the evaluation of OSDEF by comparing (instantiating) the ontology with data produced by development tools such as, e.g., static analysis tools.

Acknowledgments. NEMO (http://nemo.inf.ufes.br) is currently supported by Brazilian research funding agencies CNPq (process 407235/2017-5), CAPES (process 23038. 028816/2016-41), and FAPES (process 69382549/2015).

References

1. ISO: ISO/IEC/IEEE International Standard - Systems and software engineering - Vocabulary. Technical report, International Organization for Standardization, August 2017
2. Guimaraes, E., Garcia, A., Figueiredo, E., Cai, Y.: Prioritizing software anomalies with software metrics and architecture blueprints. In: 2013 5th International Workshop on Modeling in Software Engineering (MiSE), pp. 82–88. IEEE (2013)
3. IEEE: IEEE 1044: Standard Classification for Software Anomalies. Technical report, Technical report, Institute of Electrical and Electronics Engineers, Inc. (2009)

4. IEEE: IEEE 1028: Standard for Software Reviews and Autis. Technical report, Technical report, Institute of Electrical and Electronics Engineers, Inc. (2008)
5. Bourque, P., Fairley, R.E., et al.: Guide to the software engineering body of knowledge (SWEBOK (R)): Version 3.0. IEEE Computer Society Press (2014)
6. SEI/CMU: CMMI® for Development, Version 1.3, Improving processes for developing better products and services. no. CMU/SEI-2010-TR-033. Software Engineering Institute (2010)
7. IEEE: IEEE 1012: Standard for System, Software, and Hardware Verification and Validation. Technical report, Technical report, Institute of Electrical and Electronics Engineers, Inc. (2016)
8. Falbo, R.A.: SABiO: Systematic Approach for Building Ontologies. In: Guizzardi, G., et al. (eds.) Proceedings of the Proceedings of the 1st Joint Workshop ONTO.COM / ODISE on Ontologies in Conceptual Modeling and Information Systems Engineering, Rio de Janeiro, RJ, Brasil, CEUR, September 2014
9. Guizzardi, G.: Ontological Foundations for Structural Conceptual Models. Ph.D. thesis, University of Twente, The Netherlands (2005)
10. Guizzardi, G., Wagner, G., de Almeida Falbo, R., Guizzardi, R.S.S., Almeida, J.P.A.: Towards ontological foundations for the conceptual modeling of events. In: Ng, W., Storey, V.C., Trujillo, J.C. (eds.) ER 2013. LNCS, vol. 8217, pp. 327–341. Springer, Heidelberg (2013). https://doi.org/10.1007/978-3-642-41924-9_27
11. Falbo, R.D.A., Bertollo, G.: A software process ontology as a common vocabulary about software processes. Int. J. Bus. Process Integr. Manage. 4(4), 239–250 (2009)
12. Guizzardi, G., de Almeida Falbo, R., Guizzardi, R.S.: Grounding software domain ontologies in the unified foundational ontology (UFO): the case of the ode software process ontology. In: Proceedings of the 11th Iberoamerican Conference on Software Engineering (CIbSE), pp. 127–140 (2008)
13. Guizzardi, R.S.S., Li, F.L., Borgida, A., Guizzardi, G., Horkoff, J., Mylopoulos, J.: An ontological interpretation of non-functional requirements. In: Garbacz, P., Kutz, O. (eds.) Proceedings of the 8th International Conference on Formal Ontology in Information Systems, vol. 267, Rio de Janeiro, RJ, Brasil, pp. 344–357. IOS Press, September 2014
14. Verdonck, M., Gailly, F.: Insights on the use and application of ontology and conceptual modeling languages in ontology-driven conceptual modeling. In: Comyn-Wattiau, I., Tanaka, K., Song, I.-Y., Yamamoto, S., Saeki, M. (eds.) ER 2016. LNCS, vol. 9974, pp. 83–97. Springer, Cham (2016). https://doi.org/10.1007/978-3-319-46397-1_7
15. Borges Ruy, F., de Almeida Falbo, R., Perini Barcellos, M., Dornelas Costa, S., Guizzardi, G.: SEON: a software engineering ontology network. In: Blomqvist, E., Ciancarini, P., Poggi, F., Vitali, F. (eds.) EKAW 2016. LNCS (LNAI), vol. 10024, pp. 527–542. Springer, Cham (2016). https://doi.org/10.1007/978-3-319-49004-5_34
16. Benevides, A.B., Bourguet, J., Guizzardi, G., Peñaloza, R.: Representing the UFO-B foundational ontology of events in SROIQ. In: Proceedings of the Joint Ontology Workshops 2017 Episode 3: The Tyrolean Autumn of Ontology, Bozen-Bolzano, Italy, 21–23 September 2017
17. Guizzardi, G., Guarino, N., Almeida, J.P.A.: Ontological considerations about the representation of events and endurants in business models. In: La Rosa, M., Loos, P., Pastor, O. (eds.) BPM 2016. LNCS, vol. 9850, pp. 20–36. Springer, Cham (2016). https://doi.org/10.1007/978-3-319-45348-4_2

18. de Oliveira Bringuente, A.C., de Almeida Falbo, R., Guizzardi, G.: Using a foundational ontology for reengineering a software process ontology. J. Inf. Data Manage. **2**(3), 511 (2011)
19. Duarte, B.B., Souza, V.E.S., Leal, A.L.D.C., Guizzardi, G., Falbo, R.D.A., Guizzardi, R.S.S.: Ontological foundations for software requirements with a focus on requirements at runtime. Appl. Ontol., 1–33 (2018)
20. Guizzardi, G.: On ontology, ontologies, conceptualizations, modeling languages, and (meta) models. Front. Artif. Intell. Appl. **155**, 18 (2007)
21. de Souza, É.F., Falbo, R.D.A., Vijaykumar, N.L.: ROoST: reference ontology on software testing. Appl. Ontol., 1–32 (2017)
22. Grüninger, M., Fox, M.: Methodology for the design and evaluation of ontologies. In: Workshop on Basic Ontological Issues in Knowledge Sharing, IJCAI 1995 (1995)
23. Del Frate, L.: Preliminaries to a formal ontology of failure of engineering artifacts. In: FOIS, pp. 117–130 (2012)
24. Guizzardi, R.S.S., Franch, X., Guizzardi, G., Wieringa, R.: Ontological distinctions between means-end and contribution links in the i^* framework. In: Ng, W., Storey, V.C., Trujillo, J.C. (eds.) ER 2013. LNCS, vol. 8217, pp. 463–470. Springer, Heidelberg (2013). https://doi.org/10.1007/978-3-642-41924-9_39
25. Hogganvik, I., Stølen, K.: A graphical approach to risk identification, motivated by empirical investigations. In: Nierstrasz, O., Whittle, J., Harel, D., Reggio, G. (eds.) MODELS 2006. LNCS, vol. 4199, pp. 574–588. Springer, Heidelberg (2006). https://doi.org/10.1007/11880240_40
26. Prince, T., et al.: The common ontology of value and risk. In: Submitted to the 37th International Conference on Conceptual Modeling (ER 2018), Xi'an (2018)
27. PMI: A guide to the project management body of knowledge (PMBOK guide). Technical report, Project Management Institute (2013)
28. Chillarege, R.: Orthogonal defect classification. In: Handbook of Software Reliability Engineering, pp. 359–399 (1996)
29. Wang, X., Mylopoulos, J., Guizzardi, G., Guarino, N.: How software changes the world: the role of assumptions. In: Tenth IEEE International Conference on Research Challenges in Information Science, RCIS 2016, Grenoble, France, pp. 1–12, 1–3 June 2016
30. Fricker, S.A., Schneider, K. (eds.): REFSQ 2015. LNCS, vol. 9013. Springer, Cham (2015)
31. Leveson, N.G., Turner, C.S.: An investigation of the Therac-25 accidents. Computer **26**(7), 18–41 (1993)
32. Williams, C.: Intel's Pentium chip crisis: an ethical analysis. IEEE Trans. Prof. Commun. **40**(1), 13–19 (1997)
33. Kitamura, Y., Mizoguchi, R.: An ontological analysis of fault process and category of faults. In: Proceedings of Tenth International Workshop on Principles of Diagnosis (DX-99), pp. 118–128 (1999)
34. Avizienis, A., Laprie, J.C., Randell, B., Landwehr, C.: Basic concepts and taxonomy of dependable and secure computing. IEEE Trans. Dependable Secure Computing **1**(1), 11–33 (2004)

Schema and View Modeling

TemporalEMF: A Temporal Metamodeling Framework

Abel Gómez[1]([✉]) [iD], Jordi Cabot[1,2] [iD], and Manuel Wimmer[3] [iD]

[1] Internet Interdisciplinary Institute (IN3), Universitat Oberta de Catalunya (UOC),
Barcelona, Spain
agomezlla@uoc.edu
[2] ICREA, Barcelona, Spain
jordi.cabot@icrea.cat
[3] CDL-MINT, TU Wien, Vienna, Austria
wimmer@big.tuwien.ac.at

Abstract. Existing modeling tools provide direct access to the most current version of a model but very limited support to inspect the model state in the past. This typically requires looking for a model version (usually stored in some kind of external versioning system like Git) roughly corresponding to the desired period and using it to manually retrieve the required data. This approximate answer is not enough in scenarios that require a more precise and immediate response to temporal queries like complex collaborative co-engineering processes or runtime models.

In this paper, we reuse well-known concepts from temporal languages to propose a temporal metamodeling framework, called *TemporalEMF*, that adds native temporal support for models. In our framework, models are automatically treated as temporal models and can be subjected to temporal queries to retrieve the model contents at different points in time. We have built our framework on top of the Eclipse Modeling Framework (EMF). Behind the scenes, the history of a model is transparently stored in a NoSQL database. We evaluate the resulting *TemporalEMF* framework with an Industry 4.0 case study about a production system simulator. The results show good scalability for storing and accessing temporal models without requiring changes to the syntax and semantics of the simulator.

Keywords: Temporal models · Metamodeling
Model-driven engineering

This work was supported by the *Austrian Federal Ministry for Digital, Business and Enterprise* and by the *National Foundation for Research, Technology and Development*; the *Programa Estatal de I+D+i Orientada a los Retos de la Sociedad* Spanish program (Ref. TIN2016-75944-R); and the *Electronic Component Systems for European Leadership Joint Undertaking* under grant agreement No. 737494. This Joint Undertaking receives support from the European Union's Horizon 2020 research and innovation program and from Sweden, France, Spain, Italy, Finland & Czech Republic.

J. C. Trujillo et al. (Eds.): ER 2018, LNCS 11157, pp. 365–381, 2018.
https://doi.org/10.1007/978-3-030-00847-5_26

1   Introduction

Modeling tools and frameworks have improved drastically during the last decade due to the maturation of metamodeling concepts and techniques [9]. A concern which did not yet receive enough attention is the temporal aspect of metamodels and their corresponding models when it comes to model valid time and transaction time dimensions instead of just arbitrary user-defined times [15]. Indeed, existing modeling tools provide direct access to the most current version of a model, but very limited support to inspect the model state at specific past time periods [5,8]. This typically requires looking for a model version stored in some kind of model repository roughly corresponding to that time period and using it to manually retrieve the required data. This approximate answer is not enough in scenarios that require a more precise and immediate response to temporal queries like complex collaborative co-engineering processes or runtime models [20].

To deal with these new scenarios, temporal language support must be introduced as well as an infrastructure to efficiently manage the representation of both historical and current model information. Furthermore, query means are required to validate the evolution of a model, to find interesting modeling states, as well as execution states. Using existing technology to tackle these requirements is not satisfactory as we later discuss.

To tackle these limitations, we reuse well-known concepts from temporal languages to propose a temporal metamodeling framework, called *TemporalEMF*, that adds native temporal support. In *TemporalEMF*, models are automatically treated as temporal, and temporal query support allows to retrieve model elements at any point in time. Our framework is realized on top of the Eclipse Modeling Framework (EMF) [24]. Models history is transparently stored in a NoSQL database, thus supporting large evolving models. We evaluate the resulting framework using an Industry 4.0 case study of a production system simulator [19]. The results show good scalability for storing and accessing temporal models without requiring changes to the syntax and semantics of the simulator.

Thus, our contribution is three-fold: (*i*) we present a **light-weight extension** of current metamodeling standards to build a temporal metamodeling language; (*ii*) we introduce an **infrastructure to manage temporal models** by combining EMF and HBase [26], an implementation of Google's BigTable NoSQL storage [12]; and (*iii*) we outline a **temporal query language** to retrieve historical information from models. Please note that contributions do not change the general way how models are used: if only the latest state is of interest, the model is transparently accessed and manipulated in the standard way as offered by the EMF. Thus, all existing tools are still applicable, and the temporal extension is considered to be an add-on.

This paper is structured as follows. Section 2 presents our proposal to include temporal information in existing metamodeling standards. Section 3 presents how temporal models can be stored in a NoSQL database, and Sect. 4 presents our prototype based on EMF and HBase. Our approach is evaluated with a case study in Sect. 5. Section 6 presents related work, and Sect. 7 concludes the paper with an outlook on future work.

2   Temporal Metamodeling

In this section, we discuss how existing work on temporal modeling can be applied for temporal metamodeling. We introduce a profile for adding temporal concepts in existing metamodeling standards; and we present an Industry 4.0 case which demands for temporal metamodeling in order to realize simulation and runtime requirements.

2.1   A Profile for Temporal Metamodeling

We propose a profile for augmenting existing metamodels with information about temporal aspects. Metamodels can be regarded as just a special kind of models [7], and therefore, existing work on temporal modeling for ER [15] and UML [11] languages can be easily leveraged to specify arbitrary temporal (meta)models. Thus, we base our temporal metamodeling profile on these previous works for the static parts of the model, and extend them to cover behavioural definitions which are of particular interested if executable metamodels, i.e., executable modeling languages, are used.

Figure 1 introduces the profile for augmenting metamodels with temporal concerns in EMF Profiles notation. EMF Profiles [18] is a generalization of UML Profiles. As with UML Profiles, stereotypes are defined for predefined metaclasses and represent a way to provide lightweight extensions for modeling languages without requiring any changes to the technological infrastructure. However, in contrast to UML Profiles, EMF Profiles allow modelers to define profiles for any kind of modeling language.

The profile in Fig. 1 includes the stereotype *Temporal* (inspired from previous work on temporal UML [11]) combined with its *durability* and *frequency* properties to classify metaclasses as *instantaneous, durable, permanent* or *constant*. We also introduce novel stereotypes for annotating the operations as we consider executable metamodels. We want to define special operations for which their calls are *logged* or *vacuumed*. The former requires to keep a trace of all executions of the operation. The later forces to restart from scratch the lifespan of the modeling elements deleting their complete previous history. This should be obviously used with caution as it defeats the purpose of having the temporal annotations in the first place but it may be necessary in scenarios where runtime models are used for simulation purposes and modelers want to restart that simulation using a clean slate. Furthermore, it may help in managing the size

Fig. 1. Profile for Temporal Metamodeling.

of temporal models which may attach an extensive history where only the last periods are of interest.

Similarly, we have also adapted our previous work on the specification of temporal expressions [11] to provide temporal OCL-based query support on top of our temporal infrastructure. Before we show an application of the profile and query support, we introduce the motivating and running example of this paper.

2.2   Running Example: Transportation Line Modeling Language

The running example of the paper is taken from the CDL-MINT[1] project. The main goal is to investigate the application of modeling techniques in the domain of smart production systems. The example is about designing transportation lines made up of sets of turntables, conveyors, and multi-purpose machines. The production plant is supposed to continuously processes items by its multipurpose machines located in specific areas. Turntables and conveyors are in charge of moving items to these machines. Given a particular design for a transportation line, simulations are needed for computing different KPIs such as utilization, throughput, cycle times, etc., in order to validate if certain requirements are actually met by a particular design.

The metamodel of the transportation line modeling language is shown in Fig. 2. *System* is the root class, which is composed of several *Areas*. The system has associated a *SimConfig*, where parameters of the simulation can be specified, e.g., the simulation time or number of iterations. An area, in turn, can contain any number of *Components*. As we see, there are five types of *Components*, namely, *Conveyor*, *Machine*, *Turntable*, *StorageQueue*, and *WaitingQueue*. Items are created by *ItemGenerators* and are moved along the transportation line by starting their way in the area's associated *WaitingQueue*. On their way, they may serve as input to *Machines*. Those items that complete the transportation line successfully, should end up in a *StorageQueue*.

In Fig. 2, we also provide an application example of the introduced temporal profile. In particular, we mark the class *Item* as *Temporal* as instances of this

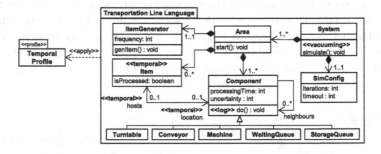

Fig. 2. Applying the Temporal Profile for a Transportation Line Modeling Language

---

[1] More information available at: https://cdl-mint.big.tuwien.ac.at.

class are created during runtime which should be tracked in the history of the model. Furthermore, not only the items, but also their assignment to particular locations should be tracked. For this, we also annotate the bi-directional reference between *Item* and *Component*. In order to understand which component is activated in a particular point in time, we annotate the *do()* operation with the *Log* stereotype. Finally, in order to create a fresh state when a simulation run is started, we annotate the *simulation()* operation with the *Vacuuming* stereotype.

Having the class *Item* marked as a temporal element as well as the involved references, we are now able to define several queries (**Qs**) to compute execution states of interest (such as those needed for provenance) and KPIs (such as utilization):

Q1 — Find all items which have been processed by machine *m*.
Q2 — Find the components which had an item assigned at a particular point in time.
Q3 — Find the components which had an item assigned within a particular time frame.
Q4 — Compute the utilization of machine *e* for the whole system execution lifecycle.

Query **Q1** retrieves the complete evolution of a structural feature, namely the hosts reference. **Q2** accesses the hosts reference for a particular point in time, while **Q3** is evaluating this reference for any particular moment between two time instants. Finally, **Q4** is performing a complex query which is also requiring the access of the time values for having items assigned and not having items assigned.

As an example of how these queries are defined using a temporal OCL [21] extension, below we find the specification for **Q2**. In particular, by using additional access methods for properties which are time sensitive (cf. *hostsAt(i:Instant)*) we are able to query the state of the hosts reference for a particular moment in the past.

```
1 Component.allInstances()->select(c | not c.hostsAt(instant).oclIsUndefined())
```

3 Approach

Enabling a temporal metamodeling language as the one discussed above requires a temporal modeling infrastructure. In this section, we introduce the core concepts of our solution, based on the use of a key-value NoSQL mechanism to store the models' historical data. Next section gives additional technical details on its design.

In a previous work [14], we discussed why NoSQL data stores and, more concretely, map-based (i.e. key-value) stores are especially well-suited to persist models managed by (meta-)modeling frameworks since map-based stores are very well aligned with the typical fine-grained APIs offered by modeling frameworks (that mostly force individual access to model elements, even when the user aims to query a large subset of the model). Alternative mechanisms, such as in-memory or XML-based, failed to scale when dealing with large models as it typically happens when working on, for instance, building information models (BIMs), modernization projects involving the model-based reengineering of legacy systems, or on simulation scenarios. This is also true for relational databases (even temporal ones, a direction they are all following in compliance with the SQL:2011 standard) mainly due to the lack of alignment with modeling tools APIs.

An interesting map-based solution is BigTable [12]. BigTable is a distributed, scalable, versioned, non-relational and column-based big data store; where data is stored in tables, which are sparse, distributed, persistent, and multi-dimensional sorted maps. These maps are indexed by the tuple **row key, column key, and a timestamp**. The native presence of timestamps and the benefits of map-based solutions to store large models make a BigTable-like solution an ideal candidate for a temporal modeling infrastructure.

Next, we describe BigTable's main concepts and how we adapt them (and, in general, similar column-based solutions) to support a temporal modeling infrastructure able to automatically persist and manage (meta)models annotated with our profile.

3.1   BigTable Basics

The top-level organization units for data in BigTable are named *tables*; and within tables, data is stored in *rows*, which are identified by their *row key*. Within a row, data is grouped by *column families*, which are defined at table creation. All rows in a table have the same column families, although may be empty. Data within a column family is addressed via its *column qualifier*; which, on the contrary, do not need to be specified in advance nor be consistent between rows. Cells are identified by their *row key*, *column family*, and *column qualifier*, do not have a data type, and store raw data which is always treated as a `byte[]`. Values within a cell are versioned. Versions are identified by their version number, which by default is the timestamp of when the cell was written. If the timestamp is not specified for a read, the latest one is returned. The number of cell value versions retained by BigTable is configurable for each column family.

3.2   Column-Based Data Model

Our proposed data model flattens the typical graph structure expressed by models into a set of key-value mappings that fit the map-based data model of BigTable. Such data model takes advantage of unique identifiers that are assigned to each model object.

Figure 3a shows a simplification of the production plant model for the case study presented in Sect. 2.2 that we will use as an example. The figure describes a production system (omitted for the sake of simplicity) with a single *Area* with *machines*, which in turn, *host* and *process* – one by one – a set of *items* that are fed into the production system. Figures 3b–d, present three instances of this

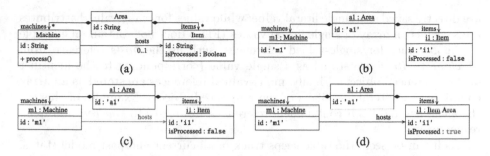

Fig. 3. Example model (a) and sample instances at t_i (b), t_{i+1} (c) and t_{i+2} (d)

model in three different consecutive instants. Changes performed at each instant are highlighted in red. Figure 3b represents an area *a1*, at a given moment in time t_i, with one machine *m1*, and one unprocessed item *i1*. Figure 3c represents the same area at time t_{i+1}, when item *i1* – which is ready to be processed – is fed into *m1*. Finally, Fig. 3d represents the area at time t_{i+2}, once *m1* has processed *i1*, thus changing the *isProcessed* status to `true`.

Our proposed data model uses a single table with three column families to store models' information: (*i*) a *property column family*, that keeps all objects' data stored together; (*ii*) a *type column family*, that tracks how objects interact with the meta-level (such as the *instance of* relationships); and (*iii*) a *containment column family*, that defines the models' structure in terms of containment references. Table 1 shows how the sample instances in Figs. 3b–d are represented using this structure.

As Table 1 shows, row keys are the objects' *unique identifiers*. The PROPERTY *column family* stores the objects' actual data. Please note that not all rows have a value for a given column (as BigTable tables are *sparse*). How data is stored depends on the *property type* and *cardinality* (i.e., upper bound). For example, values for single-valued attributes (like the *id*, which is stored in the ID column)

Table 1. Example model stored in a sparse table in BigTable

		PROPERTY					
KEY	TIMESTAMP	ROOTCONTENTS	ID	MACHINES	ITEMS	HOSTS	ISPROCESSED
'ROOT'	t_i	{ 'a1' }	—	—	—	—	—
'a1'	t_i	—	'a1'	{ 'm1' }	{ 'i1' }	—	—
'm1'	t_{i+1}	—	'm1'	—	—	'i1'	—
'm1'	t_i	—	'm1'	—	—	—	—
'i1'	t_{i+2}	—	'i1'	—	—	—	true
'i1'	t_i	—	'i1'	—	—	—	false

(continued)		CONTAINMENT		TYPE	
KEY	TIMESTAMP	CONTAINER	FEATURE	NSURI	ECLASS
'ROOT'	t_i	—	—	'http://plant'	'RootEObject'
'a1'	t_i	'ROOT'	'rootContents'	'http://plant'	'Area'
'm1'	t_{i+1}	'a1'	'machines'	'http://plant'	'Machine'
'm1'	t_i	'a1'	'machines'	'http://plant'	'Machine'
'i1'	t_{i+2}	'a1'	'items'	'http://plant'	'Item'
'i1'	t_i	'a1'	'items'	'http://plant'	'Item'

are directly saved as a single literal value; while values for many-valued attributes are saved as an array of single literal values (Fig. 3 does not contain an example of this). Values for single-valued references, such as the *hosts* reference from *Machine* to *Item*, are stored as a single value (corresponding to the identifier of the referenced object). Finally, multi-valued references are stored as an array containing the literal identifiers of the referenced objects. Examples of this are the *machines* and *items* containment references, from *Area* to *Machine* and *Item*, respectively.

As it can be seen, the table keeps track of all current and past model states. At t_i (cf. Fig. 3b), the model is stored in rows $\langle$'ROOT', $t_i\rangle$, $\langle$'a1', $t_i\rangle$, $\langle$'m1', $t_i\rangle$ and $\langle$'i1', $t_i\rangle$. After setting the *hosts* reference at instant t_{i+1} (cf. Fig. 3c), the new $\langle$'m1', $t_{i+1}\rangle$ row – which supersedes $\langle$'m1', $t_i\rangle$ – is added. When the *isProcessed* property is changed (cf. Fig. 3d), the $\langle$'i1', $t_{i+2}\rangle$ row is added; and the last model state is stored in rows $\langle$'ROOT', $t_i\rangle$, $\langle$'a1', $t_i\rangle$, $\langle$'m1', $t_{i+1}\rangle$ and $\langle$'i1', $t_{i+2}\rangle$. Note that our infrastructure is not bitemporal: we assume that valid-time and transaction-time are always equivalent.

Structurally, EMF models are trees, and thus, every non-volatile *object* (except the root *object*) must be contained within another *object* (i.e., referenced via a containment *reference*). The CONTAINMENT *column family* maintains this information for every persisted object at a specific instant in time. The CONTAINER column stores the identifier of the container object, while the FEATURE column indicates the *property* that relates the container object with the child object. Table 1 shows that, for example, the container of the *Area a1* is ROOT through the *rootContents* property (i.e., it is a root object and is not contained by any other object). In the next row we find the entry that describes that the *Machine m1* is contained in the *Area a1* through the *machines property*.

The TYPE *column family* groups the type information by means of the NSURI and ECLASS columns. For example, the table specifies the element *a1* is an instance of the *Area* class of the *Plant* metamodel (that is identified by the http://plant NSURI). Data stored in the TYPE *column family* is immutable and never changes.

3.3   Query Facilities

As mentioned in Sect. 2, several temporal query languages have been proposed before. Nevertheless, they all share the need to refer to the value of an attribute or an association at a certain (past) instant of time i in order to evaluate the temporal expressions [11] (also known as temporal interpolation functions). Based on this general requirement, we have built the generic *TObject::eGetAt(i:instant, f:feature)* method that returns the value of a feature (either an attribute or an association end) for a specific *temporal object* at a specific instant. For convenience, we also provide *TObject::eGetAllBetween(s:startInstant, e:endInstant, f:feature)*, that returns a sorted map where the key of the map is the moment when the feature was updated, and the value is the value that was set at that specific moment within the given period. In Sect. 2.2, we showed how **Q2** could be expressed in temporal OCL. As an example, below we depict how to specify such query for a specific *Area a1* in our proposed Java-based query language. This language makes use of the EMF Java API [24], taking advantage of Java streams and lambda expressions.

```
1 a1.getComponent().stream()
2 .filter(c -> c.eGetAt(instant, TllPackage.eINSTANCE.getComponent_Hosts())
     != null)
3 .collect(Collectors.toSet());
```

4 *TemporalEMF* Architecture

We have built our temporal (meta-)modeling framework on top of Apache HBase [26], the most wide-spread open-source implementation of BigTable, based on our experience on building scalable, non-temporal model persistence solutions [6].

Figure 4 shows the high-level architecture of our proposal. It consists of a *temporal model management interface – TemporalEMF –* built on top of a regular *model management interface – EMF* [24]. These interfaces use a *persistence manager* in such a way that tools built over the temporal (meta)modeling framework would be unaware of it. The persistence manager communicates with the underlying database by a *driver*. In particular we implement *TemporalEMF* as a persistence manager on top of HBase; but other persistence technologies can be used as long as a proper driver is provided.

Thanks to our identifier-based data model, *TemporalEMF* offers lightweight on-demand loading and efficient garbage collection. Model changes are automatically reflected in the underlying storage. To do so, (*i*) we decouple dependencies among objects taking advantage of the *unique identifier* assigned to all model objects. (*ii*) We implement an *on-demand loading and saving* mechanism for each live model object by creating a thin delegate object that is in charge of on-demand loading the element data from storage and keeping track of the element's state. Data is loaded/saved from/to the persistence backend by using the object's unique identifier. Finally, and thanks to the data model explained in Sect. 3.2. (*iii*) We provide a *garbage collection-friendly* implementation where no hard references among model objects are kept, so that any model object that is not directly referenced by the application can be deallocated.

TemporalEMF is designed as a simple persistence layer that adds temporal support to EMF. As in standard EMF, no thread-safety is guaranteed, and no

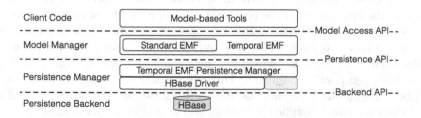

Fig. 4. Overview of the model-persistence framework

transactional support is explicitly provided, although all ACID properties [16] are guaranteed at the object level. Nevertheless, we paid special attention to keep the same semantics than in basic EMF. Thus, *TemporalEMF*, available as an open-source project [1], can be directly plugged into any EMF-based tool to immediately provide enhanced temporal support.

5   Evaluation

In this section, we perform experiments based on the guidelines for conducting empirical explanatory case studies [22]. The main goal is to evaluate the impact of the temporal extension for models on the performance as well as the capabilities of temporal queries in the context of model-based simulations. All the artifacts used in this evaluation and all the data we have gathered (either raw or processed) can be inspected at the paper web page[2].

This study aims to evaluate the possible distinct behavior of current in-memory solutions without dedicated temporal support (in the following, *StandardEMF*) and our temporal solution (in the following, *TemporalEMF*) and, more specifically, to answer the following research questions (**RQs**):

RQ1: Production Cost — Is there a significant difference of the time required for producing and manipulating temporal model elements? This question is of particular relevance for efficient model simulators which have to run for longer periods to produce a target model as well as the traces to reach such a model.

RQ2: Storage Cost — Is there a significant difference of the storage size of temporal models? This question is of particular interest since the output of a model simulation may have to be stored for provenance reasons or for comparing different variants.

RQ3: Reproduction Cost — Is there a significant difference of restoring previous versions of temporal model elements? This question is of relevance as the different properties of a simulation run have to be computed which involves accessing past states and information of the simulated model.

5.1   Case Study Setup

Next, we summarize the selected case study, the input models, the evaluation measures and the environment used to perform the evaluation.

Selected Case: Transportation Line Simulator — Section 2.2, where we exemplify the application of the temporal modeling profile, already presents the case we use in this evaluation: the Transportation Line Modeling Language. However, in order to have a reference implementation with which to compare *TemporalEMF*, we need to extend the metamodel so that *StandardEMF* can also provide temporal capabilities. Following a naive approach, we introduce additional metamodel elements (as well as listeners), so that the history of our model elements can be explicitly tracked and stored in the model itself.

---

[2]   http://hdl.handle.net/20.500.12004/1/C/ER/2018/043.

Figure 5 shows an example of such additions: the *ItemHistoryEntry* meta-class can be added to the metamodel so that everytime the *hosts* property of a *Component* changes, the corresponding *ItemHistoryEntry* element is created and/or updated. Following this naive approach, we add as many extensions to the original metamodel as many temporal properties we aim to track (in our case, the *Component::hosts* and *Item::location* properties).

Input Models — In all the experiments we have used a model with a single *System* and (as we well see later in Sect. 5.2) varying *SimConfig* parameters depending on the RQ to be answered. For reference, this *System* is composed by a single *Area*, with 1 *ItemGenerator*, 1 *WaitingQueue*, 1 *StorageQueue*, 3 *Conveyors*, 4 *TurnTables*, and 1 *Machine*.

For **RQ1**, we execute the simulation varying the processing times for the elements of the *Area* from 0 ms (no processing time), to 40 ms, following an arithmetic progression with a common difference of 5 ms. Additionally, we run the experiments in simulations whose duration (measured in the number of iterations executed) varies from 20 iterations to 40 960 iterations, following a geometric progression with a common ratio of 2. To answer **RQ2**, we only vary the duration of the simulation and the amount of memory available in the system; as well as we do for **RQ3**.

Evaluation Measures — We use different metrics depending on the nature of the research question. For **RQ1**, we measure the execution time needed for running the same simulation both using *StandardEMF* and *TemporalEMF*. We vary both the total simulation time (iterations), and the processing time in the transportation line model to evaluated how the different simulation parameters impact on the execution time. The processing time can be considered the *think time* that determines the workload we apply to the simulation execution. To evaluate **RQ2**, we measure both the used memory during the simulation as well as the storage needed to keep the simulation outcomes for both solutions. To evaluate **RQ3**, we measure the execution time for recreating previous versions of model elements for both solutions. Specifically, we execute the code implementing queries **Q1–Q4** presented in Sect. 2.2.

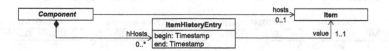

Fig. 5. Extension of the Transportation Line Modeling Language for *StandardEMF*

Environment Setup — We have executed the experiments using two Linux containers in a Proxmox VE 5.1-46 server: one for running the EMF-based code, and other for running HBase. Each one had 8 GB of RAM and 4 virtual CPUs. The actual hardware is a Fujitsu Primergy RX200 S8 server with two quad-core Intel Xeon E5-2609 v2 CPUs at 2.50 GHz, 48 GB of DDR3 RAM memory (1 333 MHz), and two hard disks (at 7 200 rpm) configured in a software-controlled RAID 1. The experiments were run using Java OpenJDK 1.8.0_162, Ant 1.9.9, Eclipse Oxygen 4.7.3a, EMF 2.13 and HBase 1.4.0.

5.2   Result Analysis

Below we summarize our experiments. For the sake of brevity, only the most significant results are shown. For a comprehensive report please refer to the paper web page.

RQ1 — Table 2 shows the results for the experiments executed to evaluate the *Production Cost*. *TemporalEMF* imposes an overhead that is especially noticeable when models are small and there is no time between model modification (i.e., processing time is zero). In those extreme cases, *TemporalEMF* is up to ~23 times slower (i.e., it has an overhead of 2291%). This is understandable, since such small models are completely loaded in memory in *StandardEMF*. On the contrary, on *TemporalEMF*, every single model operation implies a database access thus imposing a big cost.

As the simulation time increases and the model size grows, the overhead is drastically reduced: by just increasing the procesing time from 0 to 5 ms, the overhead is reduced by 10 times (i.e. only ~2 times slower). When the processing time is higher than 30 ms, the overhead is reduced to only ~0.5 times slower. These numbers remain stable with increasing simulation times. It is worth noting that the overhead is only noticeable when modifications happen in the range of ms. Thus, in activities where modifications happen in the range of seconds (e.g. collaborative modeling) the overhead is unnoticeable.

Table 2. Execution times (in seconds) for the experiments for RQ1

PROC. TIME	5120 ITERATIONS			10240 ITERATIONS			2048 ITERATIONS			40960 ITERATIONS		
	SEMF	TEMF	%	SEMF	TEMF	%	SEMF	TEMF	%	SEMF	TEMF	%
0 ms	11	261	2291%	39	588	1405%	99	1167	1078%	367	2488	578%
5 ms	102	377	268%	217	738	240%	492	1491	203%	1169	3627	210%
10 ms	193	473	145%	392	939	140%	839	1917	128%	1902	4328	128%
15 ms	282	620	120%	570	1219	114%	1206	2227	85%	2593	4752	83%
20 ms	373	711	91%	755	1301	72%	1556	2592	67%	3314	6236	88%
25 ms	461	801	74%	941	1606	71%	1939	3213	66%	4004	6616	65%
30 ms	552	841	52%	1113	1683	51%	2275	3443	51%	4738	7492	58%
35 ms	641	942	47%	1297	1880	45%	2640	3761	42%	5472	8279	51%
40 ms	731	1033	41%	1471	2084	42%	3012	4158	38%	6162	9149	48%

RQ2 — Table 3 shows the evaluation for *Storage Cost*. The table summarizes how much memory and storage space is used after running a simulation. To measure the memory consumption, the whole simulation process is executed, the resulting models are saved in disk (in XMI for *StandardEMF*, in HBase for *TemporalEMF*), and after requesting the garbage collector for three times, the actual used memory is measured. As expected, *StandardEMF* uses much more RAM than *TemporalEMF* since all model states are kept in memory; and as it can be observed, some experiments cannot be executed in *StandardEMF* because the simulation runs out of memory. On the contrary, *TemporalEMF* maintains a low memory footprint, using less than 40 MB consistently.

Table 3. Memory usage and disk usage (in MB) for the experiments for RQ2

	MEMORY USAGE (MAX HEAP 512 MB)		MEMORY USAGE (MAX HEAP 2 GB)		DISK USAGE	
Irs.	SEMF	TEMF	SEMF	TEMF	SEMF	TEMF
5 120	19	9	20	9	6	11
10 240	32	9	34	10	12	21
20 480	61	10	64	12	23	41
40 960	112	11	118	13	45	82
81 920	219	15	225	16	89	164
163 840	(*i*)	22	438	22	177	328
327 680	(*i*)	28	865	36	354	656
655 360	(*ii*)	(*ii*)	(*i*)	37	(*i*)	1 341

(*i*) Out Of Memory Error; (*ii*) Setup not executed because experiment was already failing in *StandardEMF* for smaller sizes.

Table 4. Execution times (in milliseconds) for the queries for RQ3

	QUERY EXECUTION TIME (MAX HEAP 512 MB)								QUERY EXECUTION TIME (MAX HEAP 2 GB)							
	SEMF				TEMF				SEMF				TEMF			
Irs.	Q1	Q2	Q3	Q4	Q1	Q2	Q3	Q4	Q1	Q2	Q3	Q4	Q1	Q2	Q3	Q4
5 120	22	13	8	6	145	11	17	65	13	12	8	6	107	10	16	66
10 240	23	24	20	10	225	14	29	93	23	23	15	10	157	11	19	79
20 480	37	39	29	18	1 694	17	21	138	38	36	31	18	239	10	22	172
40 960	43	49	40	35	5 492	31	32	318	47	44	35	34	1 438	23	24	257
81 920	79	63	60	45	14 453	47	35	602	69	56	50	41	11 570	28	21	417
163 840	(*i*)	(*i*)	(*i*)	(*i*)	34 035	42	23	1 510	93	79	74	52	28 913	36	30	852
327 680	(*i*)	(*i*)	(*i*)	(*i*)	68 839	34	20	2 262	184	128	122	74	61 101	33	28	2 163
655 360	(*ii*)	(*ii*)	(*ii*)	(*ii*)	(*ii*)	(*ii*)	(*ii*)	(*ii*)	(*i*)	(*i*)	(*i*)	(*i*)	135 975	36	36	3 595

(*i*) Out Of Memory Error; (*ii*) Setup not executed because experiment was already failing in *StandardEMF* for smaller sizes.

Regarding the disk usage, *TemporalEMF* requires ~2 times more storage than *StandardEMF*. However, *TemporalEMF* can take advantage of the distributed HBase infrastructure, thus allowing models to grow beyond the storage available in a single machine.

RQ3 — Table 4 shows the results for the experiments to evaluate *Reproduction Cost*. We executed **Q1–Q4** on different models for both *StandardEMF* and *TemporalEMF*. As expected *StandardEMF* outperforms *TemporalEMF* since all needed information is already in memory. However, *StandardEMF* is not scalable, and fails when models start growing or when memory is limited. On the contrary, *TemporalEMF* is able to execute all the queries in all the evaluated setups, even when memory is tightly constrained. It is worth noting that some queries are more costly than others (e.g., **Q2** and **Q3** vs **Q1** and **Q4**). In any case, most of the queries can be computed in very few milliseconds, and only **Q1** on specially big models takes several seconds (to return hundreds of thousands of elements).

5.3 Threats to Validity

Several internal and external factors may jeopardize the validity of our results. The first *internal threat* to validity is about the applied pattern for the *StandardEMF* solution. There are different patterns for keeping temporal information in object-oriented structures (even making use of external databases, thus alleviating the memory consumption). We used a standard pattern, but other well-known patterns may show a different result. The same holds for the formulation of the queries.

There are also *external threats* which may jeopardize the generalization of our results. First, we only performed one case study in the domain of model simulation domain. Other domains may show different ratios between the number of static design elements and dynamic runtime elements. Moreover, we did not allow changes to the design element during simulation. Finally, also the employed queries may not represent all possibilities on how to access a temporal model. We aimed to provide heterogeneous queries, such as provenance queries and KPI formulas. However, other queries such as retrieving full model states or a revision graph for the complete model evolution may require different capabilities and may show different runtime results.

6   Related Work

While there is abundant research work on temporal modeling and query languages to for systems data (e.g., consider [15] or [23] for a survey), ours is, as far as we know, the first fine-grained temporal metamodeling infrastructure, enabling the transparent and native tracking (and querying) of system models themselves.

Closest approaches to ours are model versioning tools, focusing on storing models in Version Control Systems (VCS) such as SVN and Git using XMI serializations [2] as well as in database technologies such as relational databases, graph databases, or tuple stores [3]. Traditionally, each version of an evolving model is stored as self-contained model instance together with a timestamp on when the instance as a whole was recorded in the VCS. There is no temporal information at the model element level, and versions are generated on demand when the designer feels there are enough changes to justify a new version (and not based on the temporal validity of the model). Therefore, reasoning on the history of specific elements with a sufficient degree of precision is barely impossible.

Trying to adapt versioning systems to mimic a temporal metamodeling infrastructure would trigger scalability issues as well. Storing full model states for each version is not efficient. E.g., several approaches use model comparison [25] to extract fine-grained historical data out of different model versions. However, these solutions are extremely costly. Just consider changing one value between two versions. This would result in mostly two identical models which have to be stored and compared. This clearly shows that historical model information is currently not well supported by existing model repositories.

A second group of related work is the family of models@run.time approaches [4]. Models@run.time refers to the runtime adaptation mechanisms that leverage software models to dynamically change the behaviour of the system based on a set of predefined conditions. While these approaches provide a modeling infrastructure to instantiate models, as we do, they do not store the history of those changes and only focus on the current state to steer the system. The only exception is the work by Hartmann et al. [17] which proposes the usage of versioning as we have seen in versioning systems for models. Instead of full models, model elements are versioned. However, the versions have to be explicitly introduced and managed as in the aforementioned versioning systems. We find a similar situation with the group of works on model execution [10,13] that focus on representing complete model states but do not keep track of the evolution of those states unless the designer manually adds some temporal patterns (e.g. the one in the previous section).

7   Conclusion

We have presented *TemporalEMF*, a temporal modeling infrastructure built on top of EMF. With *TemporalEMF*, conceptual schemas are automatically and transparently treated as temporal models and can be subject to temporal queries to retrieve and compare the model contents at different points in time. An extension to the standard EMF APIs allows modelers to easily express such temporal queries. *TemporalEMF* relies on HBase to provide an scalable persistence layer to store all past versions.

As further work, we would like to extend *TemporalEMF* in several directions. At the modeling level, we will predefine some useful temporal patterns to facilitate the definition of temporal queries and operations. At the technology level, we will explore the integration of our temporal infrastructure in other types of NoSQL backends and Web-based modeling environments to expand our potential user base. Finally, we aim to exploit the generated temporal information for a number of learning and predictive tasks to improve the user experience with modeling tools. For instance, we could classify users based on their typical modeling profile and dynamically adapt the tool based on that behaviour.

References

1. Temporal EMF. http://hdl.handle.net/20.500.12004/1/A/TEMF/001
2. Altmanninger, K., Seidl, M., Wimmer, M.: A survey on model versioning approaches. IJWIS **5**(3), 271–304 (2009)
3. Barmpis, K., Kolovos, D.S.: Comparative analysis of data persistence technologies for large-scale models. In: Proceedings of Extreme Modeling Workshop, pp. 33–38 (2012)
4. Bencomo, N., France, R., Cheng, B.H.C., Aßmann, U. (eds.): Models@run.time. LNCS, vol. 8378. Springer, Cham (2014). https://doi.org/10.1007/978-3-319-08915-7
5. Benelallam, A., et al.: Raising time awareness in model-driven engineering: Vision paper. In: Proceedings of MODELS, pp. 181–188 (2017)
6. Benelallam, A., Gómez, A., Tisi, M., Cabot, J.: Distributing relational model transformation on MapReduce. J. Syst. Softw. **142**, 1–20 (2018)
7. Bézivin, J.: On the unification power of models. Softw. Syst. Model. **4**(2), 171–188 (2005)
8. Bill, R., Mazak, A., Wimmer, M., Vogel-Heuser, B.: On the need for temporal model repositories. In: Proceedings of STAF Workshops, pp. 136–145 (2018)
9. Brambilla, M., Cabot, J., Wimmer, M.: Model-Driven Software Engineering in Practice, Synthesis Lectures on Software Engineering, 2nd edn. Morgan & Claypool Publishers, San Rafael (2017)
10. Bryant, B.R., Gray, J., Mernik, M., Clarke, P.J., France, R.B., Karsai, G.: Challenges and directions in formalizing the semantics of modeling languages. Comput. Sci. Inf. Syst. **8**(2), 225–253 (2011)
11. Cabot, J., Olivé, A., Teniente, E.: Representing temporal information in UML. In: Stevens, P., Whittle, J., Booch, G. (eds.) UML 2003. LNCS, vol. 2863, pp. 44–59. Springer, Heidelberg (2003). https://doi.org/10.1007/978-3-540-45221-8_5
12. Chang, F., Dean, J., Ghemawat, S., Hsieh, W.C., Wallach, D.A., Burrows, M., Chandra, T., Fikes, A., Gruber, R.E.: Bigtable: a distributed storage system for structured data. In: Proceedings of OSDI, pp. 15–15 (2006)
13. Ciccozzi, F., Malavolta, I., Selic, B.: Execution of UML Models: A Systematic Review of Research and Practice. Software & Systems Modeling. Springer, Heidelberg (2018)
14. Gómez, A., Tisi, M., Sunyé, G., Cabot, J.: Map-based transparent persistence for very large models. In: Egyed, A., Schaefer, I. (eds.) FASE 2015. LNCS, vol. 9033, pp. 19–34. Springer, Heidelberg (2015). https://doi.org/10.1007/978-3-662-46675-9_2

15. Gregersen, H., Jensen, C.S.: Temporal entity-relationship models - a survey. IEEE Trans. Knowl. Data Eng. **11**(3), 464–497 (1999)
16. Haerder, T., Reuter, A.: Principles of transaction-oriented database recovery. ACM Comput. Surv. **15**(4), 287–317 (1983). https://doi.org/10.1145/289.291
17. Hartmann, T., et al.: A native versioning concept to support historized models at runtime. In: Dingel, J., Schulte, W., Ramos, I., Abrahão, S., Insfran, E. (eds.) MODELS 2014. LNCS, vol. 8767, pp. 252–268. Springer, Cham (2014). https://doi.org/10.1007/978-3-319-11653-2_16
18. Langer, P., Wieland, K., Wimmer, M., Cabot, J.: EMF profiles: a lightweight extension approach for EMF models. J. Object Technol. **11**(1), 1–29 (2012)
19. Mazak, A., Wimmer, M., Patsuk-Boesch, P.: Reverse engineering of production processes based on Markov chains. In: Proceedings of CASE, pp. 680–686 (2017)
20. Mazak, A., Wimmer, M.: Towards liquid models: An evolutionary modeling approach. In: Proceedings of CBI, pp. 104–112 (2016)
21. OMG: Object Constraint Language (OCL), Version 2.3.1 (January 2012). http://www.omg.org/spec/OCL/2.3.1/
22. Runeson, P., Höst, M.: Guidelines for conducting and reporting case study research in software engineering. Empirical Softw. Eng. **14**(2), 131–164 (2009)
23. Soden, M., Eichler, H.: Temporal Extensions of OCL Revisited. In: Proceedings of ECMFA, pp. 190–205 (2009)
24. Steinberg, D., Budinsky, F., Paternostro, M., Merks, E.: EMF: Eclipse Modeling Framework 2.0. Addison-Wesley Professional, 2nd edn. (2009). ISBN 0321331885
25. Stephan, M., Cordy, J.R.: A survey of model comparison approaches and applications. In: Proceedings of MODELSWARD, pp. 265–277 (2013)
26. The Apache Software Foundation: Apache HBase (2018). http://hbase.apache.org/

A Semantic Framework for Designing Temporal SQL Databases

Qiao Gao[1]([⊠]), Mong Li Lee[1], Gillian Dobbie[2], and Zhong Zeng[3]

[1] National University of Singapore, Singapore, Singapore
{gaoqiao,leeml}@comp.nus.edu.sg
[2] University of Auckland, Auckland, New Zealand
g.dobbie@auckland.ac.nz
[3] Data Center Technology Lab, Huawei, Hangzhou, China
zengzhong4@huawei.com

Abstract. Many real world applications need to capture a mix of temporal and non-temporal entities, relationships and attributes. These concepts add complexity when designing database schemas and it is difficult to capture the temporal semantics precisely. We propose a new framework for designing SQL databases that distinguishes between temporal and non-temporal concepts while also distinguishing between entities, relationships and attributes at every step. The framework first utilizes an Entity-Relationship (ER) diagram to capture the real world semantics. Temporal constructs in the ER diagram are then annotated. Finally we map the temporal ER diagram to a normal form database schema that reduces redundant data by separating current data from historical data. We also describe how data consistency is maintained during updates. Experiment results show that we can generate database schemas that support efficient access to both current and historical information.

1 Introduction

Many organizations, especially in regulated industries such as finance and healthcare, need to manage and maintain data that changes over time. SQL:2011 [10] introduces temporal tables where a relational table can be associated with an explicit time period $[Start, End)$ to restrict the valid times of its tuples. However, designing database schemas that capture both temporal and non-temporal data with entity, relationship and attribute semantics is complex, and may lead to incorrect temporal semantics and data redundancy if the semantics are not carefully considered at every step of the design process.

A database that keeps the history of entities, relationships and their attributes will contain both current valid and historical data. Information about an entity/relationship and its temporal and non-temporal attributes may be stored over several tuples in multiple temporal and/or non-temporal tables. Temporal joins are needed to enforce constraints between the time periods of tuples from two or more temporal tables. For queries that primarily focus on

© Springer Nature Switzerland AG 2018
J. C. Trujillo et al. (Eds.): ER 2018, LNCS 11157, pp. 382–396, 2018.
https://doi.org/10.1007/978-3-030-00847-5_27

current valid data, the volume of historical data will hamper the query evaluation process. Designing database schemas that reduce temporal joins and support efficient access to both current and historical information is crucial.

Employee

	Eid	Ename	Phone	Salary	Start	End
t_{11}	e01	Alice	90000001	5,000	2000-01-01	2010-01-01
t_{12}	e01	Alice	90000001	7,000	2010-01-01	2012-01-01
t_{13}	e01	Alice	90000001	8,000	2012-01-01	9999-12-31
t_{14}	e02	Bob	90000002	6,000	2008-01-01	2012-01-01
t_{15}	e02	Bob	90000002	7,000	2012-01-01	2017-05-01

EmployeeHobby

	Eid	Hobby	Start	End
t_{21}	e01	Photography	1998-01-01	2005-01-01
t_{22}	e01	Running	2003-01-01	9999-12-31
t_{23}	e01	Swimming	2000-01-01	9999-12-31
t_{24}	e02	Basketball	2005-01-01	2017-05-01
t_{25}	e02	Running	2014-01-01	2017-05-01

Department

	Did	Dname	Start	End
t_{31}	d01	Research and Development	1995-10-01	9999-12-31
t_{32}	d01	Product	1990-01-01	9999-12-31

Project

	Jid	Tittle	Budget	Start	End
t_{51}	j01	Social Network	80,000	2007-03-01	2009-01-01
t_{52}	j02	Data Mining	50,000	2008-05-20	2010-08-01
t_{53}	j03	Keyword Search	70,000	2012-10-01	2014-01-01
t_{54}	j04	Deep Learning	80,000	2015-01-01	9999-12-31

EmpDep

	Eid	Did	Position	Start	End
t_{41}	e01	d02	Engineer	2000-01-01	2005-01-01
t_{42}	e01	d01	Engineer	2005-01-01	2012-01-01
t_{43}	e01	d01	Senior Engineer	2010-01-01	2012-01-01
t_{44}	e01	d01	Manager	2012-01-01	9999-12-31
t_{45}	e02	d01	Engineer	2008-01-01	2012-01-01
t_{46}	e02	d02	Technical Support	2012-01-01	2013-01-01
t_{47}	e02	d01	Engineer	2013-01-01	2017-05-01

EmpProj

	Eid	Pid	Start	End
t_{61}	e01	j01	2007-03-01	2009-01-01
t_{62}	e01	j02	2008-05-20	2010-08-01
t_{63}	e01	j03	2012-10-01	2014-01-01
t_{64}	e01	j04	2015-01-01	9999-12-31
t_{65}	e02	j03	2013-01-01	2014-01-01

Fig. 1. An example temporal company database with problematic design

Figure 1 shows an example company database whose data are stored using the temporal tables in SQL:2011. The schema of the database was designed using a standard technique of deriving tables from an ER diagram and adding attributes *Start* and *End* to each table to indicate the validity of the data values.

Temporal Semantics. The complexity of mixing temporal and non-temporal data may lead to difficulty in enforcing the intended temporal semantics when updates occur. Consider the temporal table *Employee* which has a non-temporal attribute *Phone* whose value may change over time but only the current value is of interest and captured, and a temporal attribute *Salary* whose changes over time are tracked. Suppose an employee Alice wants to update her phone number to *90000011* on 2017-03-01. There are two possible ways to do the update, both of which are problematic:

a. Set the valid end time of the tuple t_{13} to 2017-03-01, and insert a new tuple *<e01, Alice, 8000, 90000011, 2017-03-01, 9999-12-31>* to the table *Employee*, where the date *"9999-12-31"* means *now*. However, this violates the user requirement to keep only the latest phone number, that is, *Phone* is non-temporal, because tuples t_{11} and t_{12} still maintain the previous phone number of Alice.

b. Set the value of attribute *Phone* in the tuples t_{11}, t_{12}, t_{13} to Alice's new phone number. In this case, the tuple t_{11} becomes *<e01, Alice, 90000011, 5000, 2000-01-01, 2010-01-01>* which does not provide the correct valid time for Alice's phone number.

Data Redundancy. Having a schema where a table has both temporal and non-temporal attributes may lead to data redundancy when a temporal attribute in the table is updated. For instance, if *Alice* (*e01*) has a salary raise on 2012-01-01 from $7000 to $8000, a new tuple t_{13} will be inserted into the *Employee* table, and the end date of the tuple t_{12} is set to 2012-01-01. Note that *Alice*'s non-temporal attributes *Ename* and *Phone* are replicated, since we store them together with the temporal attribute *Salary* in one relation. Further, the temporal table *EmpDep* captures the temporal relationship between employees and departments. This table also has a temporal attribute *Position*, and any change in the job position of an employee will lead to replication of data.

Costly Query Evaluation. Processing a routine query involving temporal tables can become complex and costly. If a user wants to find the department that *Alice* currently works in, we need to carry out temporal joins of the tables *Employee*, *EmpDep* and *Department* to retrieve the current records of *Alice*. The historical records in each table participating in the temporal join will slow down the query evaluation process.

To address the above issues, we propose a framework that utilizes a high-level conceptual model, such as the Entity-Relationship (ER) model, to facilitate the database design process. We first construct a normal form ER diagram without considering the temporal aspects [12]. Then we annotate the entity types, relationship types and attributes based on the temporal aspects of the database. Finally, we map the temporal ER diagram to a set of normal form relations. We automatically generate two sets of normal form relations to increase the speed of queries, one for the current database state, and another for the historical database state. The key contributions of the mapping algorithm include:

a. Mapping each temporal and all non-temporal attributes to separate tables, thus reducing data redundancy;
b. Associating time periods with each temporal concept such as entity, relationship and attribute in the same relation instead of the tuple, further reducing data duplication;
c. Separating the current and historical data of entities and relationships, thus increasing the speed of queries.

With the separation of current and historical data, we examine how data consistency is maintained when there are updates. Experimental results show that our framework leads to temporal database schemas that reduce data redundancy, and supports efficient querying of both current and historical information.

2   Temporal ER Diagram

An ER diagram models real world entities that interact with each other via relationships, and their attributes. The works in [5,6,8] have proposed temporal ER diagrams to capture the temporal semantics in the real world. Here, we describe a temporal ER diagram (ERD^T) that contains the essential constructs required for our mapping algorithms. We also analyze the semantics when we have temporal constructs, e.g., temporal entities that participate in non-temporal relationships, or non-temporal entities that participate in temporal relationships.

Capturing both real world semantics and temporal aspects of entities and relationships typically makes an ER diagram complicated and hard to design. As such, we propose to first use an ER diagram to capture the entities, relationships and attributes of a database application, and then annotate these constructs with a superscript T to indicate that they are temporal. We have 3 types of temporal constructs depending on the changes that an application wants to keep track of.

Temporal Entity Type. A temporal entity type indicates that all the entities of this type are temporal. Each entity is associated with some implicit time period values [$Start, End$). The entity is valid within these time periods.

Temporal Relationship Type. A temporal relationship type indicates that all the relationships of this type are temporal. A relationship may be associated with more than one time period values.

Temporal Attribute. A temporal attribute indicates that the value of this attribute of an entity or relationship may change over time periods and the database keeps track of the changes.

An ERD^T can have combinations of non-temporal and temporal constructs, e.g., a temporal entity/relationship can have non-temporal attributes, and a non-temporal entity/relationship can have temporal attributes. Not all the participating entity types of a temporal relationship need to be temporal, and some participating entity types of a non-temporal relationship can be temporal. An ERD^T becomes a traditional ER diagram if all its entity types, relationship types and attributes are non-temporal. Figure 2 shows the temporal ER diagram

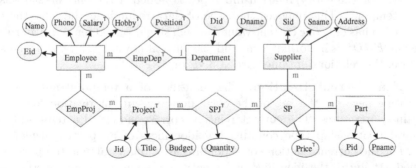

Fig. 2. A temporal ER diagram for a company database

of a company database. Department is a *non-temporal entity type*. Employee is a non-temporal entity type with *temporal attributes* Salary and Hobby. Project is a *temporal entity type*. EmpDep is a temporal relationship type with a temporal attribute Position. Note that the temporal relationship type SPJ has an aggregate entity type SP.

3   Mapping Algorithms

In this section, we present our mapping algorithms to generate the schema for a temporal database from a temporal ER diagram. The schema separates the current and historical data to facilitate the processing of queries. Depending on the workload of the application, the schema can be a set of normal form relations or a set of nested relations. The latter facilitates the frequent retrieval of all the information pertaining to some entitiy/relationship.

3.1   Current Database Schema

We can store the current state of the database in a set of normal form relations. Since the temporal ER diagram models time period attributes (*Start* and *End*) implicitly, these attributes will be added to the database schema. Details of the mapping is given in [4]. The key steps of the mapping are:

a. All non-temporal or temporal single-valued attributes of a non-temporal or temporal entity/relationship type are mapped to one relation.
b. Each non-temporal/temporal multivalued attribute together with the identifier of its entity/relationship type is mapped to a separate relation.
c. For each temporal entity/relationship type in ERD^T, we add an attribute R_Start to its corresponding current relation R_curr. This R_Start attribute indicates the start time of the temporal entities/relationships.
d. For each temporal attribute A, we add an attribute A_Start to the current relation R_curr containing this temporal attribute A to indicate the start time of the attribute values of A.

All the single valued attributes (temporal or non-temporal) are put in one relation, and a start time for each of the temporal single valued attributes, and a start time for the entity/relationship types is added if the entity/relationship type is temporal. This removes the issue of data redundancy because the current relation has only one current value for each single-valued temporal attribute. Since our ERD^T is in normal form and the current state database is a snapshot database, the relations obtained are in 3NF or 4NF [12].

Key of a Current Relation. The identifier of a temporal/non-temporal entity type becomes the key of the corresponding current relation for its single valued attributes. The current relation corresponding to a temporal/non-temporal relationship type contains the identifiers of its participating entity types. For the current relation corresponding to a temporal/non-temporal multivalued attribute, the identifier of its entity/relationship type together with this multivalued attribute form the key. Note that the *Start* time of a temporal entity/relationship/attribute is not part of the key.

Example 1 [Current Data]. Based on the ERD^T in Fig. 2, we generate some normal form relations for the current database schema.

1. Non-temporal entity type Employee with temporal attributes Salary and Hobby:
   $Employee\_curr(\underline{Eid}, Name, Phone, Salary, Salary\_Start)$
   $EmployeeHobby\_curr(\underline{Eid, Hobby}, Hobby\_Start)$
2. Temporal entity type Project with non-temporal attributes Tittle and Budget:
   $Project\_curr(\underline{Jid}, Title, Budget, Project\_Start)$
3. Non-temporal relationship type SP with temporal attribute Price:
   $SP\_curr(\underline{Sid, Pid}, Price, Price\_Start)$
4. Temporal relationship type SPJ with non-temporal attribute Quantity:
   $SPJ\_curr(\underline{Sid, Pid, Jid}, Quantity, SPJ\_Start)$
5. Temporal relationship type EmpDep with temporal attribute Position:
   $EmpDep\_curr(\underline{Eid}, Did, Position, Position\_Start, EmpDep\_Start)$

□

Note that we can have more than one start time in a relation, e.g. in relation *EmpDep_curr*, the start time for relation *EmpDep_Start* records the existence of relationship *EmpDep*, while attribute *Position_Start* records the start time of temporal attribute *Position*.

3.2 Historical Database Schema

Our mapping algorithm separates the current and historical data by generating another set of relations for the historical data. Generating the normal form relations for the historical database is similar to that for the current database. The main differences are:

a. For each temporal entity/relationship type in ERD^T, its corresponding historical relation *R_hist* contains both *R_Start* and *R_End* attributes indicating the start and end time of the temporal entities/relationships.
b. For each temporal single valued or multivalued attribute *A*, we generate a separate historical relation which contains the primary key of its original current relation together with a time period, i.e. *A_Start* and *A_End* attributes.
c. Each non-temporal multivalued attribute *A* of a temporal entity/relationship type is associated with *R_Start* and *R_End* attributes to indicate the set of attribute values for each time period of the corresponding relationship *R*.
d. The non-temporal attributes of non-temporal entity/relationship type is not stored in any historical relations.

Note that we separate temporal attributes and non-temporal attributes in the historical relations. Further, we create one historical table for each temporal attribute and each temporal entity/relationship type.

Key of Historical Relation. The start time *R_Start* for a temporal entity or relationship type *R* and the start time *A_Start* for a temporal attribute *A* are part of the key in the corresponding historical relations.

Example 2 [Historical Data]. Consider the ERD^T in Fig. 2. The relations to store the historical information are as follows:

1. Non-temporal entity type Employee with temporal attributes:
   $EmployeeSalary\_hist(\underline{Eid, Salary\_Start}, Salary\_End, Salary)$
   $EmployeeHobby\_hist(\underline{Eid, Hobby, Hobby\_Start}, Hobby\_End)$
2. Temporal entity type Project:
   $Project\_hist(\underline{Jid, Project\_Start}, Project\_End, Title, Budget)$
3. Non-temporal relationship type SP with temporal attribute Price:
   $SP\_hist(\underline{Sid, Pid, Price\_Start}, Price\_End, Price)$
4. Temporal relationship type SPJ:
   $SPJ\_hist(\underline{Sid, Pid, Jid, SPJ\_Start}, SPJ\_End, Quantity)$
5. Temporal relationship type EmpDep with temporal attribute Position:
   $EmpDep\_hist(\underline{Eid, EmpDep\_Start}, EmpDep\_End, Did)$
   $EmpDepPosition\_hist(\underline{Eid, Position\_Start}, Position\_End, Did, Position)$

   $\square$

3.3 Schema with Nested Relations

In order to minimize the cost of temporal joins, we propose to store all the attributes of an entity/relationship type in one nested relation. Since the ERD^T is in normal form, we can generate a set of normal form nested relations [13] for the ERD^T including the *Start* and *End* attributes.

Current Nested Relations. The difference between the nested relations for current data and the normal form relations lies in the mapping of multivalued attributes to a first-level repeating group inside the relation of its corresponding entity/relationship type. This removes the need to join the relations for the entity/relationship and the multivalued attribute.

Example 3 [Current Data in Nested Relations]. Consider the ERD^T in Fig. 2. *Nested* relations generated for the current state of the database include

1. Non-temporal entity type Employee with temporal attributes Salary and Hobby:
   $Employee\_currN(\underline{Eid}, Name, Phone,\quad Salary, Salary\_Start,\quad (\underline{Hobby}, Hobby\_Start)^*)$
2. Temporal entity type Project with non-temporal attributes Title and Budget:
   $Project\_currN(\underline{Jid}, Title, Budget, Project\_Start)$
3. Non-temporal relationship type SP with temporal attribute Price:
   $SP\_currN(\underline{Sid, Pid}, Price, Price\_Start)$
4. Temporal relationship type SPJ with non-temporal attribute Quantity:
   $SPJ\_currN(\underline{Sid, Pid, Jid}, Quantity, SPJ\_Start)$
5. Temporal relationship type Join with temporal attribute Position:
   $EmpDep\_currN(\underline{Eid}, Did, Position, Position\_Start, EmpDep\_Start)$

   $\square$

Note that the multivalued temporal attribute *Hobby* is stored in a first-level repeating group in relation *Employee_curr*.

Historical Nested Relations. The intuition behind historical nested relations is to put each single-valued or multivalued temporal attribute with its start and end time attributes into a repeating group, since a temporal attribute becomes a multivalued attribute in historical relation.

For each temporal entity/relationship type, all its non-temporal single-valued attributes together with attributes R_Start and R_End form the first-level repeating group of a nested relation R. The time period constrains the temporal entity/relationship as well as all its attributes. Each non-temporal multivalued attribute forms a second-level repeating group since it is constrained by the time period of the temporal entity/relationship. Each single-valued or multivalued temporal attribute A together with their time period A_Start and A_End also form a second-level repeating group.

Note that if an entity/relationship type is non-temporal but has some temporal attributes, we generate the nested relation that contains only the temporal attributes as repeating groups. Details of the mapping algorithm is given in [4].

Key for Historical Nested Relation. The start time R_Start for a temporal entity or relationship type R and the start time A_Start for a temporal attribute A is a part of the key for its corresponding historical relation or repeating group.

Example 4 [Historical Data in Nested Relations]. Consider the ERD^T in Fig. 2. The nested relations to store the historical information of temporal entities, relationships and attributes are generated as follows. "()*" denotes a repeating group, and underlined attributes in a repeating group form the key of this group.

1. Non-temporal entity type Employee with two temporal attributes:
   $Employee\_histN(\underline{Eid}, (Salary, \underline{Salary\_Start}, Salary\_End)^*,$
   $(\underline{Hobby, Hobby\_Start}, Hobby\_End)^*)$
2. Temporal entity type Project:
   $Project\_histN(\underline{Jid}, (Title, Budget, \underline{Project\_Start}, Project\_End)^*)$
3. Non-temporal relationship type SP with temporal attribute Price:
   $SP\_histN(\underline{Sid, Pid}, (Price, \underline{Price\_Start}, Price\_End)^*)$
4. Temporal relationship type SPJ:
   $SPJ\_histN(\underline{Sid, Pid, Jid}, (Quantity, \underline{SPJ\_Start}, SPJ\_End)^*)$
5. Temporal relationship type EmpDep with temporal attribute Position:
   $EmpDep\_histN(\underline{Eid, Did}, ((Position, \underline{Position\_Start}, Position\_End)^*,$
   $\underline{EmpDep\_Start}, EmpDep\_End)^*)$

□

Note that the historical relation for the non-temporal entity type Employee does not contain non-temporal attributes (e.g. *Name*), which can be obtained by joining the historical relation with the corresponding current relation in Example 3.

4 Maintaining Data Consistency

Since we separate the current and historical data, we need to ensure consistency when updates occur. Figure 3 shows the schema generated from the temporal

ER diagram in Fig. 2. We focus on updates to the current data since historical data are typically not updated but archived for subsequent analysis. Update operations on the current data include modifying the values of temporal and non-temporal attributes, and inserting or deleting tuples. These tuples may correspond to multivalued attributes or some entity/relationship.

Modify Attribute Values. If a non-temporal attribute such as an employee's phone number is modified, we simply replace the old attribute value in the current relation with the new value since the database does not keep the old values of non-temporal attributes. However, if a temporal attribute such as an employee's salary is changed, we need to insert a corresponding tuple containing the old attribute value with a valid end time to the historical relation. The start date for the new salary value in the current relation is changed to reflect the valid time of the change. Figure 4 shows the changes (in italics) in the current and historical Employee relations after updating the phone number and salary of *Alice* to 90000011 on 2017-03-01 and $10,000 on 2017-06-01 respectively.

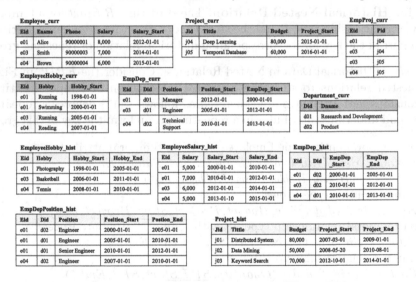

Fig. 3. Generated schema where current and historical data are separated.

Employee_curr

Eid	Ename	Phone	Salary	Salary_Start
e01	Alice	*90000011*	*10,000*	*2017-06-01*
e03	Smith	90000003	7,000	2014-01-01
e04	Brown	90000004	6,000	2015-01-01

EmployeeSalary_hist

Eid	Salary	Salary_Start	Salary_End
e01	5,000	2000-01-01	2010-01-01
e01	7,000	2010-01-01	2012-01-01
e03	6,000	2012-01-01	2014-01-01
e04	5,000	2013-01-10	2015-01-01
e01	*8000*	*2012-01-01*	*2017-06-01*

Fig. 4. Updated relations after Alice (*e01*) changes her phone number and salary.

Insert/Delete Tuples. Inserting new tuples to the current data does not affect the historical data, and will not lead to any inconsistency. However, deleting a tuple t from a current relation R_curr may require the insertion of a corresponding tuple to the historical relation with valid end time. We have 3 cases:

1. R_curr corresponds to a multivalued attribute of some entity/relationship.
   If the attribute is temporal, we insert a corresponding tuple t' to R_hist with a valid end time, and delete t from R_curr. Otherwise, we simply delete t from R_curr. For example, if we delete tuple $<e01,\ Swimming,\ 2000\text{-}01\text{-}01>$ from $EmployeeHobby\_curr$ in Fig. 3 with a valid end time 2017-08-01, we will insert a tuple $<e01,\ Swimming,\ 2000\text{-}01\text{-}01,\ 2017\text{-}08\text{-}01>$ with valid end time *2017-08-01* to the historical relation $EmployeeHobby\_hist$.
2. R_curr corresponds to an entity type t with identifier e.
   Since the information of an entity may be stored across multiple relations, its deletion may trigger the deletion of tuples that correspond to its multivalued attributes and the relationships the entity participates in.
   - Case 2(a) Entity type is non-temporal.
     We delete all tuples with identifier e from current and historical relations, i.e., remove all information about this entity from the database.
   - Case 2(b) Entity type is temporal.
     We delete tuples with identifier e from the current relations, and insert corresponding tuples with valid end times in the historical relations.
   For example, *Alice* leaves the company and we delete tuple $<e01,\ Alice,$ $90000001,\ 8000,2012\text{-}01\text{-}01>$ from $Employee\_curr$ in Fig. 3. This triggers the deletion of all tuples with $eid = e01$ from both current and historical relations. In contrast, suppose project $j04$ is completed in 2017-10-01 and we delete it from current relation $Project\_curr$. Since this is a temporal entity, the database will store its information in the historical relations. As such, we delete tuples involving $j04$ from current relations and insert the corresponding tuples with valid end times in historical relations. If employee is a temporal entity type, then we need to create historical relations to store the non-temporal single-valued and multivalued attributes of employees who have left the company, i.e., they have been deleted from current relations.
3. R_curr corresponds to a relationship type.
   If the relationship is temporal, then insert a corresponding tuple with valid end time into the historical relation R_hist. Otherwise, delete tuple t from R_curr. For example, if employee Smith ($e03$) no longer works in the department $d01$ on 2017-09-01, then we insert tuple $<e03,\ d01,\ 2012\text{-}01\text{-}01,\ 2017\text{-}09\text{-}01>$ into the historical relation $EmpDep\_hist$ and a tuple $<e03,\ d01,\ Engineer,\ 2005\text{-}01\text{-}01,\ 2017\text{-}09\text{-}01>$ is inserted into the historical relation $EmpDepPosition\_hist$. The tuple $<e03,\ d01,\ Engineer,\ 2005\text{-}01\text{-}01,\ 2012\text{-}01\text{-}01>$ is deleted from the current relation $EmpDep\_curr$.

5 Experiments

We evaluate the performance of our approach for designing temporal databases. We use Oracle 12c Enterprise Edition hosted on Solaris 10 with 2.60 GHz CPU and 128 GB RAM. Oracle implements an object-relational model and supports multi-level nested tables as objects. Three schemas are generated based on the Employee database[1] which captures employees who work for and manage departments in a company, and keeps track of the changes of the employees' salaries, job titles, departments worked for and managed.

$S1$: This is the traditional schema which mixes current and historical data.
$Dept(\underline{deptno}, name)$
$Employee(\underline{empno, Employee\_Start}, Employee\_End, birthdate, name, gender)$
$Emptitle(\underline{empno, title\_Start}, title\_End, title)$
$Empsalary(\underline{empno, salary\_Start}, salary\_End, salary)$
$Workfor(\underline{empno, Workfor\_Start}, Workfor\_End, deptno)$
$Manage(\underline{empno, Manage\_Start}, Manage\_End, deptno)$

$S2$: This is our normal form relations that separates current and historical data.
$Dept\_curr(\underline{deptno}, name)$
$Employee\_curr(\underline{empno}, birthdate, name, gender, title, title\_Start,$
$\qquad\qquad salary, salary\_Start, Employee\_Start)$
$Workfor\_curr(\underline{empno}, deptno, Workfor\_Start)$
$Manage\_curr(\underline{empno}, deptno, Manage\_Start)$
$Employee\_hist(\underline{empno, Employee\_Start}, Employee\_End, birthdate, name, gender)$
$Emptitle\_hist(\underline{empno, title\_Start}, title\_End, title)$
$Empsalary\_hist(\underline{empno, salary\_Start}, salary\_End, salary)$
$Workfor\_hist(\underline{empno, Workfor\_Start}, Workfor\_End, deptno)$
$Manage\_hist(\underline{empno, Manage\_Start}, Manage\_End, deptno)$

$S3$: This is our normal form *nested* relations that stores current and historical data separately.
$Dept\_currN(\underline{deptno}, name)$
$Employee\_currN(\underline{empno}, birthdate, name, gender, title, title\_Start,$
$\qquad\qquad salary, salary\_Start, Employee\_Start)$
$Workfor\_currN(\underline{empno}, deptno, Workfor\_Start)$
$Manage\_currN(\underline{empno}, deptno, Manage\_Start)$
$Employee\_histN(\underline{empno}, (birthdate, name, gender, (title, \underline{title\_Start}, title\_End)^*,$
$\qquad\qquad (salary, \underline{salary\_Start}, salary\_End)^*, Employee\_Start, Employee\_End)^*)$
$Workfor\_histN(\underline{empno}, deptno, (\underline{Workfor\_Start}, Workfor\_End)^*)$
$Manage\_histN(\underline{empno}, deptno, (\underline{Manage\_Start}, Manage\_End)^*)$

We have 3 datasets for each schema: original Employee dataset D_1, and two synthetically generated datasets D_2 and D_3. Table 1 shows the statistics of these datasets. In D_2, each current entity/relationship has 10 historical entities/relationships, and each temporal relationship/attribute is associated with one time period. This increases the size of historical data to evaluate our approach that separates the current and historical states of the database. In D_3, each current entity/relationship has 1 historical entity/relationship, and each temporal relationship/attribute is associated with 10 time periods. This increases the

---
[1] https://dev.mysql.com/doc/employee/en/.

number of repeating groups to evaluate our approach that stores historical data in nested relations. Table 2 gives the descriptions of our 9 test queries, and Fig. 5 shows the CPU time of SQL execution for each query on the 3 datasets.

Table 1. Statistics of Datasets

Dataset	Size	Ratio of historical to current entity/relationship	# of time periods per temporal relationship/attribute
D_1	136 MB	~1 : 3.2	~3.6 : 1
D_2	583 MB	10 : 1	1 : 1
D_3	699 MB	1 : 1	10 : 1

Table 2. Queries for Employee database

Q_1	For employees who are "Senior Engineer" now, find their start date
Q_2	Find the current number of employees in each department
Q_3	Find the maximum salary in the salary history of employees with $Eid < 100000$ who are currently working in the "Marketing" department
Q_4	Find the job titles of female employees with $Eid < 1000$ who have left the company
Q_5	Find the salary history of employee "Aris Iwayama"
Q_6	Find the last job title for all the employees who have left the company
Q_7	Find the number of departments that resigned employees (born before 1960-01-01) had joined
Q_8	Find employees who were previously "Engineer" but are now "Senior Engineer"
Q_9	Find employees who had previously managed the department "Marketing"

Q_1 and Q_2 query the current data only. These queries run much faster on all 3 datasets with schema $S2$ and $S3$ compared to $S1$. The gap in their runtime widens when the ratio of historical to current data increases (as in D_2). The query times for $S2$ and $S3$ are the same since they have the same current relations.

Q_3 to Q_5 retrieve historical information of current/resigned employees. Q_3 requires traditional join and an aggregation function on the temporal attribute. Q_4 finds all historical records for a large set of entities, while $Q5$ finds the historical records for a few entities. All these queries need to retrieve the historical record for each entity/relationship. $S3$ gives the best performance on all 3 datasets as it reduces the joins between temporal attributes and their entities/relationships.

Q_6 to Q_7 are queries on temporal relationships or constrain temporal attributes with some values. As the queries focus on either historical or current data, $S1$ is slower that $S2$. $S3$ is expensive since it is difficult to retrieve entities or relationships whose attributes satisfy some constraints from nested relations.

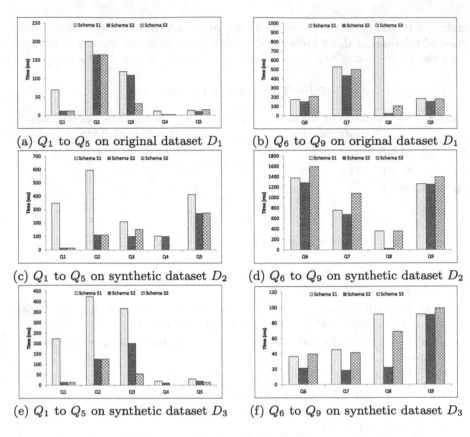

(a) Q_1 to Q_5 on original dataset D_1 (b) Q_6 to Q_9 on original dataset D_1

(c) Q_1 to Q_5 on synthetic dataset D_2 (d) Q_6 to Q_9 on synthetic dataset D_2

(e) Q_1 to Q_5 on synthetic dataset D_3 (f) Q_6 to Q_9 on synthetic dataset D_3

Fig. 5. CPU time of SQL execution for queries in Table 2

Q_8 involves a temporal join between current data and historical data. Since $S1$ combines current and historical data, we need to join two Emptitle relations and check that the time period of "Engineer" is before that of "Senior Engineer". This is time consuming. Separating current and historical data avoids such temporal joins. $S3$ does not perform as well as $S2$ because this query constrains the value of the temporal attributes, which is slow for nested relations. Q_9 retrieves both current and historical data. The runtimes for this query is similar for all three schemes since we need to scan current and historical data.

The results of our experiments demonstrate that separating the data into current and historical relations (for both flat and nested relations) improves the efficiency when queries are focused on either current or historical data. The normal form relations for storing historical data (S2) is a better design for temporal join between current and historical data, and is also good for querying historical data with some constraints on temporal attributes. The nested relations which stores historical data (S3) shows good performance for queries that retrieve two or more historical records for entities/relationships.

6   Related Work

Research in conceptual modeling captures the temporal aspects of a database by introducing new temporal constructs to the standard ER diagram [1,3,9,16–18], or extending existing constructs to include temporal semantics [2,11]. Our ERD^T is a simplified temporal ER diagram that contains the essential constructs for our mapping algorithm.

There are two main approaches to generate a temporal database schema from an ER diagram. One approach is to map a traditional ER model to a database schema, annotate it with temporal functional dependencies, and use the decomposition approach to obtain a set of normal form relations including temporal relations [14]. However, these relations do not capture the semantics of temporal entity, relationship and attribute. Further, when the database schema becomes complex, it is not easy for users to specify the temporal functional dependencies compared to annotating an ER diagram as in our proposed approach.

Another approach is to map a temporal ER model to a temporal database schema. The works in [16,18] map their temporal ER models to the standard ER model first (which may include adding explicit time period attributes) before translating the ER diagram to the database schema. However, the semantics of temporal entity, relationship and attribute are lost during the mapping since the time attributes, i.e., *Start* and *End*, are treated as regular attributes in the standard ER model. The works in [7,11,15,17] map their temporal ER models directly to a database schema. [17] generates a set of 3NF database schema that does not capture the semantics of temporal attributes since they use a temporal ER model which does not support temporal attribute and attributes of relationship type. [7] maps a temporal ER diagram to a surrogate-based relational model (RM/T model) which does not capture the semantics of temporal relationship type. [11] adds an implicit valid time period for each generated relation, thus making all the attributes in the database temporal. This may increase the data redundancy when there are one or more temporal attributes in a relation. All these methods do not separate the current data from historical data.

Although commercial databases such as Microsoft SQL server and IBM DB2 10 implemented the feature of historical and current tables, a historical table is a mirror of its current table, and the purpose is to track modifications in the current table. Further, these tables are only available to data involving transaction time, and not data involving valid time. Moreover, SQL Server and DB2 cannot distinguish between temporal attributes and temporal entities, which needs to be handled differently when updates occur. In contrast, our proposed framework provides a principled approach to generate database schema that captures these temporal semantics and supports efficient access of data involving valid time.

7   Conclusion

The requirement to capture a mix of temporal and non-temporal entities, relationships and attributes adds complexity to the design of database schemas.

We proposed a semantic framework that precisely captures temporal semantics. Each step in the framework distinguishes temporal and non-temporal entities, relationships and attributes. Our mapping algorithm generates current and historical schemas from a temporal ER diagram. This accelerates querying of current data, and provides the flexibility of mapping to a set of nested relations. We discussed how consistency between current and historical data is maintained during updates. Experiments showed that the schemas obtained provide efficient access to both current and historical information. Future work includes studying the physical design over our temporal database schema, e.g., temporal indexing.

References

1. Elmasri, R., El-Assal, I., Kouramajian, V.: Semantics of temporal data in an extended ER model. In: ER (1990)
2. Elmasri, R., Wuu, G.T.J.: A temporal model and query language for ER databases. In: IEEE ICDE (1990)
3. Ferg, S.: Modelling the time dimension in an entity-relationship diagram. In: ER (1985)
4. Gao, Q., Lee, M.L., Ling, T.W., Dobbie, G., Zeng, Z.: A semantic framework for designing temporal SQL databases. Technical report, TRB3/18, NUS (2018)
5. Gregersen, H., Jensen, C.: Conceptual modeling of time-varying information. Technical report, TimeCenter TR-35 (1998)
6. Gregersen, H., Jensen, C.S.: Temporal entity-relationship models - a survey. IEEE TKDE 11, 464–497 (1999)
7. Gregersen, H., Mark, L., Jensen, C.S.: Mapping temporal ER diagrams to relational schemas. Technical report, TimeCenter TR-39 (1998)
8. Khatri, V., Ram, S., Snodgrass, R.T., et al.: Capturing telic/atelic temporal data semantics: Generalizing conventional conceptual models. IEEE TKDE 26, 528–548 (2014)
9. Klopprogge, M.R.: TERM: an approach to include time dimension in the entity-relationship model. In: ER (1981)
10. Kulkarni, K., Michels, J.E.: Temporal features in SQL: 2011. Sigmod Rec. 41, 34–43 (2012)
11. Lai, V.S., Kuilboer, J.P., Guynes, J.L.: Temporal databases: model design and commercialization prospects. ACM SIGMIS Database 25, 6–18 (1994)
12. Ling, T.W.: A normal form for entity-relationship diagrams. In: ER (1985)
13. Ling, T.W., Yan, L.: NF-NR: a practical normal form for nested relations. J. Syst. Integr. 4, 309–340 (1994)
14. Papazoglou, M., Spaccapietra, S., Tari, Z.: Advances in Object-Oriented Data Modeling. MIT Press, Cambridge (2000). Chap. 7
15. Snodgrass, R.T.: Developing Time-Oriented Database Applications in SQL. Morgan Kaufmann Publishers, San Francisco (2000). Chap. 11
16. Tauzovich, B.: Towards temporal extensions to the entity-relationship model. In: ER (1991)
17. Theodoulidis, C., Loucopoulos, P., Wangler, B.: A conceptual modelling formalism for temporal database applications. Inf. Syst. 16, 401–416 (1991)
18. Zimanyi, E., Parent, C., Spaccapietra, S., Pirotte, A.: TERC+: a temporal conceptual model. In: International Symposium Digital Media Information Base (1997)

Nested Schema Mappings for Integrating JSON

Rihan Hai[1(✉)], Christoph Quix[1,2], and David Kensche[3]

[1] RWTH Aachen University, Aachen, Germany
{hai,quix}@dbis.rwth-aachen.de
[2] Fraunhofer Institute for Applied Information Technology FIT,
Sankt Augustin, Germany
[3] SAP Innovation Center Network, Potsdam, Germany
david.kensche@sap.com

Abstract. JSON has become one of the most popular data formats. Yet studies on JSON data integration (DI) are scarce. In this work, we study one of the key DI tasks, nested mapping generation in the context of integrating heterogeneous JSON based data sources. We propose a novel mapping representation, namely *bucket forest mappings* that models the nested mappings in an efficient and native manner. We show experimentally the practicality of our approach over six real world data sets. Moreover, via intensive experiments over synthetic scenarios we demonstrate that our approach scales well to the increasing metadata complexity of DI scenarios.

1 Introduction

JSON is gaining popularity as a universal data format. Although it enables easy interoperability with a unifying syntax, the labels and structures of JSON documents may vary between different data sources. Thus, mappings for the integration of heterogeneously structured JSON sources are necessary. In the context of data integration (DI), schema mappings are specifications that describe the relationships between schemas of data sources and the integrated schema. In virtual data integration mappings are used to rewrite queries on an integrated schema into queries over the sources.

Mappings come in different flavors. Simple element-to-element correspondences can be obtained from schema matching. From this, more expressive logical representations of mappings can be generated. Consider the mappings m_1 and m_2 shown below. They map relations from a source $Paper_s$, $Author_s$, and PA_s (representing papers, authors, and their relationships), to a nested relation Pub_t with only paper titles and a nested list of authors (AL_t). Note that only m_1 is a source-to-target tuple-generating dependency (s-t tgd), m_2 is a general first-order sentence. S-t tgds are not able to cover the semantics of this mapping [11].

© Springer Nature Switzerland AG 2018
J. C. Trujillo et al. (Eds.): ER 2018, LNCS 11157, pp. 397–405, 2018.
https://doi.org/10.1007/978-3-030-00847-5_28

$m_1 : \forall p, t\ Paper_s(p,t) \rightarrow \exists AL_t\ Pub_t(t, AL_t)$

$m_2 : \forall p, t \exists AL_t \forall a, n\ Paper_s(p,t) \wedge PA_s(p,a) \wedge Author_s(a,n) \rightarrow Pub_t(t, AL_t) \wedge AL_t(n)$

$m_3 : \forall p, t\ Paper_s(p,t) \rightarrow \exists AL_t\ (Pub_t(t, AL_t) \wedge$

$$\forall a, n\ PA_s(p,a) \wedge Author_s(a,n) \rightarrow\ AL_t(n)\ )$$

In contrast, the nested mapping m_3 integrates both basic mappings m_1 and m_2 into one logical formula. Nested mappings [6] have been proven to accurately and effectively assert the relationships between nested schemas. For hierarchical structured data such as XML and JSON, basic mappings often lead to an inefficient execution of mappings and redundant mapping assertions. For example, if we execute m_1 and m_2 over the same source, then we get duplicated values for Pub_t since it appears both in m_1 and m_2. The corresponding formalisms for nested mappings are nested tuple-generating dependencies (nested tgds) [6,11]. Intuitively, nested mappings can be considered as a nesting of basic mappings, in which the correlation of schema elements on the target side is resembled in the nesting structure of the formula.

The classical approach for nested mapping creation [6] is rather inefficient, as it first creates basic mappings (e.g., m_1 and m_2), finds their overlaps, and subsequently "nests" them to obtain nested mappings (e.g., m_3) via tableaux construction. The goal of this paper is to generate nested mappings in a more efficient manner, taking into account the desired structure on the target side. We propose a novel nested mapping representation, namely *bucket forest mappings*, as well as an approach to generate such mappings. The main intuition of our approach is to model nested mappings in a tree-like hierarchical structure, which can be generated directly without creating the basic mappings. Our approach can generate the desired nested mappings for JSON or other semi-structured data models, without transforming it into relational data.

The main contributions of our paper are:

- A novel mapping representation, namely *the bucket forest mapping*, which captures the properties of nested mappings for integrating JSON;
- Evaluation of our proposed approach with six real world data sets which shows that our approach improves the efficiency significantly;
- Proof of scalability of our approach via synthetic scenarios with varying metadata complexity factors.

Related Work: As a predominant topic in data integration, schema mapping for XML data has been intensively studied in previous works. Piazza [9] supports mappings for query reformulation using the mapping language of Piazza-XML, though the procedure of generating nested mappings is not explicitly given. Clio [6] populates nested mappings for data exchange. As discussed previously, our bucket forest approach improves the efficiency of mapping generation compared the original approach given in Clio. HePToX [4] creates mappings for query translation in P2P systems, and does not handle nested mappings. In our previous work [10] we studied modeling mappings between heterogeneous metamodels using a generic metamodel. Yet, we mainly focused on mapping composition instead of nested mapping generation. We also developed a method to generate executable queries from second-order tuple-generating dependencies, which could be also applied to the mappings presented in this paper.

Paper Structure: First we introduce preliminary concepts and provide a running example in Sect. 2. Then, we show our proposed mapping representation (Sect. 3) and generation (Sect. 4). We evaluate our solutions in Sect. 5, before we conclude the paper in Sect. 6.

2 Preliminaries

JSON Schema: We consider JSON objects as a list of key/value pairs (k, v), where k is a key (unique within the current object), and v is either an atomic value, a JSON object, or a set of atomic values or JSON objects. We are aware that JSON has list semantics for collections (arrays). We use set semantics to simplify the representation. List semantics could be simulated by using complex objects with an index. A JSON document is a collection of JSON objects. Within a JSON document, we assume that the values for the same key in different objects have the same type to simplify the schema definition.

We extend the definition of a *record schema* from [12]. A record schema of one or multiple JSON documents is a quadruple $R = (V, E, r, \tau)$. Each *path* leading to a key/value pair is represented as a node $v \in V$. There is an edge $e = (p, p.k) \in E$ if the key/value pair for the path p has a JSON object with a key k as value. The keys at the top-level of the JSON object are linked to a virtual root object r. Each path is associated with a type by the function $\tau : V \rightarrow \tau$, where $\tau = \{atomic, complex, SetOf(atomic), SetOf(complex)\}$.

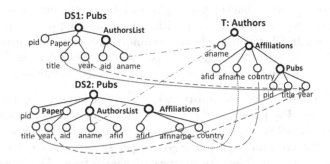

Fig. 1. Example integration scenario

Figure 1 shows the running example, extending the example from the introduction. It shows the record schemas of two publication sources $DS1$ and $DS2$. $DS1$ has a root *Pubs*, which contains one atomic subelement *pid* as the identifier, and two complex subelements *Paper* and *AuthorsList*, each with some atomic elements. $DS2$ contains another complex element *Affiliations* with the atomic elements *afid*, *afnname* and *country*. The element *afid* indicates a foreign key between *AuthorsList* and *Affiliations*. Elements with SetOf semantics are shown as bold nodes. The integrated schema (T) on the right represents a set of authors. In this work we assume that the integrated schema satisfies the partitioned normal form (PNF) [1]. To simplify the example, we assume that the author names and paper titles are unique. The source-specific IDs (*pid, aid, afid*) cannot be used to compare instances. An author may be associated with different affiliations, thus her publications may be assigned to different affiliations. Thus, in the integrated schema T, for each author, the publications *Pubs* are grouped under *Affiliations*. The connecting arrows between source and integrated schemas, indicate element-to-element correspondences.

3   Mapping Representation

In JSON-based DI scenarios often both source and integrated schemas are in hierarchical tree structures. Thus we also apply a tree-based structure for mappings, namely *bucket forest mapping*.

Definition 1 (Bucket, Bucket tree, Bucket forest). *A bucket $b = \langle p : e \rangle$ is a pair, where p is a position label and e represents a schema element (e can be an actual schema element, a position label pointing to a schema element, or a Skolem function substituting a schema element). We call a bucket empty if $e = \emptyset$. A bucket tree $T = (B, E_B)$ is a rooted tree whose nodes are a set of buckets in B; there is an edge between two nodes if there is an edge between elements of the two buckets in their record schema, and E_B is the set of such edges. A set of bucket trees $\{T_1, \ldots, T_n\}$ is a bucket forest.*

We can use a unified bucket tree to model both schemas (e is a schema element) and correspondences (e is a position label pointing to a mapped source schema element). In DI systems, schema mappings are triples of source schemas, integrated schemas, and their relationships. We use bucket trees (forests) to uniformly represent the triples, which leads to the below definition.

Definition 2 (Bucket forest mapping). *A bucket forest mapping is a triple $\mathcal{M}_{BF} = \langle \mathbf{F_S}, F_T, \mathbf{F_{CM}} \rangle$, where $\mathbf{F_S} = \{F_{DS_1}, \ldots, F_{DS_N}\}$ is the bucket forest form of the source schemas with each F_{DS_i} as a bucket forest; F_T is the bucket forest form of the integrated schema; $\mathbf{F_{CM}} = \{F_{CM_1}, \ldots, F_{CM_N}\}$ is the set of bucket forests F_{CM_i} containing the correspondences between F_{DS_i} and F_T.*

A data source may contain several JSON documents describing different entities, whose record schemas are a set of rooted trees. Thus, we use a bucket forest instead of a bucket tree to represent a single source schema F_{DS_i}, or the integrated schema F_T. In a bucket forest mapping, source schemas, integrated schemas, and their relationships are stored in $\mathbf{F_S}$, F_T, and $\mathbf{F_{CM}}$, respectively.

4   Bucket Forest Mapping Generation Procedure

The nested mapping generation problem is: given source and integrated schemas and their element-to-element correspondences, how to generate nested mappings that describe their relationship? In our approach, we recast the nested mapping generation problem into a canonical form tree construction problem. The bucket forest mapping generation procedure is divided into three steps.

Step 1. Generate $\mathbf{F_S}$ and F_T. It is desirable to obtain a unique form of source/integrated schemas. Given the record schemas of DS_1, DS_2 and T in Fig. 1, we first traverse them in the breadth-first manner from the root to the bottom, sort elements at each level lexically by their names, and assign growing non-negative integers from left to right as the values of labels (p in a bucket $b = \langle p : e \rangle$). We obtain the Breadth-First Canonical Form (BFCF) [5] bucket trees structured as shown in Fig. 2. To distinguish source and integrated schemas, we assign 0 to the root bucket of the integrated schema, and growing positive integers to the root buckets of sources schemas. For simplicity, in the rest of the paper we often use "bucket tree" to refer to a BFCF bucket tree.

Our approach supports a flexible method for adding foreign key constraints on source and integrated schemas, or finding existing ones. For instance, in Fig. 1 DS_2 has a foreign key constraint between two attributes, $DS2.Pubs.AF.afid$ (key) and $DS2.Pubs.AL.afid$ (foreign key), whose position labels are $2.0.0$ and $2.1.0$ in Fig. 2. We iterate over F_{DS_2} and find the foreign key bucket with the position label of $2.1.0$, then replace its element part with the label of the key bucket ($2.0.0$, cf. Fig. 2). To find all foreign key constraints, we just need to traverse F_{DSi} and find buckets whose elements are position labels. With $\mathbf{F_S} = \{F_{DS_1}, F_{DS_2}\}$ and F_T ready, next we generate $\mathbf{F_{CM}}$ in Steps 2 and 3.

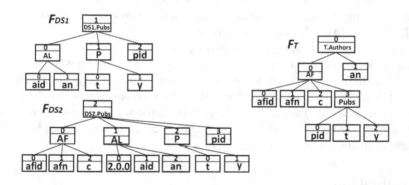

Fig. 2. F_{DS_1}, F_{DS_2}, and F_T generated via *step 1*

Step 2. Fill in correspondences in $\mathbf{F_{CM}}$. To create $F_{CM_i} \in \mathbf{F_{CM}}$, we first "copy" F_T by constructing an empty bucket tree that has the identical structure and labels as F_T. Then, we fill the element parts of the buckets with the labels of mapped source schema elements, if the element-to-element correspondences exist, resulting in the buckets with numbered elements in Fig. 3. For example, in Fig. 1 there is a correspondence between $DS2.Pubs.P.t$ and $T.A.AF.Pubs.t$, marked with a red arrow. We first retrieve their position label from F_{DS_2} and F_T, find in F_{CM_2} the bucket with the label of F_T ($0.0.3.1$), and set the element of the bucket to the position label found in F_{DS_2} ($2.2.0$). We fill the leaves of F_{CM_2} with such pairs.

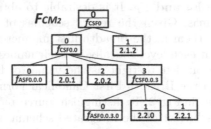

Fig. 3. F_{CM_2} for DS_2 after *step 3*

Step 3. Fill in Skolem functions. In this step, we fill in all the empty buckets in each F_{CM_i} with Skolem functions. Every Skolem function is identified by a triple, denoted as $f \langle t, n, \boldsymbol{x} \rangle$, where t is the type of the current function, which is either complex (CSF) or atomic (ASF). n is the function name; $\boldsymbol{x}$ is a vector of variables that constitute the argument list of the function. For empty buckets as inner nodes, we fill its element part with a Skolem function of complex type, while we fill empty leaf buckets with atomic type Skolem functions. We assign the position labels of the buckets as function names. Figure 3 shows the result of *step 3*. We write a Skolem function $f \langle t, n, \boldsymbol{x} \rangle$ as $f_{t,n}(\boldsymbol{x})$, e.g., $f_{CSF0}()$ with $\boldsymbol{x}$ as an empty set. The generated Skolem functions are as below.
$f_{CSF0}()$, $f_{CSF0.0}(an)$, $f_{ASF0.0.0}(an, afn, c)$,
$f_{CSF0.0.3}(an, afn, c)$, $f_{ASF0.0.3.0}(an, afn, c, t, y)$

Discussion: The final outputs of our approach are bucket forests $\mathbf{F_S}$, F_T, and $\mathbf{F_{CM}}$. With the running example we obtain $\mathbf{F_S} = \{F_{DS_1}, F_{DS_2}\}$, $F_T = F_T$, and $\mathbf{F_{CM}} = \{F_{CM_1}, F_{CM_2}\}$, as depicted in Figs. 2 and 3 respectively (we omit F_{CM_1} since it can be generated similarly to F_{CM_2}). In summary, our bucket forest mapping approach uniformly models source/target schemas (e.g., F_{DS_1}, F_{DS_2}, F_T), and their correspondences (e.g., F_{CM_1}, F_{CM_2}).

5   Evaluation

We implemented our approach in our data lake system *Constance* [7,8]. Constance loads the raw input data from data sources, extracts their schemas, performs schema matching, and generates the integrated schema as required by our algorithms. Bucket forest mappings are generated, and later used for query rewriting and formulating the final query results [8].

We first examine the practicality of our approach in real world scenarios and compare it to the classical approach [6]. Then, we study the scalability of our approach with increasing metadata complexity in synthetic scenarios. The experiments are performed on an Intel i7 2.6 GHz machine with 12 GB RAM and two cores. The results reported are the average values from fifty trials each.

Real World Scenarios: We have used six real world data sets over three integration scenarios in this experiment. In the *Publications* scenario, DBLP and Europe PubMed Central (EPMC)[1] are two online publication databases with REST APIs providing data in JSON format. In the second scenario, the Drugbank and the Protein Sequence Database (PSD)[2] share information regarding the protein name, citations, etc. Both data sets are available in XML format and converted into JSON. In the third scenario, the data sets of listed startup companies and stocks[3] are available as JSON files.

For each mapping scenario, we manually inspected the results and verified their correctness. We implemented the classical approach [6] as a baseline for performance comparison. To perform a fair comparison, we have designed the experiment such that both, the baseline approach and our approach, have the same input of record schemas, and mappings of Java objects as output. Our approach has a better performance than the baseline approach in all scenarios, with performance improvements of 96.3%, 99.2% and 52.3%, respectively. The experimental results have shown that our approach is a good fit for generating mappings during integration of JSON data in practice.

Synthetic Scenarios: To prove the applicability of our approach to more complex big data scenarios, we study the scalability of our approach and the impact of metadata complexity on the mapping generation time. To do so, we have designed a metadata generator that can populate JSON record schemas for synthetic DI scenarios. Similar to existing studies [2,3,6,13], our metadata generator requires a set of input parameters that determines the size and characteristics of the metadata. In DI scenarios, source schemas and integrated schema may have different hierarchical structure, and their schema trees may have different height (δ). We use π_δ to denote the ratio of source schema tree height $\delta(\mathbf{S})$ and integrated schema tree height $\delta(\mathbf{T})$. π_{SetOf} denotes the percentage of complex elements with SetOf semantics, among all the complex elements in the integrated schema. Moreover, we can vary the maximum length of attribute paths (L), the number of variables per level (K), and the number of sources (N). Moreover, we can also use the metadata generator to populate the element-to-element correspondences between generated source and integrated schemas.

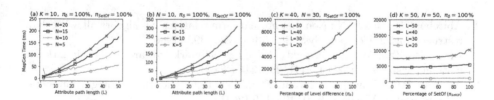

Fig. 4. (a) - (b) Scaling L; (c) Scaling π_δ; (d) Scaling π_{SetOf}

---

[1] See https://europepmc.org/ and http://dblp.uni-trier.de/.
[2] See https://www.drugbank.ca/ and http://aiweb.cs.washington.edu/research/projects/xmltk/xmldata/www/repository.html#pir.
[3] See http://jsonstudio.com/resources/.

We varied one parameter and fixed the rest, as shown in Fig. 4a–d. Foremost, we observe that our approach generates mappings in a reasonable amount of time (less than 11 s), even for high values of L, K, and N. Figure 4a shows that the mapping generation time grows slightly more than linear with increasing path lengths (L) and that a higher value for N contributes to a higher gradient. Similar observations apply to other schema complexity factors (K and N), and their combinations. In Fig. 4c, we see that with longer paths (L), the change of π_δ makes a more observable impact. Similarly in Fig. 4d, we can only observe the impact of π_{SetOf} on mapping generation time with high values for N, K, and L. It indicates that π_δ and π_{SetOf} play a less important role in the complexity compared to N, K, and L.

6   Conclusion

We have addressed the problem of nested mapping generation for integrating JSON data and proposed the bucket forest mapping approach, which generates nested mappings in triples of canonical trees representing source & integrated schemas and their relationships. Our approach can generate the nested mappings with Skolem functions without creating massive numbers of intermediate basic mappings. We have shown experimentally the practicality, efficiency and scalability of our approach over both real world and synthetic scenarios.

In the future, we plan to combine this mapping presentation with our work on mapping composition, query rewriting, and generating executable mappings [8,10], such that we are able to execute the mappings in our data lake system Constance [7].

Acknowledgements. This work has been partially funded by the German Federal Ministry of Education and Research (BMBF) (project HUMIT, http://humit.de/, grant no. 01IS14007A) and the German Research Foundation (DFG) within the Cluster of Excellence "Integrative Production Technology for High Wage Countries" (EXC 128).

References

1. Abiteboul, S., Bidoit, N.: Non first normal form relations: an algebra allowing data restructuring. J. Comput. Syst. Sci. **33**(3), 361–393 (1986)
2. Alexe, B., Tan, W.C., Velegrakis, Y.: STBenchmark: towards a benchmark for mapping systems. VLDB J. **1**(1), 230–244 (2008)
3. Arocena, P.C., Glavic, B., Ciucanu, R., Miller, R.J.: The iBench integration metadata generator. VLDB J. **9**(3), 108–119 (2015)
4. Bonifati, A., et al.: Schema mapping and query translation in heterogeneous P2P XML databases. VLDB J. **19**(2), 231–256 (2010)
5. Chi, Y., et al.: Canonical forms for labelled trees and their applications in frequent subtree mining. Knowl. Inf. Syst. **8**(2), 203–234 (2005)
6. Fuxman, A., Hernandez, M.A., Ho, H., Miller, R.J., Papotti, P., Popa, L.: Nested mappings: schema mapping reloaded. In: Proceedings of VLDB, pp. 67–78 (2006)

7. Hai, R., Geisler, S., Quix, C.: Constance: an intelligent data lake system. In: Proceedings of SIGMOD, pp. 2097–2100 (2016)
8. Hai, R., Quix, C., Zhou, C.: Query rewriting for heterogeneous data lakes. In: Benczúr, A., Thalheim, B., Horváth, T. (eds.) ADBIS 2018. LNCS, vol. 11019, pp. 35–49. Springer, Cham (2018). https://doi.org/10.1007/978-3-319-98398-1_3
9. Halevy, A.Y., Ives, Z.G., Suciu, D., Tatarinov, I.: Schema mediation in peer data management systems. In: Proceedings of ICDE, pp. 505–516 (2003)
10. Kensche, D., Quix, C., Li, X., Li, Y., Jarke, M.: Generic schema mappings for composition and query answering. Data Knowl. Eng. **68**(7), 599–621 (2009)
11. ten Cate, B., Kolaitis, P.G.: Structural characterizations of schema-mapping languages. In: Proceedings of ICDT, pp. 63–72 (2009)
12. Wang, L., et al.: Schema management for document stores. PVLDB **8**(9), 922–933 (2015)
13. Yu, C., Popa, L.: Constraint-based XML query rewriting for data integration. In: Proceedings of SIGMOD, pp. 371–382 (2004)

Languages and Models

Multi-level Conceptual Modeling: From a Formal Theory to a Well-Founded Language

Claudenir M. Fonseca[1(✉)], João Paulo A. Almeida[2], Giancarlo Guizzardi[1,2],
and Victorio A. Carvalho[3]

[1] Free University of Bozen-Bolzano, Bolzano, Italy
{cmoraisfonseca,giancarlo.guizzardi}@unibz.it
[2] Federal University of Espírito Santo (UFES), Vitória, ES, Brazil
jpalmeida@ieee.org
[3] Federal Institute of Espírito Santo (IFES), Colatina, ES, Brazil
victorio@ifes.edu.br

Abstract. Subject domains are often conceptualized with entities stratified into a rigid two-level structure: a level of classes and a level of individuals which instantiate these classes. Multi-level modeling extends the conventional two-level classification scheme by admitting classes that are also instances of other classes, a feature which is key in a number of subject domains. Despite the advances in multi-level modeling in the last decade, a number of requirements arising from representation needs in subject domains have not yet been addressed in current modeling approaches. In this paper, we tackle this issue by proposing an expressive multi-level conceptual modeling language (dubbed ML2). We follow a principled approach in the design of ML2, constructing its abstract syntax as to reflect a formal theory for multi-level modeling (termed MLT*). We show that ML2 enables the expression of a number of multi-level modeling scenarios that cannot be currently expressed in the existing multi-level modeling languages. A textual syntax for ML2 is provided with an implementation in Xtext.

Keywords: Multi-level modeling · Conceptual modeling · Modeling language

1 Introduction

A class (or type) is a ubiquitous notion in modern conceptual modeling approaches and is used in a conceptual model to establish invariant features of the entities in a domain of interest. Often, subject domains are conceptualized with entities stratified into a rigid two levels structure: a level of classes and a level of individuals, which instantiate these classes. In many subject domains, however, classes themselves may also be subject to categorization, resulting in classes of classes (or metaclasses). For instance, consider the domain of biological taxonomies [5, 8, 20]. In this domain, a given organism is classified into *taxa* (such as, e.g., Animal, Mammal, Carnivoran, Lion), each of which is classified by a biological taxonomic *rank* (e.g., Kingdom, Class, Order, Species). Thus, to represent the knowledge underlying this domain, one needs to represent entities at different (but nonetheless related) classification levels. For example, Cecil (the lion

© Springer Nature Switzerland AG 2018
J. C. Trujillo et al. (Eds.): ER 2018, LNCS 11157, pp. 409–423, 2018.
https://doi.org/10.1007/978-3-030-00847-5_29

killed in the Hwange National Park in Zimbabwe in 2015) is an instance of Lion, which is an instance of Species. A species, in its turn, is an instance of Taxonomic Rank. Other examples of multiple classification levels come from domains such as software development [13] and product types [22].

In the last decade, the importance of phenomena involving multiple levels of classification and the limitations of the fixed two-level scheme have motivated the development of a number of modeling approaches under the banner of "multi-level modeling" (e.g., [2, 18, 19, 22]). These approaches embody conceptual notions that are key to the representation of multi-level models, such as the existence of entities that are simultaneously types and instances (classes and objects), the iterated application of instantiation across an arbitrary number of (meta)levels, the possibility of defining and assigning values to attributes at the various type levels, etc.

Despite these advances, a number of requirements arising from representation needs in subject domains have not yet been addressed in current modeling approaches. For example, many approaches do not support relations between elements of different classification levels. Some others impose rigid constraints on the organization of elements into strictly stratified levels, effectively obstructing the representation of genuine domain models.

In this paper, we tackle these issues by proposing an expressive multi-level conceptual modeling language, called ML2 – Multi-Level Modeling Language. The language is aimed at multi-level (domain) conceptual modeling and is designed to cover a comprehensive set of multi-level domains. We follow a principled approach in the design of ML2, defining its abstract syntax to reflect a formal theory for multi-level modeling which we developed previously (MLT*, reported in [1]). We propose a textual syntax for ML2, which is supported by a featured Xtext-based editor in Eclipse. We show that ML2 enables the expression of a number of multi-level modeling scenarios that cannot be currently expressed in the existing multi-level modeling languages. Further, we show how ML2 incorporates rules to prevent the construction of unsound multi-level models (reflecting formal rules from MLT*).

This paper is further structured as follows: Sect. 2 briefly presents MLT*, which is the semantic foundation of ML2. Section 3 presents ML2's abstract and concrete syntax. Section 4 presents the related work, comparing ML2 to current multi-level techniques. Finally, Sect. 5 presents our final considerations.

2   MLT*: The Multi-level Theory

Types are predicative entities (e.g. "Person", "Organization", "Product") that can possibly *be applied to* a multitude of entities (including types as well). If a type t applies to an entity e then it is said that e is an *instance of* t. In contrast, *individuals* are entities that have no possible instances, i.e., they are entities that cannot *be applied to* other entities (e.g. "John", "this apple", "my cellphone"). In the philosophical literature, types are said to be repeatable, while instance are non-repeatable [14]. Since a type can be an instance of another type, it is possible to conceive of chains of instantiations (of any size), in order to represent multiple levels of classification. For instance, Fig. 1 presents an example in the biological domain with four classification levels, from the individuals "Cecil" and "Lassie", until the type "TaxonomicRank". Also, some of the types are both instances and classifiers of other entities, for example "Lion" classifies "Cecil" and is instance of (or is classified by) "Species". In the following examples, and before we arrive at the proposal of ML2, we use merely for illustrative purposes a notation inspired by the class and object notations of UML, using dashed arrows to represent relations that hold between the elements, with labels to denote the relation that applies (in this case *instance of*).

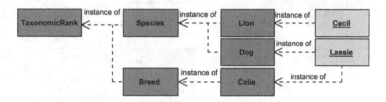

Fig. 1. A four-level instantiation chain representing a biological domain.

The theory is built up first by defining common structural relations to support conceptual modeling, starting from *specialization* between types (a transitive and reflexive relation), and *proper specialization* (which is in its turn irreflexive). Structural relations that reflect variants of the powertype pattern are also included, given the pervasiveness of this pattern in descriptions of multi-level phenomena. Based on Cardelli's notion of powertype [7], we define that t *is powertype of* t' iff every instance of t is a specialization of t' and all specializations of t' are instances of t. For example, in Fig. 2, "PersonType" is the powertype of "Person", thus every specialization of "Person" (e.g. "Man", "Woman", "Adult" and even "Person" itself) instantiates it, throughout the specialization hierarchy (e.g. "AdultMan" is also an instance of "PersonType").

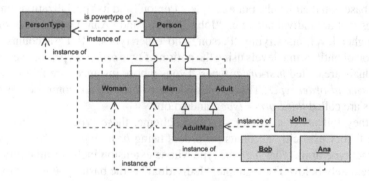

Fig. 2. PersonType and its instances.

In order to address also the notion of powertype introduced by Odell [24], MLT* also includes the so-called *categorization* relation. A type t *categorizes* a base type t' iff all instances of t are proper specializations of t'. Differently from Cardelli's powertype, in a categorization, the base type t' is not itself an instance of t. Further not all possible specializations of t' are instances of t. For instance, as presented in Fig. 3, "Employee-Type" (with instances "Manager" and "Researcher") categorizes "Person", but is not the powertype of "Person", since there are specializations of "Person" that are not instances of "EmployeeType" ("Woman" and "Man" for example).

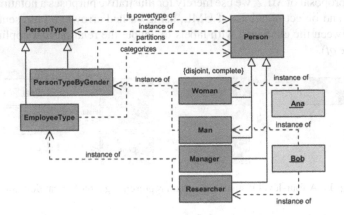

Fig. 3. Examples of the categorization and partitions relations.

The theory also defines useful variations of the categorization relation: (i) a type t *completely categorizes* a type t' iff every instance of t' is instance of at least one instance of t; (ii) a type t *disjointly categorizes* a type t' iff every instance of t' is instance of at most one instance of t; finally, (iii) t *partitions* t' iff every instance of t' is instance of exactly one instance of t. In Fig. 3, "PersonTypeByGender" partitions "Person" into "Man" and "Woman", and thus each instance of "Person" is either a "Man" or a "Woman" but not both simultaneously.

One can observe that, as presented in Fig. 2, entities in a subject domain can be organized based on their levels. For example, "Person" and its specializations only classify entities that are individuals (e.g., "John", "Bob" and "Ana"), while "PersonType" sits at a higher level, classifying "Person" and other types. MLT* accounts for this organization of entities into levels using the notion of *type order*. Types whose instances are individuals are called *first-order types*. Types whose instances are first-order types are called *second-order types*. Those types whose extensions are composed of second-order types are called *third-order types*, and so on.

Since they fall under a strictly stratified scheme, these types are called *ordered types*. The theory explicitly accounts for orders using *basic types*. A *basic type* is the most abstract type of its order, i.e., the type whose extension includes all entities in the order immediately below. For example, "Individual" is the basic type whose extension includes all individuals, "1stOT" is the basic type whose extension includes all first-order types, "2ndOT" is the one that classifies all second-order types, and so on. Due to this definition, all basic types are related in a chain of powertype relations, as presented in Fig. 4, with every ordered type specializing the basic type of the same order and instantiating the one of the order above (e.g., "Person"). The ellipsis in that figure represents that this chain of basic types can be extended as far as demanded, given the entities involved in the captured domain. However, the formalization of MLT* does not necessitate infinite chains of basic types, allowing the description of finite models (see [1] for details).

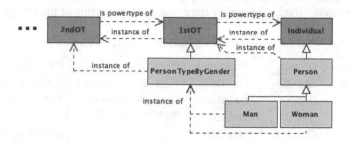

Fig. 4. Example of MLT*'s basic types.

This stratification into type orders provides for a structure useful to guide modelers in producing sound models. However, this is unable to account for types whose instances do not fall into a unique order. Examples of such include notions of "Entity", "Resource" and "Property", which are key to a number of comprehensive conceptualizations [12, 14, 21, 25]. In MLT*, types that do not conform to the stratified scheme are denominated *orderless types*. Consider the notion of "BusinessAsset" as an economic resource owned by some "Enterprise". In this context, we may say that Apple's "AppleParkMain-Building" is a business asset as well as its cellphone models, e.g., "IPhone5". Note that while the former is an individual, the latter is a first-order type (having individual cellphones as instances). Since "BusinessAsset" has instances in different orders it is an example of a domain orderless type (Fig. 5). Finally, MLT* can also be used to describe its own categories of types resulting in the model shown in Fig. 5.

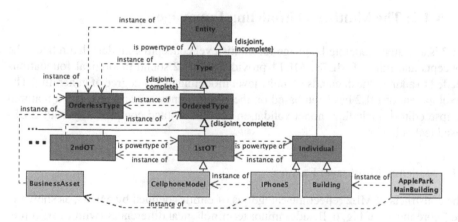

Fig. 5. MLT* basic scheme extended by a domain example.

MLT* formally defines a number of structural relations and rules to govern such relations. Some of these rules concern the nature of the instantiation relation: whenever instantiation involves solely ordered types, it is irreflexive, anti-symmetric and anti-transitive, leading to a strict stratification of types. However, when involving orderless types, there are situations in which instantiations can be reflexive, symmetric or transitive. Table 1 summarizes the logical properties and rules of types involved in MLT* structural relations, which arise out of the axioms and theorems of the theory [1].

Table 1. Summary of constraints on MLT* relations.

Relation ($t \to t'$)	Domain	Range	Constraint	Properties
Specialization	Orderless	Orderless	If t and t' are ordered types, they must be at the same type order	Reflexive, anti-symmetric, transitive
	Ordered	Orderless		
	Ordered	Ordered		
Proper specialization	Orderless	Orderless		Irreflexive, anti-symmetric, transitive
	Ordered	Orderless		
	Ordered	Ordered		
Powertype	Orderless	Orderless	t cannot be a first-order type; if t and t' are ordered types, t must be at a type order immediately above the order of t'	Irreflexive, anti-symmetric, anti-transitive
	Ordered	Ordered		
Categorization, Disjoint categorization	Orderless	Orderless	t cannot be a first-order type; if t and t' are ordered types, t must be at a type order immediately above the order of t'	Irreflexive, anti-symmetric, non-transitive
	Ordered	Orderless		
	Ordered	Ordered		
Complete categorization, Partitions	Orderless	Orderless		Irreflexive, anti-symmetric, anti-transitive
	Ordered	Ordered		

3　ML2: The Multi-level Modeling Language

ML2 is a textual modeling language for multi-level conceptual models that reflects the concepts and rules of MLT*. MLT* provides to ML2 sound theoretical foundations needed to address the demands of multi-level modeling in any degree of generality. The development of ML2 has been based on the Xtext framework and provides a featured Eclipse editor[1], including model validation capabilities and compatibility with EMF-based technologies.

3.1　Modeling Multi-level Entities

The constructs of ML2 reflect the categories of entities defined by MLT*, as shown in the Ecore model of Fig. 6. Besides minor terminological differences (with *Class* representing the notion of *type* for consistency with EMF terminology, and *EntityDeclaration* representing entities in general), there are specific constructs for every sort of entity

---

[1] The ML2 editor can be found at https://github.com/nemo-ufes/ML2-Editor.

previously presented: *Individual* representing entities with no instances; *FirstOrder-Class* representing regular classes from the two-level scheme; *HighOrderClass* representing an ordered class at a certain order; and *OrderlessClass* representing entities whose extension span across different orders. Both classes and individuals may declare the instantiation of multiple classes. Specialization (proper) and the other structural relations of the theory are considered for classes. For a class that categorizes another class, a categorization type should be defined to reflect which variant of the categorization relation applies.

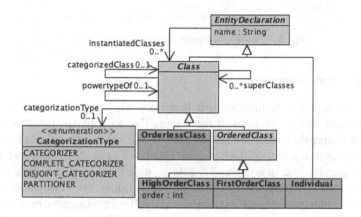

Fig. 6. Entities and classes in ML2.

Some of the rules of the theory are directly reflected in the representation of cardinality constraints. For example, given the formal definition of powertype in MLT*, a class can only be the powertype of at most one other class. All the other MLT* rules (including those that apply to the instantiation relation and those present in Table 1) are reflected in the validation functionality of the editor coded in Xtend (a high-level programming language for the Java Virtual Machine) and presented lively. The syntax of ML2[2] is inspired by traditional OO languages and applies a collection of keywords aiming at enhancing the readability of its models. The statements for entity declaration follow a common pattern, varying the available structural relations for each type of entity. Figure 7 revisits the examples presented so far using ML2. Note that a namespace mechanism is supported with modules. Unlike other multi-level languages, ML2 modules are not order bound (e.g., like MetaDepth's notion of "model" [19] that restricts the order of contained entities).

---

[2] The language's grammar is available at https://github.com/nemo-ufes/ml2-grammar.

```
module example.model {
  orderless class BusinessAsset;
  order 2 class CellphoneModel;
  class IPhone5 : CellphoneModel, BusinessAsset;
  order 2 class PersonType isPowertypeOf Person;
  order 2 class EmployeeType specializes PersonType categorizes Person;
  class Person : PersonType;
  class Manager : EmployeeType specializes Person;
  class Researcher : EmployeeType specializes Person;
  individual Bob : Person, Researcher;
}
```

Fig. 7. Examples of entity declarations in ML2.

3.2 Features and Assignments

Classes contain common features of their instances, while entity declarations contain value assignments for the features of the classes that an entity instantiates. Figure 8 presents how features are handled in the abstract syntax. ML2 distinguishes features into references and attributes (not unlike Ecore and OWL, for example). A feature has a type, which is a class in the case of references or a datatype in the case of attributes. Datatypes are first-order classes that have as instances particular values. For example, the datatype *String* is a first-order class that has as instances all well-formed sequences of characters. ML2 supports both user created datatypes and a set of primitive types, namely *String, Number* and *Boolean*. The set of primitive types covers a minimal set of data types for conceptual modeling and was inspired by JSON's specification [10].

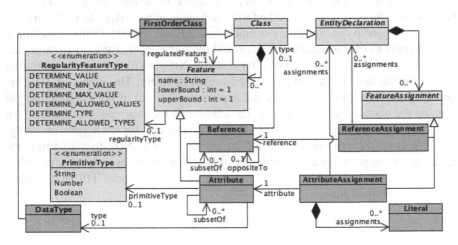

Fig. 8. Features and assignments in ML2.

Figure 9 presents an example of usage of features in an ML2 model. This model expands the one in Fig. 7 by explicitly capturing the cross-level reference *owns* between "Enterprise" and "BusinessAsset" and adding some other entities. Note that ML2 does not require exhaustive feature assignment (see "IPhone5" without an assignment for *belongsTo*). However, cardinality constraints are checked for every present assignment.

```
orderless class BusinessAsset {
    ref belongsTo : Enterprise isOppositeTo owns
};
class Enterprise {
    ref owns : [0..*] BusinessAsset isOppositeTo belongsTo
};
class Building;
order 2 class CellphoneModel categorizes Cellphone;
class Cellphone {
    screenSize : Number
};
class IPhone5 : CellphoneModel, BusinessAsset specializes Cellphone{
    ref belongsTo = Apple
};
individual AppleParkMainBuilding : Building, BusinessAsset;
individual Apple : Enterprise {
    ref owns = { AppleParkMainBuilding , IPhone5 }
};
individual MyIPhone : IPhone5{
    screenSize = 4
};
```

Fig. 9. Examples of features in ML2.

ML2 also accounts for a special kind of feature called *regularity feature* (see [8, 15]). This kind of feature has the characteristic of constraining features at a lower level. Consider the previous example of "CellphoneModel" with an *instancesScreenSize* feature that represents the specific screen size of a certain model. This feature regulates the *screenSize* feature of "Cellphone", since every cellphone will have the same screen size specified by its respective model. Instances of "CellphoneModel" such as "IPhone5" specialize "Cellphone" and determine specific value for *instancesScreenSize*, in this case, 4 in. Thereby, all instances of "IPhone5" have *screenSize* following the value assigned to *instancesScreenSize*, i.e., 4 in. Note that, in order to regulate a feature of "Cellphone", the high-order type "CellphoneModel" must categorize "Cellphone", since the regulation of a feature is defined in instances of the high-order type affecting specializations of the categorized type. Figure 10 presents a modification of Fig. 9 illustrating the usage of regularity features.

ML2 foresees six types of regularity features. In the case above, values of *instancesScreenSize* determines the exact value of *screenSize*. However, a regularity feature can also determine maximum or minimum values for a number feature (e.g., to model the maximum storage capacity of a cellphone model) and to determine the set of allowed values for a feature (e.g., to model that a phone model has either 16 or 32 GB of internal storage capacity). Additionally, a regularity feature can further constrain the type of assignment for a feature, by either determining its type(s) or determining a set of *allowed types* [14]. The specification of the regularity type can be omitted when the type of regulation does not fit one of the six foreseen types of regulation.

```
order 2 class CellphoneModel categorizes Cellphone {
    reglarity instancesScreenSize : Number determinesValue screenSize
    regularity ref compatibleProcessorModel : ProcessorModel
        determinesType installedProcessor
};
class Cellphone {
    screenSize : Number
    ref installedProcessor : Processor
};
class IPhone5 : CellphoneModel specializes Cellphone {
    instancesScreenSize = 4
    ref compatibleProcessorModel = A6
};
order 2 class ProcessorModel categorizes Processor;
class Processor;
class A6 : ProcessorModel specializes Processor;
individual Processor01 : A6;
individual MyIPhone : IPhone5 {
    screenSize = 4
    ref installedProcessor = Processor01
};
```

Fig. 10. Examples of regularity features in ML2.

Figure 10 also presents an example in which the regularity reference *compatible-ProcessorModel* of "CellphoneModel" determines the type of *installedProcessor* for instances of "Cellphone". Since "IPhone5" assigns "A6" to *compatibleProcessor-Model*, instances of "IPhone5" can only have processors that are instances of "A6". This is the case of "MyIPhone", with "Processor01" installed on it. Assignments of regulated features, if present, are checked for conformance.

4   Related Work

In this section, we position ML2 with respect to the existing work in multi-level representation approaches regarding a number of features, *which are all supported by ML2*. We consider the following multi-level modeling approaches: Melanee, M-Objects, MetaDepth and DeepTelos. Table 2 summarizes our evaluation of the various modeling approaches: a plus sign ('+') indicates support for the feature, a minus sign ('−') indicates no support, and plus/minus ('±') indicates partial support.

Melanee [2] is a tool that supports multi-level modeling founded on the notions of *strict metamodeling, clabject* and *potency*. It is based on the idea of defining *clabjects* and *fields* (attributes and slots) within the levels of a strict stratified scheme (i.e., *strict metamodeling* [4]) and assigning to both *clabjects* and *fields* a *potency*, which defines how deep the instantiation chain produced by that *clabject* or *field* can become. This allows Melanee to represent entities in multiple classification levels (F1), organizing and capturing the instantiation chains allowing an arbitrary number of levels (F2), and providing users guiding principles for the organization of models (F3). Melanee also defines *star potency* as a means to support the representation of types having instances of different potencies. While this allows the representation of types that defy a stratified

Table 2. Multi-level modeling techniques comparison.

Representation Features	Melanee	M-Objects	MetaDepth	DeepTelos
F1: represents entities of multiple classification levels	+	+	+	+
F2: allows an arbitrary number of classification levels	+	+	+	+
F3: defines guiding principles for organization of models	+	+	+	+/-
F4: represents types that defy a stratified classification scheme	+/-	–	+/-	+
F5: represents rules to govern instantiation of related types	–	–	–	+/-
F6: allows domain relations between entities in various levels	–	+	+	+
F7: represents domain features and feature assignments	+	+	+	+
F8: relates features of entities in different levels	+	–	+	–

scheme (F4), star potency does not allow self-instantiation, which is required for the abstract types we have dealt with here. Therefore, we consider that Melanee partially supports F4. In Melanee, no constructs are provided to capture rules concerning instantiation of related types at different levels (F5). For example, it is not possible to represent in Melanee that "CellphoneModel" *categorizes* (in MLT* sense) "Cellphone", and thus, it is not able to capture that every instance of "CellphoneModel" must specialize "Cellphone". Further, in Melanee, instantiations are the only relations that may cross level boundaries and, thus, it is unable to capture certain domain scenarios in which an entity is related to other entities at different instantiation levels (F6). For example, consider a scenario in which every instance of "CellphoneModel" has a "designer" being an instance of "Person" and every instance of an instance of "CellphoneModel" (i.e., every instance of "Cellphone") has an "owner" which is also an instance of "Person". Since domain relations in Melanee cannot cross levels, both "Person" and "Cellphone" must be placed in the same level to capture the "owner" relation. Because its instances are specializations of "Cellphone", "CellphoneModel" must be placed in one level higher. This makes it impossible to capture the "designer" relation, as it would cross level boundaries (which, once more, is not allowed in Melanee). Concerning domain features, Melanee supports both the representation of features of types as well as the attribution of values to those features (F7). Finally, the combination of the notions of attribute *durability* and *mutability* [9] allows one to relate features of entities in different levels (F8). For example, it allows one to capture that instances of "CellphoneModel" prescribe the exact screen size their instances must have. Note that it supports directly only one of the six types of regularity features covered in ML2 (namely, the one in which the value is fully determined).

In [22], the authors propose a multi-level modeling approach founded on the notion of *m-object*. *M-objects* encapsulate different levels of abstraction that relate to a single domain concept, and an m-object can *concretize* another m-object. The *concretize relationship* comprises indistinctive classification, generalization and aggregation relations between the levels of an m-object [22]. This approach allows the representation of entities in an arbitrary number of levels relating them through chains of *concretize relationships*, we consider that it supports F1 and F2. Given that the approach adopts a stratified scheme in which *concretize relationships* may only relate types at adjacent levels, we consider that it supports F3 and does not support F4. Further, since the *concretize relationships* are the only structural relationships that cross level boundaries, the approach fails to support F5. In [23], the authors observe that the approach was unable to capture certain scenarios in which there are domain relations between m-objects at different instantiation levels. To address this limitation, the approach was extended with the concept of Dual-Deep Instantiation, which allows the representation of relations between m-objects at different instantiation levels through the assignment of a potency to each association end, thereby supporting F6. Finally, it provides support to represent features of types (F7), but it does not include support to explain the relationship between attributes of entities in different classification levels (not supporting F8).

MetaDepth [19] is a textual multi-level modeling language founded on the same notions of clabject, potency, durability and star potency used by Melanee. Differently from Melanee, MetaDepth supports the representation of domain relationships as references, such that each reference has its own potency (a solution close to the one adopted in Dual-Deep Instantiation [23]), allowing the representation of domain relations between clabjects at different instantiation levels. Therefore, MetaDepth supports all the features Melanee supports, and further supports F6.

Finally, DeepTelos is a knowledge representation language that approaches multi-level modeling with the application of the notion of "most general instance (MGI)" [18]. In [17], the authors revisit the axiomatization of Telos and add the notion of MGI to Telos' formal principles for instantiation, specialization, object naming and attribute definition. The notion of MGI can be seen as the opposite of Odell's powertype relation. For example, to capture that "Tree Species" is a "powertype" (in Odell's sense) of "Tree", in DeepTelos it would be stated that "Tree" is the "most general instance" (MGI) of "Tree Species". Considering that the MGI construct allows representing entities in multiple classification levels and that DeepTelos allows representing chains of MGI to represent as many levels as necessary, we consider that DeepTelos supports features F1 and F2. DeepTelos builds up on Telos, whose architecture defines the notions of *simple class* and *w-class*, which are analogous to the notions of ordered and orderless types we use. Nevertheless, stratification rules for simple classes (constraining specialization and cross-level relations) are not provided. Thus, we consider that it partially supports F3 and that it supports F4 with the notion of *w-class*. Considering that DeepTelos provides only the concept of MGI to constrain the instantiation of types in different levels, not elaborating on the nuances of the relations between higher-order types and base types, we consider that it partially supports F5. It admits relations between types in different levels, thus, supporting F6. DeepTelos supports the attribution of values to features of types (F7). However, its account for attributes does not include any support to explain the relationship between attributes of entities in different classification levels, thus, not supporting F8.

5   Final Considerations

In this paper, we have presented the ML2 multi-level conceptual modeling language. We have approached the design of ML2 with a careful consideration of the conceptualization of types in classification schemes that transcend a rigid two-level structure. The language harnesses the conceptualization formalized in MLT*, reflecting the theory's definitions in its constructs and syntactical constraints. The language was designed to offer expressiveness to the modeler by addressing a comprehensive set of features for representing domains dealing with multiple levels of classification. Further, rules incorporated in ML2 have been implemented in an Eclipse-based editor that supports the live verification of models to ensure adherence to the theory. The use of a formally-verified semantic foundation is one of the distinctive features of ML2 (see [1] for axiomatization and reference to a specification in the Alloy language [16]).

During this research, we investigated a number of multi-level representation techniques reported in the literature, focusing on their capability of capturing different intended multi-level scenarios. We have observed that multi-level approaches often opt for one of two extremes: (i) to consider all classes to be orderless (or similarly to ignore the organization of elements into stratified orders, the case of DeepTelos, what is referred to as a level-blind approach in [3]), or (ii) to consider all classes to be strictly stratified (e.g., in the case of Melanee and MetaDepth). Approaches that opt for (i) are able to represent all types ML2 can capture, however, fail to provide rules to guide the use of the various structural relations (e.g., instantiation). As shown in [6], this lack of guidance has serious consequences on resulting models quality. Approaches that opt for (ii) do not support the representation of a number of important abstract notions, including those very general notions that we use to articulate multi-level domains (such as "types", "clabjects", "entities") (these abstract notions are key to ML2 being able to model the upper portion of the UFO foundational ontology [14], see ML2 models produced for it in [11] including a response to the so-called "Bicycle Challenge").

The combination of both approaches in the design of ML2 places it in a unique position in multi-level modeling approaches. On top of that, the present approaches for multi-level modeling also strive in order to accommodate the implications of the dual nature of "clabjects" regarding the representation of features and feature assignments. Besides ML2, only MetaDepth and Melanee were able to express related features in different levels and cross-level references. Although, both of them support the representation of only one of the various types of regularity attributes supported in ML2.

A few knowledge representation approaches (e.g., DeepTelos [18] and Cyc [12]) have also drawn distinctions between orderless and ordered types. However, Telos does not provide rules for the various structural relations, including instantiation and specialization. In its turn, Cyc, arguably the world's largest and most mature knowledge base nowadays, employs a conceptual architecture for types similar to MLT*'s distinctions of sorts of entities (see Fig. 5). Additionally, this architecture also includes instantiation and specialization rules [12]. However, these rules are not incorporated in some representation language and no deep characterization mechanism is provided in Cyc.

Future work concerning ML2 includes the development of transformations from ML2 into the Semantic Web approach discussed in [5], the development of an integrated constraint language to further increase the expressiveness of the language, and the investigation of a suitable visual syntax to accompany the textual syntax.

Acknowledgements. This work is partially supported by CNPq (grants number 407235/2017-5, 312123/2017-5 and 312158/2015-7), CAPES (23038.028816/2016-41), FAPES (69382549) and FUB (OCEAN Project).

References

1. Almeida, J.P.A., Fonseca, C.M., Carvalho, V.A.: A comprehensive formal theory for multi-level conceptual modeling. In: Mayr, H.C., Guizzardi, G., Ma, H., Pastor, O. (eds.) ER 2017. LNCS, vol. 10650, pp. 280–294. Springer, Cham (2017). https://doi.org/10.1007/978-3-319-69904-2_23
2. Atkinson, C., Gerbig, R.: Melanie: multi-level modeling and ontology engineering environment. Proceedings of the 2nd International Master Class on Model-Driven Engineering Modeling Wizards - MW 2012. ACM Press, New York (2012)
3. Atkinson, C., Gerbig, R., Kühne, T.: Comparing multi-level modeling approaches. In: Proceedings of the 1st International Workshop on Multi-level Modelling (2014)
4. Atkinson, C., Kühne, T.: Meta-level independent modeling. In: International Workshop "Model Engineering" (in Conjunction with ECOOP'2000), Cannes, France, p. 16 (2000)
5. Brasileiro, F., Almeida, J.P.A., Carvalho, V.A., Guizzardi, G.: Expressive multi-level modeling for the semantic web. In: Groth, P., et al. (eds.) ISWC 2016. LNCS, vol. 9981, pp. 53–69. Springer, Cham (2016). https://doi.org/10.1007/978-3-319-46523-4_4
6. Brasileiro, F., et al.: Applying a multi-level modeling theory to assess taxonomic hierarchies in Wikidata. In: Proceedings of the 25th International Conference Companion on World Wide Web, Geneva, Switzerland, pp. 975–980 (2016)
7. Cardelli, L.: Structural subtyping and the notion of powertype. In: Proceedings of the 15th ACM Symposium of Principles of Programming Languages, pp. 70–79 (1988)
8. Carvalho, V.A., Almeida, J.P.A.: Toward a well-founded theory for multi-level conceptual modeling. Softw. Syst. Model., 1–27 (2016)
9. Clark, T., Gonzalez-Perez, C., Henderson-Sellers, B.: Foundation for multi-level modelling. In: CEUR Workshop Proceedings, vol. 1286, pp. 43–52 (2014)
10. ECMA: The JSON Data Interchange Format, 1st edn. (2013). http://www.ecma-international.org/publications/files/ECMA-ST/ECMA-404.pdf
11. Fonseca, C.M.: ML2: an expressive multi-level conceptual modeling language. Dissertation (master's in informatics) - Federal University of Espírito Santo, Brazil (2017)
12. Foxvog, D.: Instances of instances modeled via higher-order classes, Foundational Aspects of Ontologies, (9–2005), pp. 46–54 (2005). http://www.uni-koblenz.de/fb4/publikationen/gelbereihe/RR-9-2005.pdf
13. Gonzalez-Perez, C., Henderson-Sellers, B.: A powertype-based metamodelling framework. Softw. Syst. Model. **5**, 72–90 (2006)
14. Guizzardi, G.: Ontological Foundations for Structural Conceptual Models, 1st edn., The Netherlands (2005)
15. Guizzardi, G., et al.: Towards an ontological analysis of powertypes. In: Proceedings of the Joint Ontology Workshops 2015, p. 1517 (2015)
16. Jackson, D.: Software Abstractions: Logic, Language and Analysis. MIT Press, Cambridge (2006)

17. Jarke, M., et al.: ConceptBase - a deductive object base for meta data management. J. Intell. Inf. Syst. **4**(2), 167–192 (1995)
18. Jeusfeld, M.A., Neumayr, B.: DeepTelos: multi-level modeling with most general instances. In: Comyn-Wattiau, I., Tanaka, K., Song, I.-Y., Yamamoto, S., Saeki, M. (eds.) ER 2016. LNCS, vol. 9974, pp. 198–211. Springer, Cham (2016). https://doi.org/10.1007/978-3-319-46397-1_15
19. de Lara, J., Guerra, E.: Deep Meta-modelling with MetaDepth. In: Vitek, J. (ed.) TOOLS 2010. LNCS, vol. 6141, pp. 1–20. Springer, Heidelberg (2010). https://doi.org/10.1007/978-3-642-13953-6_1
20. Mayr, E.: The Growth of Biological Thought: Diversity, Evolution, and Inheritance. The Belknap Press, Cambridge (1982)
21. Mylopoulos, J.: Conceptual modeling and Telos. In: Loucopoulos, P., Zicari, R. (eds.) Conceptual Modelling, Databases, and CASE: an Integrated View of Information System Development, pp. 49–68. Wiley, New York (1992)
22. Neumayr, B., Grun, K., Schrefl, M.: Multi-level domain modeling with m-objects and m-relationships. In: 6th Asia-Pacific Conference on Conceptual Modelling, vol. 96, pp. 107–116 (2009)
23. Neumayr, B., Jeusfeld, M.A., Schrefl, M., Schütz, C.: Dual deep instantiation and its ConceptBase implementation. In: Jarke, M., et al. (eds.) CAiSE 2014. LNCS, vol. 8484, pp. 503–517. Springer, Cham (2014). https://doi.org/10.1007/978-3-319-07881-6_34
24. Odell, J.: Power types. J. Object-Oriented Program. **7**(2), 8–12 (1994)
25. W3C: OWL 2 Web Ontology Language Document Overview (2009). http://www.w3.org/TR/2009/REC-owl2-overview-20091027/

Increasing the Semantic Transparency of the KAOS Goal Model Concrete Syntax

Mafalda Santos, Catarina Gralha$^{(\boxtimes)}$, Miguel Goulão, and João Araújo

NOVA LINCS, DI, FCT, Universidade NOVA de Lisboa, Lisbon, Portugal
{mcd.santos,acg.almeida}@campus.fct.unl.pt,
{mgoul,joao.araujo}@fct.unl.pt

Abstract. Stakeholders without formal training in requirements modelling languages, such as KAOS, struggle to understand requirements specifications. The lack of semantic transparency of the KAOS goal model concrete syntax is perceived as a communication barrier between stakeholders and requirements engineers. We report on a series of related empirical experiments that include the proposal of alternative concrete syntaxes for KAOS by leveraging design contributions from novices and their evaluation with respect to semantic transparency, in contrast with the standard KAOS goal model concrete syntax. We propose an alternative concrete syntax for KAOS that increases its semantic transparency (mean difference of .23, in [−1.00..1.00]) leading to a significantly higher correct symbol identification (mean difference of 19%) by novices. These results may be a stepping stone for reducing the communication gap between stakeholders and requirements engineers.

Keywords: Goal model concrete syntax · Empirical evaluation
KAOS

1 Introduction

Goal-oriented requirements models [1] are important for requirements elicitation and analysis, where communication with stakeholders plays a major role. For this to be effective, both requirements engineers and other stakeholders must have a common understanding of these models to participate in their development process. Yet, most of the stakeholders have no previous knowledge about requirements modelling languages, requiring formal training on such models. Requirements modelling languages themselves are under scrutiny, with proposals for their evolution, which include syntactic and semantic aspects. However, most of the research on these languages focuses on semantic aspects, rather than syntactic ones. Neglecting semantic transparency raises a communication barrier between requirements engineers and stakeholders. Semantic transparency has a positive impact on cognitive effectiveness, which is based on the ease, speed, and accuracy to process the information illustrated on a requirements model.

© Springer Nature Switzerland AG 2018
J. C. Trujillo et al. (Eds.): ER 2018, LNCS 11157, pp. 424–439, 2018.
https://doi.org/10.1007/978-3-030-00847-5_30

We are interested in evaluating the semantic transparency of the concrete syntaxes of goal-oriented requirements modelling languages, in particular, the KAOS goal model. We conducted a series of quasi-experiments that evaluate the cognitive effectiveness of KAOS concrete syntax. The approach followed is based on the "Physics" of Notations (PoN) [2]. We present an alternative concrete syntax for KAOS goal models proposed by novices. We evaluate it with respect to semantic transparency, in contrast with the standard KAOS goal model concrete syntax. The results suggest that the alternative concrete syntax improves cognitive effectiveness as it significantly increases semantic transparency, leading to a higher correct symbol identification by novices.

The paper is structured as follows. Section 2 discusses KAOS, PoN, and related work. Sections 4 to 7 describe the 4 studies that were undertaken to evaluate the concrete syntax of the KAOS goal model. Section 8 discusses the results and Sect. 9 draws some conclusions and discusses future work.

2   Background

The KAOS Methodology. Among the main Goal-Oriented Requirements Engineering (GORE) approaches [1,4–9], KAOS has been one of the most relevant. Its emphasis is on semi-formal and formal reasoning about behavioural goals to derive goal refinements, operationalisations, conflict management and risk analysis [10]. Goals are a prescriptive intention statement about a system whose satisfaction, in general, needs cooperation of agents that configure the system. Through *and/or* decompositions, goals can be refined into subgoals, requirements or expectations. Objects can be specified to describe the project structural model. Obstacles and goals relations can be used to identify system vulnerabilities [11]. In Fig. 1 we show 18 of the KAOS concepts that we use here.

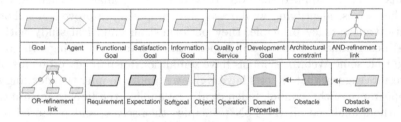

Fig. 1. Standard KAOS symbol set [1,4]

The "Physics" of Notations. Visual notations are used for communication among different kinds of stakeholders (e.g., developers, end-users, clients), where cognitive effectiveness is key to support design and problem solving. Improving the cognitive effectiveness of visual notations will promote their use and allow different stakeholders to share the same understanding of the software. The PoN theory [2] consists of a set of principles for designing cognitively effective visual notations, *optimised for processing by the human mind* [2]. These principles are related to several aspects of a modelling language such as semantic transparency, complexity management, cognitive integration, cognitive adjustment, graphic economy, semiotic clarity, visual expressiveness, double codification and perceptual discriminability. Our work focuses on the principle of *semantic transparency*, which argues that the symbols of a language must suggest their meaning: there must be a correspondence between the visual properties and the semantic properties of the represented objects [12]. Based on the notion of natural mapping, [13], a graphic language should take advantage of physical analogies, visual metaphors, common logical properties and cultural associations. Semantically transparent symbols allow to reduce the cognitive load in their recognition process as they use visual mnemonics, so their meaning can be easily deduced.

Related Work. Caire et al. [3] applied the PoN theory to i^*, and analysed the principle of semantic transparency by involving novices in the design of a new i^* concrete syntax. The results reveal that novices have a better performance than experts in creating symbols for i^*. The new symbols improved the semantic transparency compared to the standard i^* concrete syntax, suggesting that visual notations should be designed by novices, not experts. Here, we replicated this approach to KAOS, finding similar results. Our work differs from Caire et al.'s as we categorise participants (with or without knowledge on modelling languages), and include the use of the created symbols to model a small problem.

Other studies applied the PoN theory. In their seminal work, Moody et al. [25] applied the PoN principles to evaluate i^*. The authors identified flaws in the i^* concrete syntax and proposed improvements. Genon et al. [14] applied PoN to Use Case Maps. The authors analysed all the PoN principles themselves, neither proposing an alternative concrete sytax nor involving novices in the process, as we do in this paper. For each principle, they identified weaknesses and suggested improvements. The same authors identically applied the PoN to BPMN [15]. Granada et al. [16] applied the PoN to WebML. They analysed each PoN principle themselves, identifying weaknesses and proposing improvements. For the semantic transparency principle, they performed an empirical study with two groups: postgraduate students and experts in software engineering. The results showed that some WebML symbols have semantic transparency problems. Also, their results suggest that experts have a greater ability to infer the meaning from the appearance of the symbol. Saleh et al. [28] applied the PoN to misuse cases. The authors analysed all the PoN principles themselves, and proposed a new concrete syntax for misuse cases. They then compared the cognitive effectiveness of the original and the new concrete syntaxes. The results indicated that the new concrete syntax is more semantically transparent than the original one.

Matulevičius et al. [26] evaluated how KAOS and its supporting tool (Objectiver), help modellers to adhere to 9 visual modelling principles, during the modelling activity. The authors offered recommendations for modellers, language engineers and tool developers. The same authors evaluated the quality of i^* and KAOS [27]. The evaluation consisted of interviews, goal models creation, evaluation of the models and the modelling languages. The results revealed that the semantics of these languages is not defined clearly enough. Finally, Boone et al. [29] applied the PoN to CHOOSE. The authors evaluated 5 PoN principles, and created 3 alternative concrete syntax based on the results of their initial evaluation. All the concrete syntaxes were evaluated throughout an empirical study with business engineering students. One the new concrete syntaxes outperformed the others in terms of cognitive effectiveness.

3   Research Planning

Research Questions. Three research questions guided our quasi-experiments on the semantic transparency of KAOS goal models:

RQ1. Is the KAOS visual notation semantically opaque?

RQ2. Can participants with no knowledge in modelling languages design more semantically transparent symbols than participants with knowledge in modelling languages?

RQ3. Which visual notation (*standard, stereotype, or prototype*) is more semantically transparent?

Research Design. The research design consists of 4 related empirical studies, where the results of the earlier studies provide inputs to the later studies.

1. **Symbolisation experiment:** a group of 99 novice participants designed symbols for KAOS concepts, a task normally reserved for experts;
2. **Stereotyping analysis:** we identified and organised categories with the most common symbols produced for each KAOS concept. This defined the *stereotype symbol set.*
3. **Prototyping experiment:** a group of 88 novice-participants chose the symbols they consider to better represent each KAOS concept. The most voted symbols for each KAOS concept defined the *prototype symbol set.*
4. **Semantic transparency experiment:** we evaluated the ability of 52 participants to infer the meanings of novice-designed symbols (*stereotype* and *prototype symbol set*) compared to expert-designed symbols (*standard KAOS*).

For studies 1, 3 and 4, we used questionnaires, which are explained in the next sections. These questionnaires can be found at https://goo.gl/G1aDkg.

4   Study 1 – Symbolisation Experiment

Goals. The goal of this study was to **obtain candidate symbols drawn by novices to illustrate 18 KAOS goal models concepts.** We used the sign production technique [17]. This involves asking members of the target audience (i.e., those who will be interpreting the models) to generate symbols to represent a set of given concepts. The rationale is that symbols produced by members of the target audience are more likely to be understood and recognised by other members of the target audience, due to their common cognitive profile [17,18]. This approach has been used to design public information symbols [19], office equipment symbols [20], workflow modelling [21], icons for graphical user interfaces [18], and to design RE visual notations, such as i^* [3]. This type of studies has consistently shown that symbols produced in this way are more accurately interpreted by their target audience than those produced by experts.

Participants. There were 99 participants (73 males, 26 females), all students from Universidade Nova de Lisboa (UNL), from different courses (Mechanical Engineering, Industrial Engineering and Management, Environmental Engineering, Civil Engineering, and Computer Science). This diversity was deliberate, as we wanted our participants to be surrogates of stakeholders from different backgrounds who will interact with requirements engineers. We categorised the participants in: **With No Knowledge in Modelling Languages (WNKML)** and **With Knowledge in Modelling Languages (WKML)**. Participants were recruited through convenience sampling and participated voluntarily. The **WNKML** group had 53 participants (32 males, 21 females; 53 undergraduates), from courses other than Computer Science. They had no previous knowledge of modelling languages in general, or KAOS in particular, and represent stakeholders from other domains in our study. The **WKML** group had 46 participants (41 males, 5 females; 40 undergraduates, 6 MSc students), all Computer Science students. All of them had previous knowledge of modelling languages, and 37 had a brief contact with KAOS, in the context of a Software Engineering course. They are representatives of stakeholders with some technical background, but no expertise in Requirements Engineering goal models.

Experimental Material. Each participant was provided with a 6-pages questionnaire and a pen or pencil. The first page had the instructions for answering the questionnaire, which is divided into 3 parts: *Part I* provided the definition of 18 KAOS concepts. For each concept, participants were asked to create a visual representation inside a framed area, to constrain the size of the proposed symbols. *Part II* provided a requirements description. We asked participants to represent it using the visual symbols they proposed in *Part I*. *Part III* was used for collecting demographic data on participants.

Procedure. Participants were instructed to produce drawings expressing the meaning of each concept. None of the questions was mandatory. No time limit was set but, on average, participants took 45–60 min to complete the tasks.

Results. The participants produced a total of 1518 symbols, 723 of which were drawn by the WNKML and 795 by the WKML group. This corresponds to a response rate of 85.2%. There is no deviation from normality according to the Kolmogorov-Smirnov and the Shapiro-Wilk tests ($p > .05$), suggesting the response rate difference is normally distributed. There were no outliers. We conducted a paired-samples t-test to compare the response rates for each concept, from WKML and WNKML participants. This test was found to be statistically significant, $t(17) = -8.135$, $p < .001$; $d = 2.278$. The effect size for this analysis ($d = 2.378$) was found to exceed Cohen's convention for a large effect ($d = .80$). These results suggest that participants from the WKML group ($M = .960$, $SD = .033$) had a higher response rate than participants from the WNKML group ($M = .758$, $SD = .125$).

The response rate for constructing a requirements model, using a requirements description provided with the questionnaire, was considerably lower (68.6%) than the response rate for symbol proposals. The WKML group had a higher response rate (97.8%) than the WNKML group (43.4%). These results suggest that the WNKML group had more difficulty in building the KAOS model. The overall results suggest that both groups encountered more difficulties when creating the KAOS model than when proposing symbols for each concept. The WNKML group had more difficulty than the WKML, in both parts of the questionnaire.

5   Study 2 – Stereotyping Analysis

Goals. The goal of this study was to identify **the most common symbols produced by the participants** – the *stereotype symbol set* – for each KAOS concept in Study 1 (Sect. 4). This approach is based on the assumption that the most frequently produced symbol for representing a concept is also the most frequently recognised by the members of the target audience [17,18].

Procedure. The symbols produced in Study 1 were classified into symbol categories. Each symbol category represents all the symbols containing common visual features (e.g. the smiling face in Fig. 2 represents a category of similar drawings to convey the concept of *Agent*). Participants from the WKML group seem to have been influenced by the modelling languages they know. Their symbols, for each concept, were less varied, more conventional and more abstract. In contrast, the symbols created by the WNKML group were more detailed and creative, being rather varied among the participants. For both groups, we categorised the symbols based on their visual and conceptual similarity. We then combined the categories of symbols produced by both groups (naturally, some of them were the same) and counted the number of members in each category. We then selected the most representative category (i.e. the one with the highest number of members) for each concept, resulting in the *stereotype symbol set*.

Results. This study resulted in a symbol categories table, containing all the symbols produced by the participants WKML and WNKML. We extracted from it the *stereotype symbol set* (Fig. 2), with the most common symbols produced for each of the 18 KAOS concepts. The degree of stereotypy [17], or stereotype weight [18], measures the level of consensus about a concept visual representation. The average degree of stereotypy of the stereotype symbols was .212% ($SD = .128$). *Agent* is the only outlier (and the maximum value), with a stereotypy of .660. Indeed, *Agent* was the only concept with a majority, i.e., a degree of stereotypy above .5. These results confirm the inherent difficulty in representing such abstract concepts [18]. A paired samples t-test indicated that the degree of stereotypy is not significantly different for the WKML group (M = .224, SD = .148) and the WNKML group (M = .200, SD = .151), $t(17) = .637$, $p = .533$, suggesting both groups contributed similarly to the *stereotype symbol set*.

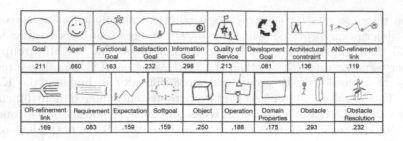

Fig. 2. KAOS stereotype symbol set with symbols degree of stereotypy

6 Study 3 – Prototyping Experiment

Goals. Stereotyping can been challenged on the grounds that the most frequently produced drawings may not necessarily be the ones that convey concepts most effectively. In fact, visual metaphors often work well as a mnemonic for the concept name, while failing to represent the concept itself [3,18]. In this study, we asked novice-participants to analyse symbols produced in Study 1 (Sect. 4) and **choose which best represents each KAOS concept**. The most frequently chosen symbol for each concept was then included in the *prototype symbol set*. This represents a consensual perception by members of the target audience about semantic transparency.

Participants. There were 80 participants (70 males and 10 females), all students from UNL. None of them had participated in Study 1, to prevent bias concerning their own proposals. Again, we categorised the participants into the WNKML and WKML groups. All students were recruited through convenience sampling and participated voluntarily. The WNKML group had 56 participants (48 males, 8 females; 56 undergraduates), all undergraduate students in Computer Science or engineering: 43 from Computer Science, 13 from other Engineering courses (Mechanical Engineering, Industrial Engineering and Management and Civil Engineering). The WKML group had 24 participants (22 males, 2 females; 8 undergraduates, 16 MSc students).

Experimental Material. Each participant was provided with a 4-page questionnaire and a pen or pencil. The questionnaire was divided into 2 parts. *Part I* provided 18 KAOS goal model concept descriptions, each with a corresponding set containing from 3 to 7 candidate symbols. The candidate symbols represented the categories corresponding to that KAOS concept with the highest stereotypy. Participants were asked to choose the symbol that represents the best visual metaphor for each concept. In *Part II*, we collected demographic data.

Procedure. Participants were verbally instructed to answer the questionnaire, by choosing the symbols that, in their opinion, best expressed the meaning of the KAOS concepts. None of the questions was mandatory. No time limit was set but, on average, participants took 5–20 min to complete the questionnaire.

Results. The result of this study is the *prototype symbol set* (Fig. 3), composed by the most voted symbol to represent each one of the 18 KAOS concepts. The overall level of consensus among judgement was lower than .5 for most symbols. Only 5 were selected by more than half of the participants. The Koklmogorov-Smirnov and Shapiro-Wilk normality tests were used to assess the normality of the distribution of the number of elements selected by each participant that made it into the prototype, for WNKML and WKML groups. The distribution among members of the WKML group departed from normality ($p =< .022$ and $p = .096$, respectively). We conducted a Welch's t-test, $t(19.442) = 9.735$, $p = .006$ to evaluate whether there were any significant differences between the number of elements selected by the WKML group ($M = 6.25$, $SD = 3.357$) and by the WNKML group ($M = 9.05$, $SD = 2.522$). On average, participants from the WKML group selected less voted elements than those from the WNKML.

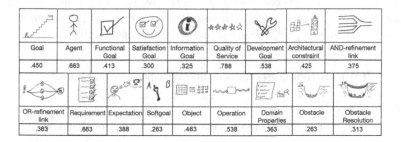

Fig. 3. KAOS prototype symbol set with level of consensus

7   Study 4 – Semantic Transparency Experiment

Goals. The goal of this study was to **evaluate the semantic transparency of 3 symbol sets for KAOS**: *standard* (Fig. 1), *stereotype* (Fig. 2) and *prototype* (Fig. 3). Semantic transparency defines the degree of association between a symbol's form and content [2]. We conducted a blind interpretation study where participants infered the concept (content) associated with each symbol (form). This method is recommended by the International Organization for Standardization (ISO) *for testing the comprehensibility of graphical symbols* [22]. ISO uses it when testing the comprehensibility of standard symbols prior to their release, which can be measured by the percentage of correct responses (i.e., hit rate).

Participants. There were 52 participants (44 males, 8 females), all students from UNL except 2 students from Universidade de Lisboa. We used different participants than in Study 1 and Study 3. We categorised the participants into the WNKML and WKML groups. All students were recruited through convenience sampling and participated voluntarily. The WNKML group had 17 participants (15 males, 2 females; 14 undergraduates, 3 MSc students) from Computer Science, Mechanical Engineering, Industrial Engineering and Management and Architecture. The WKML group had 35 participants (29 males, 6 females; 5 undergraduates, 30 MSc students), all Computer Science students.

Experimental Material. Each participant was provided with a 5-page questionnaire and a pen or pencil. The questionnaire was divided into 4 parts. In *Part I*, we provide 18 KAOS concepts and descriptions. In *Part II*, participants are asked to fill a Matching Table, by matching the symbols from each of the 3 symbol sets with each of the 18 KAOS concepts. In *Part III*, we provide a table containing the 3 symbol sets. Finally, in *Part IV* we collected demographic data.

Procedure. Participants were instructed verbally to answer the questionnaire, by selecting a symbol from each set that better described each KAOS concept. None of the questions was mandatory. No time limit was set but, on average, participants took 20–50 min do complete the questionnaire.

Hypotheses, Parameters and Variables. The independent variable is the *symbol set* (i.e., *standard, stereotype* or *prototype*). The dependent variables are the *semantic transparency coefficient* [2,3], the degree of proximity between a symbol and the semantic construct represented by it; and the *hit rate*, an indicator for measuring symbols comprehension. For each one of the dependent variables, we have defined 3 hypothesis, which we present in Table 1.

Table 1. Hypotheses for semantic transparency and hit rate

Hypotheses	Description
H_{1ST}	Stereotype is more semantically transparent than standard
H_{2ST}	Prototype is more semantically transparent than standard
H_{3ST}	Prototype is more semantically transparent than stereotype
H_{4HR}	Stereotype has a higher hit rate than standard
H_{5HR}	Prototype has a higher hit rate than standard
H_{6HR}	Prototype has a higher hit rate than stereotype

We predict that the stereotype and prototype symbol sets would outperform the standard KAOS. We also predict that the prototype would outperform the stereotype symbol set, since we consider that the most chosen symbols are easier to interpret than the most common ones, resulting in the following ordering:

prototype symbol set > stereotype symbol set > standard symbol set

Results. Table 2 shows the semantic transparency coefficient and hit rate for the 3 symbol sets. A symbol's semantic transparency coefficient is given by [3]: $\frac{maximum\ frequency - expected\ frequency}{total\ responses - expected\ frequency}$. For the semantic transparency coefficient, each cell is coloured from red (semantically opaque symbol) to green (semantically transparent symbol). For the hit rate, the highlighted values correspond to the symbols that respect the ISO threshold for comprehensibility ($\geq 67\%$) [23].

Table 2. Semantic transparency coefficient and hit rate results

Symbol	Semant. Transp. coefficient			Hit Rate		
	standard	stereotype	prototype	standard	stereotype	prototype
Agent	0.20	0.58	**0.81**	26.9%	59.9%	**81.9%**
AND-refinement link	**0.56**	0.35	0.39	**59.6%**	38.5%	42.2%
Archit. constraint	-0.09	0.05	**0.17**	0.5%	10.1%	**21.6%**
Development goal	0.01	**0.19**	0.17	9.6%	**23.1 %**	21.2%
Domain properties	0.16	-0.05	**0.23**	23.1%	0.5%	**26.9%**
Expectation	0.16	0.20	**0.44**	23.1%	24.2%	**47.3%**
Functional goal	0.16	**0.17**	0.14	**23.1%**	21.8%	18.3%
Goal	0.20	0.11	**0.24**	26.9%	16.3%	**28.1%**
Information goal	0.18	0.63	**0.75**	25.0%	65.1%	**76.0%**
Object	0.10	**0.60**	0.15	17.3%	**62.6%**	19.7%
Obstacle	0.27	0.64	**0.83**	32.7%	66.3%	**83.5%**
Obstacle resolution	0.24	0.60	**0.76**	30.8%	62.5%	**77.6%**
Operation	0.26	0.41	**0.44**	31.7%	43.9%	**46.8%**
OR-refinement link	**0.52**	0.42	0.34	**55.8%**	45.2%	37.6%
Quality of service	-0.03	0.11	**0.63**	5.8%	16.3%	**64.6%**
Requirement	**0.22**	0.04	0.18	**28.8%**	9.6%	22.2%
Satisfaction Goal	0.03	-0.04	**0.71**	11.5%	1.9%	**72.4%**
Softgoal	0.10	**0.17**	0.05	17.3%	**21.2%**	10.1%

Descriptive Statistics. Table 3 summarises the descriptive statistics for the collected metrics. Semantic transparency is defined as a scale from -1 to $+1$, and is measured by computing the semantic transparency coefficient of the symbols and the success of the participants matching the symbols from the 3 symbol sets to KAOS concepts. For each metric we present 3 rows in the table corresponding to the 3 symbol sets followed by the mean, standard deviation, skewness, kurtosis, and the p-values for the Kolmogorov-Smirnov and the Shapiro-Wilk normality tests. The *prototype* groups deviate from normality both concerning *semantic transparency* and *hit rate*, according to the Shapiro-Wilk test ($p < .05$). This is further illustrated through boxplots, presented in Fig. 4. Figure 4a presents the semantic transparency coefficient, which is higher for the prototype symbol set. Figure 4b shows the hit rate, which is also higher for the prototype symbol set.

Hypotheses Testing. We applied Levene's variance homogeneity test and found a statistically significant difference among the 3 distributions for semantic transparency coefficient ($p = .007$) and hit rate ($p = .006$). For comparing semantic transparency, we applied the Welch's t-test [24], which is robust when normality within groups and variance homogeneity among groups cannot be assumed. The semantic transparency for the 3 concrete syntaxes differs significantly according to Welch's t-test, $t(32.301) = 4.913$, $p = .014$. The hit rate also differs significantly according to the Welch's t-test, $t(32.119) = 3.857$, $p = .032$. This suggests that at least two of the concrete syntaxes differ significantly on their semantic transparency and hit rate. Post-hoc tests, using the Games-Howell post-hoc

Table 3. Descriptive statistics

Dependent variable	Symbol set	Mean	S.D.	Skew	Kurt	K-S	S-W
Semant. Transp. coefficient	standard	**.182**	.164	.845	1.256	.077	.114
	stereotype	.288	.409	.265	−1.440	.055	.057
	prototype	**.411**	.268	.386	−1.462	.110	.035
Hit rate	standard	**25.06**	15.105	.846	1.267	.091	.129
	stereotype	32.67	23.088	.269	−1.438	.051	.053
	prototype	**44.33**	25.294	.381	−1.465	.104	.034

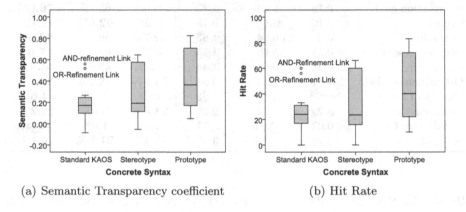

(a) Semantic Transparency coefficient   (b) Hit Rate

Fig. 4. Results for the semantic transparency coefficient and hit rate

procedure, were conducted to determine which pairs of concrete syntaxes differed significantly. This test results are outlined in Table 4 and indicate that the semantic transparency of the prototype concrete syntax ($M = .411$, $SD = .268$) is significantly higher than the one of the standard concrete syntax ($M = .182$, $SD = .164$), with $p = .12$ and $d = 1.042$. Concerning the hit rate, similar results are achieved, with the prototype concrete syntax ($M = 44.33$, $SD = 25.294$) significantly higher than the one of the standard concrete syntax ($M = 25.06$, $SD = 15.105$), with $p = .026$ and $d = .938$. The effect sizes for semantic transparency ($d = 1.042$) and hit rate ($d = .938$) exceed Cohen's convention for a large effect ($d = .80$).

Our results suggest that the prototype concrete syntax is more semantically transparent than the standard concrete syntax. We found no statistically significant differences between the prototype and the stereotype concrete syntaxes, or between the stereotype and the standard concrete syntaxes. Also, the three concrete syntaxes are semantically transparent, even if in different degrees. The standard KAOS concrete syntax differs significantly from a semantically opaque concrete syntax (which would have a mean semantic transparency score around 0). We tested this with a one-sample t-test, $t(17) = 4.708$, $p < .001$.

Table 4. Hypotheses testing for semantic transparency coefficient and hit rate

Hypotheses	Formula	Mean difference	Statistical significance (p-value)	Practical meaningfulness (d)
H_{1ST}	$stereotype > standard$	.106	.290	–
H_{2ST}	$prototype > standard$	.229	**.012***	**1.04****
H_{3ST}	$prototype > stereotype$	.123	.332	–
H_{4HR}	$stereotype > standard$	7.611	.480	–
H_{5HR}	$prototype > standard$	19.278	**.026***	**.938****
H_{6HR}	$prototype > stereotype$	11.667	.330	–

*The mean difference is statistically significant with $p < .05$
**Practical meaningfulness: $|d| > .8$, big effect size

8 Discussion

8.1 Evaluation of Results

RQ1. Is the KAOS visual notation semantically opaque? The results do not allow us to conclude that the standard KAOS symbol set is semantically opaque. 67% of the participants of the semantic transparency experiment are from the WKML group. Some of them had contact with the KAOS language as part of a Software Engineering course, which might explain the relatively high semantic transparency coefficient values for the standard KAOS symbol set.

RQ2. Can participants with no knowledge in modelling languages design more semantically transparent symbols than participants with knowledge in modelling languages? The symbols produced by the WKML group are clearly influenced by the modelling languages they know, namely UML. The symbols produced by the WNKML group are less formal, more creative and different from each other. In the prototyping experiment (Study 3), the symbols drawn by the WNKML group had more votes than the ones drawn by the WKML group. In the semantic transparency experiment (Study 4), the prototype symbol set had significantly better results, which suggests the symbols drawn by the WNKML group were more easily identifiable. We conclude that the WNKML group produced symbols that represent better visual metaphors for KAOS concepts. Some participants had a background in Computer Science, while others did not. The former were significantly more able to produce a model with their proposed symbols than the latter but were less creative in their proposals.

RQ3. Which visual notation (*standard*, *stereotype*, or *prototype*) is more semantically transparent? The results show that there is a statistically significant difference between prototype and standard KAOS in terms of semantic transparency coefficient and success rate, with a considerable *effect size* for both metrics. We conclude that the prototype symbol set is more cognitively effective than the standard KAOS in terms of semantic transparency.

8.2 Implications to Practice

The semantic transparency is only one of the 9 principles in the PoN. Improving a notation according to one particular principle does not necessarily lead to a more cognitively effective notation, as this change may have detrimental side effects with respect to other principles. For example, the ease of drawing the symbols is relevant for cognitive fitness, but is not considered here. Also, while the standard KAOS notation overloads some symbols, leading to a greater graphic economy, the prototype notation has more symbols, increasing the diagrammatic complexity. Although a symbol may be easily recognisable as mnemonic of a particular term, this may be a misrepresentation of a concept denoted by the same name, but with a significantly different semantics. For example, the symbols for obstacle and obstacle resolution were easily recognised by participants, but are not really related to the concept of obstacle in KAOS. Also, the symbols were evaluated in isolation, rather than in the context of requirements models. It may be the case that they form confusing diagrams, due to their conceptual diversity, as the metaphors were not chosen consistently from one symbol to the next.

8.3 Threats to Validity

Conclusion Validity. In the semantic transparency experiment, we used 18 candidate symbols for the stereotype and prototype concrete syntaxes, but only 12 for the KAOS standard concrete syntax, as it contains several symbols that overload different concepts [2]. This overloading introduces a bias for the smaller symbol set (standard KAOS) in terms of semantic transparency and hit rate. The probability of selecting the correct symbol by chance is higher for this set. This may have diminished the differences among the distributions of semantic transparency and hit rate. And may have hampered our ability to distinguish between the distributions of the standard and stereotype semantic transparency. The practical meaningfulness of the differences between the standard and prototype semantic transparency may be higher than measured in this experiment.

Internal Validity. We targeted RE novices, using convenience sampling. Some participants had previously contacted with RE in an academic context, but all are surrogates for non-technical stakeholders and software developers who are not experienced in RE, thus controlling expertise bias. To mitigate sequencing effects, symbols were randomly ordered in the questionnaires for each participant.

External Validity. We used novices to increase generalisability to the target population (stakeholders inexperienced with RE). As our participants are students from the same university, they share a common cultural background. Semantic transparency is often culture-specific, so their proposed and chosen concrete syntaxes were likely influenced by that background.

9   Conclusions and Future Work

We performed 3 quasi-experiments to support the evaluation of KAOS goal models semantic transparency and its improvement through the proposal of an alternative concrete syntax. We asked novices to draw candidate symbols for 18 KAOS goal model concepts. We created two alternative concrete syntaxes, based on these symbols: the stereotype and prototype symbol sets. Finally, we compared the semantic transparency of the 2 alternative concrete syntaxes and the standard KAOS, by asking a third group of novices to identify the symbol that better represents each KAOS goal model concept. The prototype's semantic transparency was significantly higher than the one in the standard KAOS concrete syntax (mean difference of .23). This suggests an opportunity for improving the communication between RE experts and other stakeholders using the prototype concrete syntax proposed in this paper. This result is in line with those obtained in similar studies for other modelling languages, as described in Sect. 2. Indeed, novices can be helpful in designing more recognisable symbols.

We plan to study other aspects of the PoN theory, such as complexity management, perceptual discriminability and cognitive fit. We also plan to assess if the prototype concrete syntax has drawbacks, in particular in model construction and model comprehension, since better symbol recognition may not necessary imply better model understanding. Moreover, since the symbols were selected independently from each other, they do not necessarily form a consistent set, in terms of the chosen visual metaphors.

Thus, further research is needed to study how an inconsistent set of symbols impacts the overall model understanding.

Acknowledgments. We thank NOVA LINCS UID/CEC/04516/2013 and FCT-MCTES SFRH/BD/108492/2015 for financial support.

References

1. van Lamswcerde, A.: Goal-oriented requirements engineering: a guided tour. In: International Symposium on Requirements Engineering, pp. 249–262 (2001)
2. Moody, D.L.: The "physics" of notations: toward a scientific basis for constructing visual notations in software engineering. IEEE Trans. Softw. Eng. **35**(6), 756–779 (2009)
3. Caire, P., Genon, N., Heymans, P., Moody, D.L.: Visual notation design 2.0: towards user comprehensible requirements engineering notations. In: RE 2013, pp. 115–124 (2013)
4. Dardenne, A., Van Lamsweerde, A., Fickas, S.: Goal-directed requirements acquisition. Sci. Comput. Program. **20**(1–2), 3–50 (1993)
5. Yu, E.: Modelling strategic relationships for process reengineering. MIT Press, Social Modeling for Requirements Engineering (2011)
6. Antón, A.I., McCracken, W.M., Potts, C.: Goal decomposition and scenario analysis in business process reengineering. In: Wijers, G., Brinkkemper, S., Wasserman, T. (eds.) CAiSE 1994. LNCS, vol. 811, pp. 94–104. Springer, Heidelberg (1994). https://doi.org/10.1007/3-540-58113-8_164

7. Castro, J., Kolp, M., Mylopoulos, J.: Towards requirements-driven information systems engineering: the TROPOS project. Inf. Syst. **27**(6), 365–389 (2002)
8. Rolland, C., Salinesi, C.: Modeling goals and reasoning with them. In: Aurum, A., Wohlin, C. (eds.) Engineering and Managing Software Requirements, pp. 189–217. Springer, Heidelberg (2005). https://doi.org/10.1007/3-540-28244-0_9
9. GRL - Goal-oriented Requirement Language Kernel Description. http://www.cs.toronto.edu/km/GRL/. Accessed Mar 2018
10. van Lamsweerde, A.: Requirements Engineering: From System Goals to UML Models to Software, vol. 10. Wiley, Chichester (2009)
11. Prisacariu, C., Schneider, G.: A formal language for electronic contracts. In: Bonsangue, M.M., Johnsen, E.B. (eds.) FMOODS 2007. LNCS, vol. 4468, pp. 174–189. Springer, Heidelberg (2007). https://doi.org/10.1007/978-3-540-72952-5_11
12. Gurr, C.A.: Effective diagrammatic communication: syntactic, semantic and pragmatic issues. J. Vis. Lang. Comput. **10**(4), 317–342 (1999)
13. Norman, D.: The Design of Everyday Things: Revised and Expanded Edition. Basic Books (AZ), New York (2013)
14. Genon, N., Amyot, D., Heymans, P.: Analysing the cognitive effectiveness of the UCM visual notation. In: Kraemer, F.A., Herrmann, P. (eds.) SAM 2010. LNCS, vol. 6598, pp. 221–240. Springer, Heidelberg (2011). https://doi.org/10.1007/978-3-642-21652-7_14
15. Genon, N., Heymans, P., Amyot, D.: Analysing the cognitive effectiveness of the BPMN 2.0 visual notation. In: Malloy, B., Staab, S., van den Brand, M. (eds.) SLE 2010. LNCS, vol. 6563, pp. 377–396. Springer, Heidelberg (2011). https://doi.org/10.1007/978-3-642-19440-5_25
16. Granada, D., Vara, J.M., Brambilla, M., Bollati, V., Marcos, E.: Analysing the cognitive effectiveness of the WebML visual notation. Softw. Syst. Model. **16**(1), 195–227 (2017)
17. Howell, W.C., Fuchs, A.H.: Population stereotypy in code design. Organ. Behav. Hum. Perform. **3**(3), 310–339 (1968)
18. Jones, S.: Stereotypy in pictograms of abstract concepts. Ergonomics **26**(6), 605–611 (1983)
19. Zwaga, H., Boersema, T.: Evaluation of a set of graphic symbols. Appl. Ergon. **14**(1), 43–54 (1983)
20. Howard, C., O'Boyle, M., Eastman, V., Andre, T., Motoyama, T.: The relative effectiveness of symbols and words to convey photocopier functions. Appl. Ergon. **22**(4), 218–224 (1991)
21. Arning, K., Ziefle, M.: "It's a bunch of shapes connected by lines": evaluating the graphical notation system of business process modeling languages. In: 9th International Conference on Work With Computer Systems, WWCS 2009, Beijing, China (2009)
22. ISO: Graphical symbols: test methods - part 1: methods for testing comprehensibility, Switzerland (2014)
23. ISO: Graphical symbols: test methods for judged comprehensibility and for comprehension, Geneva (2001)
24. Welch, B.L.: The significance of the difference between two means when the population variances are unequal. Biometrika **29**, 350–362 (1938)
25. Moody, D.L., Heymans, P., Matulevičius, R.: Improving the effectiveness of visual representations in requirements engineering: an evaluation of i* visual syntax. In: RE 2009, pp. 171–180 (2009)

26. Matulevičius, R., Heymans, P.: Visually effective goal models using KAOS. In: Hainaut, J.-L., et al. (eds.) ER 2007. LNCS, vol. 4802, pp. 265–275. Springer, Heidelberg (2007). https://doi.org/10.1007/978-3-540-76292-8_32

27. Matulevičius, R., Heymans, P.: Comparing goal modelling languages: an experiment. In: Sawyer, P., Paech, B., Heymans, P. (eds.) REFSQ 2007. LNCS, vol. 4542, pp. 18–32. Springer, Heidelberg (2007). https://doi.org/10.1007/978-3-540-73031-6_2

28. Saleh, F., El-Attar, M.: A scientific evaluation of the misuse case diagrams visual syntax. Inf. Softw. Technol. **66**, 73–96 (2015)

29. Boone, S., Bernaert, M., Roelens, B., Mertens, S., Poels, G.: Evaluating and improving the visualisation of CHOOSE, an enterprise architecture approach for SMEs. In: Frank, U., Loucopoulos, P., Pastor, Ó., Petrounias, I. (eds.) PoEM 2014. LNBIP, vol. 197, pp. 87–102. Springer, Heidelberg (2014). https://doi.org/10.1007/978-3-662-45501-2_7

Using Contextual Goal Models
for Constructing Situational Methods

Jolita Ralyté[1(⊠)] and Xavier Franch[2]

[1] University of Geneva, Geneva, Switzerland
jolita.ralyte@unige.ch
[2] Universitat Politècnica de Catalunya (UPC), Barcelona, Spain
franch@essi.upc.edu

Abstract. Situation and intention are two fundamental notions in situational method engineering (SME). They are used to assess the context of an ISD project and to specify method requirements in this context. They also allow defining the goals of the method chunks and the conditions under which they can be applied. In this way, the selection and assembly of method chunks for a particular ISD project is driven by matching situational method requirements to method chunks' goals and context descriptions. In this paper we propose the use of contextual goal models for supporting all SME steps. Our approach is based on iStar2.0 modeling language that we extend with contextual annotations.

Keywords: Situational Method Engineering · Contextual goal model
iStar2.0

1 Introduction

The mission of situational method engineering (SME) [1] consists in providing concepts and guidance for building situation-specific (i.e., situational) methods taking into account the particular context and requirements of the information systems development (ISD) project at hand. SME is founded on modularity, reusability and flexibility principles [2]: method knowledge is formalized in terms of method chunks, characterized by a set of attributes representing their reuse conditions, and stored in a method repository. Method chunks can be reused in many different method constructions. The quality of a situational method heavily depends on how well the obtained method fits the situation [3]. Arguably, selecting the right method chunks is the most challenging task in SME. It requires understanding not only functional method requirements but also the contextual ones. Functional requirements express how the method is expected to support different system engineering activities of the ISD project [4], while the contextual ones reflect the situation of the project and define the conditions in which the method will be used [5]. The SME literature exposes several different ways to deal with functional and contextual aspects in SME; some of them are discussed in the following section. Still, there is space for innovation. In this paper we propose to represent these two aspects in the same model called contextual goal model. We use these models to express situational method requirements as well as to specify method chunks. Furthermore, we introduce a systematic goal modeling in all steps of SME.

© Springer Nature Switzerland AG 2018
J. C. Trujillo et al. (Eds.): ER 2018, LNCS 11157, pp. 440–448, 2018.
https://doi.org/10.1007/978-3-030-00847-5_31

In particular, we use the iStar2.0 [6] goal modeling language that we extend with contextual elements. This type of models allows representing system engineering goals and annotating them with context criteria. To summarize, the research objective of our work is to exploit the use of contextual goal models in SME.

2   Situational and Intentional Aspects in SME

A detailed overview of the state of the art in SME can be found in [1]. In this section, we briefly present the situational and intentional aspects of SME, and the iStar2.0 goal modeling approach, which is fundamental in our contribution.

The notion of situation is often described by using a set of predefined contingency factors or context criteria that can be used at the level of a particular project, a process in the organization or the whole organization [5, 7–9]. The situation of a project is defined by giving values to a set of the most pertinent criteria, e.g. user involvement = low, time pressure = high, delivery strategy = incremental, etc. In addition, the same criteria are also used for characterising the suitability of method chunks to different situations [5].

The notion of intention or goal is mainly used to define the purpose of a process-driven method component and in particular, method chunk [5]. It allows to formalize the system engineering goal to be achieved by applying this method chunk. Goals are also used to express situational method requirements that are then matched to the method chunks' goals to find the best fitting method chunks. The goal-driven process modeling formalism Map [10] constitutes the foundation of the assembly-based [5] and the method family [11] approaches. It allows expressing methods in terms of intentions and strategies to reach the intentions. However, it has no constructs for relating context criteria to the method intentions and strategies.

Recently, iStar2.0 [6] goal models have been used in SME, mainly to formalize the content of method chunks [12] as reusable goal models. In [9] we have introduced the idea of using contextual iStar2.0 for specifying method requirements. Here we explore their usage in all steps of the assembly-based SME process.

iStar2.0 [6] is a standardized kernel goal modeling language dedicated to represent functional and quality requirements, their dependencies and their refinement. The language includes four *intentional elements*, namely *goals*, *qualities*, *tasks* and *resources*, and the notion of *actor*. Actors are autonomous entities that aim at achieving their goals in collaboration with other actors. They can be subtyped as physical agents or abstract roles and they can be related through specialization (*is-a*) or a general link called *participates-in*. Their collaboration is materialized as dependencies through intentional elements: an actor (depender) is dependent on another actor (dependee) for achieving her goal, satisfying a quality, realizing a task or obtaining a resource. All intentional elements of an actor, their refinement and interrelationships are clustered into the *actor's boundary*. View can be defined over iStar2.0 models, and in this paper we will use the Strategic Dependency (SD) and Strategic Rationale (SR) model coming from the original *i** [13]: SD models only show actors and dependencies, while SR models show the internal structure of actors.

3 SME with Contextual Goal Models

A complete SME approach has two phases [14]: (1) the construction of the contextually situated method chunks for reuse and (2) the construction of situation-specific methods by reusing and combining method chunks from the repository. Due to the space limit, we focus in this paper on the later and only briefly introduce the former.

Most of the SME approaches use the notion of goal to define method chunks and specify new method requirements. However, any of them use contextual goal models to do that. Here, we explore this idea by extending the *method chunk-based* SME approach [5] with contextual goal models, which are iStar2.0 goal models enriched with the contextual information. By relating method goals with context criteria we express not only functional but also contextual method requirements that are then used to select and assemble the most appropriate method chunks. Figure 1 depicts the process of situation method construction where contextual goal modelling is used as an underpinning technique at each of its steps.

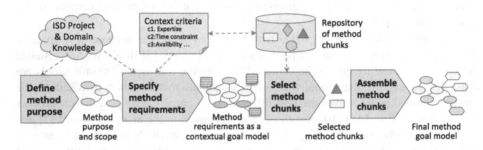

Fig. 1. The process of situational method construction with contextual goal models

3.1 Method Knowledge for Reuse

Because the selection of method chunks consists in matching method requirements with the method chunks' goals and contextual characteristics, the definition of method chunks should also be represented with contextual goal models. Which motivates us to extend the specification of method chunks [5, 14] with contextual goal models reflecting their purpose and contextual fitness. As mentioned above, we use iStar2.0 to create these goal models. The method engineer starts with developing the SD goal model of the method and the context criteria that are pertinent for defining the conditions for method application. Criteria and their values can be defined from scratch based on the related literature review and/or the method engineers' experience, or by refining some existing generic set of context criteria [5, 7, 8]. Next, for each identified method chunk the method engineer develops the SR goal model and extends it with contextual information. In order to integrate the intentional and contextual aspects of method chunks in the same model, we represent the context criteria together with their values as iStar2.0 elements decorated with context information. This representation is an extension of the standard, as introduced in [9].

3.2 Situational Method Construction by Reuse

Defining the Purpose of a Situation-Specific Method. The purpose of a situational method is closely related to the objective of the project it has to serve. For example, if the project aims at developing a web service, a situation-specific method can be required to cover the entire development process or just a part of it (e.g. service design). Therefore, defining the purpose of the method also means delimiting its scope and thus, restricting the elicitation of method requirements and the selection of method chunks. Indeed, the goal of each selected method chunk has to fit in the scope of the method, which means to be a sub-goal of the method goal.

We recommend starting with the SD goal model representing the project as a socio-technical system, where stakeholders, their roles and expectations are revealed. This model provides the context for defining the purpose of the required method, i.e. creating a goal model representing how the intended method is expected to support the achievement of different project actors' goals, resources and tasks and to deliver the expected qualities. Therefore, the intended method has to be modeled as one or several actors each of them reflecting a well-defined and autonomous part of the method (e.g. a particular platform or toolkit, an ISD process step). The different participants of the project are also represented as actors in the goal model. They depend on the method to satisfy their goals, furnish resources, and realize tasks with a given quality. On the other hand, the success of the method usually demands some expectations from the surrounding actors, therefore the actors may also act as *dependees*. The goal model is limited to the SD representation; actors' boundaries (the SR models) are not developed at this step of the approach.

Specifying Situational Method Requirements. The aim of this step is to express the requirements of the project in terms of contextual goal models. It includes two steps: (1) modeling the organizational context of the project as SR models of the actors defined in the method purpose model, and then (2) assessing the project-specific contextual conditions that should be taken into account when constructing the intended method, and in particular, when selecting the method chunks from the method repository. These conditions are formalized in terms of *context criteria*, each of them having a set of predefined values. The criteria and their values are part of the reusable method knowledge as explained in Sect. 3.1. Some criteria are uni-valued (e.g. Time pressure = Low/Medium/High), while others can be multi-valued (e.g. People per requirements elicitation session = {Individual, Group, Mass}. The assessed criteria are then added to the goal models extending them with contextual information. The obtained models express functional and contextual requirements of the project at hand.

Selecting and Assembling the Method Chunks. In this crucial step, an iterative matching process among the organizational context and the repository of method chunks is conducted by the method engineer. During this process, the models grows by: (1) recording the relevant decisions that are necessary to make in order to select the method chunk(s), (2) once selected, reflecting in the model the consequences of adopting such chunk(s).

4   Example: Requirements Elicitation Method

To illustrate our approach we propose an example of situational method construction, namely "Requirements Elicitation Method". Requirements elicitation is one of the key activities in the requirements engineering process [15] for which many different techniques exist. The example will in fact call for the selection of two different method chunks proposing such techniques for two different groups of stakeholders. In this way, we aim to demonstrate the usage of contextual goal models in method requirements specification and how these requirements depend on the contextual conditions.

4.1   Reusable Method Knowledge

Figure 2 shows the iStar2.0 model that specifies the goals of the *Requirements Elicitation Method*. This model acts as a template introducing the main elements that the forthcoming chunks need to refine. Requirements elicitation involves four actors with some mutual dependencies. The *Requirements Engineer* is in charge of delivering a *Requirements Document* to the customer *Organization* by *Conduct*ing the *Elicitation* using a *Requirements Elicitation Method*. The *Stakeholder* provides his/her *Needs* as requested by the method. These needs will be properly elicited only if the *Elicitation* is *Well-fit* and the requirements engineer commits to have the *Method Steps Followed*.

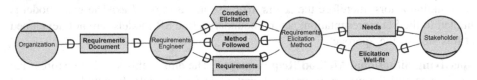

Fig. 2. The generic purpose model of the Requirements Elicitation Method

We take the systematic review and analysis of requirement elicitation techniques conducted by Carrizo et al. [16] as a source for defying the method chunks and the associated context criteria. In this paper, up to 15 different elicitation techniques are compiled and analyzed under the lenses of the context criteria. The criteria are defined in terms of four factors (Elicitor, Informant, Problem Domain, and Elicitation Process) each of them specified with a set of attributes. The fitness of method chunks to different situations is determined by assessing values of the relevant context criteria for each method chunk. In Table 1 we present an excerpt of this analysis for 6 classical elicitation techniques (considered as method chunks) and the set of selected criteria. As a result of this analysis and the representation of the criteria as iStar2.0 elements, we can build the contextual models that define the particular method chunks. Figure 3 shows an excerpt of the model for the *Elicitation by Interviews* method chunk, which is declared as a specialization of *Requirements Elicitation Method* (not shown in the figure for clarity of the drawing).

Table 1. The adequacy of requirements elicitation techniques to the selected context criteria, from [16]. Values: (+) recommended, (–) indifferent, (x) not recommended.

Context criteria		Values	Requirements elicitation techniques					
			Interviews	Questionnaire	Observation	Prototyping	Brainstorming	Scenarios
Elicitor	Elicitation experience (c_2)	High	+	+	+	+	+	+
		Medium	+	+	+	+	–	+
		Low	–	–	+	–	–	
Informant	People per session (c_5)	Individual	+	+	+	+	x	+
		Group	–	+	+	–	+	–
		Mass	x	+	–	–	–	–
	Consensus (c_6)	High	+	+	–	+	–	+
		Low	x	+	–	+	–	–
	Availability (c_{10})	High	+	+	+	+	+	+
		Low	–	+	+	–	x	–
	Location/accessibility (c_{11})	Near	+	+	+	+	+	+
		Far	–	+	–	x	x	x
Problem Domain	Type of information to be elicited (c_{12})	Strategic	+	+	–	–	+	–
		Tactical	+	+	+	+	+	+
		Basic	–	+	–	+	x	–
Elicitation process	Project time constraint (c_{15})	High	–	+	x	x	x	+
		Medium	+	+	+	+	–	+
		Low	+	+	–	–	+	+
	Process time (c_{16})	Start	+	–	+	+	+	–
		Middle	+	+	–	+	–	+
		End	+	+	x	x	x	–

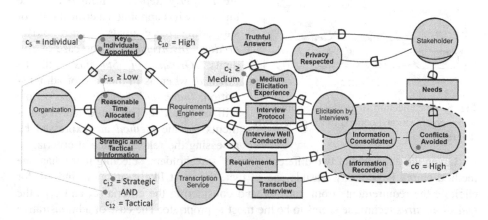

Fig. 3. The contextual goal model of the method chunk *Elicitation by Interviews*

4.2 Constructing a Situation-Specific Method

To illustrate the construction of the *Requirement Elicitation Method* in a specific setting we take the case of the University of Geneva (UniGe) and its ISD project aiming to develop a UniGe mobile application. The application should allow quickly accessing different UniGe information services, like announcement of events, school calendar, course program, access maps to different campuses, etc. The UniGe ISD department

(*UniGe ISDD*) is responsible for developing the application and needs a method for conducting requirements elicitation. *Student* and *Administrator* are two types of *Stakeholders* from whom the needs will be elicited. For the sake of brevity, the SD model of the socio-technical system of the project is not shown here. Figure 4 directly shows the SD model representing the purpose of the required project method.

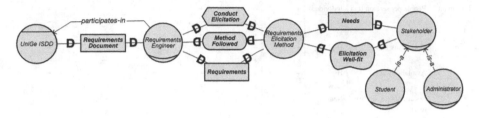

Fig. 4. The purpose of the project method

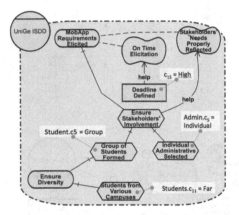

Fig. 5. Specification of method requirements in the organizational context of the project (SR diagram)

Figure 5 shows the needs of the project for the *Requirements Elicitation Method*. The main goal of *MobApp Requirements Elicited* is reinforced by two important qualities: to ensure an *On Time Elicitation* and *Stakeholders' Needs* are *Properly Reflected*. Remarkably, the university has appointed a critical task for supporting these goals: *Ensure Stakeholders' Involvement*. In the case of students, *Groups of Students* will be *Formed*, whereas in the case of administration, *Individual Administrative* people will be *Selected*. The contextual requirements for the *method* are expressed by assessing the relevant context criteria.

Given the existence of two different types of stakeholders, selecting more than one method chunk is perfectly feasible and in fact, the most likely situation. Indeed, for eliciting the requirements from students, and considering the criteria (c_5 and c_{11}), the *Questionnaires* technique seems to be the most appropriate. The case of administrative staff as stakeholders is a bit more complex. Three techniques (*Observation*, *Prototyping* and *Brainstorming*) can be discarded due to their high time constraints on the project time (c_{15}). However, it is not clear which of the others (*Interviews*, *Questionnaires* or *Scenarios*) could be applied. Therefore, the role of the method engineer as facilitator becomes crucial. Supposing that *Elicitation by Interviews* technique is the most suitable one, the two selected method chunks are then assembled into the project-specific method and applied by the project *Requirements Engineer*.

5   Conclusion

In this paper we explore contextual goal modeling as a way to deal with intentional and contextual aspects in situational method engineering (SME). In particular, we revise the method chunk-driven SME approach by introducing contextual goal models in both SME levels: (1) specification of method knowledge for reuse and (2) situation-specific method construction by reuse. For that we use iSar2.0 goal modeling language. But because this language does not include context elements, we introduce decorations in goal models to represent contextual information.

This work is a result of the OpenReq project, funded by the European Union's H2020 research and innovation programme under the grant agreement No. 732463.

References

1. Henderson-Sellers, B., Ralyté, J., Ågerfalk, P.J., Rossi, M.: Situational Method Engineering. Springer, Heidelberg (2014). https://doi.org/10.1007/978-3-642-41467-1
2. Rolland, C.: Method engineering: towards methods as services. Softw. Process Improve. Pract **14**, 143–164 (2009)
3. McBride, T., Henderson-Sellers, B.: A method assessment framework. In: Ralyté, J., Mirbel, I., Deneckère, R. (eds.) ME 2011. IAICT, vol. 351, pp. 64–76. Springer, Heidelberg (2011). https://doi.org/10.1007/978-3-642-19997-4_7
4. Ralyté, J.: Requirements definition for the situational method engineering. In: Rolland, C., Brinkkemper, S., Saeki, M. (eds.) Engineering Information Systems in the Internet Context. ITIFIP, vol. 103, pp. 127–152. Springer, Boston (2002). https://doi.org/10.1007/978-0-387-35614-3_9
5. Mirbel, I., Ralyté, J.: Situational method engineering: combining assembly-based and roadmap-driven approaches. Requir. Eng. J. **11**(1), 58–78 (2006)
6. Dalpiaz, F., Franch, X., Horkoff, J.: iStar 2.0 Language Guide. https://arxiv.org/abs/1605.07767
7. van Slooten, K., Hodes, B.: Characterizing IS development projects. In: Brinkkemper, S., Lyytinen, K., Welke, R.J. (eds.) Method Engineering. ITIFIP, pp. 29–44. Springer, Boston (1996). https://doi.org/10.1007/978-0-387-35080-6_3
8. Kornyshova, E., Deneckère, R., Claudepierre, B.: Contextualization of method components. In: RCIS 2010, pp. 235–246. IEEE Computer Society Press (2010)
9. Franch, X., et al.: A situational approach for the definition and tailoring of a data-driven software evolution method. In: Krogstie, J., Reijers, Hajo A. (eds.) CAiSE 2018. LNCS, vol. 10816, pp. 603–618. Springer, Cham (2018). https://doi.org/10.1007/978-3-319-91563-0_37
10. Rolland, C., Prakash, N., Benjamen, A.: A multi-model view of process modelling. Requir. Eng. J. **4**(4), 169–187 (1999)
11. Kornyshova, E., Deneckère, R., Rolland, C.: Method families concept: application to decision-making methods. In: Halpin, T., et al. (eds.) BPMDS/EMMSAD-2011. LNBIP, vol. 81, pp. 413–427. Springer, Heidelberg (2011). https://doi.org/10.1007/978-3-642-21759-3_30
12. López, L., Costal, D., Ralyté, J., Franch, X., Méndez, L., Annosi, M.C.: OSSAP – a situational method for defining open source software adoption processes. In: Nurcan, S., Soffer, P., Bajec, M., Eder, J. (eds.) CAiSE 2016. LNCS, vol. 9694, pp. 524–539. Springer, Cham (2016). https://doi.org/10.1007/978-3-319-39696-5_32

13. Yu, E.: Modelling strategic relationships for process reengineering. Ph.D. thesis, University of Toronto, Toronto (1995)
14. Ralyté, J., Rolland, C.: An approach for method reengineering. In: Kunii, H.S., Jajodia, S., Sølvberg, A. (eds.) ER 2001. LNCS, vol. 2224, pp. 471–484. Springer, Heidelberg (2001). https://doi.org/10.1007/3-540-45581-7_35
15. Pohl, K.: Requirements engineering: Fundamentals, Principles, and Techniques. Springer, Heidelberg (2010)
16. Carrizo, D., Dieste, O., Juristo, N.: Systematizing requirements elicitation technique selection. Inf. Softw. Technol. 56(6), 644–669 (2014)

Multi-perspective Comparison of Business Process Variants Based on Event Logs

Hoang Nguyen[1]($\boxtimes$), Marlon Dumas[2], Marcello La Rosa[3],
and Arthur H. M. ter Hofstede[1]

[1] Queensland University of Technology, Brisbane, Australia
huanghuy.nguyen@hdr.qut.edu.au, a.terhofstede@qut.edu.au
[2] University of Tartu, Tartu, Estonia
marlon.dumas@ut.ee
[3] University of Melbourne, Melbourne, Australia
marcello.larosa@unimelb.edu.au

Abstract. A process variant represents a collection of cases with certain shared characteristics, e.g. cases that exhibit certain levels of performance. The comparison of business process variants based on event logs is a recurrent operation in the field of process mining. Existing approaches focus on comparing variants based on directly-follows relations such as "a task directly follows another one" or a "resource directly hands-off to another resource". This paper presents a more general approach to log-based process variant comparison based on so-called *perspective graphs*. A perspective graph is a graph-based abstraction of an event log where a node represents any entity referred to in the log (e.g. task, resource, location) and an arc represents a relation between these entities within or across cases (e.g. directly-follows, co-occurs, hands-off to, works-together with). Statistically significant differences between two perspective graphs are captured in a so-called *differential perspective graph*, which allows us to compare two logs from any perspective. The paper illustrates the approach and compares it to an existing baseline using real-life event logs.

Keywords: Process mining · Variant analysis · Comparison · Multi-perspective

1 Introduction

The performance of a business process may vary over time, geographically, or across business units, products, or customer types. And even within a given time period, place, business unit, product, and customer type, there are usually performance variations between cases of a process. Some cases lead to a positive outcome (e.g. on-time completion), while others lead to negative outcomes. A typical question that arises in this setting is: *"What differentiates the positive and the negative cases?"*, or more broadly: *"What (statistically) significant differences exist between two variants of a process?"*

Recently, several approaches for comparing variants of a process based on their event logs have been proposed [1–4]. Given two event logs L1 and L2 corresponding to two variants of a business process, these techniques allow us to identify characteristics

© Springer Nature Switzerland AG 2018
J. C. Trujillo et al. (Eds.): ER 2018, LNCS 11157, pp. 449–459, 2018.
https://doi.org/10.1007/978-3-030-00847-5_32

that are commonly found in the cases in L1, but are rare or non-existent in L2 and vice-versa. These approaches are restricted to identifying differences in the directly-follows relations, such as "task A always directly follows task B in one variant but never in the other" or "resource X often hands-off work to resource Y in one variant but rarely in the other". However, events in a log may carry a richer set of attributes besides tasks and resources – e.g. customer attributes, location attributes, product-related attributes, etc. Differences between two event logs may be found along any of these attributes.

This paper presents an approach for comparing process variants from multiple perspectives corresponding to arbitrary sets of attributes. Specifically, the paper introduces a graph-based abstraction of an event log, namely a *perspective graph*, where a node represents any entity referenced in an attribute of the event log (task, resource, location, etc.) and an arc represents an arbitrary relation between entities (e.g. directly-follows, co-occurs, hands-off to, works-together with, etc.) within or across cases. Statistically significant differences between two perspective graphs are captured in a so-called *differential perspective graph*, which allows a user to visually compare two event logs from any given perspective using either a graphical or a matrix representation.

The proposed approach has been implemented as a proof-of-concept prototype in the ProM open-source process mining toolset. The paper illustrates the capabilities provided by differential perspective graphs using two real-life event logs, and compares them against an existing state-of-the-art approach for process variant analysis.

The paper is structured as follows. Section 2 discusses existing process variant analysis approaches. Section 3 presents the proposed approach, while Sect. 4 discusses its evaluation. Section 5 summarizes the contributions and outlines future work directions.

2   Related Work

Existing approaches to log-based process variant comparison can be classified into indicator-based, graph-based, and model-based. Indicator-based approaches extract performance indicators from two input logs and compare these indicators using visualization techniques (e.g. bar charts), e.g. risk indicators [5], performance indicators [6], and resource behavior indicators [7]. These approaches allow one to determine how two variants perform relative to each other on an aggregate basis or at a task- or resource level. For example, these techniques allow us to determine which tasks have higher cycle time in one variant than in the other. However, they do not allow us to identify behavioral differences and their impact on performance.

Graph-based approaches rely on pairwise differencing of graphs, such as event structures [2], directly-follows graphs [8], or transition systems [4]. For example, the approach in [4] abstracts a process as a transition system where each state represents an equivalence class of trace prefixes, e.g. a state may represent all prefixes that coincide on their last n events. Each transition is labeled with an event label or event attribute value. Transition systems are contrasted and differences are visually highlighted. The approach in [4] is integrated with a Process Cube approach for log slicing and dicing to generate sublogs for comparison [9]. Our technique falls into this category. In particular, the technique in [4] is used as a baseline in our evaluation.

Other related techniques take as input process models and enrich them with performance measures extracted from event logs, such as occurrence frequency or cycle time [3]. These techniques assume that a process model is available, which captures the dominant behaviors of the process variants. In contrast, in this paper we assume that no models are given a priori. Instead, the comparison of variants is conducted on an exploratory basis, from multiple perspectives, and purely based on event logs.

3 Approach

Given two event logs as input, each representing a variant of the same business process, our approach mines two perspective graphs, one from each log. Next, it compares the two graphs and visualizes their statistically significant differences using a differential graph. In the remainder, we define all ingredients of our approach: event logs, process abstraction, perspective graphs, and differential graphs.

3.1 Event Logs and Process Abstraction

An event log consists of cases where each case has a number of associated events. Cases and events can have various attributes. An example log is shown in Table 1. Rows are events and columns are event attributes. The format of event logs has been standardized in the eXtensible Event Stream by the IEEE CIS Task Force on Process Mining [10]. Table 1 provides an example of a *log schema*. Our technique can work with any log schema.

Let Ω be a universe of values, $\mathcal{E}$ be a universe of events and $\mathcal{A}$ be a universe of attribute names, where for each $A \in \mathcal{A}, A: \mathcal{E} \to \Omega$.

Definition 1 (Event Log). *An event log L over a schema $S = \{A_1, A_2, \ldots, A_n\} \subset \mathcal{A}$ is a set of events, i.e. a subset of $\mathcal{E}$.*

Assume that events in a case occur sequentially as shown in Table 1. Based on the event sequence in each case, Fig. 1 shows an example of event clusterings in each case along the time line according to the *Department* and *Location* attributes. For example, in Fig. 1a, events e_1 and e_2 share the same department attribute d_1, and the next occurring events e_3 and e_4 share the same department d_2. This is followed by event e_5 which has occurred with department attribute d_1 again. Figure 1a and b each is seen as an *abstraction* of the process in Table 1. This method of process abstraction is formalized as follows.

Table 1. Event log example

CaseID	EventID	Timestamp	Activity	Resource	Department	Location
c_1	e_1	01.10 10:00:00	a_1	r_1	d_1	l_1
c_1	e_2	02.10 10:00:00	a_2	r_2	d_1	l_1
c_1	e_3	03.10 10:00:00	a_3	r_3	d_2	l_1
c_1	e_4	04.10 10:00:00	a_1	r_3	d_2	l_2
c_1	e_5	05.10 10:00:00	a_2	r_1	d_1	l_2
c_2	e_6	02.10 10:00:00	a_3	r_1	d_1	l_1
c_2	e_7	04.10 10:00:00	a_1	r_2	d_1	l_2
c_2	e_8	06.10 10:00:00	a_3	r_4	d_2	l_2
c_2	e_9	08.10 10:00:00	a_2	r_2	d_1	l_1

Let L be a log over schema $S, CaseID$ the CaseID attribute of events, *timestamp* the timestamp attribute of events, CID_L the set of CaseIDs in L, and $T \subseteq S$ a schema with $T \neq \varnothing$ and *timestamp* $\notin T$. We assume that all timestamps are different.

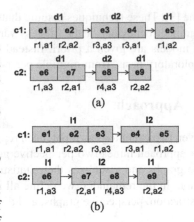

(a)

(b)

Fig. 1. Abstraction by (a) *Department*, (b) *Location*

First, we define a number of relations between events as the basis for our later formalisation. Events can be related in terms of *timestamp*, *CaseID*, and other attributes. Given two events, first they can be *ordered* based on their timestamps, i.e. $e_1 < e_2$ iff $timestamp(e_1) < timestamp(e_2)$. Second, they are *case-related* if they occur in the same case, i.e. $e_1 \doteq e_2$ iff $CaseID(e_1) = CaseID(e_2)$. Finally, in terms of a schema $T \subseteq S$, they are *T-equal* if they share values for all attributes in T, i.e. $e_1 =_T e_2$ iff $\forall t \in T [t(e_1) = t(e_2)]$; otherwise, they are *T-unequal*, i.e. $e_1 \neq_T e_2$.

Based on the above relations, two events are *case-ordered* iff they are *ordered* and *case-related*, i.e. $e_1 \lessdot e_2$ iff $e_1 < e_2 \wedge e_1 \doteq e_2$. Further, two events are *T-case-ordered* iff they are *case-ordered* and *T-unequal* or *case-ordered*, *T-equal* but separated in time from each other by another case-related but T-unequal event, i.e. $e_1 \lessdot_T e_2$ iff $(e_1 \lessdot e_2 \wedge e_1 \neq_T e_2) \vee (e_1 \lessdot e_2 \wedge e_1 =_T e_2 \wedge \exists e_3 \in L [e_1 \lessdot e_3 \wedge e_3 \lessdot e_2 \wedge e_3 \neq_T e_1])$.

The T-case-ordered relation $\lessdot_T$ can be observed in Table 1. In case c_1, in terms of schema $T = \{Department\}$, events e_1 and e_3 are *T-case-ordered* because they are ordered and T-unequal; events e_1 and e_5 are also *T-case-ordered* because they are ordered, T-equal and there is an event, e.g. e_3, that occurs between them in the same case but is T-unequal to them. The T-case-ordered relation $\lessdot_T$ forms a strict partial order over $\mathcal{E}$ for each CaseID. This can be sketched briefly. T-case-ordered is an *irreflexive* relation because $e \lessdot_T e$ implies $e \lessdot e$ hence $timestamp(e) < timestamp(e)$ which is not possible. From the definition of T-case-ordered, it can be proved that T-case-ordered is a *transitive* relation by case distinction with four cases.

Given two case-related events and a schema T, if there exists no *T-case-ordered* relation between the two events, it means that they are T-equal and non-separable in time from each other by another case-related T-unequal event. In this situation, we say that they are *T-equivalent*, i.e. $e_1 \sim_T e_2$ iff $e_1 \doteq e_2 \wedge \neg(e_1 \lessdot_T e_2) \wedge \neg(e_2 \lessdot_T e_1)$.

The T-equivalent relation $\sim_T$ on the event set $E_c = \{e \in L \mid CaseID(e) = c\}$ of a caseid c forms an equivalence relation, where the corresponding quotient set of E_c is $E_c\backslash_{\sim_T}$. Two T-equivalent events are in the same equivalence class of $E_c\backslash_{\sim_T}$. We refer to an equivalence class in $E_c\backslash_{\sim_T}$ as a *fragment*. Visually, a fragment is a row of events as shown in Fig. 1, e.g. in Fig. 1a, $\{e_1, e_2\}$ is a fragment in case c_1.

Based on the notion of fragments, we now define process abstraction.

Definition 2 (Process Abstraction). *Let L be a log, $T \subset S$ a schema, and CID_L the set of CaseIDs in L. An abstraction over T from L is defined as $\mathcal{A}_T^L = \bigcup\limits_{c \in CID_L} E_c\backslash_{\sim_T}$.*

Let $\mathcal{A}_T^L$ be an abstraction over schema T from log L. From the definition, $\mathcal{A}_T^L$ is a set of fragments where each fragment is a set of T-equivalent events. There is a *follows* relation between fragments $F_1, F_2 \in \mathcal{A}_T^L$, i.e. $F_1 \twoheadrightarrow F_2$ iff $\exists e_1 \in F_1 \; \exists e_2 \in F_2 \; [e_1 <_T e_2]$, and a *directly-follows* relation between fragments $F_1, F_2 \in \mathcal{A}_T^L$, i.e. $F_1 \rightarrow F_2$ iff $F_1 \twoheadrightarrow F_2 \wedge \not\exists F_3 \in \mathcal{A}_T^L \; [F_1 \twoheadrightarrow F_3 \wedge F_3 \twoheadrightarrow F_2]$.

3.2 Perspective Graphs

From a process abstraction as shown in Fig. 1, one can look at different relations between event attributes. For example, one can look at the co-occurrence of two attributes in the same fragment, e.g. two resources working in the same department or location. Alternatively, one can look at an inter-fragment relation where an attribute occurs in one fragment and the other attribute occurs in a directly following fragment, e.g. the flow from an activity performed in one department to another activity performed in the next department. More generally, instead of focusing on one attribute, one may focus on a number of attributes depending on the type of analysis, e.g. it may be a pair (resource, activity) representing a task assignment.

In order to represent different types of relation between event attributes, we propose two types of graphs, *intra-fragment* and *inter-fragment*, defined as follows.

Let $\mathcal{A}_T^L$ be an abstraction over schema T from log L. Let $V \subseteq S$, then $\pi_V(e)$ denotes a projection on event e of attributes in V which is defined as $\{(v, v(e)) \mid v \in V\}^1$.

Definition 3 (Intra-Fragment Graph). *Let $\mathcal{A}_T^L$ be an abstraction over schema T from log L and let $U, V \subseteq S$. An Intra-Fragment Graph $IAG_T^{U,V}(L)$ is a node and arc weighted undirected graph $G = (N, E, W_N, W_E)$, defined by:*

- $N = \{\pi_U(e) \mid e \in L\} \cup \{\pi_V(e) \mid e \in L\}$,
- $E = \{\{\pi_U(e), \pi_V(e')\} \mid e \in L \wedge e' \in L \wedge e \sim_T e' \wedge \pi_U(e) \neq \pi_V(e')\}$,
- $W_N(n) = |\{e \in L \mid \pi_U(e) = n \vee \pi_V(e) = n\}|$ *for all* $n \in N$,
- $W_E(\{n_1, n_2\}) = |\{e \in L \mid \pi_U(e) = n_1 \wedge \pi_V(e) = n_2\}| + |\{\{e_1, e_2\} \mid e_1 \in L \wedge e_2 \in L \wedge \pi_U(e_1) = n_1 \wedge \pi_V(e_2) = n_2 \wedge e_1 \sim_T e_2 \wedge e_1 \neq e_2\}|$ *for all* $\{n_1, n_2\} \in E$.

Intra-Fragment Graphs represent a *co-occurrence relation* between event attributes as they co-occur in the same fragment. For example, it can be a task assignment relation when a resource and an activity co-occur in an event, or a co-location relation when two resources co-occur in the same location.

Let $\mathcal{A}_T^L$ be an abstraction over schema T from log L. Let $\phi, \varphi: \mathcal{A}_T^L \rightarrow \mathcal{E}$ such that for all $F \in \mathcal{A}_T^L$, $\phi(F) \in \mathcal{E}$ and $\varphi(F) \in \mathcal{E}$. ϕ and φ are two choice functions on $\mathcal{A}_T^L$, i.e. choice functions can be used to extract events with certain properties from fragments. We will not specify their semantics. Just as an example, ϕ could be chosen such that it returns the latest event in a fragment, and φ could be chosen such that it returns the earliest event in a fragment.

Given two choice functions ϕ and φ, two events are in an *inter-fragment directly-follows* relation, denoted $e_1 \rightarrow_T e_2$, iff $\exists F_1, F_2 \in \mathcal{A}_T^L \mid F_1 \rightarrow F_2 \wedge e_1 = \phi(F_1) \wedge e_2 = \varphi(F_2)$. The set of all inter-fragment directly-follows event pairs in log L is $\mathcal{E}_{\rightarrow_T} = \{(e_1, e_2) \in \mathcal{E} \times \mathcal{E} \mid e_1 \rightarrow_T e_2\}$.

---

[1] $(v, v(e))$ is abbreviated to $v(e)$ when it is clear from the context for the purpose of readability.

Definition 4 (Inter-Fragment Graph). *Let $\mathcal{A}_T^L$ be an abstraction over schema T from log L and let $U, V \subseteq S$. An Inter-Fragment Graph $IEG_T^{U,V}(L)$ is a node and arc weighted directed graph $G = (N, E, W_N, W_E)$, defined by:*

- $N = \{\pi_U(e) \mid e' \in \mathcal{E} \wedge (e, e') \in \mathcal{E}_{\rightarrow_T}\} \cup \{\pi_V(e') \mid e \in \mathcal{E} \wedge (e, e') \in \mathcal{E}_{\rightarrow_T}\}$,
- $E = \{(\pi_U(e), \pi_V(e')) \mid (e, e') \in \mathcal{E}_{\rightarrow_T}\}$,
- $W_N(n) = |\{(e, e') \in \mathcal{E}_{\rightarrow_T} \mid \pi_U(e) = n\}| + |\{(e, e') \in \mathcal{E}_{\rightarrow_T} \mid \pi_V(e') = n\}|$, *for all* $n \in N$,
- *Let* $(n_1, n_2) \in E$ *and* $\mathcal{E}_{\rightarrow_T}^{n_1, n_2}$ *be the set of all inter-fragment directly-follows event pairs corresponding to* (n_1, n_2), *i.e.* $\mathcal{E}_{\rightarrow_T}^{n_1, n_2} = \{(e_1, e_2) \in \mathcal{E}_{\rightarrow_T} \mid \pi_U(e_1) = n_1 \wedge \pi_V(e_2) = n_2\}$.

$$
W_E((n_1, n_2)) = \begin{cases} |\mathcal{E}_{\rightarrow_T}^{n_1, n_2}| & \text{if frequency-based} \\ \dfrac{\sum_{(e_1, e_2) \in \mathcal{E}_{\rightarrow_T}^{n_1, n_2}} \left(timestamp(e_2) - timestamp(e_1)\right)}{|\mathcal{E}_{\rightarrow_T}^{n_1, n_2}|} & \text{if time} - based \end{cases}.
$$

Inter-Fragment Graphs represent a *flow relation* between event attributes. For example, it can be a hand-over from a resource in one department to another resource in a directly following department, or a flow from an activity executed in one location to another activity executed in a directly following location.

3.3 Comparing Perspective Graphs and Visualizing Differences

In comparing two perspective graphs, common nodes and edges on the two graphs are compared in terms of their weights. Note that the weights defined in Sect. 3.2 are computed for the whole log. Instead of comparing graphs based on these weights, this paper looks for statistically significant differences by comparing sample populations of weights obtained from log observations.

Different techniques can be used to make observations of logs. *Case-wise observations* are made on cases in the log, i.e. weights of nodes and edges are computed from events in each case. Differences determined by the tests can be understood as differences between the two variants synthesized from all cases. *Time-wise observation* allows one to see differences between logs over time. This technique uses a sliding time window starting from the earliest event in each log. Observations are made on each window, i.e. weights of nodes and edges are computed from events occurring within each window.

The result of graph comparison is a *differential graph* containing common nodes and edges and also uncommon nodes and edges that appear in one graph only. If nodes and edges are common with a statistically significant difference, their weight is the *effect size* of the difference.

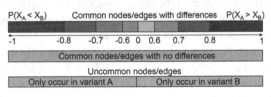

Fig. 2. Color scheme (Color figure online)

This paper chooses *common language effect size* [11] due to its interpret-ability. For example, an effect size of 80% indicates that given any random observations of the two variants, variant A has 80% chance of having a higher mean weight than variant B. If nodes or edges are common without a statistically significant difference, their weight is simply zero. Lastly, if they are uncommon, their weight is the relative weight among all uncommon nodes or edges in the graph. Differential graphs are visualised in the form of matrices (nodes are row and column headers while edges are cells). Matrices can be symmetric (for undirected graphs) as shown in Fig. 5 or asymmetric (for directed graphs) as shown in Fig. 4. Nodes and edges are color coded based on their weight and the *color scheme* shown in Fig. 2.

4 Evaluation

We implemented our approach as a ProM plugin named *Multi-Perspective Process Comparator* (MPC).[2] The plugin allows one to import two event logs in MXML or XES format as input, mine different perspective graphs and compare them to identify statistically significant differences. Using this implementation, we evaluated our approach on two real-life datasets and compared the results with the *ProcessComparator* (or PC) plugin in ProM [4].

We looked at the public real-life event logs available in the *4TU Data Center*[3] and selected two representative datasets, namely BPIC13 and BPIC15. These two datasets come with business questions that entail variants comparison, which have been posed by the process stakeholders of these datasets, as part of public contests on process mining. Due to space limits, we only report the result of our technique on the BPIC13 log focusing on aspects our technique can improve over the baseline. Detailed evaluation is documented in a technical report [12].

BPIC13[4] records cases of an IT incident handling process at Volvo Belgium. An IT ticket is raised for each incident to be investigated by various IT support teams. Teams are organized into technology-wide functions (*org:role* attribute), organization lines (*organization involved* attribute), and countries (*resource country* attribute). For our evaluation, we selected the following question from the description accompanying this dataset: "*Where do the two IT organisations (A2 and C) differ?*" where *A2* and *C* are the main organization lines responsible for most of the IT tickets.

For each dataset and business question above, we compared process variants using three sub-questions. With reference to Fig. 1, these questions are focused on two levels of granularity, event and fragment, and time-wise differences.

Q1. What are the Differences at the Event Level?

At the event level, we can look into either inter-event or intra-event relations where each event is a fragment. Regarding the former, both PC and MPC can provide the same insight. Specifically in the case of MPC, we can use the *event ID* attribute to create a process abstraction, then create an inter-fragment graph using the pair of *event name* and *status* attributes as nodes.

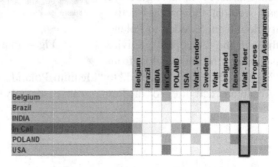

Fig. 3. Resource Country and Activity Status

[2] Executable and source code are available from http://apromore.org/platform/tools.

[3] https://data.4tu.nl/repository/collection:event_logs_real.

[4] doi:10.4121/uuid:500573e6-accc-4b0c-9576-aa5468b10cee.

However, regarding the intra-event relations, PC cannot provide a solution while MPC can investigate these relations through intra-fragment graphs. For example, on the event-based abstraction, we can use the *country* attribute as a node and the *activity status* attribute as another node. The matrix in Fig. 3 reveals that the teams in Brazil, India and the USA in the organization C choose the "Wait User" status for IT tickets more frequently than in the organization A2. This is an operational concern since IT staff can choose this status as an excuse to delay incident investigation.

Q2. What are the Differences at the Fragment Level?

In the BPIC13 dataset, we can create fragments using the *country* attribute to look into how process activities are related between IT teams from different countries. In this aspect, PC aggregates the activity flow between fragments. For example, PC shows a flow from "[Sweden]" to "[Poland]" through the "Accepted" activity meaning that this activity is performed by Sweden and then work is transferred to Poland. It may however consist of two possible flows: either "Accepted" by Sweden followed by activity "Queued" performed by Poland, or "Accepted" by Sweden followed by "Completed" performed by Poland.

Similarly to PC, MPC can look into the same flow by first using the *country* attribute to create the process abstraction, then choosing the pair of *country* and *event name* attributes as node and the *country* attribute as another node to create an inter-fragment graph. However, beyond that, MPC can elaborate the activity flow between countries by using other event attributes. Figure 4 shows an example where we chose *impact*, *country* and *event name* to represent a node. From this figure, we can see that the process activity flow from Sweden to Poland through the "Accepted" activity is actually the control flow from

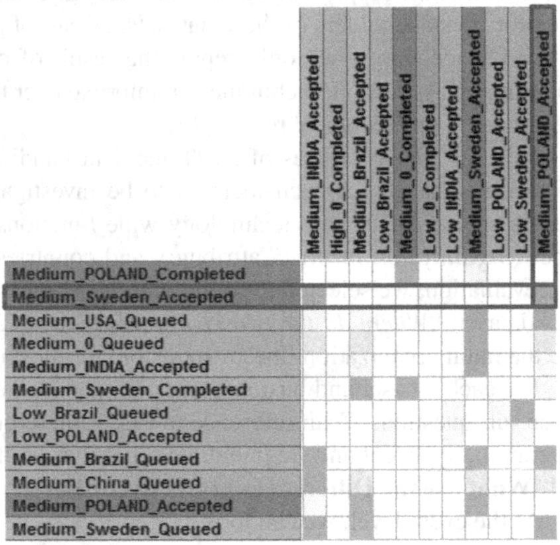

Fig. 4. Impact, Country, and Event Name

"Medium_Sweden_Accepted" to "Medium_Poland_Accepted" (i.e. from "Accepted" to "Accepted" activities for medium impact cases only).

Further, MPC can look into the differences between A2 and C within each fragment through intra-fragment graphs, while this is not possible with PC. For example, we use the *impact* attribute to create a process abstraction, and the pair of *impact* and *activity status* attributes as the node. The result is shown in Fig. 5 for medium impact incidents. Remarkably, we can see that for medium-impact incidents, most of activity statuses in A2 have approximately 60–70% chance of occurring more frequently than in C. There are no significant differences between the two organization lines in high-impact incidents.

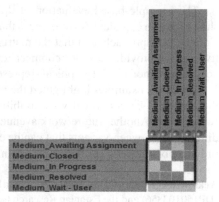

Fig. 5. Medium Impact and Activity Status

Q3. What are the Time-Wise Differences as Compared to Case-Wise Differences?

So far, the evaluation only finds case-wise differences, i.e. differences synthesized from cases in A2 and C. Time-wise observations, however, are not available in PC.

For MPC, we use time-wise observations with a sliding window set to three days as most of events in a case occur within a day. We use the *event ID* to create a process abstraction, and the pair of *event name* and *activity status* as node. In this case, the node (edge) weight captures their relative occurrence frequency in each window. The result is shown in Fig. 6. We can see that there are two remarkable differences between A2 and C over time in the node "Accepted_Wait-User" and the edge "Accepted_In Progress" →

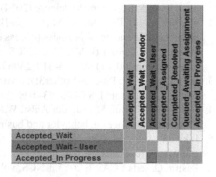

Fig. 6. Event Name and Activity Status

"Accepted_Wait-User". The difference magnitude is approximately 56%, i.e. there is a 56% probability that "Accepted_Wait-User" in A2 has lower frequency than in C. In MPC, clicking on the edge "Accepted_In Progress" → "Accepted_Wait-User" views detailed time series which shows that this difference mostly occurred between 21 Jan and 10 Mar 2012.

5   Conclusion

This paper contributes the notions of perspective graph and differential graph. A perspective graph is an abstraction of an event log in which nodes represent entities referenced by an event attribute or combination of attributes, and links refer to co-occurrence or directly-follows relations. Perspective graphs generalize directly-follows graphs and hand-off graphs, commonly supported by process mining tools. Differential perspective graphs allow us to compare two event logs (abstracted via perspective graphs) and to identify their statistically significant differences.

The example-based evaluation of differential perspective graphs on real-life logs shows that we can identify differences that are beyond the scope of the existing Process Comparator approach, and that the matrix-based representation of differential perspective graphs provides a more compact representation for displaying such differences, compared to node-link (graphical) representations used in process mining tools.

While the examples highlighted the possible advantages of the proposed approach, these need to be confirmed via a usability evaluation with end users, which is left as future work. Another future work avenue is to extend the approach in order to identify differences between variants that can be causally related to performance, e.g. structural or behavioral differences that can explain differences in cycle time between variants.

Acknowledgements. This research is partly funded by the Australian Research Council (DP150103356) and the Estonian Research Council.

References

1. Nguyen, H., Dumas, M., La Rosa, M., Maggi, F.M., Suriadi, S.: Mining business process deviance: a quest for accuracy. In: Meersman, R., et al. (eds.) OTM 2014. LNCS, vol. 8841, pp. 436–445. Springer, Heidelberg (2014). https://doi.org/10.1007/978-3-662-45563-0_25
2. van Beest, N.R.T.P., Dumas, M., García-Bañuelos, L., La Rosa, M.: Log delta analysis: interpretable differencing of business process event logs. In: Motahari-Nezhad, H.R., Recker, J., Weidlich, M. (eds.) BPM 2015. LNCS, vol. 9253, pp. 386–405. Springer, Cham (2015). https://doi.org/10.1007/978-3-319-23063-4_26
3. Wynn, M., et al.: ProcessProfiler3D: a visualisation framework for log-based process performance comparison. DSS **100**, 93–108 (2017)
4. Bolt, A., de Leoni, M., van der Aalst, W.M.P.: Process variant comparison: using event logs to detect differences in behavior and business rules. Inf. Syst. **74**, 53–66 (2018)
5. Pika, A., van der Aalst, W.M.P., Fidge, C.J., ter Hofstede, A.H.M., Wynn, M.T.: Profiling event logs to configure risk indicators for process delays. In: Salinesi, C., Norrie, M.C., Pastor, Ó. (eds.) CAiSE 2013. LNCS, vol. 7908, pp. 465–481. Springer, Heidelberg (2013). https://doi.org/10.1007/978-3-642-38709-8_30
6. Gulden, J.: Visually comparing process dynamics with rhythm-eye views. In: Dumas, M., Fantinato, M. (eds.) BPM 2016. LNBIP, vol. 281, pp. 474–485. Springer, Cham (2017). https://doi.org/10.1007/978-3-319-58457-7_35
7. Pika, A., Wynn, M.T., Fidge, C.J., ter Hofstede, A.H.M., Leyer, M., van der Aalst, W.M.P.: An extensible framework for analysing resource behaviour using event logs. In: Jarke, M., et al. (eds.) CAiSE 2014. LNCS, vol. 8484, pp. 564–579. Springer, Cham (2014). https://doi.org/10.1007/978-3-319-07881-6_38
8. Ballambettu, N.P., Suresh, M.A., Bose, R.P.J.C.: Analyzing process variants to understand differences in key performance indices. In: Dubois, E., Pohl, K. (eds.) CAiSE 2017. LNCS, vol. 10253, pp. 298–313. Springer, Cham (2017). https://doi.org/10.1007/978-3-319-59536-8_19
9. Bolt, A., van der Aalst, W.M.P.: Multidimensional process mining using process cubes. In: Gaaloul, K., Schmidt, R., Nurcan, S., Guerreiro, S., Ma, Q. (eds.) CAISE 2015. LNBIP, vol. 214, pp. 102–116. Springer, Cham (2015). https://doi.org/10.1007/978-3-319-19237-6_7
10. IEEE standard for eXtensible Event Stream (XES) for achieving interoperability in event Logs and event streams. IEEE Std 1849–2016, pp. 1–50 (2016)

11. McGraw, K.O., Wong, S.P.: A common language effect size statistic. Psychol. Bull. **111**(2), 361–365 (1992)
12. Nguyen, H., Dumas, M., La Rosa, M., ter Hofstede, A.H.M.: Multi-perspective comparison of business process variants based on event Logs (extended paper). QUT ePrints Technical report, Queensland University of Technology (2018). http://eprints.qut.edu.au/117962

NoSQL Modeling

Managing Polyglot Systems Metadata with Hypergraphs

Moditha Hewasinghage[1][✉], Jovan Varga[1], Alberto Abelló[1],
and Esteban Zimányi[2]

[1] Universitat Politècnica de Catalunya, BarcelonaTech, Barcelona, Spain
{moditha, jvarga, aabello}@essi.upc.edu
[2] Université Libre de Bruxelles, 1050 Bruxelles, Belgium
ezimanyi@ulb.ac.be

Abstract. A single type of data store can hardly fulfill every end-user requirements in the NoSQL world. Therefore, polyglot systems use different types of NoSQL datastores in combination. However, the heterogeneity of the data storage models makes managing the metadata a complex task in such systems, with only a handful of research carried out to address this. In this paper, we propose a hypergraph-based approach for representing the catalog of metadata in a polyglot system. Taking an existing common programming interface to NoSQL systems, we extend and formalize it as hypergraphs for managing metadata. Then, we define design constraints and query transformation rules for three representative data store types. Furthermore, we propose a simple query rewriting algorithm using the catalog itself for these data store types and provide a prototype implementation. Finally, we show the feasibility of our approach on a use case of an existing polyglot system.

Keywords: Metadata management · NoSQL · Polystore

1 Introduction

With the dawn of the big data era, the heterogeneity among the data storage models has expanded drastically, mainly due to the introduction of NoSQL. There are four primary data store models in NoSQL systems: (i) Key-value stores, which perform like a typical hashmap, where the data is stored and retrieved through a key and an associated value; (ii) Wide-column stores, that manage the data in a columnar fashion; (iii) Document stores, represent data in a document like structure, which can become increasingly complex with nested elements; (iv) Graph stores, are instance based and store the relationships between those instances. The heterogeneity is not only limited to the data models, but also various implementations of the same data model can be entirely different from one to another due to the lack of a standard.

Heterogeneous systems can be useful in different scenarios, because it is highly unlikely that a single data store can efficiently handle all the requirements of the end-user. Therefore, it is common to use different ones to manage

J. C. Trujillo et al. (Eds.): ER 2018, LNCS 11157, pp. 463–478, 2018.
https://doi.org/10.1007/978-3-030-00847-5_33

different portions of the data. This allows to control the storage and retrieval more efficiently for different requirements. Hence, polyglot systems were introduced, similar to traditional Federated Database Systems (FDBMS), but more complex considering the need to handle semistructured data models. Due to the heterogeneity at different levels, most of the work on polyglot systems [4, 7] suggests the implementation of wrappers or interfaces for each participating data store. However, this becomes more complex as the number of participating data store types grows.

The catalog (see [9]) maintains the meta information of the data store. Having one for a polyglot system enables end users to have a clear view of the complex system. Its metadata plays a significant role in understanding the overall picture of the underlying infrastructure and, as well, helps to improve the design of the polyglot system and determine the access patterns needed for different query requirements. It is essential to answer questions such as: What is the structure of the data being stored? Where is a piece of data stored? Is it duplicated in another store? What is the best way to retrieve this data? Nevertheless, little research addresses the managing of metadata in polyglot systems. This is mainly due to the lack of a design construct that can represent heterogeneous, semistructured data. In this paper, we address the metadata management in polyglot systems by extending an already existing NoSQL design method [2,3] and formalizing the constructs through hypergraphs.

The SOS Model [2] claims to capture the NoSQL modeling structures in data design for key-value stores, document stores, and wide-column stores utilizing three main constructs: attributes, structs, and sets. These constructs and their interactions allow to represent the physical storage of above NoSQL systems. The fact that the model is simple makes it compelling in representability, but the lack of formalization leaves space for ambiguity and hinders the automation of metadata management in such settings. Instead, it is simply used as a common programming interface for data exchange.

In this paper, we formalize SOS using a hypergraph-based representation, and extend it to maintain the catalog that manages the metadata of a polyglot system. RDF is considered to be able to represent any kind of data and is often used as a data interchange format. Therefore, we make the assumption that we have exemplars of the data in the polyglot system in RDF. Then, we build a hypergraph that maps to different data design constructs, representing the SOS model over the information. We represent the catalog of the polyglot system using these constructs, and introduce a simple query generation algorithm to show the usefulness of our approach. Next, we explore different data store models, identify their design constructs, introduce their design constraints, and define query generation rules for each of them. Finally, we show the feasibility of our approach using a use case of an existing polyglot system by representing its metadata catalog through our constructs.

The simple, yet powerful hypergraph-based approach presented in this paper is a step towards representing heterogeneous, semistructured data in a formal manner as well as managing the corresponding metadata of a polyglot system. It proves to be useful concerning (i) expressiveness: the ability to express different representations, regardless of their complexity and (ii) semantic relativism: the ability to accommodate different representations of the same data, as defined in [18].

This paper is organized as follows: First, in Sect. 2, we introduce some background. We present and formalize our data model in Sect. 3. Afterwards, we discuss the managing of the metadata through the model in Sect. 4. Then, we present an application of the constructs in Sect. 5. Finally, we introduce the related works in Sect. 6, and conclude our work with future work in Sect. 7.

2 Preliminaries

In this section, we introduce the basic concepts about RDF [14] and SOS Model [2,3] that will be used in our approach.

2.1 Resource Description Framework (RDF)

The Resource Description Framework [14] is a Wide Web Consortium (W3C) specification for representing information on the Web. It is a graph-based data model that helps to serve information by making statements about available resources.

RDF represents data as triplets consisting of subject, predicate, and object (s, p, o). These can be resources that are identified by an Internationalized Resource Identifier (IRI), which is a unique Unicode string within the RDF graph. An object can also be a literal, which is a data value. An example of RDF is shown in Listing 2.1, written in Turtle notation. The example contains information about music albums, artists, and songs and is used throughout the paper to illustrate our approach.

```
@prefix foaf: <http://xmlns.com/foaf/0.1> .
@prefix xsd: <http://www.w3.org/2001/XMLSchema#> .
@prefix mo: <http://purl.org/ontology/mo/> .
@prefix dc: <http://purl.org/dc/elements/1.1/> .
@prefix rdf: <http://www.w3.org/1999/02/22-rdf-syntax-ns#> .

<http://dbtune.org/jamendo/artist/dylan> rdf:type mo:MusicArtist;
    foaf:name "Bob Dylan"^^xsd:string;
    foaf:made <http://dbtune.org/jamendo/record/emp>.

<http://dbtune.org/jamendo/record/emp> rdf:type mo:Record;
    dc:title "Empire Burlesque"^^xsd:string;
    foaf:maker <http://dbtune.org/jamendo/artist/dylan> ;
    mo:track <http://dbtune.org/jamendo/track/Seeing>, <http://dbtune.org/
        jamendo/track/Tight> .

<http://dbtune.org/jamendo/track/Tight> rdf:type mo:Track;
    dc:title "Tight Connection to My Heart"^^xsd:string .
<http://dbtune.org/jamendo/track/Seeing> rdf:type mo:Track;
    dc:title "Seeing the Real You at Last"^^xsd:string .
```

Listing 2.1. Example RDF dataset (http://dbtune.org)

2.2   SOS Model

The high flexibility of NoSQL systems gives the freedom to have multiple designs for the same data. A particular data design built focusing on a specific scenario can result in adverse performance when applied in a different context. Most of the data design for NoSQL is carried out based on concrete guidelines for different datastores and access patterns. Nevertheless, recent approaches propose generic design constructs for NoSQL systems. For our approach, we decided to use the SOS model [2,3] as a starting point.

The SOS model introduces a basic common model (or a meta-layer) which is a high-level description of the data models of non-relational systems. This model helps to handle the vast heterogeneity of the NoSQL datastores and provides interoperability among them, easing the development process. The primary objective of the meta-layer is to generalize the data model of heterogeneous NoSQL systems. Thus, it allows standard development practices on a predefined set of generic constructs. The meta-layer reconciles the descriptive elements of key-value stores, document stores, and record stores. The different data models exposed by NoSQL datastores are effectively managed in the SOS data model with three major constructs: *Attribute*, *Struct*, and *Set* [3].

A name and an associated value characterize each of these constructs. The structure of the value depends on the type of construct. An *Attribute* can contain a simple value such as an Integer or String. *Structs* and *Sets* are complex elements which can contain multiple *Attributes*, *Structs*, *Sets* or a combination of those. SOS Model mainly addresses data design on document stores, key-value stores, and wide-column stores [2]. Each of the datastore instances is represented as a set of collections. There can be any arbitrary number of *Sets* depending on the use case. Simple elements such as key-value pairs or single qualifiers can be modeled as *Attributes* and groups of *Attributes*, or a simple entity such as a document, can be represented as a *Struct*. A collection of entities is represented in a *Set*, which can be a nested collection in a document store or a column family in a wide-column store.

3   Formalization

In this section, we introduce and formalize our data model, which is based on representative exemplars in RDF format of each kind of instances in the underlying data stores. This RDF graph contains the classes and user-defined types of the polyglot system. Previous work has shown that this global schema can be obtained by extracting the schema of each data store and reconciling them [8,20].

Building on top of the RDF data model discussed in Sect. 2.1, we introduce our design constructs based on the SOS Model. Figure 1 shows the overall class diagram of our constructs, where thicker lines represent the elements already available in the SOS (namely *Sets*, *Structs*, and *Attributes*), and the relationship multiplicities defined in SOS are preserved. On top of that, we introduce additional constructs to aid the formalization process and manage metadata.

From here on, we use letters in blackboard font to represent sets of elements (e.g., $\mathbb{A} = \{A_1, A_2 ... A_n\}$).

We rely on the concept of hypergraph, which is a graph where an edge (aka hyperedge) can relate any number of elements (not only two). This can be further generalized so that hyperedges can also contain other hyperedges (not only nodes).

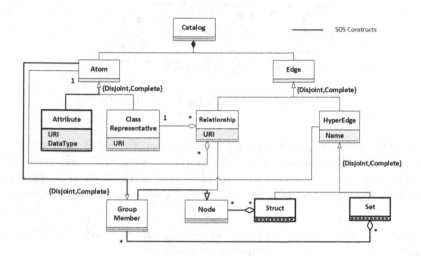

Fig. 1. Class diagram for the overall catalog

We define the overall polyglot system catalog as composed by the schema and the essential elements that support a uniquely accessible terminology for the polyglot system.

Definition 1. *A polyglot catalog $C = \langle \mathbb{A}, \mathbb{E} \rangle$ is a generalized hypergraph where $\mathbb{A}$ is a set of atoms and $\mathbb{E}$ is a set of edges.*

Definition 2. *The set of all atoms $\mathbb{A}$ is composed of two disjoint subsets of class atoms $\mathbb{A}_C$ and attribute atoms $\mathbb{A}_A$. Formally: $\mathbb{A} = \mathbb{A}_C \cup \mathbb{A}_A$*

Atoms are the smallest constituent unit of the graph and carry a name. Moreover, every A_C contains a URI that represents the class semantics, while every A_A carries the datatype and a URI for the user-defined type semantics.

Definition 3. *The set of all edges $\mathbb{E}$ composed of two disjoints subsets of relationships $\mathbb{E}_R$ that denote the connectivity between $\mathbb{A}$, and hyperedges $\mathbb{E}_H$ that denotes connectivity between other constructs of C. Formally: $\mathbb{E} = \mathbb{E}_R \cup \mathbb{E}_H$*

Definition 4. *A relationship $E_R^{x,y}$ is a binary edge between two atoms A_x and A_y and a URI u that represents the semantics of E_R. At least one of the atoms in the relationship must be an A_C. Formally: $E_R^{x,y} = \langle A^x, A^y, u \rangle | A^x, A^y \in \mathbb{A} \wedge (A^x \in \mathbb{A}_C \vee A^y \in \mathbb{A}_C)$*

This graph $G = \langle \mathbb{A}, \mathbb{E}_R \rangle$ is a representation of the available data, i.e., an RDF translation of the original representatives of the data contained in the polyglot system, that we assume to be given. This G immutable as it contains the knowledge about the data. Figure 2 shows the graph G of the original RDF example in Listing 2.1.

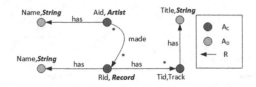

Fig. 2. Translated graph built from the RDF

We build our data design on top of G, based on the constructs introduced in SOS model. Thus, we make use of the *Hyperedges* in G and give rise to our hypergraph-based catalog C. An incidenceSet of an *Atom* or a *Hyperedge* contains the immediate set of E that *Atom* or *Hyperedge* is part of, respectively.

Definition 5. *The transitive closure of an edge E is denoted as E^+, where $E \in E^+$, $\forall e \in E^+ : e \in e'.incidenceSet \implies e' \in E^+$*

Definition 6. *A hyperedge E_H is a subset of atoms $\mathbb{A}$ and edges $\mathbb{E}$ and it cannot be transitively contained in itself. Formally: $E_H \subseteq \mathbb{A} \cup \mathbb{E} \wedge E_H.incidenceSet \cap E_H^+ = \emptyset$*

Definition 7. *A struct E_{Struct} is a hyperedge that contains a set of atoms $\mathbb{A}$, relationships $\mathbb{E}_R$, and/or hyperedges $\mathbb{E}_H$. All atoms within E_{Struct}^+ must be connected by a set of E_R that also belong to E_{Struct}^+. Formally : $E_{Struct} \subseteq \mathbb{E}_H \cup \mathbb{A} \cup \mathbb{E}_R | \forall A^x, A^y \in E_{Struct}^+ : \exists \{E_R^{x,x_1}, E_R^{x_1,x_2}, .., E_R^{x_n,y}\} \in \mathbb{E}_{Struct}^+$*

Definition 8. *A set E_{Set} is a hyperedge that contains a set of arbitrary hyperedges $\mathbb{E}_H$ or/and atoms $\mathbb{A}$. Formally: $E_{Set} \subseteq \mathbb{E}_H \cup \mathbb{A}$*

For the ease of illustration, we have also used two additional constructs in Fig. 1: A group member $M \in \mathbb{A} \cup \mathbb{E}_H$ is an element of an E_{Set}. In this context, it is an atom or a hyperedge; A node $N \in \mathbb{M} \cup \mathbb{E}_R$ is an element of an E_{Struct}, which is either an M or an E_R.

4   Metadata Management

One crucial aspect of a metadata management system is the ability to represent different data store models. In our work, we exemplify it on traditional RDBMS, document stores, and wide-column stores. Figure 3 extends our original diagram of constructs to support those.

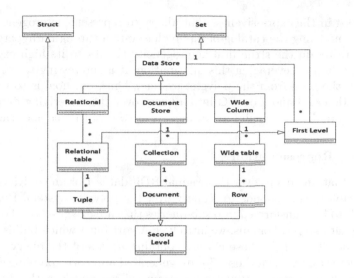

Fig. 3. Class diagram for metadata with *Hyperedge* composition

The E_{Set} are specialized into two types: *Data Store* E_D and *First Level* E_F. E_D represents a data store instance of the polyglot system. E_F denotes a collection of instances in the particular data store. All the allowed kinds of data stores are a subclass of E_D. Moreover, $E_D.incidenceSet = \emptyset$.

Thus, we define three E_D (namely *Relational* E_D^{Rel}, *Document Store* E_D^{Doc}, and *Wide-Column* E_D^{Col} in Fig. 3), which are the participants of our polyglot system. There can be multiple E_F within each E_D adhering to the number of collections that participate in the polyglot system.

All the *Atoms* and *Edges* of the polyglot system belong to the transitive closure of one or more of these E_D, $\mathbb{A} \cup \mathbb{E} = \bigcup E_D^+$. Therefore, we can deduce that the entire polyglot system catalog C can be represented by the participating E_D .

The *Second Level* (E_S), is a *Struct* that represents a kind of object residing directly in E_F. These E_S should align with the type of E_D where it is contained. It is a tuple for E_S^{Rel}, a document stored directly in the collection for E_S^{Doc}, and a row for E_S^{Col}.

Each E_H carries a name which is interpreted depending on the context. In E_D, it can represent the physical location of the underlying data store. In E_F, it can be the collection name or table name. Depending on the type of E_D that represents the data store, we can identify specific constraints and transformation rules for the queries over the representatives.

The *Edges* of the catalog can carry much more information than just the name. For example, an E_R can indicate the multiplicity between *Atoms*. Likewise, an E_H can carry information like the size of a collection, percentage of null values, maximum, minimum, and average of values. However, in this work, we focus mainly on the structural metadata. This catalog differs from a relational catalog in the expressiveness that allows to represent heterogeneous data models. By modeling the catalog of a polyglot system through a hypergraph, it is possible to retain the structural heterogeneity thanks to its high expressiveness and flexibility. Leveraging this information, it is interesting to see how we can retrieve the data from the polyglot system. Thus, our goal is to transform the formulation of a query over G into a query over the underlying data stores. Inter-data store data reconciliation or merges are out of the scope of this paper.

4.1 Query Representation

We assume that any query over the original RDF dataset or an equivalent query over the graph G corresponds to a query over the polyglot system. Hence, this query needs to be transformed into sub-queries that are executed on the relevant underlying data stores. For this, we introduce Algorithm 1 which builds an adjacency list for all the E_H whose closure contains $Atoms$ of the query, aided by the incidence sets. Once all those E_H are generated and corresponding E_D identified, a simple projection query can be composed recursively with Algorithm 2 for each of the E_F according to different rules depending on the kind of data store. Algorithm 2 uses a prefix, a suffix and the path relevant for each of the constructs of the data stores.

We are only generating projection queries, but selections can be considered a posteriori by pushing down the predicates over the query path. Also, there can be cases where the same information is available in multiple data stores. If this happens, this overlap can be easily identified as the considered $Atom$ will be contained in more than one E_D^+. Then a join should be performed in the corresponding mediator.

Algorithm 1. Query over polyglot system algorithm

Input: A Subgraph of G including the query Atoms and Relationships
Output: A set of multi language queries $\mathbb{Q}$ corresponding to data store queries
1: $\mathbb{Q} \leftarrow \emptyset$
2: $M \leftarrow newHashmap() < N, Set >$
3: $Q \leftarrow newQueue()$
4: **for each** $Atom\ a \in G$ **do**
5:     **for each** $E_H\ i \in a.incidenceSet$ **do**   // hyperedges containing an Atom
6:         $Q.enqueue(< i, a >)$
7:     **end for**
8: **end for**
9: **while** $Q \neq \emptyset$ **do**
10:     $temp \leftarrow Q.dequeue$
11:     $current \leftarrow temp.first$
12:     $M.addToSet(current, temp.second)$ // adds the second parameter to the set
13:     **for each** $E_H\ j \in current.incidenceSet$ **do**
14:         $Q.enqueue(< j, current >)$
15:     **end for**
16: **end while**
17: **for each** $E_F\ f \in M.keys$ **do**
18:     $\mathbb{Q}.add(CreateQuery(f, ""))$
19: **end for**
20: **return** $\mathbb{Q}$

Algorithm 2. Create Query algorithm

Input: *source E_H, path of E_H* (adjacency list M from Algorithm 1 is also available)
Output: A data store query q
1: $q \leftarrow prefixOf(source, path)$
2: **for each** *child* $\in M.get(source)$ **do**
3:    $q \leftarrow q + CreateQuery(child, pathOf(source))$
4: **end for**
5: $q \leftarrow q + suffixOf(source)$
6: **return** q

4.2 Constraints and Transformation Rules on Data Stores

Considering the constructs and the query generation algorithm mentioned earlier, each of the data store models would have its own rules and constraints on
the data. Therefore, in this section, we analyze the constraints and transformation rules for 3 of them: relational stores, document stores, and wide-column
stores.

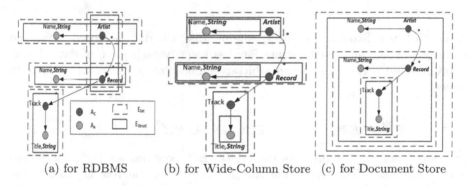

(a) for RDBMS   (b) for Wide-Column Store   (c) for Document Store

Fig. 4. Alternative data design representations

Relational Database Management Systems. A typical example of the type
of data design in RDBMS is shown in Fig. 4a. The constraints and the mappings
on E_H can be represented in a grammar as follows:

$$E_D^{Rel} \implies E_F^{Rel} *, \ E_F^{Rel} \implies E_S^{Rel}, \ E_S^{Rel} \implies A_C A *$$

A traditional RDBMS data storage system consists of tables, tuples, and simple
attributes. The data store can have multiple tables, which are represented by
E_F^{Rel}. Within a table, the schema of the tuple E_S^{Rel} is fixed. Therefore, there
can only be a single E_S^{Rel} inside a E_F^{Rel}. Finally, the tuple contains at least
one A_C, which is the primary key, or multiple A_C in case of compound keys.
The E_H containing an E_R that crosses two E_F^{Rel} (only one of both) is the one
corresponding to the relation that has the attribute.

The RDBMS design of Fig. 4a represents the following tables:

$Artist[A\_id, name], Record[R\_id, name], Artist\_Record[A\_id(FK), R\_id(FK)],$
$Track[T\_id, title, R\_id(FK)].$

The following prefix and suffix are used in Algorithm 2 to generate the corresponding queries. Note that $deleteComma()$ is an operation that deletes the trailing comma of a string (Table 1).

Table 1. Symbols for Algorithm 2 in RDBMS

Symbol	Prefix	Suffix	Path
E_F^{Rel}		$"FROM" + E_F^{Rel}.name$	
E_S^{Rel}	$"SELECT"$	$deleteComma()$	
A	$A.name$	$","$	

Wide-Column Stores. In a wide-column store, the data is stored in vertical partitions. A key and fixed column families identify each piece of data. Inside a column family, there can be an arbitrary number of qualifiers which identify values.

$$E_D^{col} \implies E_F^{col} {}^*, \; E_F^{col} \implies E_S^{col}, \; E_S^{col} \implies A_C E_{Struct}^{col} {}^+, \; E_{Struct}^{col} \implies A^+$$

The outer most E_F^{col} represents the tables. E_F^{col} contains an E_S^{col}, which represents the rows. The A_C inside this E_S^{col} becomes the row key. E_S^{col} contains several E_{Struct}^{col}, which represent the column families. The A inside the E_{Struct}^{col} represents the different qualifiers. The relationships between A_C can be represented as reverse lookups. In our example scenario, the hypergraph in Fig. 4b contains one column family per table. This can be mapped into $Artist[A\_id, [name, \{R\_id\}]], Record[R\_id, [name]], Track[T\_id[title, R\_id]]$.

Wide-column stores generally support only simple get and put queries and require the row key to retrieve the data. We use HBase query structure to demonstrate the capability of simple query generation. Table 2 depicts the translation rules for simple queries in wide-column stores used in Algorithm 2.

Document Stores. Document stores have the least constraints when it comes to the data design. They enable multiple levels of nested documents and collections within. Figure 4c shows a document data store design of our example scenario. The constraints and mappings in a document store design are as follows:

$$E_D^{Doc} \implies E_F^{Doc} {}^*, \; E_F^{Doc} \implies E_S^{Doc} {}^+, \; E_S^{Doc} \implies A_C (A|E_{Set}^{Doc}|E_{Struct}^{Doc})^*,$$
$$E_{Set}^{Doc} \implies E_{Struct}^{Doc} {}^+, \; E_{Struct}^{Doc} \implies (A|E_{Set}^{Doc}|E_{Struct}^{Doc})^+$$

Table 2. Symbols for Algorithm 2 in Wide-Column Stores

Symbol	Prefix	Suffix	Path
E_F^{Col}	"$scan'$ " $+ E_F^{Col}.name +$ " ', $\{COLUMNS => [$"		
E_S^{Col}		$deleteComma() + $ "$]\}$"	
E_{Struct}^{Col}			$path + E_{Struct}^{Col}.name +$ "$.$"
A	"'" $+ path + A.name + $ "'"	","	

E_F^{Doc} represents the collections of the document store. E_S^{Doc} inside the E_F^{Doc} represents the documents within the collection, which must have an identifier A_C. Apart from that, E_F^{Doc} can have *Atoms*, or E_S^{Doc}, which represents nested collections, or E_{Struct}^{Doc}, or a combination of any of them. E_{Set}^{Doc} represents a nested collection which contains documents E_{Struct}^{Doc}.

The design in Fig. 4c can be mapped into a document store design as $Artist\{A\_id :, name :, Records : [\{R\_id :, name :, Tracks : [\{T\_id :, title :\}]\}]\}$.

We use MongoDB syntax for the queries as it is one of the most popular document stores at the moment. Table 3 identifies the symbols for the queries.

Table 3. Symbols for Algorithm 2 in Document Stores

Symbol	Prefix	Suffix	Path
E_F^{Doc}	"$db.$" $+ E_F^{Doc}.name +$ "$.find(\{\}$",$\{$	"$\})$"	
E_S^{Doc}		$deleteComma()$	
$E_{Struct/Set}^{Doc}$			$path +$ $E_{Struct/Set}^{Doc}.name + $ "$.$"
$A(path \neq \varnothing)$	"'" $+ path + A.name + $ "' : 1"	","	
$A(path = \varnothing)$	$A.name + $ "$: 1$"	","	

Other Data Stores. As discussed above, we have managed to model and infer constraints and transformation rules for RDBMS, wide-column stores, and document stores which cover most of the use cases. However, it is also interesting to see the capability of the approach to represent other data stores. Since our model is based on graphs, we can simply conclude that it can express graph data stores. We only need to map the data into G, and define a single set with all *Atoms*. The key-value stores do not have sophisticated data structures, and it can be considered as a single column in a column family. Thus, since we are disregarding the storage of complex structures in the values that are not visible to the data store itself, we can state that our model covers key-value stores as well.

5   Use Case

In this section, we showcase our technique applied on an already available polyglot system. We base the example on the scenario used for ESTOCADA [5]. This involves a typical transportation data storage for a digital city open data warehousing. It uses RDBMS, document stores, and key-value stores. Figure 5 shows the graph G corresponding to ESTOCADA use case.

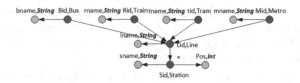

Fig. 5. Graph representation of ESTOCADA

The ESTOCADA system is used to store train, tram, and metro information in a RDBMS, the train and metro route information in a document store, and bus route together with the buses information in a key-value store (see [5] for more details). This information can be represented in our polyglot catalog[1] as follows (shown as containment sets):

$$C = \{E_D^{Rel}, E_D^{Kv}, E_D^{Doc}\}$$

$$E_D^{Rel} = \{E_{F\_Train}^{Rel}, E_{F\_Merto}^{Rel}, E_{F\_Tstat}^{Rel}, E_{F\_Mstat}^{Rel}, E_{F\_Station}^{Rel}\},$$

$$E_{F\_Train}^{Rel} = \{E_{S\_Train}^{Rel}\}, \quad E_{S\_Train}^{Rel} = \{A_{C\_rid}, A_{A\_rname}\},$$

$$E_{F\_Metro}^{Rel} = \{E_{S\_Metro}^{Rel}\}, \quad E_{S\_Metro}^{Rel} = \{A_{C\_mid}, A_{A\_mname}\},$$

$$E_{F\_Tstat}^{Rel} = \{E_{S\_Tstat}^{Rel}\}, \quad E_{S\_Tstat}^{Rel} = \{A_{C\_rid}, A_{C\_sid}, A_{A\_pos}\},$$

$$E_{F\_Mstat}^{Rel} = \{E_{S\_Mstat}^{Rel}\}, \quad E_{S\_Mstat}^{Rel} = \{A_{C\_mid}, A_{C\_sid}, A_{A\_pos}\},$$

$$E_{F\_Station}^{Rel} = \{E_{S\_Station}^{Rel}\}, \quad E_{S\_Station}^{Rel} = \{A_{C\_sid}, A_{A\_sname}\},$$

$$E_D^{Doc} = \{E_{F\_Metros.Trams}\}, \quad E_{F\_Metros.Trams}^{Doc} = \{E_{S\_Metros.Trams}^{Doc}\},$$

$$E_{S\_Metros.Trams}^{Doc} = \{A_{C\_mtid}, A_{A\_lname}, E_{Set\_route}^{Doc}\},$$

$$E_{Set\_route}^{Doc} = \{E_{Struct\_Station}^{Doc}\}, E_{Struct\_Station}^{Doc} = \{A_{A\_sname}\},$$

$$E_D^{Kv} = \{E_{F\_Station}^{Kv}, E_{F\_Bus}^{Kv}\}, \quad E_{F\_Station}^{Kv} = \{E_{S\_Station}^{Kv}\},$$

$$E_{S\_Station}^{Kv} = \{A_{A\_sname}, E_{Set}^{Kv\_loc}\}, \quad E_{Set}^{Kv\_loc} = \{A_{A\_lname}\}, E_{F\_Bus}^{Kv} = \{E_{S\_Bus}^{Kv}\},$$

$$E_{S\_Bus}^{Kv} = \{A_{A\_lname}, E_{Set}^{Kv\_st}\}, \quad E_{Set}^{Kv\_st} = \{A_{A\_sname}\}$$

---

[1] The implementation of the catalog is available in https://git.io/vxyHO.

Our goal was to store the metadata of the ESTOCADA polyglot system with a hypergraph. Thus, we used HyperGraphDB[2] to save the entire catalog information including the *Atoms*, *Relationships*, and *Hyperedges* for the structures. With this catalog, one can quickly detect where each fragment of the polyglot system lies by merely referring to *Hyperedges* and the content within.

6   Related Work

There are few polyglot systems already available to support heterogeneous NoSQL systems. In BigDAWG [7] different data models are classified as islands. Each of the islands has a language to access its data, and the data stores provide a shim to the respective islands it supports for a given query. Cross-island queries are also allowed, provided appropriate query planning and workload monitoring. Contrastingly, ESTOCADA [4,5] enables the end user to pose queries using the native format of the dataset. In this case, the storage manager fragments and stores the data in different underlying data stores by analyzing the access patterns. These fragments may overlap, but the query executor decides the optimal storage to be accessed. ODBAPI [19] introduces a unified data model and a general access API for NoSQL and heterogeneous NoSQL systems. This approach supports simple CRUD operations over the underlying systems, as long as they provide an interface adhering to the global schema. SQL++ [16] introduces a unifying query interface for NoSQL systems as an extension of SQL to support complex constructs such as maps, arrays, and collections. This is used as the query language for the FORWARD middleware that unifies structured and non-structured data sources. [21] uses a similar approach to SQL++. Katpathikotakis et al. [13] introduces a monoid comprehension calculus-based approach which supports different data collections and arbitrary nestings of them. Monoid calculus allows transformations across data models and optimizable algebra. This enables the translation of queries into nested relational algebra, that can be executed in different data stores through native queries.

Using different adapters or drivers for heterogeneous data stores is a common approach used in polyglot systems. The aforementioned systems use the same principal. Liao et al. [15] uses adapter-based approach for RDBMS and HBase. The authors introduce a SQL interface to RDBMS and NoSQL system, a DB converter that transforms the information with table synchronization, and three mode query approach which provides different policies on how applications access the data. The Spring framework [12] is one of the most popular softwares used to access multiple data stores, as it supports different types by using specific drivers and a common access interface. Apache Gremlin [17] and Tinkerpop[3] follow a similar approach but particularly for graph data stores. The main drawback in this approach is that each and every implementation needs to adhere to a common interface, which is difficult due to the vast number of available data stores.

---

[2] http://www.hypergraphdb.org.
[3] http://tinkerpop.apache.org.

Standalone data stores have their own metadata catalogs. For example, HBase uses HCatalog[4] (for hive) to maintain the metadata. They are strictly limited to the respective data models involved. In our work, we introduce a catalog to handle heterogeneous data models. MongoDB, on the contrary, does not maintain any metadata but rather handles the documents themselves (without any schema information). But some work has been carried out in managing schema externally [22].

Several work has been carried out on data design methodologies for NoSQL systems. NoSQL abstract model (NoAM) [1,6] is designed to support scalability, performance, and consistency using concepts of collections, blocks and entries. This model organizes the application data in aggregates. It defines the four main activities: conceptual modeling, aggregate design, aggregate partitioning, and implementation. The aggregate storage in the target systems is done depending on the data access patterns, scalability and consistency needs. Similarly, a general approach for designing a NoSQL system for analytical workloads has been presented in [10]. It adapts the traditional 3-phase design methodology of conceptual, logical, and physical design, and integrates the relational and co-relational models into a single quantitative method. At the conceptual level, the traditional ER diagram is used and transformed into an undirected graph. Nodes denote the entities, and the edges represent their relationships, tagged with the relationship type (specialization, composition, and association). In cases where an entity can become a part of several different hypernodes, it is replicated in each of them. The Concept and Object Modeling Notation (COMN) introduced in [11] covers the full spectrum of not only the data store, but also the software design process. COMN is a graphical notation capable of representing the conceptual, logical, physical and real-world design of an object. This helps to model the data in NoSQL systems where the traditional Entity-Relationship (ER) diagrams fail in representing certain situations, such as nesting.

7   Conclusions and Future Work

In this paper, we introduced a hypergraph-based approach for managing the metadata of a polyglot system. We based our work on an already existing data exchange model (SOS). First, we formalized the design constructs, and extended them to support metadata management through a catalog. Next, we defined the constraints and the rules for simple query generation on heterogeneous data store models. Finally, we implemented a simple use case to showcase our approach on an existing polyglot system. We showed the expressiveness of hypergraphs, which is the essential feature of a canonical model of a federated system [18]. Then, effectively introduced a metadata management approach for polyglot systems leveraging this expressiveness.

In addition to the structural metadata, the statistical metadata can be easily included by extending the information kept in the *Hyperedges*. Moreover, this work allows us to identify restrictions and limitations of different underlying

---

[4] https://cwiki.apache.org/confluence/display/Hive/HCatalog.

data store models that can aid in data design decisions. Thus, this formalization of the data design can be extended to make data design decisions on NoSQL systems as well as optimizing an existing design by transforming the data using the hypergraph representation.

Acknowledgments. This research has been funded by the European Commission through the Erasmus Mundus Joint Doctorate Information Technologies for Business Intelligence - Doctoral College (IT4BI-DC)

References

1. Atzeni, P., Bugiotti, F., Cabibbo, L., Torlone, R.: Data modeling in the NoSQL world. Comput. Stand. Interfaces (2016)
2. Atzeni, P., Bugiotti, F., Rossi, L.: Uniform access to non-relational database systems: the SOS platform. In: International Conference on Advanced Information Systems Engineering, CAiSE (2012)
3. Atzeni, P., Bugiotti, F., Rossi, L.: Uniform access to NoSQL systems. Inf. Syst. **43** (2014)
4. Bugiotti, F., Bursztyn, D., Deutsch, A., Manolescu, I., Zampetakis, S.: Flexible hybrid stores: constraint-based rewriting to the rescue. In: IEEE 32nd International Conference on Data Engineering ICDE (2016)
5. Bugiotti, F., Bursztyn, D., Diego, U.C.S., Ileana, I.: Invisible glue: scalable self-tuning multi-stores. In: CIDR (2015)
6. Bugiotti, F., Cabibbo, L., Atzeni, P., Torlone, R.: Database design for NoSQL systems. In: International Conference on Conceptual Modeling ER (2014)
7. Duggan, J., et al.: The BigDAWG polystore system. ACM SIGMOD Rec. **44**(2) (2015)
8. Euzenat, J., Shvaiko, P., et al.: Ontology Matching, vol. 18. Springer, Heidelberg (2007). https://doi.org/10.1007/978-3-642-38721-0
9. Garcia-Molina, H., Ullman, J., Widom, J.: Database Systems: The Complete Book. Pearson Education India (2008)
10. Herrero, V., Abelló, A., Romero, O.: NoSQL design for analytical workloads: variability matters. In: 35th International Conference Conceptual Modeling ER (2016)
11. Hills, T.: NoSQL and SQL Data Modeling. Technics Publications (2016)
12. Johnson, R., Hoeller, J., Donald, K., Pollack, M., et al.: The spring framework-reference documentation. Interface **21**, 27 (2004)
13. Karpathiotakis, M., Alagiannis, I., Ailamaki, A.: Fast queries over heterogeneous data through engine customization. Proc. VLDB Endowment **9**(12), 972–983 (2016)
14. Klyne, G., Carroll, J.J.: Resource description framework (RDF): concepts and abstract syntax. http://www.w3.org/TR/2004/REC-rdf-concepts-20040210/. Accessed 16 Feb 2018
15. Liao, Y.T., et al.: Data adapter for querying and transformation between SQL and NoSQL database. Fut. Gener. Comput. Syst. **65**, 111–121 (2016)
16. Ong, K.W., Papakonstantinou, Y., Vernoux, R.: The SQL++ semi-structured data model and query language: a capabilities survey of sql-on-hadoop, NoSQL and newsql databases. CoRR abs/1405.3631 (2014)
17. Rodriguez, M.A.: The Gremlin graph traversal machine and language. In: 15th Symposium on Database Programming Languages. ACM (2015)

18. Saltor, F., Castellanos, M., García-Solaco, M.: Suitability of data models as canonical models for federated databases. ACM SIGMOD Rec. **20**(4), 44–48 (1991)
19. Sellami, R., Bhiri, S., Defude, B.: Supporting multi data stores applications in cloud environments. IEEE Trans. Serv. Comput. **9**(1), 1 (2016)
20. Shvaiko, P., Euzenat, J.: Ontology matching: state of the art and future challenges. IEEE Trans. Knowl. Data Eng. **25**(1), 158–176 (2013)
21. Vathy-Fogarassy, Á., Hugyák, T.: Uniform data access platform for SQL and NoSQL database systems. Inf. Syst. **69**, 93–105 (2017)
22. Wang, L., et al.: Schema management for document stores. PVLDB **8**(9), 922–933 (2015)

Renormalization of NoSQL Database Schemas

Michael J. Mior[✉] and Kenneth Salem

Cheriton School of Computer Science, University of Waterloo, Waterloo, Canada
{mmior,kmsalem}@uwaterloo.ca

Abstract. NoSQL applications often use denormalized databases in order to meet performance goals, but this introduces complications as the database itself has no understanding of application-level denormalization. In this paper, we describe a procedure for reconstructing a normalized conceptual model from a denormalized NoSQL database. The procedure's input includes functional and inclusion dependencies, which may be mined from the NoSQL database. Exposing a conceptual model provides application developers with information that can be used to guide application and database evolution.

Keywords: Renormalization · NoSQL · Database design
Conceptual modeling

1 Introduction

NoSQL databases, such as Apache Cassandra, Apache HBase, and MongoDB, have grown in popularity, despite their limitations. This is due in part to their performance and scalability, and in part because they adopt a flexible approach to database schemas. Because these systems do not provide high-level query languages, applications must denormalize and duplicate data across physical structures in the database to answer complex queries. Unfortunately, the NoSQL database itself typically has no understanding of this denormalization. Thus, it is necessary for applications to operate directly on physical structures, coupling applications to a particular physical design.

Although NoSQL systems may not require applications to define rigid schemas, application developers must still decide how to store information in the database. These choices can have a significant impact on application performance as well as the readability of application code [6]. For example, consider an application using HBase to track requests made to an on-line service. The same requests may be stored in multiple tables since the structures available determine which queries can be asked. The choice of data representation depends on how the application expects to use the table, i.e., what kinds of queries and updates it needs to perform. Since the NoSQL system itself is unaware of these application decisions, it can provide little to no help in understanding what is being represented in the database.

J. C. Trujillo et al. (Eds.): ER 2018, LNCS 11157, pp. 479–487, 2018.
https://doi.org/10.1007/978-3-030-00847-5_34

Together, the lack of physical data independence and the need for workload-tuned, denormalized database designs creates challenges for managing and understanding physical schemas, especially as applications evolve. In our on-line service example, request information might be stored twice, once grouped and keyed by the customer that submitted the request, and a second time keyed by the request subject or the request time. If the application updates a request, or changes the information it tracks for each request, these changes should be reflected in both locations. Unless the application developer maintains external documentation, the only knowledge of this denormalization is embedded within the source code. We aim to surface this knowledge by generating a useful conceptual model of the data.

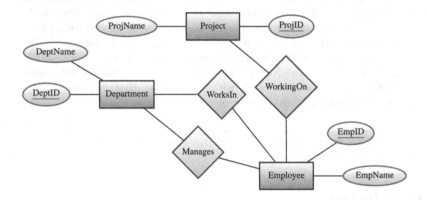

Fig. 1. Schema example after renormalization

We refer to this surfacing task as *schema renormalization*. This work addresses the schema renormalization problem through the following technical contributions:

- We present a semi-automatic technique for extracting a normalized conceptual model from an existing denormalized NoSQL database. It produces a normalized conceptual model for the database, such as the one shown in Fig. 1.
- We develop an normalization algorithm in Sect. 5 which forms the core of our approach. This algorithm uses data dependencies to extract a conceptual model from the NoSQL system's physical structures. Our algorithm ensures that the resulting model is free from redundancy implied by these dependencies. To the best of our knowledge, this is the first normalization algorithm to produce a schema in interaction-free inclusion dependency normal form [9].
- Finally, Sect. 6, presents a case study which shows the full schema renormalization process in action for a NoSQL application. We use this case study to highlight both the advantages and the limitations of our approach to renormalization.

The conceptual data model that our algorithm produces can serve as a simple reference, or specification, of the information that has been denormalized across the workload-tuned physical database structures. We view this model as a key component in a broader methodology for schema management for NoSQL applications.

2   Renormalization Overview

We renormalize NoSQL databases using a three step process. The first step is to produce a *generic physical schema* describing the physical structures that are present in the NoSQL database. We describe the generic physical model in more detail in Sect. 3, and illustrate how it can be produced for different types of NoSQL systems. The second step is to identify dependencies among the attributes of the generic model. The dependencies, which we discuss in Sect. 4, can be provided by a user with understanding of the NoSQL system's application domain or automatically using existing mining techniques. We provide a brief overview of these steps in the following sections. More detail is available in an extended technical report [14].

The final step in the renormalization process is to normalize the generic physical schema using the dependencies, resulting in a logical schema such as the one represented (as an ER diagram) in Fig. 1. This step is automated, using the procedure described in Sect. 5. Our algorithm ensures that redundancy in the physical schema captured by the provided dependencies is removed.

3   The Generic Physical Schema

The first step in the renormalization process is to describe the NoSQL database using a generic schema. The schemas we use are relational. Specifically, a generic physical schema consists of a set of *relation schemas*. Each relation schema describes a physical structure in the underlying NoSQL database (e.g., a document collection). A relation schema consists of a unique relation name plus a set of attribute names. Attribute names are unique within each relation schema.

If the NoSQL database includes a well-defined schema, then describing the physical schema required for renormalization is a trivial task. The generic schema simply identifies the attributes that are present in the table, and gives names to both the attributes and the table itself. For example, Cassandra stores table definitions which can directly provide the generic schema.

In general, we anticipate that the definition of a generic physical schema for an application will require user involvement. However, there are tools that may assist with this process. For example, Izquierdo et al. [7] have proposed a method for extracting a schema from JSON records in a document store, which could be applied to extract the generic physical schema required for renormalization.

4   Dependency Input

The second step of the renormalization process is to identify dependencies among attributes in the generic physical schema. Our algorithm uses two types of dependencies: functional dependencies (FDs) and inclusion dependencies (INDs). These two forms of dependencies are easy to express and are commonly used in database design [11].

For input to our algorithm, we require that all INDs are superkey-based. That is, for an IND $R(A) \subseteq S(B)$, B must be a superkey of S. We do not believe that this is a significant restriction since we intend for INDs to be used to indicate foreign key relationships which exist in the denormalized data. Indeed, Mannila and Räihä [11] have previously argued that only key-based dependencies are relevant to logical design.

5   Normalization Algorithm

Levene and Vincent [9] define a normal form for database relations involving FDs and INDs referred to as inclusion dependency normal form (IDNF). They have shown that normalizing according to IDNF removes redundancy from a database design implied by the set of dependencies. However, one of the necessary conditions for this normal form is that the set of INDs is non-circular. This excludes useful schemas which express constraints such as one-to-one foreign key integrity. For example, for the relations $R(\underline{A}, B)$ and $S(\underline{B}, C)$ we can think of the circular INDs $R(A) = S(B)$ as expressing a one-to-one foreign key between $R(A)$ and $S(B)$.

Levene and Vincent also propose an extension to IDNF, termed *interaction-free inclusion dependency normal form* which allows such circularities. The goal of our normalization algorithm is to produce a schema that is in interaction-free IDNF. This normal form avoids redundancy implied by FDs and INDs while still allowing the expression of useful information such as foreign keys. As we show in Sect. 6, this produces useful logical models for a real-world example.

Data: A set of relations **R**, FDs **F**, and INDs **I**
Result: A normalized set of relations $\mathbf{R}'''$ and new dependencies $\mathbf{F}'$ and $\mathbf{I}^{+'''}$
begin
    $\mathbf{F}', \mathbf{I}^+ \leftarrow$ Expand $(\mathbf{F}, \mathbf{I})$;   // Perform dependency inference
    $\mathbf{R}', \mathbf{I}^{+'} \leftarrow$ BCNFDecompose $(\mathbf{R}, \mathbf{F}', \mathbf{I}^+)$;   // BCNF normalization
    $\mathbf{R}'', \mathbf{I}^{+''} \leftarrow$ Fold $(\mathbf{R}', \mathbf{F}', \mathbf{I}^{+'})$;   // Remove attributes/relations
    $\mathbf{R}''', \mathbf{I}^{+'''} \leftarrow$ BreakCycles $(\mathbf{R}'', \mathbf{I}^{+''})$;   // Break circular INDs
end

Fig. 2. Algorithm for normalization to interaction-free IDNF

Figure 2 provides an overview of our normalization algorithm, which consists of four stages. In the reminder of this section, we discuss the normalization algorithm in more detail. We will make use of a running example based on the simple generic (denormalized) physical schema and dependencies shown in Fig. 3.

Physical Schema

EmpProjects(<u>EmpID</u>, EmpName, <u>ProjID</u>, ProjName)

Employees(<u>EmpID</u>, EmpName, DeptID, DeptName)

Managers(<u>DeptID</u>, EmpID)

Functional Dependencies

EmpProjects : EmpID → EmpName   Employees : EmpID → EmpName, DeptID
EmpProjects : ProjID → ProjName   Employees : DeptID → DeptName
Managers : DeptID → EmpID

Inclusion Dependencies

EmpProjects (EmpID, EmpName) ⊆ Employees (...)

Employees (DeptID) ⊆ Managers (...)

Managers (EmpID) ⊆ Employees (...)

When attributes have the same names, we use ... on the right.

Fig. 3. Example generic physical schema and dependencies.

5.1 Dependency Inference

To minimize the effort required to provide input needed to create a useful normalized schema, we aim to infer dependencies whenever possible. Armstrong [1] provides a well-known set of axioms which can be used to infer FDs from those provided as input. Similarly, Mitchell [15] presents a similar set of inference rules for INDs.

Mitchell further presents a set of inference rules for joint application to a set of FDs and INDs. We adopt Mitchell's rules to infer new FDs for INDs and vice versa. The pullback rule enables new FDs to be inferred from FDs and INDs. The collection rule allows the inference of new INDs. These new dependencies allow the elimination of attributes and relations via the Fold algorithm (see Sect. 5.3) to reduce the size of the resulting schema while maintaining the same semantic information.

There is no complete axiomatization for FDs and INDs taken together [3]. Our Expand procedure, which uses Mitchell's pullback and collection rules for inference from FDs and INDs, is sound but incomplete. However, it does terminate, since the universe of dependencies is finite and the inference process is purely additive. Although Expand may fail to infer some dependencies that are implied by the given set of FDs and INDs, it is nonetheless able to infer dependencies that are useful for schema design.

5.2   BCNF Decomposition

The second step, BCNFDecompose, is to perform a lossless join BCNF decomposition of the physical schema using the expanded set of FDs. When relations are decomposed, we project the FDs and INDs from the original relation to each of the relations resulting from decomposition. In addition, we add new INDs which represent the correspondence of attributes between the decomposed relations. For example, when performing the decomposition $R(ABC) \rightarrow R'(AB), R''(BC)$ we also add the INDs $R'(B) \subseteq R''(B)$ and $R''(B) \subseteq R'(B)$. In our running example, we are left with the relations and dependencies shown in Fig. 4 after the Expand and BCNFDecompose steps.

Physical Schema

Employees $\left(\underline{\text{EmpID}}, \text{EmpName}, \text{DeptID}\right)$ Departments $\left(\underline{\text{DeptID}}, \text{DeptName}\right)$

EmpProjects $\left(\underline{\text{EmpID}}, \underline{\text{ProjID}}\right)$ EmpProjects' $\left(\underline{\text{EmpID}}, \text{EmpName}\right)$

Managers $\left(\underline{\text{DeptID}}, \text{EmpID}\right)$ Projects $\left(\underline{\text{ProjID}}, \text{ProjName}\right)$

Functional Dependencies

Employees : EmpID $\rightarrow$ EmpName, DeptID   Departments : DeptID $\rightarrow$ DeptName

Projects : ProjID $\rightarrow$ ProjName   Managers : DeptID $\rightarrow$ EmpID

EmpProjects' : EmpID $\rightarrow$ EmpName

Inclusion Dependencies

Projects (ProjID) = EmpProjects (...)   EmpProjects (EmpID) $\subseteq$ Employees (...)

EmpProjects' (EmpID) = EmpProjects (...)   Managers (DeptID) $\subseteq$ Departments (...)

EmpProjects' (EmpID, EmpName) $\subseteq$ Employees (...)   Managers $\subseteq$ Employees (...)

Fig. 4. Relations and dependencies after BCNF decomposition. Note that = is used to represent bidirectional inclusion dependencies.

5.3   Folding

Casanova and de Sa term the technique of removing redundant relations *folding* [2]. A complete description of our algorithm, Fold, is given in an extended technical report [14]. Fold identifies attributes or relations which are recoverable from other relations. Specifically, folding removes attributes which can be recovered by joining with another relation and relations which are redundant because they are simply a projection of other relations. Fold also identifies opportunities for merging relations sharing a common key.

Consider the EmpProjects' relation which contains the EmpName attribute. Since we have the IND EmpProjects'(EmpID,EmpName) $\subseteq$ Employees(...) and the FD Employees: EmpID $\rightarrow$ EmpName we can infer that the EmpName attribute in EmpProjects' is redundant since it can be recovered by joining with the Employees relation.

5.4   Breaking IND Cycles

Interaction-free IDNF requires that the final schema be free of circular INDs. Mannila and Räihä [11] use a technique, which we call `BreakCycles`, to break circular INDs when performing logical database design. We adopt this technique to break IND cycles which are not proper circular.

5.5   IDNF

The goal of our normalization algorithm is to produce a schema that is in interaction-free IDNF with respect to the given dependencies. The following conditions are sufficient to ensure that a set of relations $\mathbf{R}$ is in interaction-free IDNF with respect to a set of FDs $\mathbf{F}$ and INDs $\mathbf{I}$: (1) $\mathbf{R}$ is in BCNF [5] with respect to $\mathbf{F}$, (2) all the INDs in $\mathbf{I}$ are key-based or proper circular, and (3) $\mathbf{F}$ and $\mathbf{I}$ do not interact. A set of INDs is proper circular if for each circular inclusion dependency over a unique set of relations $R_1(X_1) \subseteq R2(Y_2), R_2(X_2) \subseteq R_3(Y_3), \ldots, R_m(X_m) \subseteq R_1(Y_1)$, we have $X_i = Y_i$ for all i. The schema produced by the normalization algorithm of Fig. 2 is in interaction-free IDNF. We provide a proof in an extended technical report [14].

6   RUBiS Case Study

In previous work, we developed a tool called NoSE [13], which performs automated schema design for NoSQL systems. We used NoSE to generate two Cassandra schemas, each optimized for a different workload (a full description is given in an extended technical report [14]). In each case, NoSE starts with a conceptual model of the database which includes six types of entities (e.g., users, and items) and relationships between them. The two physical designs consist of 9 and 14 Cassandra column families.

Our case study uses NoSE's denormalized schemas as input to our algorithm so that we can compare the schemas that it produces with the original conceptual model. For each physical schema, we tested our algorithm with two sets of dependencies: one manually generated from the physical schema, and a second mined from an instance of that schema using techniques discussed in an extended technical report [14]. The first set of dependencies resulted in a conceptual model that was identical (aside from names of relations and attributes) to the original conceptual model used by NoSE, as desired.

For the second set of tests, renormalization produced the original model for the smaller Cassandra schema. The mining process identified 61 FDs and 314 INDs. Ranking heuristics were critical to this success. Without them, spurious dependencies lead to undesirable entities in the output schema. For the larger schema, mining found 86 FDs and 600 INDs, many of them spurious, resulting in a model different from the original. No set of heuristics will be successful in all cases and this is an area for future work.

These examples show that FDs and INDs are able to drive meaningful denormalization. Runtime for the normalization step of our algorithm was less than one second on a modest desktop workstation in all cases.

7   Related Work

Much of the existing work in normalization revolves around eliminating redundancy in relational tables based on different forms of dependencies. However, it does not deal with the case where applications duplicate data across relations. Inclusion dependencies are a natural way to express this duplication. Other researchers have established normal forms using inclusion dependencies [9–11] in addition to FDs. Our approach borrows from Manilla and Räihä, who present a variant of a normal form involving inclusion dependencies and an interactive normalization algorithm. However, it does not produce useful schemas in the presence of heavily denormalized data. Specifically, their approach is not able to eliminate all data duplicated in different relations.

A related set of work exists in database reverse engineering (DBRE). The goal of DBRE is to produce an understanding of the semantics of a database instance, commonly through the construction of a higher level model of the data. Unlike our work, many approaches [4,17] present only an informal process and not a specific algorithm.

There is significant existing work in automatically mining both functional [12] and inclusion [8] dependencies from both database instances and queries. These approaches complement our techniques since we can provide the mined dependencies as input into our algorithm. Papenbrock and Naumann [16] present of heuristics for making use of mined dependencies to normalize a schema according to BCNF. We leverage these to incorporate mining into our algorithm as discussed in an extended technical report [14].

8   Conclusions and Future Work

We have developed a methodology for transforming a denormalized physical schema in a NoSQL datastore into a normalized logical schema. Our method makes use of functional and inclusion dependencies to remove redundancies commonly found in NoSQL database designs. We further showed how we can make use of dependencies which were mined from a database instance to reduce the input required from users. Our method has a variety of applications, such as enabling query execution against the logical schema and guiding schema evolution as application requirements change.

References

1. Armstrong, W.W.: Dependency structures of data base relationships. In: IFIP Congress, pp. 580–583 (1974)
2. Casanova, M.A., de Sa, J.E.A.: Mapping uninterpreted schemes into entity-relationship diagrams: two applications to conceptual schema design. IBM J. Res. Develop. 28(1), 82–94 (1984)
3. Casanova, M.A.: Inclusion dependencies and their interaction with functional dependencies. J. Comput. Syst. Sci. 28(1), 29–59 (1984)

4. Chiang, R.H., et al.: Reverse engineering of relational databases: extraction of an EER model from a relational database. Data Knowl. Eng. **12**(2), 107–142 (1994)
5. Codd, E.F.: Recent investigations into relational data base systems. Technical report RJ1385, IBM, April 1974
6. Gómez, P., Casallas, R., Roncancio, C.: Data schema does matter, even in NoSQL systems! In: RCIS 2016, June 2016
7. Cánovas Izquierdo, J.L., Cabot, J.: Discovering implicit schemas in JSON data. In: Daniel, F., Dolog, P., Li, Q. (eds.) ICWE 2013. LNCS, vol. 7977, pp. 68–83. Springer, Heidelberg (2013)
8. Kantola, M., et al.: Discovering functional and inclusion dependencies in relational databases. Int. J. Intell. Syst. **7**(7), 591–607 (1992)
9. Levene, M., Vincent, M.W.: Justification for inclusion dependency normal form. IEEE TKDE **12**(2), 281–291 (2000)
10. Ling, T.W., Goh, C.H.: Logical database design with inclusion dependencies. In: ICDE 1992, pp. 642–649, Feburary 1992
11. Mannila, H., Räihä, K.J.: Inclusion dependencies in database design, pp. 713–718. IEEE Computer Society, February 1986
12. Mannila, H., Räihä, K.J.: Algorithms for inferring functional dependencies from relations. DKE **12**(1), 83–99 (1994)
13. Mior, M.J., Salem, K., Aboulnaga, A., Liu, R.: NoSE: Schema design for NoSQL applications. In: ICDE 2016, pp. 181–192 (2016)
14. Mior, M.J., Salem, K.: Renormalization of NoSQL database schemas. Technical report CS-2017-02, University of Waterloo (2017)
15. Mitchell, J.C.: Inference rules for functional and inclusion dependencies. In: PODS 1983, pp. 58–69. ACM (1983)
16. Papenbrock, T., Naumann, F.: Data-driven schema normalization. In: Proceedings of EDBT 2017, pp. 342–353 (2017)
17. Premerlani, W.J., Blaha, M.R.: An approach for reverse engineering of relational databases. Commun. ACM **37**(5), 42–49 (1994)

Modeling Strategies for Storing Data in Distributed Heterogeneous NoSQL Databases

Moditha Hewasinghage[1](✉), Nacéra Bennacer Seghouani[2],
and Francesca Bugiotti[2]

[1] Universitat Politcnica de Catalunya, Barcelona, Spain
moditha@essi.upc.edu
[2] LRI, CentraleSupélec, Paris-Saclay University, Paris, France

Abstract. In this work, we propose HerM (Heterogeneous Distributed Model), a NoSQL data modeling approach which supports the use of multiple heterogeneous NoSQL systems in a distributed environment. We define the conceptual elements necessary for data modeling, and we identify optimized data distribution patterns. We implemented a flexible framework, where we deployed our proposed modeling strategies and that we evaluated comparing our approach against native the NoSQL data distribution methodology provided by the NoSQL databases MongoDB.

Keywords: Conceptual modeling · Heterogeneity · Data distribution

1 Introduction

How to store and use Big Data to extract information efficiently and effectively has become crucial in many fields. To support this, the industry has started moving towards new systems called non-relational or NoSQL. These systems are highly scalable, can efficiently handle large amounts of data, and support flexible, semi-structured data. NoSQL systems are mostly based on simple read-write operations where performance is a crucial concern and where ACID properties can be traded off [6,11]. Currently, there are more than one hundred NoSQL systems available with different characteristics of the data model, different data access APIs, and different consistency and durability guarantees [14]. The various data models have been categorized into four main categories namely the key-value, the document, the column-family (or extensible record), and the graph [6]. This vast heterogeneity has opened up new problems. NoSQL data-stores are claimed to be flexible, but research [1,4,5] has shown that even in the NoSQL world data design requires significant modeling decisions that impact on considerable quality requirements. Moreover a complete and fully formalized data modeling procedure for NoSQL systems does not exist. The distribution of data plays a significant role in NoSQL systems [6] and many research approaches have

---

This work was achieved at LRI Paris-Saclay University.

© Springer Nature Switzerland AG 2018
J. C. Trujillo et al. (Eds.): ER 2018, LNCS 11157, pp. 488–496, 2018.
https://doi.org/10.1007/978-3-030-00847-5_35

introduced efficient methods and algorithms for distributing data [3, 9]. In this paper we define modeling strategies that support the usage of multiple heterogeneous data-stores in a distributed environment. Our approach enables modeling heterogeneous data stores and provides a configurable data distribution methodology, achieved thanks to the usage of a general conceptual model composed by a set of general constructs. We also identify data distribution strategies at the conceptual level. Our approach provides a flexible mechanism to deploy the conceptual model that is implemented through a framework that allows the users access data efficiently and transparent of the underlying data stores. We based our data model, called HerM (**He**terogeneous **D**istributed **M**odel) on an existing general data model for heterogeneous NoSQL systems: the SOS Model [2]. HerM extends SOS to support distributed environment and can be used to evaluate the best data distribution strategy for different use cases. We map HerM into a logical model with a step by step methodology, mapping the relations and distributing the data among the physical cluster nodes. We use MongoDB and Oracle NoSQL as two heterogeneous data models. Next, we implement a flexible framework that provides transparent access to the data with a familiar REST API and a JSON based, easy-to-learn query interface. We offer the ability in configuring the data model for the system through simple configuration based on HerM giving much more portability and flexibility in supporting different use cases. Finally, we evaluate our framework against a native sharded MongoDB instance on a large dataset based on Twitter, for different data storage and retrieval scenarios. With this comparison, we show that HerM provides useful distribution strategies and that implemented model performs on par with native MongoDB. Moreover, the system is exceptionally stable and give constant performance throughout the experiments.

This paper is organized as follows: in Sect. 2 we explore the related work, then we define our data model in Sect. 3. The framework we implemented is introduced in Sect. 4. We finally describe our experiments in Sect. 5.

2   Related Work

Only a few research has been conducted on systematically modeling NoSQL data-stores. In [2] the authors introduce a high-level description of an interface for NoSQL systems based on a generic model. They propose a meta-modeling approach that inspired our research where a common underlying structure is defined with the methods to access the system. [5] introduces a NoSQL abstract model called NoAM which is designed to support scalability, performance, and consistency but does not take into account the data distribution process. Another general approach for designing a NoSQL system focused on analytical workloads have been discussed in [7]. Different works have analyzed the problem of selecting a target cloud data-store. Among those [10] defines a general XML schema that proposes a multi-criteria optimization model for calculating the optimal data distribution. Some other works have addressed the problem of querying data without proposing a specific general data model. DBMS+ is a new notion introduced in [8] where a DBMS encapsulates the different execution and storage engines. OPEN-PaaS-DataBase API (ODBAPI) [12] defines a unified REST-based API for executing queries over NoSQL data-stores. This approach is extended to a concept of a virtual data-store which enables complex queries over hetero-geneous data-stores [13]. Finally ESTOCADA [4] is a constraint-based query rewriting system for heterogeneous NoSQL data-stores that chooses the optimal data-storage methodology for the required scenario and the queries are executed depending on how they are interacting with the different data-stores.

3 Heterogeneous Distributed Model (HerM)

In this section, we discuss our general distribution modeling strategy on hetero-geneous NoSQL systems. This modeling strategy is based on the definition and on the usage of HerM (**He**terogeneous **D**istributed **M**odel), a general conceptual data model.

3.1 HeRM Data Model

The starting point of our conceptual model is the SOS data model [2]. The SOS data model can represent the main categories of NoSQL systems (key-value, documents, and column) but does not offer data distribution facilities. Therefore, we extend it and propose HerM. SOS data model defines three major constructs: *Attribute*, *Struct*, and *Set*. Simple elements of entities such as key-value pairs can be modeled as *Attributes* and groups of heterogeneous *Attributes* can be represented as a *Struct*. A collection of entities are represented in a *Set*. A name and an associated value characterize each construct. The structure of the value depends on the type of the construct. An *Attribute* can contain a simple value such as an integer or string. *Structs* and *Sets* are complex elements which can contain multiple *Attributes*, *Structs*, *Sets* or combination of those. Different kind of binary relations can relate entities, but many-to-many relations are not taken into account. A database is a set of collections of these constructs. Our data distribution approach considers the conceptual model as a hierarchy of related entities. This hierarchy is mapped into a data distribution strategy. The most important entity of the hierarchical representation is the *Root* that follows this rule: "The *Root Struct* should not have any direct or indirect incoming connections from another *Struct* with many-to-many cardinality." An incoming relation for an entity means that the entity is nested into the referring one.

In Fig. 2 shows an example of the use of our data model for storing Twitter data. Three entities that are candidate to be elected as the *Root* entity: USER, TWEET, and HASHTAG. In the given example the only *Struct* that satisfies this requirement is the USER. If there are multiple candidates for the *Root* the end user could choose one of them and have the others as child entities. The pos-sibility of having more than one distribution is beyond our scope. The entities are represented as *Structs* in the model. In the example, it is the USER_ID. All the other *Structs* that represent an entity should have an *Attribute* that helps to identify the entity uniquely, referred as Key (K).

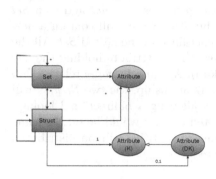

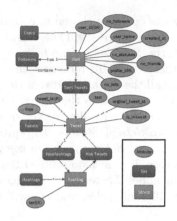

Fig. 1. HerM meta model

Fig. 2. HerM model of the Twitter example

The *Distribution Key* (DK) denotes the distribution field. The data is stored in the different available physical nodes of a NoSQL distributed data-store handled by the framework according to the DK. To guarantee an even and correct data distribution, the distribution key must be chosen among the candidate attributes of the *Root*. The HerM meta model is shown in Fig. 1.

3.2   From HerM to the Logical Model

After identifying the *Distribution Key* of the *Root* and the keys for other *Struct*s the next step is to determine the collections of our entities.

Identifying the Collections. Collections are the independent sets in HerM model. An independent set is a *Set* which does not have any incoming connections and is directly related to a *Struct*. In Fig. 2 there are three independent *Set*s USERS, TWEETS, and HASHTAGS.

Data Distribution. In our approach, the data distribution mechanism is guided by the Distribution Key. The different parts of data lie on different data-stores as well as different physical nodes. Data should be distributed *in a logical way*: data locality is important and minimizes the joins and the data duplication.

Storing different entities in different data-store collections allow effective data manipulation concerning the usage of a single data-store. The root entity is the main entry point of the data distribution. The distribution key defines where the root entity instances will lie on the physical node. It is impossible to have a perfect distribution mainly because the related entities are of different sizes for a root entity instance. According to the hierarchy fixed from the data model, the entities related to the *Root* entity are stored in the same physical node. In HerM model an entity is represented as a *Struct*. The relationships between the entities are expressed differently depending on the cardinality between the structs that is specified in the data model. A *One-to-one* relationship relates two structs, one of the *Struct*s could be chosen to contain the attributes of the other *Struct* into it. This merging process will then result in a single *Struct* having all the attributes of both the *Struct*s. A *One-to-many* relationship relates a *Struct* and to a *Set* of another *Struct*. To represent this relationship, the *Struct* will contain a new *Set* with all the key attributes of the *Struct*s contained in the nested *Set*. All the *Struct*s contained into the nested set will have a new attribute holding the key of the father *Struct* (allowing the reverse lookup). A *Many-to-many* relationship relates two *Set*s of *Struct*s. To represent this relationship, the two *Struct*s will contain a new *Set* with all the key attributes allowing a bidirectional lookup. In this case, it is not guaranteed that the nested entities would lie on the same physical node. Therefore, both many relationship entities will be distributed among different nodes, and one will be repeated.

4   HerM Framework

The overall architecture of the HerM Framework is shown in Fig. 3. It is a master-slave architecture and consists of one main physical node and multiple physical data nodes. The main node is responsible for managing the overall data distribution, handling the queries, and returning the results. Each data node contains the NoSQL data-store instances which are responsible for the data storage and managing the queries (and the joins if necessary). The framework is able to take into account the following aspects: (*i*) Providing a transparent query interface; (*ii*) Dynamically modeling the meta-layer; (*iii*) Handling the distribution of a single entity and the joins among related entities. The meta-layer plays a significant role in HerM framework as it contains all the necessary information about the relationships and the key information of the entities defined in the meta-model. It keeps the name of the entity, whether it is the root entity, the parent entity (if it is not the root entity), primary key of the entity, reverse index to the parent entity (if it is a child entity), the type of the relationship with the parent (one or many) for all the entities. This information helps to identify and execute queries throughout the HerM Framework. It also carries the information related to the

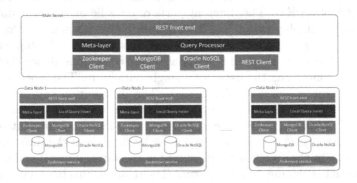

Fig. 3. Overall architecture of HerM Framework

physical data distribution and the target data-store parameters. This information needs to be available and updated throughout all the data nodes. Therefore we used Apache Zookeper which helps to manage configuration information in a distributed environment. The meta-data is stored in JSON format, and the individual data nodes have their own Zookeeper instance.

Handling Data Distribution and Joins of Related Entities. The data distribution is handled in two steps. The first one consists of the initial distribution of the root entity among the data nodes and the second one is the distribution of the related entities stored in separate data stores according to the distribution key. In our example the USER_ID is selected as the distribution key, and the user's data is split among the available number of data nodes with ranges. When insertion of a USER is sent as a query the HerM Framework analyses the data and determine that it is the root element then according to the distribution key USER_ID defines the data node for the data as well as the datastore and the collection. Handling joins, or related entities is a challenging task since most of the NoSQL systems do not provide the join operation. Moreover, HerM framework uses different data stores, which means that there are no pre-existing tools that support this. We propose to execute the joins locally on the data node where possible. By doing this, we expect the system to be more stable and the workload on the central location to be less, resulting in better overall performance. The performance also increases because our framework distributes data trying to put related entities in the same physical node.

5   Experiments

In this section, we describe the different experiments we run to evaluate the performances of HerM Framework. We studied different INSERT and SELECT queries and compared to MongoDB, analyzing the performance and the stability. The experiments were carried out in a cluster environment using MongoDB 3.4 and Oracle NoSQL 4.3.11. The data was collected from Twitter using the Twitter REST API containing 412814 users 275139 tweets and 8203 unique hashtags that occur in these tweets. The data was distributed among four data nodes using the distribution of the USER_ID as distribution key as described in Fig. 2. The quartiles of the USER_ID were used to distribute the data. The users were stored in MongoDB and the tweets of the user are stored in the Oracle NoSQL instance in the same data node as well as the HASHTAGs. The main HerM Framework instance was running on a separate node. The same nodes were used as the MongoDB shard instances. USER is stored as the main collection of documents with embedded TWEETs and HASHTAGs.

The experiments evaluated four different scenarios: *(i)–(ii)* The system and MongoDB without indexes, *(iii)* The system with an index on USER_ID in MongoDB and *(iv)* native MongoDB with index on USER_ID and TWEET_ID.

Table 1 shows the runtime in milliseconds for inserting users, tweets, querying a user, tweet, and followers of a given user. The average insertion time of USER in native MongoDB is faster compared to the system in both with and without indexes. This is expected because the introduction of the REST API introduces additional overhead.

Table 1. Average query run times

Query/System	INSERT user	SELECT user	INSERT tweet	SELECT tweet	SELECT followers
HerM Framework	1.45	80.05	4.11	31.1	147.92
Native MongoDB	1.22	108.90	101.33	134.74	107.84
HerM Framework + index	2.14	33.82	4.13	29.80	39.10
Native MongoDB + index	1.43	12.64	10.10	21.70	10.76

The performance of the system for querying users is nearly 30% faster than querying without indexes the native MongoDB shard instance. Querying without an index produces a full scan on the documents and in our system we have to do the scan only on the relevant node where MongoDB has to do a scan on all the nodes. Since the tweets are stored as nested elements inside the user document in MongoDB, in each insert a push operation to the nested collection is needed. Indexing on native MongoDB improved the insertion time by improving the finding of user. The next experiment was to query a tweet by the ID. In our system, the reverse index for the USER_ID, TWEET_ID stored in Oracle NoSQL is referred to make the major-minor key combination to retrieve a particular tweet. In MongoDB the tweets are stored as nested objects making retrieval of an inner element costly. When retrieving the followers, in our system, the user is retrieved at first. Once the USER_IDs of the followers are retrieved from the user, the IDs are divided between the nodes as it is the distribution key. Then separate nodes will retrieve a set of USER_IDs that lies in the node. Finally, the users from all the nodes are merged. In native MongoDB, the user followers are retrieved directly given a list of ids. MongoDB shard with the indexes has the best performance closely followed by our system with the index. We also measured the percentage of outliers using Tukey's method for each of the scenarios in both systems. For inserting a user, the percentage of outliers is less than 5% which can be considered as usual. In the inserts for MongoDB, we are inserting close to 300000 tweets, and almost 14% of outliers mean that around 42000 of them have taken quite more than the average time and the maximum value was about 3 s for an insertion. This performance issue could be a result of the complex nested structure of the data model. The results of the experiments show that the performance and the stability of the system are better compared to the native MongoDB almost all scenarios. The performance measured in the above experiments are based on the total time taken from the moment the client sends the REST request to the API until the server sends the response. This total time contains certain additional overheads which are beyond our control. Mainly the REST client and the server implementations in jersey. Figure 4 shows the internal average query runtime of the system compared to the total time of the system with indexes and the average time taken by the native MongoDB cluster with indexes for different queries. The internal time of the system is quite low in all the cases. In almost all the cases the difference between the total time and the internal time has a difference of around 25–30 ms. This constant time is for the additional requirements added by the REST API: the time to create the REST client by the end user, create the request, time for the request to reach the server, server processing the request, time for the server to serialize the response and the time for the response to reach the server. The results shows that the average internal time is not only better than the overall time but also better than the native MongoDB shard as well.

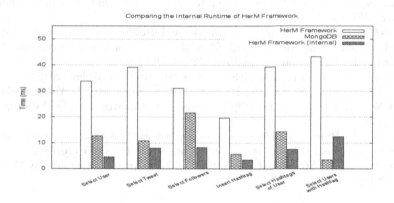

Fig. 4. Internal average query time comparison

Our experiments show that our modeling approach has better or comparable performance with respect to the native MongoDB implementation. The performance of the REST API is beyond our control: we define our framework without relying on third party software or optimizing the implementation of the APIs. Given that the aim of this paper is to show that our approach of data modeling in NoSQL system has its advantages those factors are beyond our scope.

References

1. Atzeni, P., Bellomarini, L., Bugiotti, F., Celli, F., Gianforme, G.: A runtime approach to model-generic translation of schema and data. Inf. Syst. **37**(3), 269–287 (2012)
2. Atzeni, P., Bugiotti, F., Rossi, L.: Uniform access to NoSQL systems. Inf. Syst. **43**, 117–133 (2014)
3. Basani, V.R., et al.: Method and apparatus for reliable and scalable distribution of data files in distributed networks, US Patent 6,718,361, 6 Apr 2004

4. Bugiotti, F., Bursztyn, D., Deutsch, A., Manolescu, I., Zampetakis, S.: Flexible hybrid stores: constraint-based rewriting to the rescue. In: 32nd IEEE International Conference on Data Engineering, pp. 1394–1397 (2016)
5. Bugiotti, F., Cabibbo, L., Atzeni, P., Torlone, R.: Database design for NoSQL systems. In: Yu, E., Dobbie, G., Jarke, M., Purao, S. (eds.) ER 2014. LNCS, vol. 8824, pp. 223–231. Springer, Cham (2014). https://doi.org/10.1007/978-3-319-12206-9_18
6. Cattell, R.: Scalable SQL and NoSQL data stores. SIGMOD Rec. **39**(4), 12–27 (2010)
7. Herrero, V., Abelló, A., Romero, O.: NOSQL design for analytical workloads: variability matters. In: Comyn-Wattiau, I., Tanaka, K., Song, I.-Y., Yamamoto, S., Saeki, M. (eds.) ER 2016. LNCS, vol. 9974, pp. 50–64. Springer, Cham (2016). https://doi.org/10.1007/978-3-319-46397-1_4
8. Lim, H., Han, Y., Babu, S.: How to fit when no one size fits. In: The Conference on Innovative Data Systems Research (CIDR) (2013)
9. Pokorny, J.: NoSQL databases: a step to database scalability in web environment. Int. J. Web Inf. Syst. **9**(1), 69–82 (2013)
10. Ruiz-Alvarez, A., Humphrey, M.: A model and decision procedure for data storage in cloud computing. In: 12th IEEE/ACM International Symposium on Cluster, Cloud and Grid Computing, pp. 572–579. IEEE (2012)
11. Sadalage, P.J., Fowler, M.J.: NoSQL Distilled. Addison-Wesley (2012)
12. Sellami, R., Bhiri, S., Defude, B.: ODBAPI: a unified REST API for relational and NoSQL data stores. In: IEEE International Congress on Big Data, pp. 653–660 (2014)
13. Sellami, R., Bhiri, S., Defude, B.: Supporting multi data stores applications in cloud environments. IEEE Trans. Serv. Comput. **9**(1), 59–71 (2016)
14. Stonebraker, M.: Stonebraker on NoSQL and enterprises. Commun. ACM **54**, 10–11 (2011)

How the Conceptual Modelling Improves the Security on Document Databases

Carlos Blanco[1,3](✉), Diego García-Saiz[1], Jesús Peral[2], Alejandro Maté[2], Alejandro Oliver[2], and Eduardo Fernández-Medina[3]

[1] ISTR Research Group, Department of Computer Science and Electronics,
University of Cantabria, Santander, Spain
{Carlos.Blanco,Diego.Garcia}@unican.es
[2] LUCENTIA Research Group, Languages and Computing Systems Department,
University of Alicante, Alicante, Spain
{jperal,amate}@dlsi.ua.es, aor14@alu.ua.es
[3] GSyA Research Group, Institute of Information Technologies and Systems,
Information Systems and Technologies Department, University of Castilla-La
Mancha, Toledo, Spain
Eduardo.Fdezmedina@uclm.es

Abstract. Big Data is becoming a prominent trend in our society. Ever larger amounts of data, including sensitive and personal information, are being loaded into NoSQL and other Big Data technologies for analysis and processing. However, current security approaches do not take into account the special characteristics of these technologies, leaving sensitive and personal data unprotected, thereby risking severe monetary losses and brand damage. In this paper, we focus on assuring document databases, proposing a framework that considers three stages: (1) The source data set is analysed by using Natural Language Processing techniques and ontological resources, in order to detect sensitive data. (2) We define a metamodel for document databases that allows designers to specify both structural and security aspects. (3) This model is automatically implemented into a specific document database tool, MongoDB. Finally, we apply the proposed framework to a case study with a data set from the medical domain. The great advantages of our framework are that: (1) the effort required to secure the data is reduced, as part of the process is automated, (2) it can be easily applied to other NoSQL technologies by adapting the metamodel and transformations, and (3) it is aligned with existing standards, thus facilitating the application of recommendations and best practices.

Keywords: Conceptual modelling · Security · Big data
Document databases · Natural language processing

1 Introduction

Nowadays, the exchange of great volumes of data by means of Big Data technologies is a usual activity in several domains (scientific, medical, citizens, etc.).

© Springer Nature Switzerland AG 2018
J. C. Trujillo et al. (Eds.): ER 2018, LNCS 11157, pp. 497–504, 2018.
https://doi.org/10.1007/978-3-030-00847-5_36

These data include critical enterprise information and personal data which could be exposed if they are not correctly assured [9]. Unfortunately, although some security approaches such as UMLSec [3] exist for traditional technologies, there are no well-known mechanisms to define security policies that maintain the confidentiality, privacy, integrity and other characteristics required for assuring critical data within Big Data environments [6,11].

One of the main challenges is that, while new technologies such as document, columnar, and graph databases have appeared, they have focused mainly on dealing with Big Data characteristics (volume, variety, velocity and complexity). Security and privacy constraints in these technologies have been relegated to a secondary place [4,11], thus leading to information leaks, scandals and millionaires loss.

The main objective of our research deals with incorporating security in Big Data technologies, focusing on document databases as a starting point. We consider that to obtain a robust secure solution, security constraints need to be included at early development stages in order to be considered in design decisions. Then, they can be correctly incorporated into the final implementation. In this way, this paper presents a framework for modeling and implementing document databases taking into account security issues and applies it to a case study within the medical domain.

The remainder of this paper is organized as follows. In Sect. 2 the related work about security in Big Data is shown. Following this, in Sect. 3, our framework for modelling secure document databases including data set analysis, modelling, and implementation is defined. In Sect. 4, our framework is applied to a case study within the medical domain. Finally, Sect. 5 presents the conclusions and sketches future works.

2   Related Work

Nowadays, the exchange of a great volume of data is an usual activity in several domains (scientific, medical, citizens, etc.). But current tools and technologies are not designed for managing the variety, velocity and complexity of these massive data sets [4,7], neither to incorporate adequate security and privacy constraints [8,10,11]. Actually, we do not have (or we do not know how to correctly apply) security policies to assure (confidentiality, privacy, integrity, etc.) in Big Data domains [6,9,11,12].

Furthermore, the current proposals provide isolated security mechanisms, such as the possibility of anonymizing and encrypting the data [5], the definition of graph-based reputation models [14], the signature encryption [13] or the creation of specific signatures for Big Data [2]. These are not enough since security concepts should be considered from design in order to be considered in design decisions and finally implement robust solutions that perfectly fits them.

3   Framework Proposed

The framework proposed in this paper is composed of three stages. First, we carry out an analysis of the data set in order to automatically detect and tag sensitive data by applying NLP and ontological techniques. Then, designer use that information as recommendations to model the structural and security aspects of the database. In this stage, designer decides which NoSQL database technology to use (document, columnar or graph) and uses the metamodel proposed. Finally, the implementation of the system into a specific database tool is automatically generated from that model by applying the mappings defined in our methodology.

In this paper we have defined a path for document databases, composed of: data set analysis, modelling for document database technology and implementation into MongoDB. In further works we plan to expand our framework including other technologies (columnar, graph, etc.) and destination tools. The implementation of our proposal (metamodels, transformation rules, validation example, etc.) has been carried out by using Eclipse modeling tools (with Epsilon languages) and can be public accessed[1].

3.1   Data Set Analysis

Firstly, starting from a data set (for instance in a CSV format), each field is analysed searching for sensitive information. This phase of analysis is reusable, regardless of the technology (for instance, document, columnar or graph databases) used in the following phases.

We use the lexical resource WordNet 3.1[2] that a domain expert enriches by including a security level for the concepts which contains. By default, all the concepts have the lowest security level (SL = 1) and he increases them to sensitive information (SL = 2) and very sensitive information (SL = 3). These concepts are searched in WordNet, the new security level is modified and it is propagated to all their children.

It is noteworthy that the NLP techniques are suitable to be applied on fields that contain different concepts and they are expressed in natural language (for instance the diagnosis or the medical history of a patient). After undertaking a lexical-morphological analysis of the field content (POS tagging) and a partial syntactic analysis (partial parsing) the main concepts of the text are detected. In addition, these techniques can solve specific problems of language, such as ellipses, anaphoric references, ambiguities, etc. present in the text.

The analysis process is carried out by assigning a security level to each text field and all the information obtained is used as recommendations for the designer to model the data set.

---

[1] https://github.com/GSYAtools.
[2] http://wordnetweb.princeton.edu/perl/webwn (visited on April, 2018).

3.2   Conceptual Modelling

In this stage the designer models the database according to a metamodel proposed in this paper (Fig. 1) and taking into account the recommendations obtained in the previous stage. This metamodel allows the specification of both structural and security aspects related with this document databases.

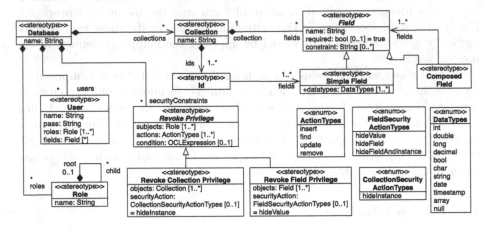

Fig. 1. Metamodel for secure document databases

It permits to model structural aspects such as *Databases* (as *Packages*), *Collections* and *Fields*: that can be simple fields or fields composed of several simple or composed fields. Furthermore, it allows to specify identificators for collections, data types, constraints, etc.

The security configuration of the system is defined by using a Role-Based Access Control (RBAC) policy in which users are classified into a hierarchical structure of security roles (*Role*). Once the security configuration has been established, we define security constraints by considering an open world approach in which users have privileges over any object if there are not any specific rule for revoking them. Then, we have to revoke permissions by defining security rules (*Revoke Privilege*) that indicates the actions (insert, find, update and remove) that certain subjects (*Role*) cannot carry out over certain objects (*Collections* or *Fields*). Depending on the objects a different set of security actions such as hiding values, fields, instances, etc. can be selected. Furthermore, security constraints can include a condition to be evaluated at execution time, for instance to check if the value of certain field is different of a given one.

3.3   Implementation

The last stage is the implementation of the modelled database into a specific document database management tool (such as MongoDB, CouchDB, etc.). Our model represents all the concepts needed for its implementation in different tools, but we have to define how to map the model to each destination tool. Once we have defined these mappings, they are implemented as model-to-text transformations and integrated in a Model Driven Engineering approach that allows us to automatically obtain the final implementation.

In this paper, MongoDB has been considered as the destination tool and the necessary transformation rules has been also defined identifying two stages:

1. **Database structure implementation.** The creation of the database, followed by the definition of the collections which store the data, by using a JSON schema validator for each one. With this JSON schema, it is defined the collection's simple fields along with their constrains, data types and if they are required. Also, it includes the definition of the collection's composed fields. Finally, a unique index is created over every ID field or set of fields in a collection.

2. **Security constraints implementation.** In this second stage, a user and an associated role is created for every role defined in the model. Then, for each role, all the security constraint affecting them are checked. If there is a *Revoke field privilege* constraint defined for a given role over a collection, it means that this role has its access to the affected collection limited. In this scenario, it is needed to create a view over the collection with the properly conditions, in order to filter or hide the necessary fields of the collection for the role. On the other hand, if for the role it is defined a *Revoke collection privilege* constraint, it would also be needed to create a view that, in this case, filters the instances that can no be seen by this role in the properly collection, given a set of conditions. Moreover, if the role has different *Revoke collection or field privilege* constraints over the same collection, it would be needed to create a view that implements all the filters at once. Finally, if the *Revoke collection privilege* constraint has not an *Action Type*, it means that the role has not the properly permissions over any of the instances of the collection, so it will be enough to revoke them directly over the collection, without the need of creating a new view. Finally, the strategy for revoking privileges depends of the security action selected: hide values is implemented inserting null values, hide fields with projections and hide instances with matches.

4   Case Study

The data set used in this experimentation is a synopsis that represents the clinical care at 130 hospitals between the years 1999 and 2008. This data set is available in the UCI Machine Learning Repository [1].

4.1   Data Set Analysis

In this case, the domain expert identify as sensitive ($SL = 2$) the concepts related to treatments, medicaments and medical specialty, and very sensitive ($SL = 3$) some specialties (Oncology, etc.).

After carrying out the analysis we obtain the following recommendations:

Patient Fields. Race: Level 3; Address: Level 2; Remaining fields: Level 1.

Admission Related Fields. Admission type and treatment: Level 2; Medical specialty: Level 2 (and if it is oncology: Level 3); Diagnosis: Level 2; Remaining fields: Level 1.

4.2   Conceptual Modelling

Figure 2 shows the model designed for our case study. It presents two collections: *Patient* that manages information about patient id, name, surname, race, etc. with their data types and constraints associated and *Admission* that focuses on information related with encounters, diagnosis and treatments, that are pairs of medicaments and doses embedded data of in *Admission* collection.

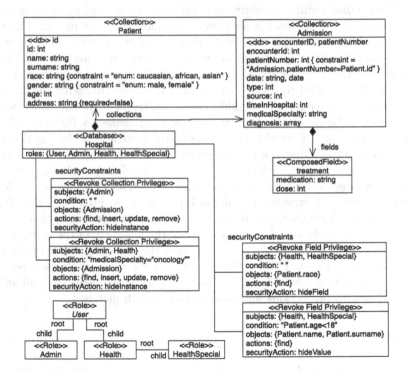

Fig. 2. Case study: model

Next, the security aspects are modeled creating several roles (*Admin, Health* and *HealthSpecial*) and security constraints. For instance, a security constraint over *Admission* collection is defined according to the recommendation of the previous stage. It represents that admissions with a medical specialty related with oncology will be only granted to the role *HealthSpecial*. In this way, the permission defined revokes the execution of any action to the roles *Admin* and *Health* if the medical specialty is oncology. To achieve this goal applies a security action that hides to these unauthorized roles all the instances with oncology values.

4.3 Implementation

The system is automatically implemented into MongoDB by applying the set of transformation rules defined in our proposal. As it was mentioned before, first the database structure is generated and then, the security constraints.

For instance, the *Health* role will be limited to execute the find operation over the view *Patients_View* and has a *Revoke collection privilege* constraint in the *Admission* collection which indicates that it can't access to instances in this collection that have the value *oncology* in the *medical_speciality* field. That means that a new view that filters these instances must be created for this role, in order to have permission to read data from the rest of instances.

5   Conclusions

Currently, large amounts of data handled by Big Data technologies can expose sensitive enterprise and personal information if they are not correctly assured. This is often the case due to the focus on functionality first, security later of existing Big Data technologies, such as document, columnar, or graph databases, combined with an absence of security approaches tailored to the Big Data domain.

In order to tackle this problem, we have presented a framework to assure document databases. The framework is composed of three stages: (1) analysis of the data set in order to automatically detect and tag sensible data by applying NLP and ontological techniques. (2) Modelling the structural and security aspects of the database; in this stage, the NoSQL database technology to use (document, columnar or graph) is defined and the metamodel proposed in our framework for document databases is used. (3) Implementation of the database into a specific tool. This implementation is automatically generated from the previous model according to the mappings defined in our framework. In this paper we have consider MongoDB as the final tool.

Our framework presents several advantages: (1) the effort required to protect the data is reduced, as part of the process is automated, (2) it can be easily applied to other NoSQL technologies by adapting the metamodel, and (3) it is aligned with existing standards, thus facilitating the application of recommendations and best practices.

The applicability and benefits of our proposal have been shown by presenting a case study involving a data set from the medical domain in which: the analysis of the data is carried out to establish security recommendations; the conceptual model is specified according to our metamodel; the implementation is generated for MongoDB.

As part of our future work, we want to automate our process for document databases extracting database model from the data set to use it as base model, that will be modified by the designer in order to include security rules. We would like also to include in this base model, the output of the data set analysis. Furthermore, we plan to extend our framework to columnar and graph database technologies.

Acknowledgments. This work has been developed within the SEQUOIA Project, funded by Fondo Europeo de Desarrollo Regional FEDER and Ministerio de Economía y Competitividad, (TIN2015-63502-C3-1-R) (MINECO/FEDER).

References

1. Andrew, F., Arthur, A.: UCI machine learning repository. http://archive.ics.uci.edu/ml. irvine, ca: University of california. School of Information and Computer Science, 213 (2010)
2. Hou, S., Huang, X., Liu, J.K., Li, J., Xu, L.: Universal designated verifier transitive signatures for graph-based big data. Inf. Sci. **318**, 144–156 (2015)
3. Jurjens, J.: Secure Systems Development with UML. Springer, Heidelberg (2004)
4. Kshetri, N.: Big data's impact on privacy, security and consumer welfare. Telecommun. Policy **38**(11), 1134–1145 (2014)
5. La Fuente, G.: The big data security challenge. Netw. Secur. **1**, 12–14 (2015)
6. Michael, K., Miller, K.: Big data: new opportunities and new challenges [guest editors' introduction]. Computer **46**(6), 22–24 (2013)
7. NIST: Big Data Interoperability Fremework, Security and Privacy. Big Data Public Working Group, vol. 4 (2017)
8. Okman, L., Gal-Oz, N., Gonen, Y., Gudes, E., Abramov, J.: Security issues in NoSQL databases. In: Proceedings of 10th IEEE International Conference on Trust, Security and Privacy in Computing and Communications (TrustCom) (2011)
9. RENCI/NCDS. Security and privacy in the era of big data. White paper (2014). http://www.renci.org/wp-content/uploads/2014/02/0313WhitePaper-iRODS.pdf
10. Saraladevi, B., Pazhaniraja, N., Paul, P., Basha, M.S., Dhavachelvan, P.: Big data and hadoop-a study in security perspective. Procedia Comput. Sci. **50**, 596–601 (2015)
11. Thuraisingham, B.: Big data security and privacy. In: 5th ACM Conference on Data and Application Security and Privacy, pp. 279–280. ACM (2015)
12. Toshniwal, R., Dastidar, K., Nath, A.: Big data security issues and challenges. Int. J. Innov. Res. Adv. Eng. (IJIRAE) **2**(2), 15–20 (2015)
13. Wei, G., Shao, J., Xiang, Y., Zhu, P., Lu, R.: Obtain confidentiality or/and authenticity in big data by ID-based generalized signcryption. Inf. Sci. **318**, 111–122 (2015)
14. Yan, S.R., Zheng, X.L., Wang, Y., Song, W.W., Zhang, W.Y.: A graph-based comprehensive reputation model: exploiting the social context of opinions to enhance trust in social commerce. Inf. Sci. **318**, 51–72 (2015)

Conceptual Modeling for Machine
Learning and Reasoning I

Leveraging the Dynamic Changes from Items to Improve Recommendation

Zongze Jin[1,2], Yun Zhang[1,2(✉)], Weimin Mu[1,2], Weiping Wang[2,4], and Hai Jin[2,3]

[1] School of Cyber Security, University of Chinese Academy of Sciences, Beijing, China
[2] Institute of Information Engineering, Chinese Academy of Sciences, Beijing, China
{jinzongze,zhangyun,muweimin,wangweiping}@iie.ac.cn
[3] School of Computer Science and Technology, Huazhong University of Science and Technology, Wuhan, China
hjin@hust.edu.com
[4] National Engineering Research Center Information Security Common Technology, NERCIS, Beijing, China

Abstract. User-generated reviews contain rich information, which has been ignored by most of recommender systems. Recently, some recommender systems using reviews with deep learning techniques have demonstrated that they can potentially alleviate the sparsity problem and improve the quality of recommendation. However, they only consider the dynamic interests from users but ignoring the changed properties of items. In this paper, we present a deep model which can capture not only the common users behaviors, the changed users interests and fundamental item properties, but also the changed properties of items. Experimental results conducted on a variety of datasets demonstrate that our model significantly outperforms all baseline recommender systems.

Keywords: Recommender system · Dynamic item reviews
Deep learning

1 Introduction

Recently, the variety and the number of products and services have increased dramatically. With the development of recommender systems, numerous approaches have been proposed by researchers. The most prominent one is the Collaborative Filtering (CF) techniques [22], which use sharing as the main idea. Generally speaking, recommender systems recommend customers products and services based on their preferences, needs, and history behaviors. Many CF approaches are based on matrix factorization (MF), which can not only find out the hidden factors, but also learn their importance for each user and how each of them satisfies each factor.

© Springer Nature Switzerland AG 2018
J. C. Trujillo et al. (Eds.): ER 2018, LNCS 11157, pp. 507–520, 2018.
https://doi.org/10.1007/978-3-030-00847-5_37

Although CF techniques have shown the good performance, they still have limitations. First of all, CF techniques cannot solve the sparsity problem very well. Another one is the poor interpretability of CF techniques.

The approach that uses reviews can address the above issues [17, 18]. In many recommender systems, other than the numeric ratings, users can write reviews for the products. Reviews contain rich information, such as users' sentiments, reasons, interests and so on. The existing CF techniques focus on numeric ratings but ignoring the abundant information in the review text.

Recently, some studies [17, 18] have shown that using review text can improve the prediction accuracy, especially for items and users with few ratings [28]. The typical task with reviews is Hidden Factors and Hidden Topics (HFT) [18]. With the development of deep learning techniques, some deep learning based methods [23, 24, 35] with reviews have been proposed. However, Deep Cooperative Neural Networks (DeepCoNN) [35] and User Word Composition Vector Model (UWCVM) [24] ignore users' changed interests and item properties. Temporal Deep Semantic Structured Model (TDSSM) [23] only takes the changed interests from users into account, but it neglects the changed items.

In this paper, we propose a novel method, named Static-Dynamic Features Capture Networks (SDFCN), for item recommendation that considers the changed interests of users and the changed properties of items. For example, you give a good review to a hotel at first, because the ordered room is quiet and clean. However, when you come back the second time, you order a room near the elevator which is much smaller and noisier than the one you ordered last time. In this case, your reviews to the hotel will be changed.

To capture the dynamic reviews, we propose to model the users and the items separately: the aggregated reviews from a user are used to build a user-specific model and the aggregated reviews on an item are used to build an item-specific model. We use convolutional neural networks (CNN) to extract the static features which are unchanged. For a user, the static features contain age, sex and so on. And for an item, the static features are the description of items including color, materials and so on. Besides, we use long short term memory (LSTM) to extract the dynamic features. For a user, the dynamic features are his changed interests, habits and so on. For an item, the dynamic features are changed shape, changed taste and so on.

In this paper, we propose a model named SDFCN based on CNN [4] and LSTM [10], our contributions can be summarized as follows:

- We find a new phenomenon that the reviews can be changed with time, and the phenomenon can affect the recommendation accuracy.
- We present a deep model, SDFCN, which can capture not only the common users' behaviors, the changed users' interests and fundamental item properties, but also the changed properties of items.
- We compare SDFCN with the state-of-the-art methods on three real-world datasets to demonstrate its effectiveness. Our method also shows a good interpretive ability with reviews.

The rest of the paper is organized as follows. In Sect. 2, we represent a short review of the related work. Besides, we describe SDFCN in detail in Sect. 3. In Sect. 4, we present our experiments. Meanwhile, we analyze SDFCN and demonstrate its effectiveness. Finally, conclusions are presented in Sect. 5.

2 Related Work

In early studies, review rating prediction is a fundamental task in sentiment analysis. Pang and Lee [5] pioneer this field by regarding review rating prediction as a classification problem. Using reviews in rating prediction task is the first study that focuses on how to use reviews to get ratings [1, 30]. Most researches on using reviews for recommendation focus on topic modeling for items from review text, such as HFT and ratings meet reviews (RMR) [17]. HFT adopts a LDA-like topic model on review text for items, and uses matrix factorization to fit the ratings. RMR is different from HFT, which uses Gaussian mixtures to model the rating instead of MF techniques. TopicMF [2] is better than HFT, the former one learns the topics for each review and the latter one learns the topics for each item.

But these studies have one limitation that the textual similarity of these approaches is simply based on lexical similarity. They ignore the important semantic meaning. Besides, many researchers use bag-of-word (BOW) [5] to represent reviews. Then, the methods which employ topic modeling techniques are proposed, which suffer from the scalability problem and also cannot deal with cold start.

With the development of recommender systems, some researches [11, 23] leverage temporal modeling with reviews to improve the performance of recommendations. Temporal modeling is an important element in many tasks and it has been shown to improve results over no-temporal model. In earlier researches, Koren et al. [13] proposes that the matrix factorization model is extended to allow each user to have a base latent vector and another set of time dependent vectors. Xiong [32] proposes a bayesian extension to this work, which adopts regularization to replace prior distributions. By treating matrix factorization as an autoregressive model, Yu et al. [33] has adopted a temporal matrix factorization to predict next steps from historical data. Yuan et al. [34] uses collaborative filtering technique to compute explicit user similarity function and extends the similarity function to incorporate temporal pattern similarities between users. Li et al. [15] leverages long-term user history to provide coarse grain news recommendation for certain news groups.

Recently, more and more researchers [6, 9, 16, 19, 21, 23, 27, 29, 35] have adopted deep learning techniques to improve the performance of recommendation tasks. In music recommendation, many studies use deep models of CNN and DBN to learn the latent factors from music data [9, 27]. These models only consider the items' latent factors from items' content, but they ignore reviews. Besides, some studies focus on sentiment prediction and word embedding with

deep learning techniques [6,8,19] to improve the performance of recommendations. Besides, Zheng et al. [35] adopt DeepCoNN to learn hidden latent features for users and items jointly using two coupled neural networks. It verifies
that reviews can improve recommendation, but it ignores the dynamic features
of reviews. Tang et al. [24] use reviews to represent users, but ignoring user's
changed interests and item properties. However, using deep learning for temporal recommendation has not yet been extensively studied. Hidasi et al. [11] uses
Recurrent Neural Networks (RNN) for recommending shopping items to users
based on the users current session history. Song et al. [23] adopt TDSSM to
consider the changed interest from users, but it ignores some items which will
change.

3   Algorithm

3.1   Architecture

Figure 1 shows the architecture of the proposed model, SDFCN, which contains
two parts: user networks and item networks. User networks contain two tightly
coupled networks which are the CNN network to capture the static or long-
term feature of users and the LSTM network to gain the dynamic or short-term
property of users. Same with the user networks, item networks also consist of
two networks, which are the CNN network to capture the item attributes and
the LSTM network to capture the changed feature of items.

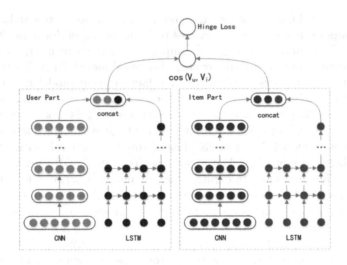

Fig. 1. The architecture of SDFCN

In the first layer, which is the look-up layer, we can gain the semantic feature
from matrices of word embeddings with reviews. In user networks, we use CNN

to capture the static feature of users and use LSTM to gain the changed users' interest. In item networks, we use CNN to learn the static properties of items and use LSTM to gain the changed properties. In the next layer, we calculate the cosine distance, and the top layer is the objective function, hinge loss function, which measures whether the users are interested or not.

3.2 Word Embedding

The embedding layer maps each non-zero feature into a dense vector representation. We consider that reviews contain both user information and item information. Firstly, we pre-train the dictionary of words such as word2vec. In current researches, some NLP applications, such as [4,7], demonstrate word embedding can enhance the performance. The document matrix D in our approach is:

$$D = \begin{pmatrix} & | & | & | & \\ \cdots & w_{i-1} & w_i & w_{i+1} & \cdots \\ & | & | & | & \end{pmatrix} \tag{1}$$

For user networks, we merge all the reviews written by user u into a single sequence S_n^u. In this look-up layer, we find the corresponding vectors and concatenate them. Then we write the input matrix V_{CNN}^u of CNN layers as:

$$V_{CNN}^u = concat(v(s_1^u), v(s_2^u), \cdots, v(s_k^u)) \tag{2}$$

where $v(s_k^u)$ denotes the corresponding c-dimensional word vector for the word s_k^u, which denotes the k-th word of the sequence S_n^u.

Same as the user networks, we merge all the reviews about item i into a single sequence S_n^i. The input matrix V_{CNN}^i of CNN layers in item networks is:

$$V_{CNN}^i = concat(v(s_1^i), v(s_2^i), \cdots, v(s_k^i)) \tag{3}$$

where $v(s_k^i)$ denotes the corresponding c-dimensional word vector for the word s_k^i, which is the k-th word of the sequence S_n^i. For LSTM, given a set of sentences $s = \{s_1, s_2, ..., s_{l_i}\}$, $s_i \in s$ is a sentence with l_{s_i} words $s_i = (w_1, w_2, ..., w_{l_{s_i}})$. L is a lookup table to encode words w_i into a distributed vector representation $x_i = L(w_i)$ $x_i \in \mathbb{R}^{n_I}$, where n_I indicates the dimension of the distributed vector. For the user part, we use V_{LSTM}^u to represent the input of LSTM. For the item part, we use V_{LSTM}^i to represent the input of LSTM. Using word embeddings, we can get the semantics of reviews, and it should be better than the bag-of-words techniques.

3.3 User Networks and Item Networks

Our model contains user networks and item networks. Each of them includes a CNN model and a LSTM model. In this section, we present the CNN layers, LSTM layers and the output layers of the user networks, and the item networks are similar with the user networks.

CNN Layers. A convolutional neural network extracts the static feature in our approach, which includes convolutional layers, max pooling and fully connected layers. We use the CNN model proposed in [14]. For each convolutional kernel K_j, we can get the output c_i^j from the convolutional layer. In our model, we adopt ReLUs (Rectified Linear Units) [20] as our activation function to avoid the problem of vanishing gradient. In [14], they demonstrate that ReLU can get better performance than other non-linear activation functions. Then, the convolutional feature c^j of a sequence is constructed. The pooling layers extract representative features from convolutional layers. Our task follows [7], and we can use max pooling operation to get the reduced fixed size vector. The results from the max-pooling layer are passed to a fully connected layer with a weight matrix W. Last we obtain the output X_{cu}. The details are as follows:

$$c_i^j = ReLU(V_{CNN}^u * K_j + b_j) \tag{4}$$

$$c^j = [c_1^j, c_2^j, \cdots, c_i^j, \cdots, c_k^j] \tag{5}$$

$$m = [max(c^1), max(c^2), max(c^3), ..., max(c^n)] \tag{6}$$

$$X_{cu} = f(W * m + b_{cu}) \tag{7}$$

where K_j is convolutional kernel, $*$ is the convolutional operator, b_j is a bias for V_{CNN}^u and b_{cu} is the bias term for X_{cu}.

LSTM Layers. The other part of user network is long short-term memory (LSTM). Both feed-forward neural network and recurrent neural network have been frequently used in text categorization, sequential word generation and so on. LSTM is also attracting more and more attention especially in speech recognition.

We adopt the LSTM model in our user network to capture the dynamic feature of the users. Each LSTM unit at time t contains a memory cell c_t, an input gate i_t, a forget gate f_t and the output gate o_t. We use the previous hidden state h_{t-1} and the current input z_t to compute these gates:

$$[f_t, i_t, o_t] = \sigma[W_{lstm}[h_{t-1}, z_t] + b_{lstm}] \tag{8}$$

The memory cell c_t is updated by partially forgetting the existing memory and adding a new memory content l_t:

$$l_t = tahn[V[h_{t-1}, z_t] + d] \tag{9}$$

$$c_t = f_t * c_{t-1} + i_t * l_t \tag{10}$$

Once the memory content of the LSTM unit is updated, the hidden state at time step t is:

$$h_t = o_t * tahn(c_t) \tag{11}$$

The update of the hidden states of LSTM at time step t is denoted as $h_t = LSTM(h_{t-1}, z_t)$. The dynamic feature output by user network X_{lu} is $X_{lu} = h_t$.

The Output Layer. We use concatenation operator to connect the static feature of CNN and the dynamic feature of LSTM. Note that the computing process of the static feature of items X_{ci} and the dynamic feature of items X_{li} are the same as that of X_{cu} and X_{lu}, respectively. The output vector of the user network is $V_U = concat(X_{cu}, X_{lu})$ and that of item network is $V_I = concat(X_{ci}, X_{li})$.

We use the cosine distance between V_U and V_I as the similarity between users and items.

$$cos(V_U, V_I) = \frac{\sum_{i=1}^{n}(V_U * V_I)}{\sqrt{\sum_1^n(V_U)^2} * \sqrt{\sum_1^n(V_I)^2}} \tag{12}$$

Finally, we use hinge loss [31] as the loss function to train our model.

$$L = max(0, 1 - t_h \cdot cos(V_U, V_I)) \tag{13}$$

where $t_h = \pm1$ represents whether the users are interested in the items ($t_h = 1$) or not ($t_h = -1$).

3.4 Network Training

We take the derivative of the loss with respect to the whole set of parameters through back-propagation, and use stochastic gradient descent with mini-batch to update the parameters. Word vectors are learned with word2vec[1]. We empirically set vector dimension d as 512. We use dropout to avoid the neural network being over-fitting.

4 Experiments

4.1 Data Sets

In order to evaluate the effectiveness of our model, we conduct experiments on three real datasets. We show the statistical information in Table 1.

Yelp16: It is a large-scale dataset consisting of restaurant reviews from Yelp. It is released by the seventh round of the Yelp Dataset Challenge in 2017. There are more than 1M ratings and reviews in Yelp2016[2].

Skytrax: It is an air travel dataset consisting of user reviews from Skytrax[3]. This dataset can be downloaded at Github[4].

Trip: It is one of the world's top travel sites[5]. We collect the dataset from 2015 to 2016.

---

[1] https://code.google.com/p/word2vec.
[2] http://www.yelp.com.
[3] http://www.airlinequality.com.
[4] https://github.com/quankiquanki/skytrax-reviews-dataset.
[5] http://www.trip.com.

Table 1. The statistic of datasets

Dataset	#users	#reviews	Scale	len_{avg}	V
Yelp16	70817	335018	1–5	75.1	137816
Skytrax	8741	41396	1–10	41.4	81241
Trip	65712	254142	1–5	67.8	136012

In this table, #users and #reviews are the number of users and reviews, respectively, len_{avg} is the average length of reviews in each dataset, V is the vocabulary size of words. In all datasets, each review consists of less than 80 words on average.

4.2   Comparison Methods

In the experiments, we evaluate our model by comparing with several state-of-the-art approaches on review recommendation:

LDA [3]: Latent Dirichlet Allocation is a well-known topic modeling algorithm presented in [3].

HFT [18]: Hidden Factors and Hidden Topics is proposed to employ LDA to learn a topic distribution from a set of reviews for each item. By treating the learned topic distributions as latent features for each item, latent features for each user are estimated by optimizing rating prediction accuracy with gradient descent.

CNN [4,6]: Convolution Neural Network is a state-of-the-art performer on sentence-level sentiment analysis tasks.

LSTM [10]: Long Short-Term Memory is an application in sentence-level sentiment analysis with RNN.

DeepCoNN [35]: Deep Cooperative Neural Networks (DeepCoNN) learns hidden latent features for users and items jointly using two coupled neural networks such that the rating prediction accuracy is maximized.

TDSSM [23]: Temporal Deep Semantic Structured Model shows that deep learning can be effective in learning temporal recommendation models by combining traditional feedforward networks (DSSM) with recurrent networks (LSTM).

4.3   Metrics

In our experiments, we use 3 popular metrics for evaluation: mean squared error (MSE), precision at position 1($P@1$) and 5($P@5$) and Mean Reciprocal Rank (MRR).

$$MSE = \frac{1}{N} \sum_{i=1}^{N} (r_i - \hat{r_i})^2 \tag{14}$$

$$P@N = \frac{\sum_{i=1}^{|N|} hasTrue(N_i)}{|N|} \tag{15}$$

where $hasTrue(N_i) = 1$ means the i-th item N_i should be recommended, and vice versa.

$$MRR = \frac{1}{|Q|} \sum_{i=1}^{|Q|} \frac{1}{rank_i} \tag{16}$$

where $|Q|$ is the number of query and $rank_i$ refers to the rank position of the first relevant document for the i-th query.

4.4 Settings

We divide each dataset shown in Table 1 into three sets of training set, validation set and test set. We split the original corpus into train, val and test sets with a 80:10:10 split.

For our model, SDFCN, we present some parameters in Fig. 2. We show the performance of SDFCN on the validation set of Yelp2016, latent factors n from 5 to 100, K_j from 10 to 400 and l from $\{7, 14, 21, 30, 60, 120, 180\}$ to investigate its sensitivity. We can easily get the conclusion that the model is stable when the number of latent factors n ranges from 40 to 80, the number of kernels c_k ranges from 50 to 200 and the length of LSTM l ranges from 60 to 180, respectively. So we use 50 latent factors, $c_k = 100$ and $l = 30$. We use word2vec as the pre-trained word embeddings.

We conduct ten-fold cross-validation on training set and validation set to tune the best hyper-parameter of each baseline. For LDA, we set $K = 10$ for LDA. For HFT, we utilize HFT-10 and HFT-50 to show the impact of the number of latent factors. Then we set $\alpha = 0.1$, $\lambda_u = 0.02$ and $\lambda_v = 10$ for HFT on validation set.

In addition, we list the important parameters for deep learning approaches. For CNN and LSTM, we set word embedding size as 512. We choose the previous one-month reviews as short-term history. Thus we set the length of the LSTM to 30. For DeepCoNN [35], we set $|x_u| = |y_i| = 50$ and $n_1 = 100$. And other hyper-parameters: t, c, λ and batchsize are set as 3,300, 0.002 and 100, respectively. We used word2vec as the pre-trained word embeddings. For TDSSDM, we use the implementation of Sent2Vec from [12]. The output embedding is set to 300. Same as LSTM, we set 30 as the length of the LSTM for TDSSM. Optimizations terminate until convergence or 200 learning epochs.

Our models are implemented in Keras[6]. The CNNs and RNNs in our models are implemented using CNN and LSTM. All models are trained and tested on an NVIDIA GeForce GTX1080 GPU.

4.5 Experimental Results

Performance Evaluation. In order to evaluate effectiveness of our method, we first show our experimental results with MRR in Fig. 3 and we provide a range of recommendation list sizes ranging from 5 to 25 with 5 as the increment.

---

[6] http://keras.io.

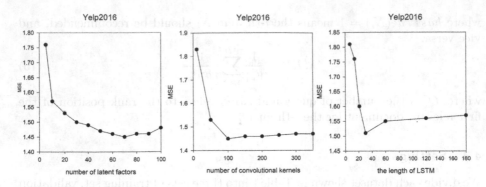

Fig. 2. The impact of the number of latent factors, convolutional kernels of CNN and the length of LSTM on the performance of SDFCN in terms of MSE (*Yelp*16 *Dataset*)

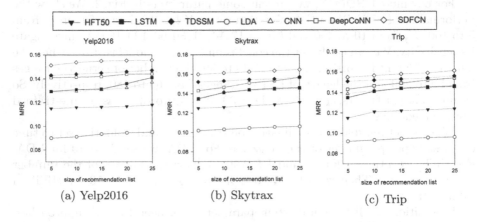

Fig. 3. Experimental results of the compared methods using MRR

From Fig. 3, we have the following observations: (1) Using CNN or LSTM with review text is better than LDA. (2) Both the deep model DeepCoNN and TDSSM perform better than single deep learning techniques, like CNN or LSTM. The result demonstrates that the deep model with review text is better than the approaches without it. (3) SDFCN improves the performance of each review prediction task significantly on the big dataset like Yelp16 and small dataset like Skytrax.

The performance of SDFCN and the baselines are reported in terms of $P@1$, $P@5$ and MRR in Table 2. Table 2 shows the results on the three datasets. The experiments are repeated 5 times, and the averages are reported with the best performance shown in bold.

In Table 2, all models perform better on *Skytrax* than on the others. It is mainly due to the sparsity of *Yelp16* and *Trip*.

From the results we can see that, the LDA method is the worst among all the methods. Both HFT-10 and HFT-50 perform better than LDA. In addition, we can observe the deep learning techniques perform better than the traditional methods. Using review text with deep learning can achieve better performance.

Finally, SDFCN achieves the best performances among most methods in terms of all the measures. The results demonstrate that by using dynamic item reviews, we can capture more information to achieve better performance.

Table 2. P@1, P@5 and MRR comparison with baselines

Method	Yelp16			Strytrax			Trip		
	P@1	P@5	MRR	P@1	P@5	MRR	P@1	P@5	MRR
LDA	0.157	0.072	0.092	0.178	0.091	0.104	0.152	0.082	0.095
HFT-10	0.156	0.072	0.112	0.179	0.092	0.127	0.161	0.084	0.121
HFT-50	0.168	0.079	0.118	0.184	0.095	0.13	0.167	0.091	0.129
CNN	0.201	0.112	0.131	0.210	0.121	0.142	0.198	0.101	0.142
LSTM	0.199	0.097	0.133	0.211	0.106	0.142	0.201	0.107	0.143
DeepCoNN	0.227	0.131	0.143	0.240	0.139	0.150	0.226	0.125	0.149
TDSSM	0.231	0.134	0.145	0.242	0.145	0.151	0.230	0.129	0.153
SDFCN	**0.239**	**0.141**	**0.154**	**0.250**	**0.156**	**0.162**	**0.236**	**0.137**	**0.158**

The Effect of Word Embedding. In our experiments, we investigate the effect of word embedding on review rating prediction. We try the randomly initialized word vectors (Random), the word vectors learned from SkipGram and the sentiment-specific word embeddings learned from SSWE [26] and SSPE [25].

From Fig. 4, we find that all pre-trained word vectors outperform randomly initialized word vectors. Compared with SkipGram, SSWE does not yield improvements. The assumption is reasonable for tweets as they are short, but it is unsuitable for the document-level reviews where negation and contrast phenomenons are frequently appeared. And SSPE performs slightly better than others, as it optimizes the word vectors by using a global document vector to predict the sentiment of a review. Nevertheless, sentiment embeddings do not obtain obvious performance boost than word2vec in our experiments.

Comparing LSTM with GRU. In order to measure the impact of choosing different recurrent neural networks on the model, we show the results which compare LSTM with GRU in Fig. 5. The length of RNN is from $\{7, 15, 30, 60, 120, 180\}$. From Fig. 5, by comparing three different datasets, we can conclude that GRU is significantly worse than LSTM. Meanwhile, we observe that the model gains the best performance when we set 30 as the length of RNN. In order to ensure the best performance of our model, we adopt LSTM to capture the dynamic features.

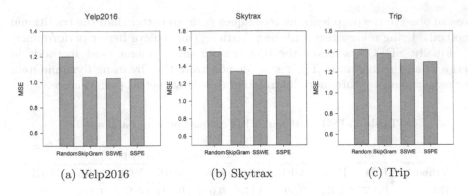

Fig. 4. Comparing different word embeddings on the datasets

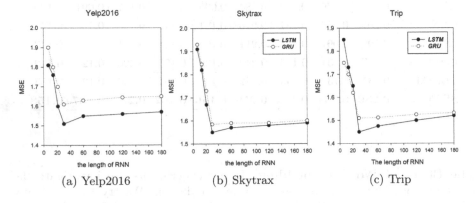

Fig. 5. Comparing LSTM with GRU

5 Conclusion

In this paper, we propose a novel temporal model based on deep learning, Static-Dynamic Features Capture Networks (SDFCN). The SDFCN model contains two key parts: the User Part and the Item Part. SDFCN can capture not only the common users behaviors, the changed users interests and fundamental item properties, but also the changed properties of items. The reviews of some datasets (e.g., hotels, trips or restaurants) will change according to the user's different experience. Thus, our model can improve the preformance of recommendations. However, the users usually do not change their attitudes on some reviews like products. In this situation, our model can improve the performance weakly. At last, experiments on real-world datasets demonstrate that our model can outperform state-of-the-art baselines consistently under different evaluation metrics.

Acknowledgment. This work was supported by National Key Research and Development Plan (2016QY02D0402).

References

1. Baccianella, S., Esuli, A., Sebastiani, F.: Multi-facet rating of product reviews. In: Boughanem, M., Berrut, C., Mothe, J., Soule-Dupuy, C. (eds.) ECIR 2009. LNCS, vol. 5478, pp. 461–472. Springer, Heidelberg (2009). https://doi.org/10.1007/978-3-642-00958-7_41
2. Bao, Y., Fang, H., Zhang, J.: TopicMF: simultaneously exploiting ratings and reviews for recommendation. In: Twenty-Eighth AAAI Conference on Artificial Intelligence, pp. 2–8 (2014)
3. Blei, D.M., Ng, A.Y., Jordan, M.I.: Latent Dirichlet allocation. J. Mach. Learn. Res. 3(Jan), 993–1022 (2003)
4. Blunsom, P., Grefenstette, E., Kalchbrenner, N.: A convolutional neural network for modelling sentences. In: Proceedings of the 52nd Annual Meeting of the Association for Computational Linguistics (2014)
5. Bo, P., Lee, L.: Seeing stars: exploiting class relationships for sentiment categorization with respect to rating scales. In: Meeting on Association for Computational Linguistics, pp. 115–124 (2005)
6. Chen, Y.: Convolutional neural network for sentence classification. Master's thesis, University of Waterloo (2015)
7. Collobert, R.: Natural language processing from scratch. J. Mach. Learn. Res. 12, 2393–2537 (2011)
8. Dai, H., Wang, Y., Trivedi, R., Song, L.: Deep coevolutionary network: embedding user and item features for recommendation. In: KDD (2017)
9. Dieleman, S., Schrauwen, B.: Deep content-based music recommendation. In: International Conference on Neural Information Processing Systems, pp. 2643–2651 (2013)
10. Graves, A.: Supervised Sequence Labelling with Recurrent Neural Networks. Springer, Heidelberg (2012). https://doi.org/10.1007/978-3-642-24797-2
11. Hidasi, B., Karatzoglou, A., Baltrunas, L., Tikk, D.: Session-based recommendations with recurrent neural networks. Computer Science (2015)
12. Huang, P.S., He, X., Gao, J., Deng, L., Acero, A., Heck, L.: Learning deep structured semantic models for web search using clickthrough data. In: ACM International Conference on Conference on Information and Knowledge Management, pp. 2333–2338 (2013)
13. Koren and Yehuda: Collaborative filtering with temporal dynamics. Commun. ACM 53(4), 89–97 (2009)
14. Krizhevsky, A., Sutskever, I., Hinton, G.E.: Imagenet classification with deep convolutional neural networks. In: International Conference on Neural Information Processing Systems, pp. 1097–1105 (2012)
15. Li, L., Zheng, L., Yang, F., Li, T.: Modeling and broadening temporal user interest in personalized news recommendation. Expert Syst. Appl. 41(7), 3168–3177 (2014)
16. Li, S., Kawale, J., Fu, Y.: Deep collaborative filtering via marginalized denoising auto-encoder. In: Proceedings of the 24th ACM International on Conference on Information and Knowledge Management, pp. 811–820. ACM (2015)
17. Ling, G., Lyu, M.R., King, I.: Ratings meet reviews, a combined approach to recommend. In: Proceedings of the 8th ACM Conference on Recommender Systems, pp. 105–112. ACM (2014)
18. Mcauley, J., Leskovec, J.: Hidden factors and hidden topics: understanding rating dimensions with review text. In: ACM Conference on Recommender Systems, pp. 165–172 (2013)

19. Mitchell, J., Lapata, M.: Composition in distributional models of semantics. Cogn. Sci. **34**(8), 1388 (2010)
20. Nair, V., Hinton, G.E.: Rectified linear units improve restricted boltzmann machines. In: International Conference on International Conference on Machine Learning, pp. 807–814 (2010)
21. Salakhutdinov, R., Mnih, A., Hinton, G.: Restricted boltzmann machines for collaborative filtering. In: Proceedings of the 24th International Conference on Machine Learning, pp. 791–798. ACM (2007)
22. Sarwar, B., Karypis, G., Konstan, J., Riedl, J.: Item-based collaborative filtering recommendation algorithms. In: International Conference on World Wide Web, pp. 285–295 (2001)
23. Song, Y., Elkahky, A.M., He, X.: Multi-rate deep learning for temporal recommendation. In: International ACM SIGIR Conference on Research and Development in Information Retrieval, pp. 909–912 (2016)
24. Tang, D., Qin, B., Liu, T., Yang, Y.: User modeling with neural network for review rating prediction. In: International Conference on Artificial Intelligence, pp. 1340–1346 (2015)
25. Tang, D., Wei, F., Qin, B., Zhou, M., Liu, T.: Building large-scale twitter-specific sentiment lexicon: a representation learning approach. In: Proceedings of the 25th International Conference on Computational Linguistics (COLING 2014), pp. 172–182 (2014)
26. Tang, D., Wei, F., Yang, N., Zhou, M., Liu, T., Qin, B.: Learning sentiment-specific word embedding for twitter sentiment classification. In: Meeting of the Association for Computational Linguistics, pp. 1555–1565 (2014)
27. Wang, H., Wang, N., Yeung, D.Y.: Collaborative deep learning for recommender systems. In: ACM SIGKDD International Conference on Knowledge Discovery and Data Mining, pp. 1235–1244 (2015)
28. Wang, H., Lu, Y., Zhai, C.: Latent aspect rating analysis on review text data: a rating regression approach. In: ACM SIGKDD International Conference on Knowledge Discovery and Data Mining, pp. 783–792 (2010)
29. Wu, Y., DuBois, C., Zheng, A.X., Ester, M.: Collaborative denoising auto-encoders for top-n recommender systems. In: Proceedings of the Ninth ACM International Conference on Web Search and Data Mining, pp. 153–162. ACM (2016)
30. Wu, Y., Ester, M.: Flame: a probabilistic model combining aspect based opinion mining and collaborative filtering. In: Eighth ACM International Conference on Web Search and Data Mining, pp. 199–208 (2015)
31. Yichao, W., Liu, Y.: Robust truncated hinge loss support vector machines. Publ. Am. Stat. Assoc. **102**(479), 974–983 (2007)
32. Xiong, L., Chen, X., Huang, T.K., Schneider, J., Carbonell, J.G.: Temporal collaborative filtering with Bayesian probabilistic tensor factorization. In: SIAM International Conference on Data Mining, SDM 2010, 29 April–1 May 2010, Columbus, Ohio, USA, pp. 211–222 (2010)
33. Yu, H.F., Rao, N., Dhillon, I.S.: Temporal regularized matrix factorization, Computer Science (2015)
34. Yuan, Q., Cong, G., Ma, Z., Sun, A., Thalmann, N.M.: Time-aware point-of-interest recommendation. In: International ACM SIGIR Conference on Research and Development in Information Retrieval, pp. 363–372 (2013)
35. Zheng, L., Noroozi, V., Yu, P.S.: Joint deep modeling of users and items using reviews for recommendation. In: Tenth ACM International Conference on Web Search and Data Mining, pp. 425–434 (2017)

Mining Rules with Constants from Large Scale Knowledge Bases

Xuan Wang, Jingjing Zhang, Jinchuan Chen$^{(\boxtimes)}$, and Ju Fan

School of Information, Renmin University of China, Beijing, China
jcchen@ruc.edu.cn

Abstract. Rules or constraints can be used to clean a knowledge base, or find new facts which should have been included. Recently there are many efforts on automatically mining rules from large scale knowledge bases. However, these rules usually contain no constants. In practice, we often need some detailed rules, for example, rules restricted to a special country or a special profession. One major challenge of appending constants lies in that there are large amount of constants, each of which can generate a new rule. Moreover, we have to choose appropriate granularity in order to trade off between the applicability and precision (or support and confidence in traditional rule mining terminology). In this paper, we propose a Spark based solution to mine rules with constants, a taxonomy based approach to control the granularity, and several techniques to improve the efficiency. We also conduct extensive experiments to evaluate the efficiency and effectiveness of our solution with comparison with the state of the art works.

Keywords: Knowledge base · Rule mining · SPARK · Big data

1 Introduction

Recent years, large-scale RDF knowledge bases have been constructed [1], such as Freebase, YAGO and DBLP. These knowledge bases store structured information about real-world entities and the relationships between entities, which are formed as RDF triple (*subject, predicate, object*). RDF is an imporant tool for representing conceptual models like UML. There are also many applications built based on RDF knowledge bases, such as automated customer service, structured search and semantic graph search etc.

However, these knowledge bases are far from complete, missing many facts and also containing some incorrect data [2,3]. Thus recently there are lots of research on mining rules or constraints in order to complete knowledge bases [4–8]. These rules can be in different forms like function dependencies [9], denial constraints [7], and first-order logic formulas [10]. The rationale is that it should be a common rule for the whole knowledge base if it is correct for most triples. For example, suppose we find that the nationality of one person is probably the same as his/her birthplace, we may conclude the following rule ϕ:

$$isCitizenOf(x,y) \leftarrow wasBornIn(x,y) \tag{1}$$

© Springer Nature Switzerland AG 2018
J. C. Trujillo et al. (Eds.): ER 2018, LNCS 11157, pp. 521–535, 2018.
https://doi.org/10.1007/978-3-030-00847-5_38

Applying this rule to the knowledge base, we may find some missing facts about the nationality information.

Since the scale of these knowledge bases are quite large[1], traditional rule mining approaches like FOIL [11] or ILP [12] have limited scalability and cannot be applied to large-scale knowledge bases. The work in [10] proposed a Spark-based rule-mining approach and illustrates nice scalability. However, [10] can only obtain rules without constants. This may miss many high-quality rules. For example, the rule ϕ in Eq. 1 is not good enough because in some countries nationality does not depend on birthplaces. But apparently we can have some qualified rules if we restrict y to some countries like USA. That is, we refine ϕ to ϕ':

$$isCitizenOf(x,'USA') \leftarrow wasBornIn(x,'USA') \tag{2}$$

The rule ϕ' covers fewer facts than ϕ, but is obviously more precise. In the tasks of knowledge completion, we need rules with high precision. Thus it is necessary to introduce constants into rules.

The works in [7,8] are able to obtain rules with constants. But the approach proposed in [8] is not designed for the distributed computing environment and is hard to scale out for large data sets. The work in [7] focuses on denial constraints only. The denial constraints can only identify that some combinations of facts are incorrect. They are useful for detecting errors but not able to infer new facts.

The target of this paper is to mine Horn-clause formed rules with constants. Note that allowing constants improve the complexity greatly. Each rule (without constants) may have thousands of extended sub-rules by appending different constants. Similar to [10], we propose a Spark-based approach to obtain scalability. We find that the performance is quite poor when directly apply the approach of [10] to deal with rules with constants. The reason lies in that it takes too much time for loading and exchanging the large data set in the Spark cluster when evaluating candidate rules. We try to solve this problem by designing a novel method, which reduces the chances of data loading as much as possible.

We further find that it is necessary to finely control the granularity of the rules. In the process of learning rules, we can split a rule ϕ into sub-rules $\phi_1, \cdots, \phi_k$. Each $\phi_i(i = 1, \cdots, k)$ covers a sub-set of facts that are covered by ϕ, while the confidence[2] of ϕ_i may increase. Note that when we append a condition $x =' c'$ to ϕ, we restrict x in the most strict way. According to our experiments, we can only find a few rules when directly restricting variables to constants, because these rules are too strict to cover enough number of facts.

With this observation, we propose a taxonomy based approach to trade off between the granularity and the quality. Instead of restricting a variable x to constants, we try to restrict x to some domains. The domains are hierarchical and form a taxonomy tree. Hence we can control the granularity by restricting x to an appropriate node (domain) in the tree.

---

[1] Yago3 has more than 10 million facts, and Freebase has about 2.4 billion facts.
[2] We will define this in Sect. 2.3.

The contributions of this paper are listed as follows.

- We propose a scalable solution for mining rules with constants from large-scale knowledge bases.
- We design a series of techniques for reducing the chances of data loading and improve the efficiency.
- We conduct extensive experiments to evaluate the efficiency and effectiveness of our approach. We also utilize our solution for completing a real data set and report some interesting findings.

The remaining parts of this paper are organized as follows. We first explain the learning model and propose the taxonomy based approach in Sect. 2. Then we propose our solution in Sect. 3 and report our experimental results in Sect. 4. Section 5 summarizes the related works and finally we conclude this paper in Sect. 6.

2   Learning Model

In this section, we will discuss the model utilized when mining rules from knowledge bases, including language bias and the quality metrics. We will also propose a taxonomy-based model to refine rules by appending constants.

2.1   Language Bias

There are many different types of rules or constraints, which can be applied to RDF knowledge bases. In this paper, we focus on Horn clauses, which is a special kind of logic formulas with at most one positive atom in the head.

Horn clauses. A *Horn clause* consists of a head and a body, where the head is a single atom and the body is a set of atoms.

$$H(\bar{X}) \leftarrow \overrightarrow{B}(\bar{X})$$

where $H(\bar{X})$ is an atomic formula, $\overrightarrow{B}(\bar{X})$ is a conjunction of atomic formulas, and $\bar{X}$ is a vector of variables.

As an example, the rule in Eq. 1 specifies that we can infer that x is a citizen of y if x was born in y. Here, both *isCitizenOf* and *wasBornIn* are predicates and x,y are variables. When learning rules from RDF knowledge bases, all RDF predicates will be adopted as predicates such as *livesIn, wasBornIn, isLocatedIn and human* etc. Some of them receive only one argument, meaning that the argument variable has a property, e.g. $human(x)$ specifies that x is a human. The others receive two arguments, specifying that the two argument variables have a special relationship, e.g. $wasBornIn(x, y)$.

We will also introduce a special predicate, *equals*, in order to append constants into rules. This predicate receives two arguments. Specifically, $equals(x,' c')$ means that a variable x equals to a constant $'c'$. In this paper, we mark constants by circling them with quotes.

Furthermore, we only consider *connected* and *closed* Horn clauses to ensure that the result rules do not contain irrelative atoms. Two atoms are *connected* if they share at least one variable. A rule is *connected* if every two atoms in the rule are connected directly or transitively. A rule is *closed* when every variable appears more than once.

We also limit the length of all rules to be less than or equal to three. As reported by [10], 90.3% rules with length longer than or equal to four can be reduced to shorter rules. Hence these long rules provide little knowledge than short ones.

Take all the circumstances of permutations into account, all the rules (without constants) satisfying the above constraints can be grouped into the following six types.

1. $p(x, y) \leftarrow q(x, y)$
2. $p(x, y) \leftarrow q(y, x)$
3. $p(x, y) \leftarrow q(z, x) \wedge r(z, y)$
4. $p(x, y) \leftarrow q(x, z) \wedge r(z, y)$
5. $p(x, y) \leftarrow q(z, x) \wedge r(y, z)$
6. $p(x, y) \leftarrow q(x, z) \wedge r(y, z)$

2.2   Refining Rules with Constants

One of the major tasks of this paper is to refine rules with constants. An intuitive way of introducing constants is to replace some variables in rules by constants or append an atom with the *equals* predicate. For example, $isCitizenOf(x, y) \leftarrow equals(y,'USA'), wasBornIn(x, y)$. Note that this rule is logically the same as $isCitizenOf(x,'USA') \leftarrow wasBornIn(x,'USA')$. Thus we will not count in the atoms with the predicate *equals* when calculating the length of a rule.

Appending the *equals* predicate looks like a matter of course for introducing constants. However, we find that there are only a few qualified rules with the *equals* predicate. When the support and confidence[3] thresholds are set to 100 and 0.7 respectively, there are only 26 rules can be mined from knowledge base[4].

Note that, when appending an atom with the *equals* predicate into a rule ϕ, we exactly replace ϕ with a set of sub-rules, each of which is bound with a different constant. For example, the rule $isCitizenOf(x, y) \leftarrow wasBornIn(x, y)$ will generate more than two hundred thousand sub-rules if we append an atom $equals(y,'c')$ because there are 281,036 possible choices for $'c'$ in the knowledge base. It seems that the rules with the *equals* predicate are too refined and we need to adjust the granularity.

We observe that an argument of every predicate is bound with a specific domain defined in the schema of a knowledge base. For example, the x in $wasBornIn(x, y)$ must be a *person*. Moreover, a domain has sub-domains and sub-domains may also have sub-domains, which finally builds a taxonomy tree.

---

[3] We will explain the two metrics soon.

[4] In this paper, without otherwise specified, the default knowledge base is Yago3.

For example, the type *person* can be classified into *slave, worker* and so on. Therefore, instead of restricting x to some constant like 'Donald_Trump', it looks more meaningful to restrict x to some domains, e.g. 'USA_People'.

With this observation, we propose another way of refining rules with constants. In order to refine a rule ϕ, we append an atom with predicate $rdftype(x,'c')$, where $'c'$ can be the domain of x as defined by the schema or any sub-domain of it.

According to our experiments, the taxonomy-based approach obtains many more meaningful rules than constants-based approach, i.e. appending atoms with *equals*. Another merit of the taxonomy-based approach lies in that it is easy to control the granularity. We can trade off between the accuracy and coverage by choosing appropriate node in the taxonomy tree.

In the following, if a rule contains atoms of *equals* or *rdftype*, we call it *a rule with constants*. Otherwise, we call it a rule without constants.

2.3 Quality Metrics

We adopt the quality metrics in [8], i.e. the support (denoted by *supp*) and confidence (denoted by *conf*).

The support of each rule is defined as the number of distinct subject and object pairs in the head atom of all instantiations satisfying the rules in the knowledge base.

$$supp(H(x,y) \leftarrow \overrightarrow{B}(\bar{X})) := \#(x,y) : \exists z_1, ..., z_m : \overrightarrow{B}(\bar{X}) \wedge H(x,y) \qquad (3)$$

Here H(x,y) denotes the head atom and $\overrightarrow{B}(\bar{X})$ is the conjunction of a set of atoms in the body atom, and $z_1, ..., z_m$ are all the variables appearing in rule $\varphi(\bar{X})$ except x and y.

The confidence of each rule is defined to be the ratio of rule predictions in the knowledge base. It is obtained by dividing the support value by the number of distinct subject and object pairs in the head atom from all the instantiations satisfying the body atom in the knowledge base.

$$conf(H(x,y) \leftarrow \overrightarrow{B}(\bar{X})) := \frac{supp(H(x,y) \leftarrow \overrightarrow{B}(\bar{X}))}{\#(x,y) : \exists z_1, ..., z_m : \overrightarrow{B}(\bar{X})} \qquad (4)$$

Lemma 1. *Monotonity of Supp.* *Suppose ϕ' is a rule refined from ϕ by appending an atom of $rdftype(x,'c')$ (x can be any variable appearing in ϕ), then $supp(\phi') \leq supp(\phi)$.*

Proof. Without loss of generalization, suppose ϕ is $H(x,y) \leftarrow \overrightarrow{B}(\bar{X})$, and ϕ' is $H(x,y) \leftarrow \overrightarrow{B}(\bar{X}) \wedge rdftype(x,'c')$. It is not hard to see that for every valuation $v' \vDash \phi'$, there must exist a valuation v which is a sub-set of v' and $v \vDash \phi$. Since the head variables $\{x,y\}$ are contained by both ϕ and ϕ', we can obtain that $v' \upharpoonright_{\{x,y\}} = v \upharpoonright_{\{x,y\}}$. Therefore, $\{v \upharpoonright_{\{x,y\}} |v \vDash \phi'\} \subseteq \{v \upharpoonright_{\{x,y\}} |v \vDash \phi\}$.

Here, $v \vDash \phi$ means that ϕ is satisfied by the assignment v, and $v \upharpoonright_{\bar{x}} = < a_1, \cdots, a_n >$, if $a_i = v(\bar{x}[i])$ ($i = 1, \cdots, n$).

3   The Approaches

In this section, we will illustrate how to learn rules from knowledge bases. We will first explain how to learn rules based on the taxonomy tree in Sect. 3.1. Then we will analyze the bottleneck and propose two methods to improve the efficiency in Sects. 3.2 and 3.3.

3.1   Mining and Refining (MR)

Our first approach is called *mining and refining* (MR in short). As implied by the name, it contains two phases. In the first phase, we learn rules without constants from the knowledge base. Next, we try to refine the rules by appending the *rdftype* predicate.

The mining phase adopts the algorithm in [10]. The refining phase tries to extend each candidate rule ϕ to a set of sub-rules with constants. For each variable x in ϕ, we first load the domain c of x from the KB's schema. Then we traverse the taxonomy tree rooted at c. The traverse process is listed in Algorithm 1. It is basically a DFS. The lines 2–5 in Algorithm 1 specify the conditions of splitting a domain. We split a domain c if and only if the support score of the current rule (bound with domain c) is high enough and the confidence score is not qualified. Note that by splitting a domain, we may obtain rules with higher confidence scores, but the support scores cannot be improved (Lemma 1).

Algorithm 1. traverse(ϕ, c)

1 $\phi' \leftarrow rdftype(x,' c') \wedge \phi$;
2 **if** $supp(\phi') < \alpha$ **then**
3 $\quad\mid$ Return;
4 **else if** $conf(\phi') > \beta$ **then**
5 $\quad\mid$ Return;
6 **else**
7 $\quad\mid$ **for** *each sub-domain s of c* **do**
8 $\quad\mid\quad\mid$ traverse(ϕ, s);

In the experiments, we find that MR is quite slow. The reasons are two folds. Firstly, MR needs to extend each candidate rule ϕ to a set of sub-rules with constants by DFS, and for each sub-rule, we need to load the dataset to calculate its support and confidence score, which is very time-consuming. Secondly, each rule obtained in the mining phase has many candidate sub-rules, each of which requires an independent Spark job containing many unnecessary tasks. Next we will discuss how to improve the efficiency of MR.

Algorithm 2. MR-Batch

require : $facts1 = \{(pred, sub, obj)\}, facts2 = \{(pred, sub, obj)\}$
require : $rules = \{(ID, head, body)\}, types = \{(entity, type)\}$
1 **Join** facts1, facts2 and types on facts1.pred=rules.head and
facts2.pred=rules.body and facts1.sub=facts2.obj and facts1.obj=facts2.sub
and facts1.sub=types.entity;
2 **Distinct** the $\{(rules.head, rules.body, facts1.sub, facts1.obj, types.type)\}$;
3 **GroupBy** types.type, yielding a list of
$\{(rules.head, rules.body, facts1.sub, facts1.obj)\}$ pairs for each type;
4 **Count** $\{(rules.head, rules.body, facts1.sub, facts1.obj)\}$ pairs for each type,
yielding a list of $\{(type, supp)\}$ pairs;
5 **Join** supp and denominator(the denominator is calculated similar to the way of
calculating supp), yielding a list of $\{(type, supp, deno)\}$;
6 **Map** $\{(type, supp, deno)\}$ to $\{(type, supp, supp/deno)\}$;

3.2 MR-Batch

The MR algorithm needs to evaluate the quality of each refined rule, which is very inefficient because there are millions of refined rules and each of which requires an independent Spark job for quality evaluation. In this section, we propose an approach named MR-Batch, which evaluates the quality scores of all rules refined from one rule in the same Spark job.

Like MR, the MR-Batch approach also contains two phases, mining and refining. The difference lies in that now we replace Algorithm 1 with Algorithm 2. In Algorithm 1, we need to invoke a Spark job each time when we evaluate a rule. However, in Algorithm 2, we invoke only one Spark job if we want to refine a rule ϕ.

As illustrated in Algorithm 2[5], in Step 1 and Step 2, we join the tables and use the Distinct operator to deduplicate (x, y) pairs. Then, we group all tuples with the same key $(types.type)$ into one set and count $\{(rules.head, rules.body, facts1.sub, facts1.obj)\}$ pairs for each type, yielding a list of $\{(type, supp)\}$ pairs. The next step is to calculate the denominator of confidence and join the support and denominator scores. Finally, map each $\{(type, supp, deno)\}$ to $\{(type, supp, supp/deno)\}$ to yield a list of $\{(type, supp, conf)\}$.

3.3 Loading Data only once (LDOO)

MR-Batch is much more efficient than MR. Actually, for the tasks of mining rules from Yago3, MR cannot finish in two days, while the MR-Batch takes only about two hours. But MR-Batch still wastes too much time on repeatedly loading data and can be further improved.

---

[5] For ease of illustration, both Algorithms 2 and 3 are tailored for the second rule type.

We design a new algorithm named LDOO (Loading Data Only Once). The algorithm loads the whole data set and mines all rules through a single Spark task.

The basic idea of LDOO is simple. For a given rule type, like $p(x,y) \leftarrow q(y,x)$, we join facts (i.e. triples) according to the join conditions specified in the rule type. We also make another join with the $rdftype$ triples with the join condition x or y. Then we divide the whole join results into groups by the predicates and the constants in $rdftype$. We utilize the taxonomy information to generate the $rdftype$ triples. For example, if $rdftype(x, 'worker')$, we will also have $rdftype(x, 'person')$ since $'worker'$ is a sub-type of $'person'$.

Algorithm 3. LDOO

> **require** : $facts1 = \{(pred, sub, obj)\}, facts2 = \{(pred, sub, obj)\}$
> **require** : $types = \{(entity, type)\}$
> 1 **Join** facts1, facts2 and types on facts1.sub=facts2.obj and facts1.obj=facts2.sub and facts1.sub=types.entity;
> 2 **Distinct** the $\{(facts1.pred, facts2.pred, facts1.sub, facts1.obj, types.type)\}$;
> 3 **GroupBy** (facts1.pred, facts2.pred, types.type), yielding a list of $\{(facts1.sub, facts1.obj)\}$ pairs for each key;
> 4 **Count** $\{(facts1.sub, facts1.obj)\}$ pairs for each key, yielding a list of $\{((p, q, t), supp)\}$ pairs;
> 5 **Join** supp and denominator(the denominator is calculated similar to the way of calculating supp), yielding a list of $\{((p, q, t), supp, deno)\}$;
> 6 **Map** $\{((p, q, t), supp, deno)\}$ to $\{((p, q, r, t), supp, supp/deno)\}$;

The details of the LDOO algorithm are illustrated in Algorithm 3 with Spark primitives. In Step 1 and Step 2, we join tables and deduplicate the pairs. In Step 3 and Step 4, we group tuples with same key $\{(facts1.pred, facts2.pred, types.type)\}$ into one set and count the number of pairs for each key, yielding a list of $\{((p, q, t), supp)\}$ pairs. Then, calculate the denominator of confidence. Finally, we join support and denominator scores and map each $\{((p, q, t), supp, deno)\}$ to $\{((p, q, r, t), supp, supp/deno)\}$ to yield a list of $\{((p, q, t), supp, conf)\}$.

Note that the MR and MR-Batch algorithm are based on the taxonomy structure and can only mine rules with the $rdftype$ predicate. But LDOO can learn rules with $rdftype$ or $equals$. To do this, we just need to remove the join conditions with $rdftype$ and add a GroupBy parameter x or y. In our experiments, we implement LDOO for both $rdftype$ and $equals$.

Aggregation Pushdown. LDOO worked well for types 1 to 5, but it failed in mining rules for the six-th rule type. By checking the logs, we found that some workers crashed due to huge intermediate results. The six-th rule type, i.e. $p(x,y) \leftarrow q(x,z), r(y,z)$, requires a join on the *object* attribute as indicated by the variable z. However, many object values have high appearance frequency which results in large number of join results. In Yago, there are 10,502 objects which appear more than 100 times, 25 objects more than 10,000 times, and 2 objects (*i.e.* $'female'$ and $'male'$) even more than 100,000 times.

Based on this observation, we try to improve the performance of LDOO by pushing down the aggregation operator. The idea is to perform the COUNT operator before JOIN, in order to reduce the amount of intermediate results.

Figure 1 illustrates the original query plan of the mining task for the six-th rule type, and Fig. 2 shows the rewritten plan by pushing down the COUNT operators. According to [19, 20], the two plans are equivalent.

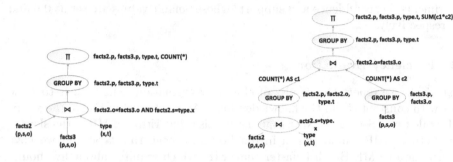

Fig. 1. Original plan

Fig. 2. Plan after pushing down COUNT

4   Experimental Study

In this section, we first summarize the settings of the experiments in Sect. 4.1. Then we report the efficiency and effectiveness results in Sects. 4.2 and 4.3 respectively.

4.1   Settings

We conduct all experiments on a 18-nodes cluster, consisting of one master and 17 workers. All these nodes are hosted on a cloud platform. Each node is equipped with one 2.4 GHz processor and 500G hard disk space. The master has 8G RAM and while each worker node has 4G RAM. All the approaches are implemented in scala 2.11.8 and java 1.8.0. The jobs run on the Spark 2.1.0 platform, and the OS is CentOS 6.5.

All the experiments are conducted on two real datasets, YAGO 3.0.0[6] and DBPedia[7]. The statistics are listed in the following (Table 1).

Table 1. Statistics of the datasets

	Yago	DBPedia
# of Triples	10,400,678	52,680,098
# of Entities	4,464,016	14,592,204
# of Predicates	125	59,149
# of Types	569,312	385

[6] https://www.mpi-inf.mpg.de/departments/databases-and-information-systems/research/yago-naga/yago/.

[7] http://wiki.dbpedia.org/develop/datasets.

Compared Methods. We implement the three methods proposed in Sect. 3, i.e. MR, MR-Batch and LDOO. We also compare our methods with the one in [10]. Note that we also implement LDOO for mining rules with the *equals* predicate. However there are only a few rules can have enough *supp* scores, because the granularity of these rules are too small. When *conf* is set as 0.6, we can obtain only 26 rules. Therefore, we only report the results of mining rules with the *rdftype* predicate in the following. There are two parameters in the experiments, i.e. confidence and support, whose default values are set as 0.6 and 100 respectively.

4.2   Efficiency Evaluation

First of all, we report the results of efficiency evaluation. The task is to mine rules with constants from the knowledge base. As illustrated in Fig. 3, the x-axis is the rule type (Sect. 2.1). Note that the y-axis is logarithmic. MR is the slowest one. Actually MR cannot finish in ten hours for each rule type and we have to terminate it. MR-Batch is faster than MR, which requires about five hours. LDOO is the best, which costs about thirty minutes. As explained in Sect. 3.3, LDOO saves the expensive cost of reloading data from disk and moving data among the network. There is an exceptional case in Rule-Type 5, where MR-Batch beats LDOO. The reason is there are only a few (less than 10) rules to be refined for this type, which greatly reduces the workload of MR-Batch. Figure 4 illustrates the mining time on DBPedia. Note that we only report the result of LDOO here because neither MR nor MR-Batch can finish the mining job on DBPedia in 24 h.

Next we take a breakdown analysis for MR-Batch and show the results in Fig. 5. As discussed in Sect. 3.2, MR-Batch contains two phases, mining and refining. As illustrated in Fig. 5, the refining phase dominates the time cost. The reason is that we need to invoke a SPARK job for each rule obtained in the mining phase in order to evaluate the candidate sub-rules.

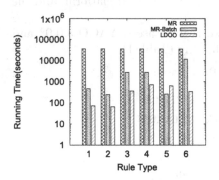
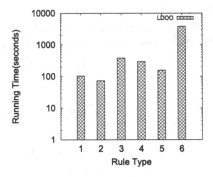

Fig. 3. Rule mining time (Yago)   **Fig. 4.** Rule mining time (DBPedia)

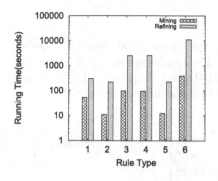

Fig. 5. Breakdown of MR-Batch (Yago)

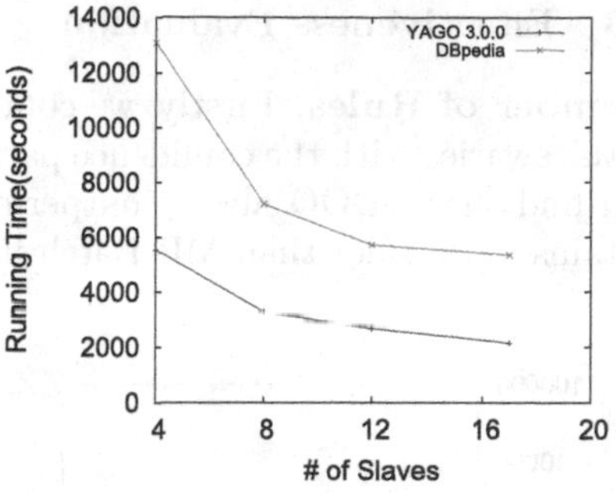

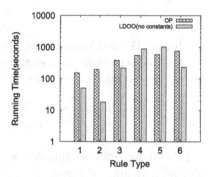

Fig. 6. Scalability (Yago & DBPedia)

We also conduct an experiment to evaluate the scalability of LDOO. As shown in Fig. 6, when the number of worker nodes increases, the time cost of LDOO decreases. For example, it takes 5,532 s when there are only 4 nodes and 2,149 s for 17 nodes, when mining rules from Yago.

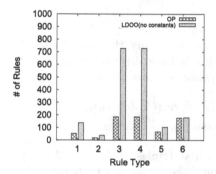

Fig. 7. Comparison of # of rules (Yago)

Fig. 8. Comparison of time (Yago)

Comparison with Ontological Pathfinding (OP). The OP approach in [10] depends on the domains of predicates when generating candidate rules. For example, when generating a rule $p(x, y) \leftarrow q(x, y)$, OP requires that the subject/object domains of predicates p, q must overlap because they share the same variables x, y. However, the schema data of Yago are not complete. Therefore, OP misses many meaningful rules.

As illustrated in Fig. 7, LDOO obtains several times more rules than OP, because LDOO does not have this limit of domain overlapping. Moreover, the time cost of LDOO is similar to OP (Fig. 8). Note that here we restrict LDOO to find rules without constants in order to make a fair comparison. Hence in the figure, the label is LDOO(no constants).

4.3   Effectiveness Evaluation

Number of Rules. Firstly, we conduct an experiment about how the number of rules varies with the confidence parameter. The results are shown in Fig. 9. We can find that LDOO always outperforms MR-Batch. Therefore, LDOO always obtains more rules than MR-Batch in spite of the confidence threshold.

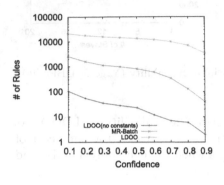

Fig. 9. # of rules (Yago)   **Fig. 10.** # of new facts (Yago)

Knowledge Base Completion. In this task, we apply the mined rules to automatically replenish knowledge base, which is basically to infer new facts based on the qualified rules. For example, suppose we have the following rule:

$$hasChild(x, y) \leftarrow isMarriedTo(x, z), hasChild(z, y),$$
$$rdftype(x, Russian\_grand\_dukes)$$

The rule above means: suppose we have three triples, $isMarriedTo(x, z)$, $hasChild(z, y)$ and $rdftype(x, Russian\_grand\_dukes)$, we can then infer that the triple $hasChild(x, y)$ must be correct and put it into the knowledge base if it is missing.

First of all, we did several experiments to find an appropriate value for the $conf$ parameter. We randomly select 50 new facts for each $conf$ setting and manually check the correctness. The results are $0.92, 0.92, 0.96, 1.0$ for $conf = 0.5, 0.6, 0.7, 0.8$ respectively.

Figure 10 illustrates how the number of new facts varies with the confidence parameter. We can find that LDOO outperforms LDOO(no constants). When the confidence is set as 0.7, we can obtain more than 200,000 new facts by the rules with constants (about 96% of these new facts are correct). However, if there are no constants in the rules, we can only obtain about 3,000 new facts for the same confidence. This result confirms the effectiveness of our solution for completing knowledge bases.

Table 2 shows several example rules mined by LDOO. The third column of Table 2 compares the confidence values of the rule and its parent rule (i.e. the corresponding rule without constants). Clearly the refined rules have higher confidence values than their parent rules. We also list several inferred new facts in Table 3.

Table 2. Example rules

Rule	rdftype of x	Conf.
$playsFor(x, y) \leftarrow isAffiliatedTo(x, y)$	$football\_player$	0.78(0.59)
$isLocatedIn(x, y) \leftarrow wasBornIn(z, x), isCitizenOf(z, y)$	$Cities\_in\_California$	0.87(0.56)
$hasFamilyName(x, y) \leftarrow hasChild(z, x), hasFamilyName(z, y)$	$lawyer$	0.72(0.51)

Table 3. New facts inferred by rules

New facts
playsFor(Oleksiy_Antonov,Ukraine_national_football_team)
hasFamilyName(Sam_Houston, "Kaufman" @eng)
isLocatedIn(Moscow,Russian_Federation)

We put the whole set of rules and inferred new facts in a public web site. Interested readers can find them at the following url:

https://pan.baidu.com/s/1oAqprqa   (pass: 8m8d)

5 Related Works

Rule learning is a classical AI task which aims at learning first-order logic formulas from given data sets [13]. Classical rule learning methods [11,12,14] rely on some traverse strategies to search and evaluate all candidate rules, which are hard to deal with large data sets. Some recent works [8,15] improve the efficiency by running mining tasks in parallel, but it is not a distributed method and hard to scale out.

Chen et al. [4,10] proposed an ontology-based routing algorithm OP [16]. The whole task is divided into many sub-tasks, which run in parallel on a Spark platform. The limitation of OP lies in that it excludes constants and misses many valuable rules.

The works [7,17] admit constants, but both have limitations. [7]'s language model is quite limited and not able to find new facts which should be included in knowledge bases, while [17]'s implementation is for single machine and hard to be applied in large data sets.

In this paper, we design and implement our approaches based on the Spark platform [18]. Spark can provide an efficient and general data processing platform. Its core consists of a set of powerful and high-level libraries for large-scale parallel and distributed data processing.

6 Conclusion

Mining rules with constants is important for knowledge base construction. This paper proposes a scalable solution based on the Spark platform, including a series of approaches to improve both the efficiency and quality. In the future work, we will study how to generalize the learning model, e.g. introducing negative atoms. We will also study how to further improve the efficiency by optimizing the mining process.

Acknowledgments. This work was supported by the National Key Research & Develop Plan (No. 2016YFB1000702), National Science Foundation of China (No. 61602488), and Talent Training Fund at RUC.

References

1. Mahdisoltani, F., Biega, J., Suchanek, F.M.: YAGO3: a knowledge base from multilingual wikipedias. In: Seventh Biennial Conference on Innovative Data Systems Research, CIDR 2015, Asilomar, CA, USA, 4–7 January 2015, Online Proceedings (2015)
2. Niu, F., Zhang, C., Ré, C., Shavlik, J.W.: DeepDive: web-scale knowledge-base construction using statistical learning and inference. In: Proceedings of the Second International Workshop on Searching and Integrating New Web Data Sources, Istanbul, Turkey, 31 August 2012, pp. 25–28 (2012)
3. Shin, J., Wu, S., Wang, F., De Sa, C., Zhang, C., Ré, C.: Incremental knowledge base construction using DeepDive. PVLDB **8**, 1310–1321 (2015)
4. Chen, Y., Wang, D.Z., Goldberg, S.: ScaLeKB: scalable learning and inference over large knowledge bases. VLDB J. **25**, 893–918 (2016)
5. Lao, N., Mitchell, T., Cohen, W.W.: Random walk inference and learning in a large scale knowledge base. In: Proceedings of the Conference on Empirical Methods in Natural Language Processing, pp. 529–539 (2011)
6. Chen, Y., Wang, D.Z.: Knowledge expansion over probabilistic knowledge bases. In: Proceedings of the 2014 ACM SIGMOD International Conference on Management of Data, pp. 649–660 (2014)
7. Chu, X., Ilyas, I.F., Papotti, P.: Discovering denial constraints. Proc. VLDB Endow. **6**, 1498–1509 (2013)
8. Galárraga, L.A., Teflioudi, C., Hose, K., Suchanek, F.: AMIE: association rule mining under incomplete evidence in ontological knowledge bases. In: Proceedings of the 22nd International Conference on World Wide Web, pp. 413–422 (2013)
9. Fan, W., Geerts, F., Li, J., Xiong, M.: Discovering conditional functional dependencies. IEEE Trans. Knowl. Data Eng. **23**, 683–698 (2011)
10. Chen, Y., Goldberg, S., Wang, D.Z., Johri, S.S.: Ontological pathfinding: mining first-order knowledge from large knowledge bases. In: Proceedings of the 2016 International Conference on Management of Data, pp. 835–846 (2016)
11. Quinlan, J.R.: Learning logical definitions from relations. Mach. Learn. **5**, 239–266 (1990)
12. Muggleton, S., De Raedt, L.: Inductive logic programming: theory and methods. J. Log. Program. **19**, 629–679 (1994)
13. Frnkranz, J., Gamberger, D., Lavrac, N.: Foundations of Rule Learning. Springer, Heidelberg (2012). https://doi.org/10.1007/978-3-540-75197-7

14. Savasere, A., Omiecinski, E., Navathe, S.B.: An efficient algorithm for mining association rules in large databases. In: Proceedings of the 21st International Conference on Very Large Data Bases, pp. 432–444 (1995)
15. Galárraga, L., Teflioudi, C., Hose, K., Suchanek, F.M.: Fast rule mining in ontological knowledge bases with AMIE$$+$$+. VLDB J. **24**, 707–730 (2015)
16. Agrawal, R., Srikant, R.: Fast algorithms for mining association rules in large databases. In: VLDB 1994, Proceedings of 20th International Conference on Very Large Data Bases, 12–15 September 1994, Santiago de Chile, Chile, pp. 487–499 (1994)
17. Zeng, Q., Patel, J.M., Page, D.: QuickFOIL: scalable inductive logic programming. Proc. VLDB Endow. **8**, 197–208 (2014)
18. Dean, J., Ghemawat, S.: MapReduce: simplified data processing on large clusters. Commun. ACM **51**, 107–113 (2008)
19. Yan, W.P., Larson, P.: Eager aggregation and lazy aggregation. VLDB **31**(12), 345–357 (1995)
20. Harinarayan, V., Gupta, A.: Generalized projections: a powerful query-optimization technique (1995)

Concepts in Quality Assessment for Machine Learning - From Test Data to Arguments

Fuyuki Ishikawa[✉] [iD]

National Institute of Informatics, Tokyo, Japan
f-ishikawa@nii.ac.jp

Abstract. There have been active efforts to use machine learning (ML) techniques for the development of smart systems, e.g., driving support systems with image recognition. However, the behavior of ML components, e.g., neural networks, is inductively derived from training data and thus uncertain and imperfect. Quality assessment heavily depends on and is restricted by a test data set or what has been tried among an enormous number of possibilities. Given this unique nature, we propose a MLQ framework for assessing the quality of ML components and ML-based systems. We introduce concepts to capture activities and evidences for the assessment and support the construction of arguments.

Keywords: Machine learning · Assurance cases · Arguments
Test management · Artificial intelligence

1 Introduction

Practical and industrial applications of machine learning (ML), or, in a broader term, artificial intelligence, are actively investigated, and it is highly demanded that they be dependable. However, the majority of research on ML techniques has targeted algorithmic aspects for performance, typically some types of accuracy and learning costs. Research on engineering disciplines is still limited.

The behavior of ML components, e.g., neural networks, is inductively derived from training data and in uncertain. It is an unexplainable blackbox [5], imperfect and potentially affected by slight perturbations of inputs (adversarial examples) [4], and sensitive to changes [11]. In addition, ML-based systems often target the open real world, e.g., driving support, or fuzzy user satisfaction, e.g., recommendation. It is intrinsically impossible to declare a clear boundary regarding what an ML component or whole system should/can do. Quality assessment for ML-based systems heavily depends on and is restricted by test data or what has been tried among an enormous number of possibilities.

© Springer Nature Switzerland AG 2018
J. C. Trujillo et al. (Eds.): ER 2018, LNCS 11157, pp. 536–544, 2018.
https://doi.org/10.1007/978-3-030-00847-5_39

The ML community has focused on metrics such as accuracy, precision, and recall to measure the quality of ML components. However, these metrics are relative to a target set of test data. For example, it is nonsense to claim dependability with a very high accuracy of 99% in a test data set that contains only red cars as obstacles to be recognized in a driving support system. This ML component may fail to recognize white cars. As a means of engineering, we need a framework for arguing quality by explicitly annotating test data sets.

In this paper, we propose a framework for assessing the quality of ML components and ML-based systems (MLQ framework for ML-based system quality). Our focus is test data, or, more generally, empirical evidences, given the aforementioned nature lf ML. We define concepts for capturing activities and evidences for quality assessment. We provide a template for arguments that describe how the quality of ML components or ML-based systems is assessed and how evidences support it in Goal Structuring Notation (GSN) [9]. Thus, we support the construction and continuous management of arguments regarding ML-specific aspects. We demonstrate the proposed framework with a case study on a driving support system with image recognition.

In the remainder of the paper, we first discuss the background and related work in Sect. 2. We then present the proposed framework with a case-study scenario in Sect. 3. We give concluding remarks in Sect. 4.

2   Background and Related Work

2.1   Terminology and Scope

When we use ML techniques, we construct a program that implements a *ML algorithm*, such as random forest and backpropagation, by using a set of *Training Data* as the input. We deploy the output of the program as a *ML Component*, e.g., Decision Tree and Neural Network[1]. A ML component is contained in a whole system or *(ML-based) System*, such as a driving support system. In this paper, we focus on the quality of ML components and ML-based systems in terms of how well the expected functionality is realized. We do not discuss the non-functional quality aspects or quality of ML algorithm code.

2.2   Testing and Verification of ML

A notable movement is the emergence of automated testing methods for ML components and ML-based systems.

One direction is systematic and comprehensive testing. Covering various inputs has been considered by using image synthesis [2] or formal verification [6]. Specific coverage for white-box testing was also proposed ("neuron coverage" for deep neural networks) [10]. These approaches are aimed at a kind of exhaustiveness and help detect adversarial examples or unexpected wrong classifications.

---

[1] We avoid the confusion by calling this as a "model" as in the ML community.

Another direction is applying metamorphic testing [8,13]. It is difficult or practically impossible to define an oracle or expected output for arbitrary inputs. In metamorphic testing, we check the expectation that "changing the input in this way should change the output in that way" (metamorphic relation).

The work in [1] extracts and focuses on the possibility that the failure of an ML component will lead to system failures. This direction is very significant, in contrast with most efforts in the ML community, to linking system-level requirements and environmental assumptions to the quality of ML components.

In this paper, we additionally consider human-designed testing by preparing data sets with certain characteristics to reflect requirements and environmental assumptions. Thus, our work in this paper provides a comprehensive viewpoint of testing rather than focusing on one specific aspect or technique.

2.3   Arguments

Argument models have been widely investigated and used for assuring system quality (e.g., assurance cases) [9]. The essence is to explicitly describe how goals are decomposed into sub-goals and finally linked to solutions or evidences. A tree structure, such as GSN, is typically used, the root node is the top goal, and the leaf nodes are solutions or evidences that work as grounds. The constructed models can be used for validation and discussion with various stakeholders. They also allow for continuous tracing and management, e.g., updating evidences with runtime monitoring information [12]. The work in [3] discusses how to argue risk mitigation for ML. Our previous work in [7] focused on extending GSN and its usages under uncertainty in ML-based systems.

Our focus in this paper is to provide the foundation to capture essential concepts for quality assessment of ML components and ML-based systems as well as to link arguments with state-of-the-art testing techniques.

3   MLQ Framework

3.1   Overview

We now present our framework for assessing the quality of ML components and ML-based systems (MLQ framework for ML-based system quality). The objective of the framework is to support quality assessment via arguments over relevant activities and obtained evidences. Our key focus is evidences or descriptions of what we have tried, given the uncertainty of ML.

In the framework, we provide concepts that describe relevant activities and evidences as in Fig. 1. These concepts are used to describe and record the result of quality assessment, including the obtained evidences. Thus, instances of the concepts appear in leaf nodes of a tree structure for arguments. In this paper, we illustrate the concepts only informally, whereas actual models based on the concepts can be represented in any usable notation, e.g., spreadsheets. We also provide a top-down viewpoint to position each of testing techniques, in the form of an argument template.

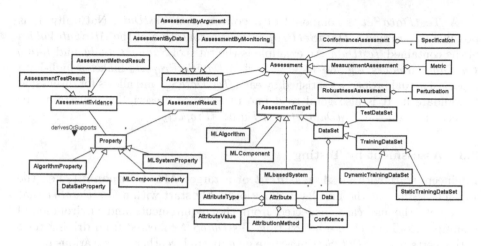

Fig. 1. Concepts in proposed MLQ framework

We use a case-study scenario to demonstrate the significance of the concepts and assessment activities to be proposed. Specifically, we discuss a driving support system that contains a ML component for image recognition. The component judges obstacles ahead of a car, such as other cars and pedestrians. According to the information from the component (as well as other kinds of sensors), the system automatically controls the throttle and brake as necessary to mitigate the risk and damage of collisions.

We focus only on quality assessment activities motivated by the unique nature of ML. Quality assessment at the ML component level is completely within this scope. The system level is also affected by the unique nature, too, though there are points that can be handled as before. For example, an argument about safety in the case of ML component failures will be a common one that reasons about alternative means or fail-safe measures.

3.2   Test Data Attributes and Test Data Set Properties

The concepts in Fig. 1 contains how we describe each piece of test data and each test data set. Each **TestData**, such as a *CarCameraImage*, has a set of **Attribute**s. Each *Attribute* has an **AttributeValue** and belongs to a certain **AttributeType**, such as *Sunny* in *WeatherCondition*. Besides this very common aspect, we explicitly include information on how each *AttributeValue* is obtained and validated as **AttributionMethod**s. For example, suppose that the *AttributeValue* for *WeatherCondition* is obtained by the *WeatherReport* method, recording the weather announced at the time each *CarCameraImage* is taken. As the weather is reported for a wide area of a city, the result can be different from the actual weather at a specific local point. Then, the *WeatherMLCheck* method is also used to deploy a ML system to calculate the confidence on the correctness of the *WeatherCondition* by analyzing the *CarCameraImage* and then conduct filtering. The **Confidence** on the attribution is given such as *middle*.

A *TestDataSet* is composed of a collection of *TestData*. Naturally, it is possible to describe a *DataSetProperty* as a collection of the *AttributeValue*s of the contained *TestData*. An example is the rates of *Sunny*, *Cloudy*, and *Rainy CarCameraImages*, e.g., 30-40-30%. Such a *DataSetProperty* makes a high-level report of actual properties satisfied by each *TestDataSet* and allows for discussion to validate it. It is also good to define in advance prescriptive constraints or expectations on the *TestDataSet* as a *DataSetProperty*.

3.3   Assessment by Testing

Engineers start with an *Assessment* of a target *ML Component*, i.e., the *ImageClassifier*, at the unit level. They probably start with a *TestDataSet* that represents the use cases expected from the requirements and environmental assumptions. They choose an existing *TestDataSet* collected from driving tests with a certain set of *DataSetProperty*, e.g., expected weather rates, various points for different times of day, and different seasons given the target area of operation.

We distinguish between *MeasurementAssessment*, where quality is measured with target *Metric*s, and *ConformanceAssessment*, where quality is verified with target *Specification*s. Suppose that, for the unit level, *MeasurementAssessment* is conducted. Engineers run a *MeasurementAssessment* with the target *Metric*s, e.g., *Accuracy*, and obtains a set of *MLComponentProperties*, e.g., *Accuracy = 78%*, as the *AssessmentResult*. The *AssessmentTestResult* holds the details of the result, such as with which *TestData* recognition failed. In the case of *ConformanceAssessment*, the target is *Specifications*, e.g., *Accuracy > 85%*, and the obtained *MLComponentProperty* looks like "*Accuracy > 85% is not satisfied*". *AssessmentMethod*, the means of *Assessment*, is *AssessmentByData* in this example.

Detailed analysis on the *AssessmentResult* will lead to further *MeasurementAssessment*s or improvement to the implementation. For example, thanks to the *AttributeValue*s of failed *TestData*, it may turn out that the *Accuracy* of *Rainy* images is very low, so an intensive *MeasurementAssessment* is made with a new *TestDataSet* that contains only an increased number of *Rainy* images.

System-level testing can be done for designed *TestDataSet*s, similarly. For example, engineers may combine a Simulink model for the controller of a *MLbasedSystem*, *DrivingSupportSystem* in this example, combined with the *ML* component (*ImageClassifier*). *Assessment*s are made similarly, but here, a *MLSystemProperty* is targeted. For example, a *Metric* is used in a *MeasurementAssessment* to record the *DistanceToCollision*, which is the minimum distance between the car and an obstacle during a simulation.

Evidences obtained by other kinds of testing or verification techniques can be argued. Robustness testing against adversarial examples can be explained as *RobustnessAssessment*, another type for *Assessment*. As the evaluation criterion, a set of *Perturbations* are defined: under which a *ML Component* or *MLbasedSystem* is tested. The *Perturbations* define the space for a search together with the base *TestDataSet*. The means of *Assessment*, that is, *AssessmentMethod*, is often different from the standard one of *AssessmentByTestData* (preparing a certain *TestDataSet* beforehand), e.g., *RandomTesting* by picking from the space of possible *Perturbations*. A kind of *Assessment* using expected constraints or relations, such as metamorphic testing, is covered in our modeling framework as *ConformanceAssessment* with a *Specification*.

3.4   ML Algorithms and Static/Dynamic Training Data

It is possible to ensure that some properties hold by reasoning about how a ML component is created, i.e., the properties of ML algorithms and training data. First, we give *Assessments* on a set of *Properties* of the **TrainingDataSet** and **MLAlgorithm**. Then, we assess the target *Property* on the *MLComponent*. Typically this process makes an **AssessmentByArgument**, which considers the decomposition of a *Property* into a set of sub-*Properties*.

A *TrainingDataSet* is classified into a **StaticTrainingDataSet**, prepared and used at development time, and a **LiveTrainingSet**, given to a system at runtime. In the latter case, **AssessmentByMonitoring** is a means of assessing what *Property* holds or not to detect or prevent undesirable learning.

3.5   Argument Construction

Figure 2 shows a simplified version of an argument for the case-study scenario in the form of GSN. The top side box (module) is a template we provide, which summarizes how we capture the whole picture of quality assessment for ML components. Arguments for ML-based systems will include further aspects irrelevant to the specific natures of ML, which are out of the scope of this paper. The presented models, such as test data sets, are referred to in the evidence or solution part at the leaf nodes. As presented, our framework focuses on this part to reflect the intrinsically empirical nature of quality assessment for ML components and ML-based systems.

3.6   Continuous Validation and Management

Explicitly representing and recording the essence of ML support continuous validation given frequent updates under the increasing uncertainty.

One example is updating a test data set given the data collected during operation. The properties are checked against the updated data sets. This may lead to an update of the concrete goals to reflect requirements and environmental conditions. As another example, suppose that additional weakness of the image classifier is discovered during operation. This would lead to a demand for an additional test data set to intensively assess the quality of the ML component regarding the weakness (and then improve it). This update is made to the argument model. Then, knowledge about the weakness and the demand for the test data set is documented and effectively used in future activities.

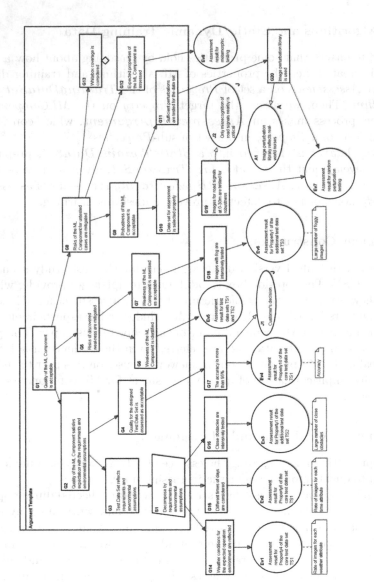

Fig. 2. Example argument derived in proposed MLQ framework

4 Concluding Remarks

In this paper, we proposed a framework for assessing the quality of ML components and ML-based systems (MLQ Framework). The objective of the framework is to support quality assessment via arguments over relevant activities and obtained evidences. Our key focus is evidences or descriptions of what we have tried, given the uncertainty of ML.

Our future work includes primarily two directions. One is the continuous validation and refinement of the presented framework with further experiences, especially using different kinds of applications. The other is the implementation of supporting tools that link test management and arguments.

Acknowledgments. This work is partially supported by the ERATO HASUO Meta-mathematics for Systems Design Project (No. JPMJER1603), JST. We are thankful to the industry researchers and engineers who gave deep insights into the difficulties in the engineering of ML and practices in the case-study scenario.

References

1. Dreossi, T., Donzé, A., Seshia, S.A.: Compositional falsification of cyber-physical systems with machine learning components. In: Barrett, C., Davies, M., Kahsai, T. (eds.) NFM 2017. LNCS, vol. 10227, pp. 357–372. Springer, Cham (2017). https://doi.org/10.1007/978-3-319-57288-8_26
2. Dreossi, T., Ghosh, S., Seshia, S., Sangiovani-Vincentelli, A.: Systematic testing of convolutional neural networks for autonomous driving. In: ICML 2017 Workshop on Reliable Machine Learning in the Wild, August 2017
3. Burton, S., Gauerhof, L., Heinzemann, C.: Making the case for safety of machine learning in highly automated driving. In: Tonetta, S., Schoitsch, E., Bitsch, F. (eds.) SAFECOMP 2017. LNCS, vol. 10489, pp. 5–16. Springer, Cham (2017). https://doi.org/10.1007/978-3-319-66284-8_1
4. Goodfellow, I., Shlens, J., Szegedy, C.: Explaining and harnessing adversarial examples. In: International Conference on Learning Representations (ICLR), May 2015
5. Gunning, D.: Explainable artificial intelligence (XAI). In: IJCAI 2016 Workshop on Deep Learning for Artificial Intelligence (DLAI), July 2016
6. Huang, X., Kwiatkowska, M., Wang, S., Wu, M.: Safety verification of deep neural networks. In: Majumdar, R., Kunčak, V. (eds.) CAV 2017. LNCS, vol. 10426, pp. 3–29. Springer, Cham (2017). https://doi.org/10.1007/978-3-319-63387-9_1
7. Ishikawa, F., Matsuno, Y.: Continuous argument engineering: Tackling uncertainty in machine learning based systems. In: The 6th International Workshop on Assurance Cases for Software-Intensive Systems (ASSURE 2018), September 2018
8. Jarman, D.C., Zhou, Z.Q., Chen, T.Y.: Metamorphic testing for Adobe data analytics software. In: The 2nd International Workshop on Metamorphic Testing, pp. 21–27, May 2017
9. Kelly, T., Weaver, R.: The goal structuring notation - a safety argument notation. In: Dependable Systems and Networks 2004 Workshop on Assurance Cases, July 2004
10. Pei, K., Cao, Y., Yang, J., Jana, S.: DeepXplore: automated whitebox testing of deep learning systems. In: The 26th Symposium on Operating Systems Principles (SOSP 2017), pp. 1–18, October 2017

11. Sculley, D., et al.: Machine learning: the high interest credit card of technical debt. In: NIPS 2014 Workshop on Software Engineering for Machine Learning (SE4ML), December 2014
12. Tokuda, H., Yonezawa, T., Nakazawa, J.: Monitoring dependability of city-scale IoT using D-case. In: 2014 IEEE World Forum on Internet of Things (WF-IoT), pp. 371–372, March 2014
13. Xie, X., Ho, J.W., Murphy, C., Kaiser, G., Xu, B., Chen, T.Y.: Testing and validating machine learning classifiers by metamorphic testing. J. Syst. Softw. 84(4), 544–558 (2011)

Inductive Discovery by Machine Learning for Identification of Structural Models

Wolfgang Maass and Iaroslav Shcherbatyi[(⊠)] [iD]

German Research Center for Artificial Intelligence (DFKI), Kaiserslautern, Germany
{wolfgang.maass,iaroslav.shcherbatyi}@dfki.de
http://www.dfki.de

Abstract. Automatic extraction of structural models interferes with the deductive research method in information systems research. Nonetheless it is tempting to use a statistical learning method for assessing meaningful relations between structural variables given the underlying measurement model. In this paper, we discuss the epistemological background for this method and describe its general structure. Thereafter this method is applied in a mode of inductive confirmation to an existing data set that has been used for evaluating a deductively derived structural model. In this study, a range of machine learning model classes is used for statistical learning and results are compared with the original model.

Keywords: Structural models · Machine learning · SEM
Data science.

1 Introduction

Structural equation models (SEM) are means for assessing and modifying theoretical models as often used in Social Sciences and Information Systems research in particular [13]. SEM combines two models: (1) a measurement model that relates observed features (aka measures) to assumed constructs by factor analysis and (2) a structural model that specifies directed relations between constructs as derived from some theory [2].

The underlying method consists of two model selection steps: (1) selection of a local model space from theory and (2) selection of the best fitting model by statistical analysis and heuristics. With the uprise of big data, data science and artificial intelligence, this general method becomes the target of reevaluation [3]. Many thousands of expert players of the game of Go have developed strategies and tactics over hundreds of years. Nonetheless they were surprised when the AI-based system AlphaGo found new strategies, tactics and moves that were first considered a failure but later proved to be the cornerstone of success. The defeated world-champion Lee Sedol learned these new moves and became successful in the next 21 games himself, i.e. he extracted knowledge from patterns learned by AI technologies [28]. Translated into the scientific realm, this opens the question how AI-based research can be done for the first step of SEM,

© Springer Nature Switzerland AG 2018
J. C. Trujillo et al. (Eds.): ER 2018, LNCS 11157, pp. 545–552, 2018.
https://doi.org/10.1007/978-3-030-00847-5_40

i.e. selection of appropriate models from theory. In this paper, we will discuss this in the context of confirmation of structural models, i.e. an existing traditionally developed model is compared with models derived by machine learning algorithms.

In detail, we focus on a covariance-based structural equation model (SEM) that is used as a standard tool for research in social sciences, and the information systems community in particular. Assuming measurement models with strong ties (covariance-based or PLS-based) with the latent structural variables they depend on [8], we investigate whether structural models can be automatically extracted from data instead of deriving them from theory alone. In the following, we discuss our research question by using an existing deductive study as a baseline. We show how data is analyzed by various methods of statistical learning resulting in a set of structural relations as input for a structural model. Resulting structural models are compared with the original model.

2   Research Using Structural Equation Modeling

2.1   Standard Approach

Structural equation modeling (SEM) is used for evaluating parameters defined by a hypothesized underlying model by statistical analysis of empirical measurements [15] with an emphasis on analyzing covariance [14] or component-based [9] structures between observed and latent variables [8]. Conventionally theory drives model specification that is assessed by statistical analysis of empirically assessed data [15]. By theoretical considerations the latent variable model consisting of exogenous and endogenous variables is described in closed form as follows: $\eta = B\eta + \Gamma\xi + \zeta$ with η a vector of latent endogenous variables to be explained, ξ a vector of latent exogenous variables, ζ capturing disturbances and B and Γ are coefficient matrices [15]. Similarly, measurement models are described in closed form as follows: $x = \Lambda_x \xi + \delta$ and $y = \Lambda_y \eta + \epsilon$ [15].

With a given data set, measurement models of SEM are typically respecified and reestimated of both models untill an acceptable fit has been achieved [2]. Evaluation is done by cross-validation on another data sample drawn from the population resulting in values for free model parameters. Thus, model fitting is assessed and used as positive or negative support of an initially hypothesized model [17].

From a pure mathematical perspective, for a graph with p independent variables (concepts of a structural model) $2^{p(p-1)}$ different directed graphs can be derived with each of them being a candidate for explaining the data [10]. For instance, for 6 independent variables this results in $2^{30} \approx 1$ billion models. Most of these models are trivial or nonsense but it exceeds researchers' mental capacities to think through all models that might make sense. A brute force approach evaluates all feasible models and selects best fitting models from which researchers select those models that are aligned with theory.

Taking these two perspectives together, the search space for structural models in SEM is spanned between models derived from theory alone and all combinations of mathematically possible models.

2.2 Inductive Discovery for Structural Models

We ask how many potentially meaningful structural models are possible if n variables and p paths are given that exhibit equally likely interpretations of data [18] as a similarity class of models, so-called confounds [6]. The search model approach resembles and is even anchored in the discussion currently conducted around the term Big Data [1,7,16]. A key claim of proponents of big data research is that patterns can be extracted automatically from large data sets [23]. In contrast, it is argued that also big data is subject to sampling bias, dependent on viewpoints, tools for collecting data, and data ontologies and the need for epistemological interpretations by domain experts [16].

In the field of pattern recognition, a pattern is considered as a characteristic regularity in some data discovered by the use of computer algorithms [5]. In the area of conceptual modelling, pattern recognition has rarely been used for discovery of structural models. Marcen et al. use machine learning by encoding ontological models into feature vectors for discovery of most relevant features used for concept location and traceability link retrieval [19]. Nalchigar et al. are looking into the design of data analytical systems by proposing a modelling framework for requirements analysis [21]. Sunkle et al. discuss a data-driven approach for semi-automatic extraction of regulatory rules from texts by utilization of domain models [29].

The proposed method for *inductive discovery for structural models* consists of the following phases:

1. specification of a measurement model and data collection
2. discovery of patterns between variables
3. assessment of the quality of selected models by SEM analysis, and
4. confirmation or non-confirmation of selected models

For the evaluation of this method, we use a published SEM model (M_O) and its data set [30]. Verniers and Vala compared M_O with two alternative models $M_{D(MS)}$ und $M_{D(OM)}$. These three models are compared with models automatically extracted by data analytical methods with machine learning $(M_{D(ML)})$.

3 Computational Methods for Inductive Discovery

We focus on path analysis in structural models and abstain from using fit estimates in SEM for comparing models. This is done for the same reason as discussed by [6], i.e. we presuppose particular measurement models but leave freedom for choosing structural models randomly [6]. Instead, we apply and compare results of different data-driven statistical learning models by introducing a targeted accuracy metric. Resulting models are tested and evaluated by separate test and evaluation data sets. We assume that we are given a set of concepts, and we want to establish whether there exist directed relational dependencies between different concepts.

We concentrate on establishing all pairwise dependencies between concepts. For every pair of concepts, we solve a supervised learning problem [4] for prediction of values describing observation of concept Y given the values of corresponding observation of the concept X and vice versa. We use a value of a coefficient of determination R^2 of machine learning model on hold-out set (20% of data) as a value that describes the strength of a relation between two concepts. This value allows to establish an ordering of all n^2 one-to-one relations by strength, and among them we select $m \leq n^2$ strongest ones, where m is provided by the user. The resulting relation set is used as input for evaluation methods for structural models, such as covariance-based SEM. We utilize a range of ML models, such as Lasso, Artificial Neural Networks, K Neighbors and Gradient Boosted models, and select their hyperparameters using cross-validation [4].

4   Structural Model Selection and Performance

The starting point for our analysis is a study that investigates gender discrimination in the workplace [30]. This study investigated whether the myth that women's work threatens children and family life mediates the relationship between sexism and opposition to a mother's career. The study is based on a large survey of attitudes towards women in workforce; In total there are more than 47000 responses available to the survey questions. Three concepts are considered, such as "Motherhood myths" (2 indicators), "Opposition to womens career" (2 indicators) and "Sexism" (1 indicator). Additionally, for every survey response, data is available about gender, age, year of survey, and some other indicators; see [30] for details. We apply our statistical learning approach as given in the previous section, and obtain a relationship strength coefficients, shown in the Fig. 1. For extraction of relations using ML methods, we use a randomly selected partition of 80% of the total dataset (37219 observations). The rest of

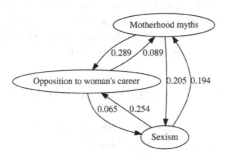

Fig. 1. The relationship strength values, derived according to our methodology. Higher values indicate stronger directed relationship. The values are average test R^2 for prediction of indicators of a target variable given indicators of an input variable.

the data (9187 observations) is used for evaluation of SEM models. Note that this is disjoint set from testing set for ML model evaluation. Such split is used in order to avoid biasing the results of our approach on the same dataset.

4.1 Structural Model Selection

Obtained models are evaluated using R and the Lavaan package with a test data set with 9187 cases that have not been used for training and validating the extracted model. Due to the fact that the measurement was re-used from the original model, we discuss results for the structural variable model alone.

Directed binary relations between any structural variables A and B ($A \rightarrow B$) of the model [30] with high and significant R^2 are selected, to match size of original model. Higher R^2 values stand for better linear prediction of B using A.

In the original paper, Verniers and Vala compared M_O with two models: (1) $M_{O(MS)}$: reversal of the relation Sexism $\rightarrow$ Myth into Myth $\rightarrow$ Sexism and (2) $M_{O(OM)}$ reversal of the relation Myth $\rightarrow$ Opposition into Opposition $\rightarrow$ Myth. These models were only compared on the country level where it showed some differences. When we evaluated all three models on the whole hold-out test data set, we found no differences for model performance measures (cf. Table 1). We note that M_O, $M_{D(MS)}$ and $M_{D(OM)}$ include a mediating variable, i.e. *Myth*, *Sexism*, and *Opposition* respectively.

These models are compared with a model $M_{D(ML)}$ that emerged purely by a data analytical approach with machine learning (ML) that reverses *Sexism* $\rightarrow$ *Opposition* into *Opposition* $\rightarrow$ *Sexism*.

5   Structural Model Performance Analysis

For overall performance measures of a fitted model, a baseline model is usually defined that assumes all observed variables being uncorrelated [25]. It is compared with the fitted model and used for computation of relative indexes of model fit CFI and TLI. A CFI of .955 is often considered "good". Keep in mind, however, that both TLI and CFI are relative indexes of model fit—they compare the fit of the model to the fit of the (worst fitting) null model. Hu & Bentler [11] recommend reporting both a relative and an absolute index of model fit. Absolute indexes of model fit compare the fit of a model to a perfect fitting model, i.e. RMSEA and SRMR.

All four models that we consdier yielded the same results for CFI (0.994), TLI (0.979), RMSEA (0.057 and p-value 0.138) and SRMR (0.012). The explanation for this result is that only one relation was turned for each model compared to M_O. The same holds for information criteria, i.e. AIC (93434), BIC (93517) and adjusted BIC (93479). Therefore, overall performance measures of fitted models and information criteria do distinguish between any model. In the original paper, Verniers and Vala gained differences only by comparing M_O with $M_{D(MS)}$ and $M_{D(OM)}$ on the country level. Overall countries these models cannot be

Table 1. Standardized maximum likelihood coefficients for the structural models tested with 20% hold-out data set values and overall model performance measures

	S → M	M → S	M → O	O → M	S → O	O → S
M_O	0.426(0.010)	na	0.277(0.011)	na	0.085(0.007)	na
$M_{D(MS)}$	na	0.754(0.018)	0.277(0.011)	na	0.085(0.006)	na
$M_{D(OM)}$	0.240(0.012)	na	na	0.916(0.035)	0.203(0.005)	na
$M_{D(ML)}$	0.330(0.013)	na	0.296(0.010)	na	na	0.764(0.052)

distinguished. Fitted values for latent variables based on the measurement model are the same for all models.

All four models show substantial path coefficients. Only the coefficients for $S \rightarrow O$ in model M_O and $M_{D(MS)}$ are small, i.e. below 0.1 while the same coefficient for model $M_{D(OM)}$ is above 0.2. Verniers and Vala found support for selecting M_O over the other two models by applying these models on country level. Because national data is much smaller, it cannot be ruled out that these effects are related to sample size. Therefore we used the whole data set for our study.

The question is whether the model purely derived from data shows any properties that are similar to the theory-driven approach. We already mentioned that overall performance measures are exactly the same as for the other three models which is not surprising because it only changed the direction of one relation. On the level of single relations, $M_{D(ML)}$ exhibits significant (0.1) for each coefficient. Absolute values for coefficients between $S \rightarrow M$ and $M \rightarrow O$ are close to the other models, i.e. a change in the independent variable has a similar effect on the dependent variable. The reversal of $S \rightarrow O$ into $O \rightarrow S$ results in a strong coefficient that is significant at a 10% level.

5.1   Discussion of Results

Two issues come up by our analysis. First, we found that all three models presented by Verniers and Vala [30] exhibit almost the same overall performance indicators and path coefficients when fitted to the whole data set. Thus quantitative arguments given in the original paper are not sufficient for ruling out two of three models with respect to the whole data set.

Second, the model ($M_{D(ML)}$) derived purely by data analytics with machine learning exhibits the same overall performance indicators and significant path coefficients as well. This brings the argumentation back to a theoretical discussion between *Sexism* and *Opposition*. Going back to Verniers and Vala's argumentation, it appears that little prior knowledge exists for the causal relationship from *Sexism* to *Opposition*: "We assume that an equivalent mediational process underlies the justification of gender discrimination in the workplace or, put differently, that the sexism-opposition to womens career relationship is mediated by legitimizing myths." Unfortunately, this conclusion is not supported by [8].

Other research shows ambivalent views on the relationship between *opposition* and *sexism* [27].

These results call for stronger evidence of relationship between variables as provided by [30]. First, it was found that the originally evaluated models are not as differentiating as reported. Secondly, another model was extracted by data analytical processes alone that cannot be rejected by overall performance measures or detailed analysis of relations. Therefore, our approach does not fully support the model proposed by Verniers and Vala, and asks for more scrutiny.

6 Summary and Outlook

We have presented an approach for extracting structural relationships by data-driven, statistical learning methods. We described a method consisting of the following steps: (1) specification of a measurement model and data collection, (2) discovery of patterns between variables by applying data analytical methods with machine learning, (3) assessed the quality of selected models by SEM analysis, and (4) confirmation or non-confirmation of selected models.

A limitation is that the proposed method has been applied to one original model only. Studies for applying this method to a set of structural models including cross-evaluation are a subject of ongoing work. Furthermore, a statistical method for comparing structural models is needed as stressed in [6]. Recent work presented some initial methods for comparing models based on statistically sound similarity measures [17].

References

1. Anderson, C.: The end of theory: The data deluge makes the scientific method obsolete. Wired magazine **16**(7) (2008). 16–07
2. Anderson, J.C., Gerbing, D.W.: Structural equation modeling in practice: a review and recommended two-step approach. Psychol. Bull. **103**(3), 411 (1988)
3. Berente, N., Seidel, S., Safadi, H.: Data-driven computationally-intensive theory development. Inf. Syst. Res. (forthcoming)
4. Bishop, C.M.: Pattern Recognition and Machine Learning (Information Science and Statistics). Springer, New York, Secaucus (2006)
5. Christopher, M.B.: Pattern Recognition And Machine Learning. Springer, New York (2016)
6. Fife, D.A., Rodgers, J.L., Mendoza, J.L.: Model conditioned data elasticity in path analysis: assessing the "confoundability" of model/data characteristics. Multivar. Behav. Res. **49**(6), 597–613 (2014)
7. Floridi, L.: Big data and their epistemological challenge. Philos. Technol. **25**(4), 435–437 (2012)
8. Glick, P., Fiske, S.T.: The ambivalent sexism inventory: differentiating hostile and benevolent sexism. J. Pers. Soc. Psychol. **70**(3), 491 (1996)
9. Hair, J.F., Ringle, C.M., Sarstedt, M.: PLS-SEM: indeed a silver bullet. J. Mark. Theory Pract. **19**(2), 139–152 (2011)
10. Harary, F., Palmer, E.M.: Graphical Enumeration. Elsevier (2014)

11. Hu, L., Bentler, P.M.: Cutoff criteria for fit indexes in covariance structure analysis: conventional criteria versus new alternatives. Struct. Equ. Model.: Multidiscip. J. **6**(1), 1–55 (1999)
12. James, G., Witten, D., Hastie, T., Tibshirani, R.: An Introduction to Statistical Learning, vol. 112. Springer, New York (2013). https://doi.org/10.1007/978-1-4614-7138-7
13. Jöreskog, K.G.: Structural analysis of covariance and correlation matrices. Psychometrika **43**(4), 443–477 (1978)
14. Jöreskog, K.G.: Lisrel. Wiley Online Library (2006)
15. Kaplan, D.: Structural Equation Modeling: Foundations and Extensions, vol. 10. Sage Publications, Newbury Park (2008)
16. Kitchin, R.: Big data, new epistemologies and paradigm shifts. Big Data Soc. **1**(1), 2053951714528481 (2014)
17. Lai, K., et al.: Assessing model similarity in structural equation modeling. Struct. Equ. Model.: Multidiscip. J. **23**(4), 491–506 (2016)
18. MacCallum, R.C., Wegener, D.T., Uchino, B.N., Fabrigar, L.R.: The problem of equivalent models in applications of covariance structure analysis. Psychol. Bull. **114**(1), 185 (1993)
19. Marcén, A.C., Pérez, F., Cetina, C.: Ontological evolutionary encoding to bridge machine learning and conceptual models: approach and industrial evaluation. In: Mayr, H.C., Guizzardi, G., Ma, H., Pastor, O. (eds.) ER 2017. LNCS, vol. 10650, pp. 491–505. Springer, Cham (2017). https://doi.org/10.1007/978-3-319-69904-2_37
20. Meseguer-Artola, A., Aibar, E., Lladós, J., Minguillón, J., Lerga, M.: Factors that influence the teaching use of wikipedia in higher education. J. Assoc. Inf. Sci. Technol. **67**(5), 1224–1232 (2016)
21. Nalchigar, S., Yu, E., Ramani, R.: A conceptual modeling framework for business analytics. In: Comyn-Wattiau, I., Tanaka, K., Song, I.-Y., Yamamoto, S., Saeki, M. (eds.) ER 2016. LNCS, vol. 9974, pp. 35–49. Springer, Cham (2016). https://doi.org/10.1007/978-3-319-46397-1_3
22. Popper, K.: The Logic of Scientific Discovery. Routledge, New York (2005)
23. Prensky, M.: H. sapiens digital: from digital immigrants and digital natives to digital wisdom. Innov. J. Online Educ. **5**(3), 1 (2009)
24. Rodgers, J.L.: The epistemology of mathematical and statistical modeling: a quiet methodological revolution. Am. Psychol. **65**(1), 1 (2010)
25. Rosseel, Y.: Lavaan: an R package for structural equation modeling and more. version 0.5-12 (beta). J. Stat. Softw. **48**(2), 1–36 (2012)
26. Schölkopf, B., Janzing, D., Peters, J., Sgouritsa, E., Zhang, K., Mooij, J.: On causal and anticausal learning. arXiv preprint arXiv:1206.6471 (2012)
27. Sibley, C.G., Perry, R.: An opposing process model of benevolent sexism. Sex Roles **62**(7–8), 438–452 (2010)
28. Silver, D., et al.: Mastering the game of go without human knowledge. Nature **550**(7676), 354 (2017)
29. Sunkle, S., Kholkar, D., Kulkarni, V.: Comparison and synergy between fact-orientation and relation extraction for domain model generation in regulatory compliance. In: Comyn-Wattiau, I., Tanaka, K., Song, I.-Y., Yamamoto, S., Saeki, M. (eds.) ER 2016. LNCS, vol. 9974, pp. 381–395. Springer, Cham (2016). https://doi.org/10.1007/978-3-319-46397-1_29
30. Verniers, C., Vala, J.: Justifying gender discrimination in the workplace: the mediating role of motherhood myths. PloS one **13**(1), e0190657 (2018)

Conceptual Modeling for Machine Learning and Reasoning II

Realtime Event Summarization from Tweets with Inconsistency Detection

Lingting Lin[1], Chen Lin[1($\boxtimes$)], and Yongxuan Lai[2]

[1] Department of Computer Science, Xiamen University, Xiamen, China
chenlin@xmu.edu.cn
[2] School of Software, Xiamen University, Xiamen, China

Abstract. The overwhelming amount of event relevant tweets highlights the importance of realtime event summarization systems. When new information emerges, former summaries should be updated accordingly to deliver most recent and authoritative information. Existing studies couldn't preserve the integrity of a realtime summary. For example, for an ongoing earthquake event, existing studies might generate a summary including new and old estimates of number of injuries, which are inconsistent. In this contribution we present a realtime event summarization system with explicit inconsistency detection. We model the realtime summarization problem as multiple integer programming problems and solve the relaxed linear programming form by an improved simplex update method. To reduce the storage and computational cost of expensive inconsistency detection, we embed a novel fast inconsistency detection strategy in the simplex update algorithm. We conduct comprehensive experiments on real twitter sets. Compared with state-of-the-art methods, our framework produces summaries with higher ROUGE scores and lower inconsistency rates. Furthermore our framework is more efficient.

Keywords: Realtime summarization · Inconsistency detection
Simplex

1 Introduction

Emergence of microblogging platforms, such as Twitter [5], has resulted in a large community of microblogging users and a massive amount of instant reports about live events all over the world. Event summarization system is needed to facilitate knowledge management and improve user experiences. We have witnessed rapidly increasing popularity of research efforts in event summarization from tweets [3,6,7,12–15,17]. Most previous studies are based on extractive methods, i.e. they extract a smallest set of representative tweets to form a brief summary. Extractive methods are easy to implement and have shown to perform well [3,6,7,12,14,15,17]. Our arguments in this paper are founded on extractive summarization methods.

© Springer Nature Switzerland AG 2018
J. C. Trujillo et al. (Eds.): ER 2018, LNCS 11157, pp. 555–570, 2018.
https://doi.org/10.1007/978-3-030-00847-5_41

Beyond the usual requirements for text summarization systems, such as coverage and representativeness, event summary must also be **realtime**. On one hand, the response must be fast. The summary must be efficiently updated as new tweets arrive. On the other hand, the summary must report the current status of the event. As an ongoing event often involves changing information, the old summary must be updated to include new information when it emerges.

During the update process, the integrity of the summary must be preserved. An outdated tweet report must be replaced if it leads to **inconsistency** in the summary. An inconsistent summary is harmful for most live events because it is confusing and misleading. Examples include (1) in an earthquake the number of injuries is increasing over time. Thus the numbers in previous summaries become obsolete and they should be replaced by the most up-to-date numbers. (2) The number of injuries is estimated by several parties, such as bystanders, hospitals and so on. When a more authoritative source, such as the local government announces the new estimate, the old estimates in previous summaries are no longer credible and must be replaced.

Realtime event summarization from tweets is still an open problem. In the literature, most of previous works treat the problem as producing different forms of summaries from a static set of tweets [3,6,7,12,15]. A few recent research works focused on efficient algorithms to summarize the tweet streams [14,17]. Their summarization systems are based on coarse grained semantic analysis, and thus are not able to detect inconsistency. Though we have shown that integrity of the event summary is crucial, to the best of our knowledge, none of the previous works is able to produce a realtime event summarization which is guaranteed to exclude inconsistent information.

Two challenges arise in producing realtime event summarization without inconsistent information.

The first challenge lies in the macro-level algorithm. Realtime summarization requires an efficient algorithm to analyze the streaming tweets. As the amount of available tweets constantly increases to infinity, re-computation based on a complete set of all tweets up to the current timestamp is infeasible. The ideal algorithm is to incorporate new tweets as they become available, and discard old tweets when possible to limit the storage and speed up processing.

The second challenge is related to the micro-level analysis to detect inconsistency. Inconsistency detection is based on pair-wise similarity. Coarse grained semantic analysis, such as the cosine similarity measure based on the bag of words representation in previous works [3,6,7,12–15,17] is suitable to capture topic similarity in a summarization. However, it is not able to detect inconsistent information. We need to design new similarity metrics for this purpose. Furthermore, inconsistency detection is computationally expensive. It is important to avoid unnecessary pair-wise comparisons.

Our goal in this paper is to design a system that delivers realtime summary with integrity from tweets. To address the first challenge, we assume that, the realtime summarization problem given a small batch of new tweets can be modeled as two integer programming problems, one of which on the old tweets, and another on the new batch. Both integer programming problems can be relaxed to linear programming problems and be solved by the simplex method. In each update we first optimize the problem on the new batch. We use the solution on the new batch to modify the problem on the old tweets and incrementally update the summary. In this manner, we do not need to store or operate on the complete tweet set and the full similarity matrix.

To address the second challenge, we propose skeleton similarity: a new similarity metric to assess information similarity between any pair of tweets. An inconsistency detection strategy, which is a combination of the skeleton similarity and authority estimation heuristics, is then adopted in the pivoting operation in the simplex method. It has two advantages in embedding the skeleton similarity computation in the simplex algorithm. (1) It significantly reduces the number of information similarity comparisons. (2) It ensures that the former summary will be replaced by most up-to-date and authoritative information.

Our contributions are three folds. (1) The integrity of event summary is a relatively unexplored area in Tweet summarization. Because different data sources may provide different data for the same concept, resulting in inconsistent data. We propose to improve the integrity of event summarization by explicit inconsistency detection. (2) Our system is targeted towards text streams. We model the realtime summarization problem as integer programming problems in small batches. We differ from existing work in that we enable incremental updates in the simplex framework. (3) We propose a novel skeleton similarity to efficiently and effectively capture inconsistency, which is beneficial for a wide range of applications, including conceptual modeling.

This paper is organized as follows. We briefly survey the related work in Sect. 2. In Sect. 3, we first introduce the idea of modeling a summarization problem as an integer programming problem and the standard simplex procedure to solve the relaxed linear programming problem. In Sect. 4, we give the problem definition for realtime event summarization given a small batch of new tweets and the improved simplex solution. The inconsistency detection strategy is a component of the simplex update algorithm. We present and analyze the experimental results on a real data set in Sect. 5. We conclude our work and suggest future directions in Sect. 6.

2   Related Work

Multi-document summarization conveys the main and most important meaning of several documents. There are generally two types of summarization techniques. One type is extraction-based summarization, which extracts objects from the entire collection and combines the objects into a summary without modifying the objects themselves. The other type is abstraction-based summarization which rephrases the source document. The majority of summarization systems are extractive. The extracted objects are often sentences. The selection is usually based on the representativeness of sentences, i.e. with significant frequency [16], or is a structural centroid in a sentence graph [6], or is considered important by a submodularity function [22].

The emergence of Twitter motivates recent research works on summarizing microblogging contents. Tweet summarization systems are successfully applied in entity-centric opinion summarization [9], personal summarization of interesting content [1,10], search results grouping [8], and summarizing tweets for natural or social events [3,6,7,12,14,15,17].

At the algorithm level, tweet summarization also uses extractive and abstractive methods. Except a few work based on abstractive methods [13], most tweet summarization methods are extractive, i.e. they rank and select the most representative tweets. A few recent works start to improve general multi-document summarization methods for better efficiency. In [14], an incremental clustering method is presented. In [17] the selection range is shrinked by detecting sub-events and selecting one sentence with maximal similarity to any new sub-event.

To achieve a better performance, the noisy and social nature of microblogs must be taken into consideration. Most tweet summarization systems identify influential tweets [4], promote recent tweets [2], and circumnavigate spam and conversational posts [3]. However, the integrity of summary has not yet been fully studied. The work that is most related to ours is the classification and summarization of situational information in [11,12]. However they do not explicitly identify inconsistency as they simply provide all versions of inconsistent information. Furthermore, they do not accelerate algorithms for realtime responses.

3   Static Summarization

The standard summarization task is conducted on a static set of tweets. We refer this task as a static summarization. In this section, we first model the static summarization problem as an integer programming problem. We then present a high level explanation about the simplex method. We finally give an outline for the algorithm to solve static summarization.

3.1   Problem Definition

Suppose that we have a universe of N *candidate* tweets, within which M tweets are credible and relevant. The credible and relevant tweets are important so they are considered to be the *seeds* of the summary. The extractive method for any static summarization is to select a few representative tweet *reports*[1] from the tweet universe to form the summary. To model this problem, we use a vector $\mathbf{x} \in \mathcal{R}^N$, where each element $\mathbf{x}_j \in \{0,1\}$ is a binary variable. If a candidate i is chosen to be a report in the summary, the corresponding $x_i = 1$. Otherwise, we set $x_i = 0$. We use another N-dimensional vector $\mathbf{c} \in \mathcal{R}^N$ to describe the loss of choosing each candidate as a report. $\mathbf{A} \in \mathcal{R}^{M \times N}$ is a similarity matrix, where $A_{i,j}$ is the similarity between a seed i and a candidate j in the tweet universe.

---

[1] To distinguish the three types of tweets, we will refer the tweets to be summarized as candidates, the tweets in the summary as reports, and the credible and relevant tweets as seeds.

$\mathbf{b} \in \mathcal{R}^M$ is a weight vector, where $b_i > 0$ indicates the importance of seed i being covered in the summary. Our objective is to:

$$\min \mathbf{c}^T \mathbf{x} \text{ subject to } \mathbf{Ax} \geq \mathbf{b}, \forall i \ x_i \in \{0, 1\}. \tag{1}$$

The optimization problem in Eq. 1 aims to find a minimal set of reports that delivers all credible and relevant information. Thus the result summary achieves brevity, coverage and representativeness. The decision of b, c and the M indices affects the performance of summarization. We allow the flexibility of choosing any sounding technique to determine the credible and relevant set of tweets, their importance values and the loss of choosing any candidate. For example, in the experiment, we choose the top M results from a search engine, determine $\mathbf{b}$ by estimating the authority of each tweet account, and assign $\mathbf{c}$ by assessing the textual quality of each candidate. The technical details are discussed in Sect. 5.

3.2   The Simplex Method

We transform the integer programming problem in Eq. 1 to a bounded linear programming problem by making the following adjustments: $\tilde{\mathbf{c}} = [\mathbf{c}, \mathbf{0}], \tilde{\mathbf{x}} = [\mathbf{x}^T, \mathbf{z}]^T, \tilde{\mathbf{A}} = [\mathbf{A}, -\mathbf{I}]$, where $\mathbf{z} \in \mathcal{R}^M$, $\mathbf{I}$ is the $M \times M$ identity matrix. Therefore we have the following objective:

$$\min \tilde{\mathbf{c}}^T \tilde{\mathbf{x}} \text{ subject to } \tilde{\mathbf{A}}\tilde{\mathbf{x}} = \mathbf{b}, \forall 1 \leq j \leq N, \ \tilde{x}_j \leq 1, \tilde{\mathbf{x}} \geq \mathbf{0}. \tag{2}$$

The linear programming problem in Eq. 2 can be solved by the simplex method. Each iterate in the simplex method is a *basic feasible point* that (1) it satisfies $\tilde{\mathbf{A}}\tilde{\mathbf{x}} = \mathbf{b}, \forall 1 \leq i \leq M + N, \tilde{x}_i \geq 0, \forall 1 \leq j \leq N, \tilde{x}_j \leq 1$ and (2) there exists three subsets $\mathcal{B}, \mathcal{U}, \mathcal{L}$ of the index set $\mathcal{A} = \{1, 2, \cdots M + N\}$ such that $\mathcal{B}$ contains exactly M indices, $\mathcal{A} = \mathcal{B} \cup \mathcal{U} \cup \mathcal{L}$ and $i \in \mathcal{U} \Rightarrow \tilde{x}_i = 1, i \in \mathcal{L} \Rightarrow \tilde{x}_i = 0$.

The major issue at each simplex iteration is to decide which index to be removed from the basis $\mathcal{B}$ and replaced by another index outside the basis $\mathcal{B}$. As the optimal is achieved when the KKT conditions are satisfied, in each simplex iteration first a **pricing** step is conducted to check on the KKT conditions. If the KKT conditions are satisfied then the problem is solved. Otherwise a **pivoting** operation is implemented to select entering and leaving indices.

To obtain an easy start of basic feasible points, we solve the following linear programming problem.

$$\min \mathbf{e}^T \mathbf{s} \text{ subject to } \tilde{\mathbf{A}}\tilde{\mathbf{x}} + \mathbf{Is} = \mathbf{b}, \forall 1 \leq j \leq N, \ \tilde{x}_j \leq 1, \tilde{\mathbf{x}} \geq \mathbf{0}, \mathbf{s} \geq \mathbf{0}. \tag{3}$$

where $\mathbf{e}$ is the vector of all ones, $\mathbf{I}$ is a diagonal matrix whose diagonal elements are $I_{ij} = 1$. $\mathbf{s}$ are called artificial variables. It is easy to see that the solution to Eq. 3 is a basic feasible point for Eq. 2, if the objective $\mathbf{e}^T \mathbf{s} = 0$.

The remaining problem is that the solution we obtained for Eq. 2 are not integers. As our interest is on the variables $\mathbf{x}$ in Eq. 1, we can interpret the values obtained for Eq. 2 $\tilde{x}_j, \forall 1 \leq j \leq N$ as the probability of choosing j as a response. For indices that are in the upper bound set $j \in \mathcal{U}$, the corresponding candidates are deemed in the summary. For indices that are in the lower bound set $j \in \mathcal{L}$, the corresponding candidates are filtered. For the remaining candidates, we perform a **rounding** algorithm. We first sort the solutions we obtained for Eq. 2 in descending order of their values $0 < \tilde{x}_j < 1, \forall 1 \leq j \leq N$ and then sample x_j based on the probability of assigned value $\tilde{x}_j$.

$$p(x_j = 1) = \tilde{x}_j, \forall 1 \leq j \leq N \tag{4}$$

To conclude this section, we present Algorithm 1. The framework of static summarization includes three stages. The first stage (step 1 to step 2) is to obtain a basic feasible point for the linear programming problem Eq. 2, the second stage (step 3 to step 4) is to optimize the linear programming problem in Eq. 2, and the third stage (step 5) is to obtain the integer solutions for Eq. 1.

Input: $\mathbf{c}, \mathbf{b}, A$
Output: $\mathbf{x}$
1 Initialize $\tilde{\mathbf{x}} = \mathbf{0}, \tilde{\mathbf{A}} = [\mathbf{A}, -\mathbf{I}], \mathbf{s} = \mathbf{b}$;
2 Solve $\min \mathbf{e}^T \mathbf{s}$ subject to $\tilde{\mathbf{A}}\tilde{\mathbf{x}} + \mathbf{Is} = \mathbf{b}, \forall 1 \leq j \leq N, \tilde{x}_j \leq 1, \tilde{\mathbf{x}} \geq \mathbf{0}, \mathbf{s} \geq \mathbf{0}$ by simplex method;
3 Initialize $\tilde{\mathbf{x}}$ by the solution above, set $\tilde{\mathbf{c}} = [\mathbf{c}, \mathbf{0}]$;
4 Solve $\min \tilde{\mathbf{c}}^T \tilde{\mathbf{x}}$ subject to $\tilde{\mathbf{A}}\tilde{\mathbf{x}} = \mathbf{b}, \forall 1 \leq j \leq N, \tilde{x}_j \leq 1, \tilde{\mathbf{x}} \geq \mathbf{0}$ by simplex method;
5 $\tilde{\mathbf{x}} = [\mathbf{x}, \mathbf{z}]^T$, Rounding $\mathbf{x}$;

Algorithm 1. The framework for static summarization

---

4   Dynamic Summarization

For streaming texts, the realtime summarization system must be able to update the summary as new contents arrive. The summary is dynamic because it changes over time. We here consider a batch setting: the summary is updated when a predefined number T of tweets are received. Note that if we set $T = 1$ we enable fully online update. In this section, we first model the dynamic summarization problem and then we present the improved simplex method.

4.1   Problem Definition

When a small batch of tweets arrive, the candidate universe will be a new set of N_1 candidates with importance weights denoted by $\mathbf{c_1}$ and the former candidates. We keep previous reports $\mathbf{x_0}$ with $x_i = 1$ in the former summary and their loss value $\mathbf{c_0}$ and discard everything else from former corpus to reduce storage space.

To produce a summary that covers the demanded information, the model in Sect. 3 requires a predefined set of credible and relevant seed tweets and their importance weights $\mathbf{b}$. It is crucial here to mention that in the dynamic setting the seed set is priorly unknown due to inconsistency. For example, if a new credible tweet is found to conflict with a former credible tweet, then the former one must be excluded. Suppose we have M seeds, but the integrity is not guaranteed. It is possible to check on inconsistency to prune the M seeds to identify a consistent new set. However, the inconsistency detection algorithm on $M \times (M-1)$ pairs is expensive. Henceforth, our idea is to call the inconsistency detection algorithm only when it is necessary.

Suppose we segment the timeline of tweets into two divisions. M_0 represents the set of seeds that are covered by the former summary, M_1 is the set of seeds published after the former summary was generated. M_1 is generated as follows. We first retrieve R relevant tweets published after the summary. Then we can adopt any heuristic to fast prune inconsistent seeds in M_1. We can also adopt any external strategy to exclude seeds that are less credible than any seed in M_0. Therefore the seeds in M_1 must be covered with priority if they are inconsistent with the seeds in M_0, because any seed in M_1 is newer and at least as credible as any seed in M_0. To deliver an up-to-date summary, we set $\mathbf{c}$ so that the loss for new candidates is smaller than the old candidates.

Below we give the architecture of the dynamic summarization system. We model the dynamic summarization as sequential integer programming problems. Given the whole set of candidates $\mathbf{x_0}, \mathbf{x_1}$ and the associated loss $\mathbf{c_0}, \mathbf{c_1}$, the importance of possible seeds $\mathbf{b_1}$ in M_1, at stage I, we first solve the optimization problem for the higher division of seeds.

$$\min \mathbf{c_0}^T \mathbf{x_0} + \mathbf{c_1}^T \mathbf{x_1} \text{ subject to } S\mathbf{x_0} + B\mathbf{x_1} \geq \mathbf{b_1}, \forall i, \forall j \in \{0,1\} \ x_{j,i} \in \{0,1\}. \tag{5}$$

We adopt an incremental method to update $\mathbf{x_0}$ given its original assignment in the previous time slot. We will show later that in solving Eq. 5 we also fast prune inconsistent seeds in $\mathbf{b_0}$. As we emphasize on new and credible tweets, we will preserve $\mathbf{b_1}$ whenever possible and obtain the refined seed set $\mathbf{b_0}$. In stage II we solve the optimization problem based on the refined $\mathbf{b_0}$:

$$\min \mathbf{c_0}^T \mathbf{x_0} + \mathbf{c_1}^T \mathbf{x_1}$$
$$s.t. \begin{bmatrix} F & D \\ S & B \end{bmatrix} \begin{bmatrix} \mathbf{x_0} \\ \mathbf{x_1} \end{bmatrix} \geq \begin{bmatrix} \mathbf{b_0} \\ \mathbf{b_1} \end{bmatrix},$$
$$\forall i, \forall j \in \{0,1\} \ x_{j,i} \in \{0,1\}, \tag{6}$$

where $F \in \mathcal{R}^{M_0 \times N_0}$ is the similarity matrix between former reports and old seeds, $D \in \mathcal{R}^{M_0 \times N_1}$ is the similarity matrix between new candidates and old seeds, $S \in \mathcal{R}^{M_1 \times N_0}$ is the similarity matrix between old reports and new seeds, $B \in \mathcal{R}^{M_1 \times N_1}$ is the similarity matrix between new candidates and new seeds.

4.2 Simplex Update Method

As in Sect. 3, we introduce new variables. $\tilde{\mathbf{c}} = [\mathbf{c_0}, \mathbf{c_1}, \mathbf{0}], \tilde{\mathbf{x}} = [\mathbf{x_0}, \mathbf{x_1}, \mathbf{z}]^T, \tilde{\mathbf{A}} = [S, B, -\mathbf{I}], \mathbf{s}$, where $\mathbf{z} \in \mathcal{R}^{M_1}$, $\mathbf{I}$ is the $M_1 \times M_1$ identity matrix. To initialize, we set $\mathbf{x_0} = 1$ and keep $\mathbf{x_0}$ fixed while solving Eq. 5. We use the solution as the initialized guess for the second stage optimization.

In each iterate of simplex, we first operate the pricing step. Let's compute $\bar{c}_j = \tilde{c}_j - y^T \tilde{\mathbf{A}}_{.,j}$ for every $\tilde{x}_j$ not in the basis, where $\mathbf{y} = \mathcal{B}^{T-1} \mathbf{c}_{\mathcal{B}}$. If $\tilde{x}_j = 0, \bar{c}_j > 0$ and $\tilde{x}_j = 1, \bar{c}_j < 0$, then the objective function is optimized. Otherwise we implement the pivoting operation.

We modify the conventional pivoting operation in the simplex method. In the initialization we keep $\mathbf{x_0} = 1$, which indicates that the constraint is satisfied with all the former reports kept. If the pricing step suggests the objective function is not optimized, there are usually multiple indices which violate the KKT conditions. Hence in selecting the entering index, we prefer a z index, because selecting an auxiliary variable leads to minimal modification of the summary. However, if an index from $\mathbf{x_0}$ is to leave and an index from $\mathbf{x_1}$ is to enter, a new tweet is able to cover the same set of seeds with smaller cost (given that we assign higher loss c_0 to old tweets). We can not rule out the possibility of inconsistency. In fact, inconsistency only happens when two tweets are semantically closely related. Therefore we conduct inconsistency test between the entering index and the leaving index. If the leaving index is found to be inconsistent of the new report, we delete the corresponding component $x_{0,i}$ from the problem. This strategy ensures that inconsistent candidates do not repeatedly enter and leave the summary. We remove all seeds that are related to $x_{0,i}$ in $\mathbf{b_0}$. The pivoting operation is described in Algorithm 2.

4.3   Inconsistency Detection

To effectively deal with tweets which contain abbreviations, numerals, foreign words and symbols, we perform case-folding, lemmatization, stop-word removal, marker deletion, POS tagging [12] and Named Entity Recognition [24]. We ignore the Twitter-specific words (assigned by tag "G"). We replace all numerals by a special token "numeral". The named entity phrases, such as person, organization location are replaced by special tokens. A tweet is transformed to a sequence of tokens after pre-processing.

The skeleton of a pair of tweets is the longest common sequence (LCS) of tokens. Note that we do not consider the position of special tokens in finding the LCS. Special tokens, despite of their values, are considered to be identical and a component of the LCS. After the LCS is found, we append the specific tokens to the LCS. As shown in Table 1, the LCS for the given pair of tweets is "least numeral location" which we do not consider the position of "location".

Table 1. Illustrative example of a pair of inconsistent tweets

Original tweet 1	At least 38 killed in Pakistan by blast outside Lahore park
Processed tweet 1	least (numeral) killed (location) blast (location)
Original tweet 2	TTP claims Lahore park explosion that killed at least 65
Processed tweet 2	(organization) claim (location) explosion killed least (numeral)

Two tweets are inconsistent, if they are similar at the sentence level, yet they are different in some of the key components. In a real event, numerals and named entities are often key components because they provide situational information about an event. Therefore we identify inconsistency if (1) the length ratio of LCS v.s. the shortest tweet in the pair is equal to or larger than 0.5 and (2) the special tokens, such as numerals, persons, organizations and locations have different values in the two tweets. For example, the two tweets in Table 1 are inconsistent, because the LCS has length 3 and the shortest tweet has length 6 and the numerals and locations are different (i.e. 38 v.s. 65). According to Algorithm 2 if tweet 2 is newer (thus in x_1), we will remove tweet 1 from the candidate index and delete indices in $\mathbf{b}_0$ which are covered by tweet 1.

---

1   $q = \arg\max_j \left\{ 10 \times \bar{c}_j \; \forall j \in \mathbf{z}_{\mathcal{N}}, \bar{c}_j \; \forall j \in \mathbf{x}_{\mathcal{L}}, -\bar{c}_j \; \forall j \in \mathbf{x}_{\mathcal{U}} \right\};$

2   $d = \mathcal{B}^{-1}\tilde{\mathbf{A}}_{,q};$

3   **if** $q \in \mathbf{z}_{\mathcal{N}}$ **then**

4      $x_q^{new} = \min_i \left\{ \{ \frac{x_{B_i}}{d_i}, \frac{s_{B_i}}{d_i}, \frac{z_{B_i}}{d_i} \} \forall d_i > 0, \{ \frac{x_{B_i} - 1}{d_i} \} \forall d_i < 0 \right\};$

5      $\mathbf{x}_{Bold}^{new} = \mathbf{x}_{Bold}^{old} - dx_q^{new};$

6      $p = \arg\min_i;$

7   **end**

8   **if** $q \in \mathbf{x}_{\mathcal{L}}$ **then**

9      $x_q^{new} = \min_i \left\{ \{ \frac{x_{B_i}}{d_i}, \frac{s_{B_i}}{d_i}, \frac{z_{B_i}}{d_i} \} \forall d_i > 0, \{ \frac{x_{B_i} - 1}{d_i} \} \forall d_i < 0, 1 \right\};$

10     $\mathbf{x}_{Bold}^{new} = \mathbf{x}_{Bold}^{old} - dx_q^{new};$

11   **if** $x_q^{new} \neq 1$ **then**

12       $p = \arg\min_i;$

13   **end**

14   **end**

15   **if** $q \in \mathbf{x}_{\mathcal{U}}$ **then**

16     $x_q^{new} = 1 - \min_i \left\{ \{ \frac{x_{B_i}}{-d_i}, \frac{s_{B_i}}{-d_i}, \frac{z_{B_i}}{-d_i} \} \forall d_i < 0, \{ \frac{1 - x_{B_i}}{d_i} \} \forall d_i > 0, 1 \right\};$

17     $\mathbf{x}_{Bold}^{new} = \mathbf{x}_{Bold}^{old} + d(1 - x_q^{new});$

18   **if** $x_q^{new} \neq 0$ **then**

19       $p = \arg\min_i;$

20   **end**

21   **end**

22   **if** $1 \leq p \leq N_0, N_0 + 1 \leq q \leq N_0 + N_1$ **then**

23   Inconsistency detection on (p, q);

24   **if** p, q *are inconsistent* **then**

25   Refine $\mathbf{b}_0$;

26   Remove p from index set $\mathcal{A}$;

27   **end**

28   **end**

29   $\mathcal{B}^{new} \leftarrow \mathcal{B}^{old} - \{p\} \bigcup \{q\};$

Algorithm 2. Pivoting in the simplex update method

5 Experiment

5.1 Experimental Setup

We use the twitter event data set [18]. The corpus includes tweets for 10 real world events, collected between 2014 and 2016, through the Twitter API using keyword matching. Details of the data set, including the description of each event, the number of tweets related to each event are shown in Table 2.

Table 2. Statistics of the events and related tweets

Abbrevation	♯ Tweet	Time period (start end)	Event description
EOutbreak	12983	2014/07/01 2014/07/31	Large-scale virus outbreaks in West Africa
GUAttack	21000	2014/06/02 2014/07/17	Military operation launched by Israel in the Hamas-ruled Gaza Strip
HProtest	27000	2014/09/26 2014/10/17	A series of civil disobedience campaigns in Hong Kong
THagupit	10315	2014/12/05 2014/12/11	A strong cyclone code-named Typhoon Hagupit hit Philippines
CHShoot	15000	2015/01/07 2015/01/07	Two brothers forced their way into the offices of the French satirical weekly news-paper Charlie Hebdo in Paris
HPatricia	9288	2015/10/24 2015/12/08	The second-most intense tropical cyclone on record worldwide
RWelcome	19725	2015/09/02 2015/11/24	Rising numbers of people arrived in the European Union because of European refugee crisis
BAExplossion	15000	2016/03/22 2016/03/22	Three coordinated suicide bombings occurred in Belgium
HPCyprus	14917	2016/03/29 2016/03/30	A domestic passenger flight was hijacked by an Egyptian man in Cyprus
LBlast	13423	2016/03/27 2016/03/30	A suicide bombing that hit the main entrance of Gulshan-e-Iqbal Park

Different types of tweets are contained in the data set, including replies and retweets. The corpus is multilingual, including English, Japanese and so on. In pre-processing, we first filter non-English tweets. We use the Bloom Filter algorithm [19] to filter duplicate tweets. Emoji expressions, http links and mentions (@somebody) are removed from the vocabulary.

5.2 Summarization Performance

We first evaluate the summarization performance of the proposed algorithm. For each event, we chronologically order the tweets and segment the timeline.

Each time segment contains 3000 tweets. We run the search engine Terrier[2] which choose the BM25, PL2 and TF_IDF models to calculate the weight of keywords about event in the tweet seperately, and then get the weight of each tweet on average, and return the top 100 related posts as the seeds of the summarys. We assign $b = 0.2$ for tweets published by any user with more than 5 tweets in the data set, and $b = 0.1$ for other tweets. We assign $c_i = 0.25 \times I_i(u_i) + 0.25 \times \frac{nr_i}{5000} + 0.5 \times \frac{nt_i}{500}$, where $I_i(u_i)$ is an identity function which returns whether the tweet user u_i is a certified account, nr_i is the number of retweets of i, nt_i is the number of thumbs up i receives.

We compare our summarization scheme with four state-of-the-art summarization methods. (1) LPR [21]: summarization based on a sentence connectivity matrix. (2) MSSF [22]: multi-document summarization based on submodularity hidden in textual-unit similarity property. (3) SNMF [23]: summarization based on symmetric non-negative matrix decomposition. (4) Sumblr [14]: online tweet summarization based on incremental clustering.

The ground truth of summaries for each time segment is manually generated. For each time segment of each event, three human volunteers review the summaries generated by the above mentioned methods and individually select the tweets that can summarize the events of this time period. The final gold standard summary is selected by taking a simple majority vote on each tweet.

The measurement is ROUGE (Recall-Oriented Understudy for Gisting Evaluation). ROUGE is a standard evaluation toolkit for document summarization [20] which automatically determines the quality of a summary by comparing it with the human generated summaries through counting the number of their overlapping textual units (e.g., n-gram, word sequences, and etc.). There are four different ROUGE measures: ROUGE-N, ROUGE-L, ROUGE-W, and ROUGE-S, depending on the textual units to be compared.

We report the ROUGE F-score for all versions of ROUGE measures, averaged over the 10 events in our data set, by our simplex algorithm and the comparative state-of-the-art methods. As shown in Fig. 1, our system (marked as Simplex) outperforms state-of-the-art methods in terms of all ROUGE metrics. Our method outperforms Sumblr, which is also a realtime tweet summarization system, by more than 20% in terms of all ROUGE-F scores. It is clear that, despite of the nature of events, the proposed system which is based on integer programming problems is able to model the summarization task well. Furthermore, by explicitly detecting for inconsistency, the quality of summary is significantly enhanced.

5.3   Inconsistency Detection Performance

Next we study the integrity of summaries generated by different methods. As events in our dataset vary in the length of their lifespan, we select the first 4 time segments for every event, and compute the inconsistency rate of the summary produced by each method. We first adopt the inconsistency detection strategy in

---

[2] www.terrier.org.

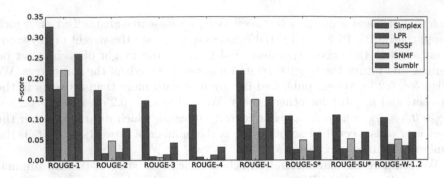

Fig. 1. Average ROUGE F-score on 10 events by different methods.

Sect. 4 to identify inconsistent reports in each summary, and then we manually check the inconsistent reports. We finally obtain the inconsistency rate, which is the proportion of the number of inconsistent reports to the number of reports in a summary.

As shown in Fig. 2, the proposed method preserves the integrity of the summary. At each time segment, the median, 25th and 75th percentile of the inconsistency rate over all event summaries produced by the proposed methods are all near zero. On the contrary, the state-of-the-art comparative methods can not generate an consistent summary. They do not have consistency check between the tweets, hence it is possible that some inconsistent tweets got selected into the summary at some timepoints and are never replaced later. LPR and SNMF are better than MSSF and Sumblr, possibly due to the fact that LPR and SNMF tend to select a smaller number of reports. Sumblr and MSSF tend to select more redundant reports, thus the possibility of inconsistency is increased.

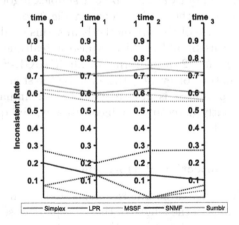

Fig. 2. Inconsistentcy rate, including median (solid line), the 25th percentile and the 75th percentile (dashed line) over ten events on first four time segments for each method.

5.4 Efficiency Study

We now analyze the efficiency of the various techniques. We implement the proposed algorithm in java and use the open source code for LPR, MSSF, SNMF and Sumblr. All comparative methods are given the set of candidates in each time segment. All experiments have been executed on a server with an Intel(R) Xeon(R) CPU E5-1660 3.20 G Hz (4 cores) and main memory 64 G bytes.

As some methods adopt incremental updating (i.e. Sumblr and ours) while some do not (i.e. LPR, MSSF, SNMF), efficiency analysis must be separated for the first summary and the summaries afterwards. We report execution times of each method in generating summary in the first time segment, averaged over all events in Fig. 3(a). The graph y-axes has logarithmic scale to illustrate the difference more clear. We can see that the proposed algorithm is the fastest. We emphasize here that the proposed algorithm is 10 times faster than the second fast method Sumblr, which is also an incremental updating summarization system. This highlights the potential of our proposed method, since generating realtime summaries is critical during disaster events.

We then report the average execution times by different methods, over all events and all follow-up time segments in Fig. 3(b). We see that again our proposed method is the fastest. Compared with Fig. 3(a), the execution times for LPR, MSSF and SNMF to generate follow-up summaries differ by two orders of magnitude. This is expected as LPR, MSSF and SNMF are not incremental methods. Thus they are not suitable for delivering real-time summaries. For any ongoing event their computation time will increase exponentially with the size of accumulated tweets. However, Sumblr and our proposed method are more efficient in updating the summaries, as re-computation is limited to a small range of reports. Moreover, our method requires significantly less execution time (smaller median of the boxplot) than Sumblr.

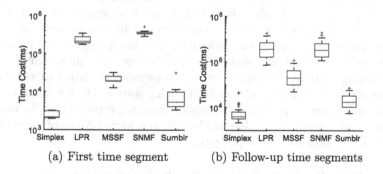

(a) First time segment   (b) Follow-up time segments

Fig. 3. Average execution times to generate a summary by various methods.

5.5 Effects of Parameters

Finally we study the effect of the size of credible and relevant tweets (the number of seeds M in Sect. 3) in Fig. 4. The execution time is the time cost to generate the first summary. We set $M = \{50, 100, 150, 200, 250, 300\}$. We can see that the execution time grows linearly with the size of credible and relevant tweets. However when the number of seeds is too big, the time cost is significantly higher. In order to generate realtime summary, it is better to limit the size of seeds.

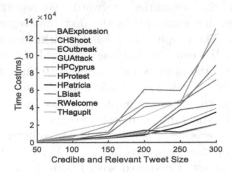

Fig. 4. Execution times to generate summary for each event by the proposed method, given different size of credible and relevant tweets.

6 Conclusion

In this paper we study the problem of realtime event summarization. We implement inconsistency detection in the update procedure to preserve integrity of the summary. We model the realtime summarization problem as integer programming problems and solve the relaxed linear programming form by simplex method. We present a novel inconsistency detection strategy and embed it in the simplex algorithm. Our work sheds insights in intellectual information management in social media platforms, and is especially important for disaster monitoring systems. Our future direction includes exploring more advanced inconsistency detection strategies and exploiting state-of-the-art numerical optimization techniques to solve the integer programming problems. We also plan to incorporate concept extraction to identify and merge coherent concepts and find an optimal summary concept map.

Acknowledgments. The work is supported by the Natural Science Foundations of China (under grant No.61472335, 61672441).

References

1. Chin, J.Y., Bhowmick, S.S., Jatowt, A.: Totem: Personal tweets summarization on mobile devices. In: SIGIR, pp. 1305–1308 (2017)
2. Efron, M., Golovchinsky, G.: Estimation methods for ranking recent information. In: SIGIR, pp. 495–504 (2011)
3. Gillani, M., Ilyas, M.U., Saleh, S., et al.: Post summarization of microblogs of sporting events. In: WWW, pp. 59–68 (2017)
4. Hannon, J., Bennett, M., Smyth, B.: Recommending twitter users to follow using content and collaborative filtering approaches. In: RecSys, pp. 199–206 (2010)
5. Kwak, H., Lee, C., Park, H., et al.: What is twitter, a social network or a news media? In: WWW, pp. 591–600 (2010)
6. Lin, C., Lin, C., Li, J., et al.: Generating event storylines from microblogs. In: CIKM, pp. 175–184 (2012)
7. Liu, Z., Huang, Y., Trampier, J.R.: LEDS: local event discovery and summarization from tweets. In: GIS, pp. 53:1–53:4 (2016)
8. Mathioudakis, M., Koudas, N.: Twittermonitor: trend detection over the twitter stream. In: SIGMOD, pp. 1155–1158 (2010)
9. Meng, X., Wei, F., Liu, X., et al.: Entity-centric topic-oriented opinion summarization in twitter. In: KDD, pp. 379–387 (2012)
10. Ren, Z., Liang, S., Meij, E., et al.: Personalized time-aware tweets summarization. In: SIGIR, pp. 513–522 (2013)
11. Rudra, K., Banerjee, S., Ganguly, N., et al.: Summarizing situational tweets in crisis scenario. In: HT, pp. 137–147 (2016)
12. Rudra, K., Ghosh, S., Ganguly, N., et al.: Extracting situational information from microblogs during disaster events: a classification-summarization approach. In: CIKM, pp. 583–592 (2015)
13. Sharifi, B., Hutton, M.-A., Kalita, J.: Summarizing microblogs automatically. In: HLT, pp. 685–688 (2010)
14. Shou, L., Wang, Z., Chen, K., et al.: Sumblr: continuous summarization of evolving tweet streams. In: SIGIR, pp. 533–542 (2013)
15. Takamura, H., Yokono, H., Okumura, M.: Summarizing a document stream. In: ECIR, pp. 177–188 (2011)
16. Yih, W., Goodman, J., Vanderwende, L., et al.: Multi-document summarization by maximizing informative content-words. IJCAI 7, 1776–1782 (2007)
17. Zubiaga, A., Spina, D., Amigó, E., et al.: Towards real-time summarization of scheduled events from twitter streams. In: HT, pp. 319–320 (2012)
18. Zubiaga, A.: A longitudinal assessment of the persistence of twitter datasets. arXiv preprint arXiv:1709.09186 (2017)
19. Bloom, B.: Space/time tradeoffs in hash coding with allowable errors. Communi. ACM 13(7), 422–426 (1970)
20. Lin, C.Y.: ROUGE: a package for automatic evaluation of summaries. In: Text Summarization Branches Out (2004)
21. Erkan, G., Radev, D.R.: LexPageRank: Prestige in multi-document text summarization. In: EMNLP, pp. 365–371 (2004)

22. Li, J.X., Li, L., Li, T.: MSSF: a multi-document summarization framework based on submodularity. In: SIGIR, pp. 1247–1248 (2011)
23. Wang, D., Li, T., Zhu, S., et al.: Multi-document summarization via sentence-level semantic analysis and symmetric matrix factorization. In: SIGIR, pp. 307–314 (2008)
24. Finkel, J. Rose., Grenager, T., Manning, C.: Incorporating non-local information into information extraction systems by gibbs sampling. In: ACL, pp. 363–370 (2005)

CRAN: A Hybrid CNN-RNN Attention-Based Model for Text Classification

Long Guo[1(✉)], Dongxiang Zhang[2], Lei Wang[2], Han Wang[2], and Bin Cui[1]

[1] Peking University, Beijing, China
{guolong,bin.cui}@pku.edu.cn
[2] University of Electronic Science and Technology of China, Chengdu, China
{zhangdo,201621060133,2015200101009}@uestc.edu.cn

Abstract. Text classification is one of the fundamental tasks in the field of natural language processing. The CNN-based approaches and RNN-based approaches have shown different capabilities in representing a piece of text. In this paper, we propose a hybrid CNN-RNN attention-based neural network, named CRAN, which combines the convolutional neural network and recurrent neural network effectively with the help of the attention mechanism. We validate the proposed model on several large-scale datasets (i.e., eight multi-class text classification and five multi-label text classification tasks), and compare with the state-of-the-art models. Experimental results show that CRAN can achieve the state-of-the-art performance on most of the datasets. In particular, CRAN yields better performance with much fewer parameters compared with a very deep convolutional networks with 29 layers, which proves its effectiveness and efficiency.

Keywords: Text classification · Convolutional neural network
Recurrent neural network · Attention mechanism

1 Introduction

Text classification is one of the fundamental tasks in the field of natural language processing where one needs to assign a single or multiple predefined labels to a sequence of text. It has attracted significant attention from both academic and industry communities due to its broad applications, such as topic modeling [21], sentiment classification [16], and spam detection [2]. Traditional approaches of text classification generally consist of a feature extraction stage, where some sparse lexical features, such as n-grams, are extracted to represent a document, followed by a classification stage, where the features are sent to a linear or kernel classifier. More recent approaches start using deep neural networks, such as convolutional neural networks [11,12] and recurrent neural networks based on long short-term memory (LSTM) [9] or gated recurrent unit (GRU) [5], to jointly perform feature extraction and classification for document classification.

© Springer Nature Switzerland AG 2018
J. C. Trujillo et al. (Eds.): ER 2018, LNCS 11157, pp. 571–585, 2018.
https://doi.org/10.1007/978-3-030-00847-5_42

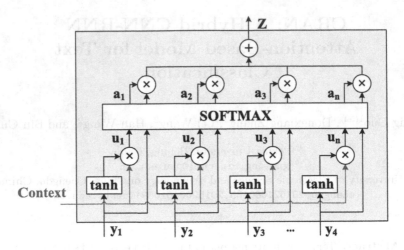

Fig. 1. Attention model

Although neural-network-based approaches to text classification have been proved quite effective, the CNN-based and the RNN-based approaches have shown different capabilities in representing a piece of text. On the one hand, the CNN-based approaches utilize layers with convolutional filters that are applied to local features, and attempt to extract effective text representation by identifying the most influential n-grams of different semantic aspects. However, these methods can only learn local patterns, and it is challenging to deal with the order-sensitive and long distance dependency patterns contained in the sentences. On the other hand, the RNN-based approaches are particularly good at modeling sequential data, and capable of building effective text representation by learning temporal features and long-term dependencies between nominal pairs. However, these methods treat each word in the sentences equally and thus cannot distinguish between the key words that contribute more to the classification and the common words. While the convolutional neural networks and the recurrent neural networks can complement each other for text classification, existing methods [20,22] make a straightforward attempt to combine CNN and RNN with a serial structure by directly applying RNN on top of CNN. This kind of serial structure would suffer from some feature loss since the top module cannot be applied directly on the original text. To exploit the full advantages of these two kinds of models, new effective combination method needs to be designed.

Recently, attention-based neural networks have demonstrated great success in a wide range of tasks ranging from image captioning, speech recognition and question answering to machine translations [1,6,8,19,23]. In terms of natural language classification, Att-BLSTM [28] and HAN [24] can achieve the state-of-the-art performance for relation classification and multiple sentences classification, respectively. Both models apply an attention model on top of a bidirectional RNN model. The basic architecture of the attention model is shown in Fig. 1, which takes n arguments $y_1, ..., y_n$ and a context and returns a vector Z which is supposed to be the summary of the input y. The intuition underlying the attention mechanism is that not all parts of a document are equally relevant for answering a query. Therefore, the vector Z is a weighed arithmetic mean of the input y, where the weights are chosen according to the relevance of each y_i given the context. However, in Att-BLSTM and HAN, the context is initialized randomly, where the role of the context is weakened significantly.

Inspired by the role of the context in the attention model, we propose a hybrid parallel CNN-RNN framework, named CNN-RNN attention-based neural network (CRAN), to model the local patterns and long-distance dependency patterns in the text simultaneously. In our model, CNN and RNN are first applied to the original text in parallel. Then, the attention model takes the output of the RNN as the input and the output of the CNN as the context. Therefore, the returned vector Z can pick the useful local features from the sequences generated by the RNN according to the context generated by the CNN.

The key difference to previous work on attention mechanism is that our model uses the output of the CNN as the *context* to discover when a sequence of token is relevant rather than simply ignoring the context or initializing the context randomly. Experimental results show that CRAN can achieve the state-of-the-art performance on most of the datasets, which demonstrates that the attention mechanism is an effective framework to unify CNN and RNN, and the CNN layer serves as a meaningful context for the attention mechanism. In addition, CRAN yields better performance with much fewer parameters compared with a very deep convolutional networks (VD-CNN) with 29 layers, which proves its effectiveness and efficiency.

2 Model

In this section we describe our proposed CRAN model in detail. As shown in Fig. 2, the model proposed in this paper contains the following components.

2.1 Input Layer

In our model, word embedding is used to map words from a vocabulary to a corresponding vector of real values. In more detail, given a document consisting of n words $T = \{w_1, w_2, ..., w_n\}$, every word w_i is converted into a real-valued vector e_i. For each word in T, we first look up the embedding matrix $\mathbf{W} \in \mathbb{R}^{d \times |V|}$, where $|V|$ is a fixed-sized vocabulary and d is the dimension of the word embedding vector. We transform a word w_i into its word embedding e_i by using the matrix-vector product:

$$e_i = \mathbf{W}v_i, \tag{1}$$

where v_i is a vector of size $|V|$ with the element corresponding to e_i set to 1 and other elements set to 0. In this way, the document can be represented as real-valued vectors $\mathbf{e} = \{e_1, e_2, ..., e_n\}$ and fed into the next layer, i.e., the CNN layer and RNN layer.

2.2 CNN Layer

The CNN layer is used to extract the most influential n-grams of different semantic aspects from the text. CNN performs a discrete convolution on the input embedding matrix with a set of different filters. In this paper, we adopt a CNN structure similar with the model proposed in [12]. Th convolutional layer consists of two stages. The first stage is called "convolution", where a set of k filters are applied to the input sequence $\mathbf{e}$. Each filter $\mathbf{F} \in \mathbb{R}^{h \times d}$ with size h is applied to a window of h embedding words $\mathbf{e}_{i:i+h-1}$ to produce a new feature c_i as follows:

$$c_i = f(\mathbf{F}\mathbf{e}_{i:i+h-1} + b), \qquad (2)$$

where f is a non-linear function such as tanh or a rectifier. The filter $\mathbf{F}$ is applied to each possible window of words in the embedding sequence to produce a feature map $\mathbf{c} = [c_1, c_2, ..., c_{n-h+1}]$. With k filters, the first stage produces k feature maps. The second stage is called "pooling", where a max-over-time

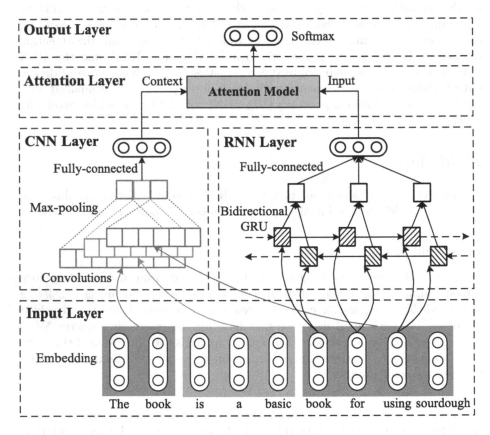

Fig. 2. The hybrid CNN-RNN attention-based model. The shaded rectangles represent the filters in the CNN layer with different sizes. Please refer to Fig. 1 for more details of the attention model.

pooling operation is applied to each feature map and the maximum value $\hat{c} = \max\{c\}$ is taken as the feature corresponding to $\mathbf{F}$. As a result, k features are extracted from the feature maps to form the penultimate layer $\mathbf{p} = \{\hat{c}_1, \hat{c}_2, ..., \hat{c}_k\}$ and are passed to a fully connected layer to reshape the output of the CNN layer, i.e., $\mathbf{p} \in \mathbb{R}^{d_a}$.

2.3   RNN Layer

The RNN layer is used to extract the temporal features and long-term dependencies from the text sequences. RNN is a class of neural network that maintains internal hidden states to model the dynamic temporal behaviour and long-distance patterns of sequences through directed cyclic connections between its units. Take the standard RNN as example. Given the word embedding vectors $\{e_1, e_2, ..., e_n\}$, the word vectors are put into the recurrent layer step by step. For each step t, the network accepts the word vector e_t and the output of the previous step h_{t-1} as the input and produces the current output h_t by a linear transformation followed by a non-linear activation function (e.g., tanh), as follows:

$$h_t = \tanh(\mathbf{W}e_t + \mathbf{U}h_{t-1} + b). \tag{3}$$

To avoid the vanishing gradient problem suffered by the standard RNN, GRU is proposed to control the update of the information via two types of gates: the reset gate r_t and the update gate z_t, which is defined by the following equations:

$$
\begin{aligned}
r_t &= \sigma(\mathbf{W}_{er}e_t + \mathbf{W}_{hr}h_{t-1} + b_r) \\
z_t &= \sigma(\mathbf{W}_{ez}e_t + \mathbf{W}_{hz}h_{t-1} + b_z) \\
\tilde{h}_t &= \tanh(\mathbf{W}_{eh}e_t + \mathbf{W}_{hh}(r_t \odot h_{t-1}) + b_h) \\
h_t &= z_t \odot h_{t-1} + (1 - z_t) \odot \tilde{h}_t.
\end{aligned}
\tag{4}
$$

One property of the one-directional recurrent layer is that there is imbalance in the amount of information seen by the hidden states at different time steps. This can be easily alleviated by having a bidirectional recurrent layer which is composed of two GRU layers working in opposite directions, as shown in Fig. 2. We obtain the representation of the word w_t by concatenating the forward hidden state $\overrightarrow{h}_t$ and the backward hidden state $\overleftarrow{h}_t$, i.e., $h_t = [\overrightarrow{h}_t, \overleftarrow{h}_t]$. In this way, the document can be represented as $\mathbf{h} = \{h_1, h_2, ..., h_n\}$. Before feeding $\mathbf{h}$ into the next attention layer, we use a fully-connected layer to reshape the dimension of $\mathbf{h}$, in order to be aligned with the output of the CNN layer, i.e., $\mathbf{h} \in \mathbb{R}^{n \times d_a}$.

2.4   Attention Layer

The key component of CRAN is the attention layer which combines the output of the CNN layer and RNN layer effectively. In CRAN, the attention mechanism is first applied on top of the output of RNN, i.e., $\mathbf{h}$. For each time step t, we first feed h_t through a fully-connected network to get u_t as a hidden representation of

h_t, and then measure the importance of the word w_t as the similarity of u_t with a context vector u_w, followed by a softmax function to get a normalized importance weight a_t. After that, we compute the text vector $\mathbf{d}$ as a weighed arithmetic mean of $\mathbf{h}$ based on the weights $\mathbf{a} = \{a_1, a_2, ..., a_n\}$. The above procedure is defined as:

$$u_t = \tanh(\mathbf{W}_w h_t + b_w)$$
$$a_t = \frac{\exp(u_t^T u_w)}{\sum_t \exp(u_t^T u_w)} \tag{5}$$
$$\mathbf{d} = \sum_t a_t h_t.$$

The context vector u_w contains useful information to guide the attention model to locate the informative word from the input sequences, and thus plays an important role in the attention mechanism. However, previous works either ignore the context vector u_w or initialize u_w randomly, which weakens the role of the context significantly. Inspired by the observation and the capability of the CNN model which is effective at exploring regional syntax of words, we propose to use the output of the CNN layer as the context of the attention model. As a result, the new attention layer can be defined as follows:

$$u_t = \tanh(\mathbf{W}_w h_t + b_w)$$
$$a_t = \frac{\exp(u_t^T \mathbf{p})}{\sum_t \exp(u_t^T \mathbf{p})} \tag{6}$$
$$\mathbf{d} = \sum_t a_t h_t = \mathbf{a}\mathbf{h}.$$

where $\mathbf{p} \in \mathbb{R}^{d_a}$ is the output of the CNN layer and $\mathbf{h} \in \mathbb{R}^{n \times d_a}$ is the output of the RNN layer.

Note that different from traditional attention mechanism where the context vector is the same to all the samples, each sample in CRAN has a unique context vector (i.e., the output of CNN), which provides more flexibility and potential to achieve better performance. By merging the CNN layer and RNN layer using the attention mechanism, the unified model can pick the useful local features from the sequences generated by the RNN model according to the context generated by the CNN model, and thus reserve the merits of both models.

2.5   Output Layer

Our model is designed to support both the multi-class text classification and multi-label text classification.

Multi-class Classification. When used for multi-class classification, we use a softmax classifier to predict a single label y from a discrete set of classes Y for a text T. The classifier takes the output of the attention layer as the input and computes the predictive probabilities for all the labels. Then the label with the

Table 1. Large-scale text classification data sets. #Labels represents the average number of labels per question.

(a) Multi-class classification

Dataset	Train	Test	Classes	Classification Task
AG's news	120k	7.6k	4	English news categorization
Sogou news	450k	60k	5	Chinese news categorization
DBPedia	560k	70k	14	Ontology classification
Yelp Review Polarity	560k	38k	2	Sentiment analysis
Yelp Review Full	650k	50k	5	Sentiment analysis
Yahoo! Answers	1400k	60k	10	Topic classification
Amazon Review Full	3000k	650k	5	Sentiment analysis
Amazon Review Polarity	3600k	400k	2	Sentiment analysis

(b) Multi-label classification

Subject	Train	Test	Classes	#Labels
English	16k	1.8k	236	1.25
Chinese	22k	2.4k	72	1.10
Chemistry	320k	35k	536	1.47
Physics	380k	40k	430	1.49
Mathematics	480k	53k	782	1.35

highest probability is chosen as the label of T.

$$p(y|T) = softmax(\mathbf{W}_{as}\mathbf{d} + b_s)$$
$$y = \arg\max_y p(y|T) \tag{7}$$

The cost function is the categorical cross-entropy loss function:

$$J(\theta) = -\frac{1}{m}\sum_{i=1}^{m}\bar{y}_i \log(y_i) + \lambda \,||\theta||_2^F \tag{8}$$

where $\bar{y}_i$ is the ground truth of text t_i, y_i is the estimated probability for each class, m is the size of the dataset, and λ is an L2 regularization hyper-parameter.

Multi-label Classification. When used for multi-label classification, we add a fully connected output layer with a sigmoid activation function to predict k labels y from Y for a text T, which produces a probability for each of the labels. Then the k labels with the highest probabilities are chosen as the labels of T.

$$p(y|T) = sigmoid(\mathbf{W}_{as}\mathbf{d} + b_s)$$
$$y = \arg\operatorname*{sort}_{0:k} p(y|T) \tag{9}$$

The cost function is the binary cross-entropy loss function:

$$J(\theta) = -\frac{1}{m} \sum_{i=1}^{m} \sum_{j=1}^{K} (\bar{y}_{ij} \log(y_{ij}) + (1 - \bar{y}_{ij}) \log(1 - y_{ij})) + \lambda \|\theta\|_2^F \qquad (10)$$

where $\bar{y}_{ij}$ is the ground truth for label l_j of text t_i, y_{ij} is the estimated probability for l_j, and K is the number of classes.

3 Experiments

3.1 Datasets

We conduct extensive experiments on several datasets for both the multi-class classification and multi-label classification, which are introduced as follows.

Multi-class Classification. We evaluate the effectiveness of our model on eight freely available large-scale datasets introduced by [26] which cover several classification tasks such as news categorization, sentiment analysis and topic classification. A summary of the statistics for each dataset is listed in Table 1(a). Please refer to [26] for more detailed information of the datasets.

Multi-label Classification. Since most datasets for multi-label classification are quite small and contain small number of classes, we build several real-world datasets to fully evaluate the effectiveness of our model, which contain large number of questions collected from the exercises and exams of senior high school in China. Table 1(b) shows the statistics for each dataset. The datasets cover several subjects such as mathematics and physics, and the classes represent the knowledge points corresponding to each subject.

3.2 Baselines

We compare CRAN with several baseline methods, which cover most of the existing CNN-based approaches and RNN-based approaches[1]. For fairness, we do not compare with the models which utilize the hierachical structure of the documents [20,24], where our model can be adopted as a component in their models. In our experiments, all the models take the whole document as a single sequence. Since the baselines are designed for multi-class classification task, we modify their output layer to support multi-label classification and conduct new experiments on the multi-label classification datasets. In the following, we introduce the baseline methods briefly.

- **CNN-word** represents the word-based CNN model [12], where a CNN model with one convolutional layer is applied to the word embeddings of the documents.

---

[1] We ignore some relevant models because of the unavailability of their codes.

- **CNN-char** represents the character-based CNN model [26], where a CNN model with six convolutional layers is applied to the character representation of the documents.
- **LSTM** treats the whole document as a single sequence and the average of the hidden states of all words are used as feature for classification.
- **Att-BLSTM** represents the attention-based bidirectional LSTM [28], where the attention model is adopted on top of a bidirectional LSTM layer. Att-BLSTM is similar with our CRAN model without the CNN layer.
- **CNN-RNN** represents a hybrid model that processes the input sequence of characters with a number of convolutional layers followed by a single recurrent layer [22].
- **VD-CNN** represents a very deep character-based CNN model [7], where a CNN model with up to 29 convolutional layers is applied to the character representation of the documents.

3.3 Model Settings and Training

The word embeddings can be pretrained or directly trained together with the model. In our experiments, we adopt the pretrained word embeddings. We obtain the word embeddings by training an unsupervised word2vec [17] model on the training and testing datasets. We set the wording embedding dimension to 300.

Since there is no official validation set, we randomly select 10% of the training samples as the validation set. The hyper parameters of the models are tuned on the validation set. In terms of the CNN layer, we use filters of $3, 4, 5$ with 256 feature maps each and reshape the dimension of penultimate vector p to 200. In terms of the RNN layer, we set the GRU dimension to 100 and thus the bidirectional GRU layer gives us the representation vector h_t with 200 dimensions. Dropout is an effective way to regularize deep neural networks. We apply dropout after the CNN layer as well as after the RNN layer, where the dropout rate is set as 0.3. Other parameters in our model are initialized randomly.

Our model was trained using Adam [13] with a learning rate of 0.001 and a mini-batch of size 256. The input text is padded to a fixed size of 300, where longer text is truncated and masks are generated to help identify the padded region. The implementation is done using Keras with Theano as the backend. All experiments are performed on a single GeForce GTX TITAN X GPU.

3.4 Experimental Results and Analysis

The experimental results of all the models on the multi-class classification datasets and multi-label classification datasets are shown in Tables 2 and 3, respectively. In the following, we offer some empirical analysis of our CRAN model with respect to different baseline models.

CRAN is an Effective Hybrid Framework. Experimental results show that the attention mechanism is an effective framework to unify CNN and RNN. As shown in Tables 2 and 3, CRAN performs much better than the standalone CNN and RNN. In particular, CNN-RNN is a rather coarse combination manner for unifying CNN and RNN. Although it can improve performance compared with CNN or RNN, it performs much worse compared with CRAN.

The CNN Layer Serves as a Meaningful Context for the Attention Mechanism. Although Att-BLSTM performs as a rather strong model which can beat most of the existing methods, our CRAN model can still perform better than Att-BLSTM. Remember that Att-BLSTM is similar to our CRAN model except that the context of Att-BLSTM is ignored. This fact proves the effectiveness of the CNN layer as a context of the attention model. The CNN layer can provide useful information for CRAN to pick the important words from the sequences generated by the RNN layer.

CRAN can Achieve Better Performance than the State-of-the-Art Model VD-CNN While with Much Fewer Parameters. Specifically, CRAN performs much better than CNN-char with 6 convolutional layers and the VD-CNN with 17 convolutional layers. Even though VD-CNN increases its number of layers to 29, CRAN can still beat VD-CNN on most of the datasets for multi-class classification and all the datasets for multi-label classification. In particular, CRAN performs much better than VD-CNN on multi-label classification which is a significantly harder task than multi-class classification. In addition, Table 4 shows the number of parameters for the models. As shown, CRAN has significantly fewer parameters compared with CNN-char and VD-CNN.

3.5   Visualization of Attention

To further show the effectiveness of our model in selecting informative words in a document, we visualize the attention layers of Att-BLSTM and CRAN in Figs. 3 and 4. We pick two typical examples from the AG's news and Yahoo! Answers, respectively. We group the words into three categories according to their weights, which is denoted as the blue rectangles with different color depth.

Figure 3 shows a document picked from the AG's news corpus. As shown, CRAN predicts the document to be 4, which represents "Sci/Tech", while Att-BLSTM labels the document with 3, which represents "Business". The reason that CRAN can predict the correct label is that it selects the words carrying strong information, i.e., "information technology", which are highly related to "Sci/Tech". On the other hand, Att-BLSTM focuses on the words "strategy business" which are more related to "Business" and mislead the judgement. A similar phenomenon can be observed in Fig. 4. The document in Fig. 4 is more complicated than that in Fig. 3. Att-BLSTM distributes more weights on the words "screen" where "screen" is a rather common words in the field of "Computers". "screen" also obtains a rather high weight in CRAN. However, CRAN also gives high weights for the words "movie" and "theater", which helps the model to recognize that "screen" represents the screen of movie theater instead of computer.

Table 2. Testing errors of all the models for multi-class text classification, where numbers are in percentage. For VD-CNN, we report the results of the model with different number of layers.

Method	Depth	AG	Sogou	DBP	Yelp P	Yelp F	Yah. A	Amz. F	Amz. P
CNN-word	1	9.92	4.39	1.42	4.60	40.16	31.97	44.40	5.88
CNN-char	6	9.85	4.88	1.66	5.25	38.40	29.55	40.53	5.50
LSTM	-	13.94	4.82	1.45	5.26	41.38	29.16	40.57	6.10
Att-BLSTM	-	8.98	4.10	1.33	4.98	36.77	27.25	38.05	4.95
CNN-RNN	2-3	8.64	4.83	1.43	5.51	38.18	28.26	40.77	5.87
VD-CNN	9	9.17	3.58	1.35	4.88	36.73	27.60	37.95	4.70
VD-CNN	17	8.88	3.54	1.40	4.50	36.07	27.35	37.50	4.41
VD-CNN	29	8.67	3.18	1.29	4.28	35.28	26.57	**37.00**	**4.28**
CRAN	1	**8.30**	**3.15**	**1.05**	**4.00**	**34.69**	**26.28**	37.28	4.35

Table 3. F1 scores of all the models for multi-label text classification, where numbers are in percentage. For VD-CNN, we only report the best result when the number of layers is 29.

	English			Chinese			Chemistry			Physics			Mathematics		
	k=1	k=2	k=3	k=1	k=2	k=3	k=1	k=2	k=3	k=1	k=2	k=3	k=1	k=2	k=3
CNN-word	54.42	45.41	38.54	79.22	62.94	48.66	55.41	49.27	43.24	61.94	57.80	49.87	54.56	50.52	43.31
CNN-char	42.46	37.23	33.64	78.07	61.98	48.83	50.41	46.52	40.58	57.68	54.74	48.11	49.11	46.63	40.81
LSTM	53.24	46.27	37.23	81.41	62.76	47.66	54.24	48.49	42.51	60.24	57.75	50.01	53.28	49.79	42.27
Att-BLSTM	58.21	49.02	42.61	85.34	63.76	49.74	58.74	52.11	45.72	63.96	60.23	51.42	56.66	53.18	46.09
CNN-RNN	59.89	47.18	41.61	84.44	63.30	49.08	56.91	51.22	45.67	63.67	59.28	52.44	56.51	52.95	45.83
VD-CNN	53.09	43.50	37.74	82.30	63.00	48.89	57.05	51.90	44.92	63.79	59.33	51.57	56.91	53.31	46.10
CRAN	62.05	51.79	44.49	86.19	64.15	50.33	60.54	54.81	47.17	65.66	61.16	52.82	57.02	53.32	46.29

Table 4. Number of parameters per model. Note that CNN-char adopts the convolutional layer with more filters than that of VD-CNN and thus has more parameters than VD-CNN.

Method	Depth	AG	Sogou	DBP	Yelp P	Yelp F	Yah. A	Amz. F	Amz. P
CNN-char	6	27M	27M	27M	27M	27M	27M	2.7M	2.7M
VD-CNN	9	2.2M	2.2M	2.2M	2.2M	2.2M	2.2M	2.2M	2.2M
VD-CNN	17	4.3M	4.3M	4.3M	4.3M	4.3M	4.3M	4.3M	4.3M
VD-CNN	29	4.6M	4.6M	4.6M	4.6M	4.6M	4.6M	4.6M	4.6M
CRAN	1	**1.2M**	**1.2M**	**1.2M**	**1.2M**	**1.2M**	**1.2M**	**1.2M**	**1.2M**

From the above examples, we can see CNN indeed serves as a meaningful context for the attention mechanism. CNN can help the attention mechanism to focus on more correct informative words to help the model make the right decision.

Ground Truth: 4

Att-BSLTM: Prediction 3
does nick carr matter strategy business concludes that a controversial new book on the strategic value of information technology is flawed but correct

CRAN: Prediction 4
does nick carr matter strategy business concludes that a controversial new book on the strategic value of information technology is flawed but correct

Fig. 3. Documents from AG's news. Label 3 means Business while label 4 means Sci/Tech.

Ground Truth: 8

Att-BSLTM: Prediction 5
h ow many holes does a movie theater screen have every movie theater screen ive ever seen has had a bunch of holes in it i think its so they can have speakers behind the screen is there a standard for the size and spacing of the holes is there a certain horizontal and vertical count

CRAN: Prediction 8
h ow many holes does a movie theater screen have every movie theater screen ive ever seen has had a bunch of holes in it i think its so they can have speakers behind the screen is there a standard for the size and spacing of the holes is there a certain horizontal and vertical count

Fig. 4. Documents from Yahoo! Answers. Label 5 means Computers & Internet while label 8 means Entertainment & Music.

4   Related Work

In this section, we review the existing solvers for text classification according to the evolving of their two primary components: feature representation and classifier design. The early methods [10,18,21] on text classification used bag-of-words or n-grams to construct features and adopted traditional classifiers.

In recent years, various deep learning based models have been proposed and achieved state-of-the-art performance. We group them into the following categories and conduct a brief review in the following.

Convolution Neural Networks. Since 2014, the CNN-based and RNN-based approaches have emerged and become the trend for text classification. Kim in [12] proposed convolutional neural networks (CNN) for sentence classification. By word embedding, a sentence with length n is converted into $n \times k$ matrix representation. Then, a convolution operation with a filter window of length h words, together with a max-over-time pooling layer is adopted. In DCNN proposed in [11], Kalchbrenner et al. applied dynamic k-max pooling over time to generalize the original max pooling in traditional CNN. There are also several convolutional networks proposed to work at character level [7,26] and very deep models (up to 29 layers) can be designed with small convolutions and pooling operations.

Recurrent Neural Networks. Carrier and Cho in [4] gave a tutorial on using a recurrent neural network for sentiment analysis which is one type of text classification. Their model adopts a single long short-term memory layer followed by a mean pooling over time. Lai et al. in [14] introduced a bidirectional recurrent neural network for text classification, which adopts a bidirectional long short-term memory layer followed by a max pooling over time. A similar structure is proposed in [25] for relation classification. The goal of the recurrent structure is to capture contextual information as far as possible when learning word representations.

Hybrid CNN-RNN Models. Tang et al. in [20] proposed a hierachical neural network model named Conv-GRNN for hierarchical processing of a document. Their model first learns sentence representation with convolutional neural network or long short-term memory. Then the representation of the sentences is sent to a gated recurrent neural network. Xiao and Cho [22] proposed a hybrid model named ConvRec that processes an input sequence of characters with a number of convolutional layers followed by a single long short-term memory layer. Other CNN-RNN models including [3,15,27] apply a similar serial structure with ConvRec. The major difference between Conv-GRNN and ConvRec lies in the purpose of combining the convolutional network and recurrent network. In conv-GRNN, the convolutional network is constrained to model each sentence, and the recurrent network to model inter-sentence structures. While in ConvRec, the recurrent layer is applied to the output of the convolutional layers to capture the long-term dependencies, where the whole document is viewed as a single sentence.

Different from ConvRec, our proposed CRAN model combines the convolutional network and the recurrent network with the help of the attention mechanism. In addition, CRAN is orthogonal to the hierarchical model Conv-GRNN and can be plugged in Conv-GRNN as a sentence feature extraction module.

Attention-Based Models. In [28], Att-BLSTM is proposed to capture the most import semantic information in a sentence for relation classification, where the attention model is applied on top of a bidirectional LSTM layer. In [24], bidirectional GRU with two levels of attention mechanisms named HAN were proposed for text classification. The former is used to encode the sentences and the latter to capture importance of the sentences. Att-BLSTM can be viewed as an component in HAN, where Att-BLSTM is used to capture the important words in the word attention encoder and the important sentences in the sentence attention encoder of HAN.

The difference between CRAN and Att-BLSTM is that Att-BLSTM adopts a randomly initialized vector as the context of the attention mechanism while CRAN utilizes the output of the CNN model as the context which can provide more instructive information. Again, CRAN is orthogonal to HAN and can be adopted as a component in HAN similar to Att-BLSTM.

5 Conclusion

In this paper, we propose a hybrid CNN-RNN attention-based neural network, named CRAN, which combines the convolutional neural network and recurrent neural network effectively with the help of the attention mechanism. The proposed model is able to pick the useful local features from the sequences generated by the RNN layer according to the context generated by the CNN layer, and thus reserve the merits of both models in representing a piece of text. We validate the proposed model on several large scale datasets for both multi-class and multi-label text classification. Experimental results show that CRAN can achieve state-of-the-art performance on most of the datasets. The proposed model is a general hybrid architecture that is not limited to text classification or natural language inputs. It will be interesting to see future research on applying the architecture on other applications.

Acknowledgments. This work is supported in part by the National Natural Science Foundation of China under Grant No. 61702016, 61602087, 61632007, 61572039, the China Postdoctoral Science Foundation under Grant No. 2017M610019, 973 program under No. 2014CB340405, Shenzhen Gov Research Project JCYJ20151014093505032, the Fundamental Research Funds for the Central Universities under grants No. ZYGX2016J080, ZYGX2014Z007.

References

1. Bahdanau, D., Cho, K., Bengio, Y.: Neural machine translation by jointly learning to align and translate. CoRR abs/1409.0473 (2014)
2. Blanzieri, E., Bryl, A.: A survey of learning-based techniques of email spam filtering. Artif. Intell. Rev. **29**(1), 63–92 (2008)
3. Cai, R., Zhang, X., Wang, H.: Bidirectional recurrent convolutional neural network for relation classification. In: Proceedings of the 54th Annual Meeting of the Association for Computational Linguistics, ACL 2016, 7–12 August 2016, Berlin, Germany, vol. 1, Long Papers (2016). http://aclweb.org/anthology/P/P16/P16-1072.pdf
4. Carrier, P.L., Cho, K.: LSTM networks for sentiment analysis. Deep Learning Tutorials (2014)
5. Cho, K., van Merrienboer, B., Gülçehre, Ç., Bougares, F., Schwenk, H., Bengio, Y.: Learning phrase representations using RNN encoder-decoder for statistical machine translation. CoRR abs/1406.1078 (2014). http://arxiv.org/abs/1406.1078
6. Chorowski, J., Bahdanau, D., Serdyuk, D., Cho, K., Bengio, Y.: Attention-based models for speech recognition. In: Advances in Neural Information Processing Systems, pp. 577–585 (2015)
7. Conneau, A., Schwenk, H., Barrault, L., LeCun, Y.: Very deep convolutional networks for natural language processing. CoRR abs/1606.01781 (2016)
8. Hermann, K.M., et al.: Teaching machines to read and comprehend. CoRR abs/1506.03340 (2015)
9. Hochreiter, S., Schmidhuber, J.: Long short-term memory. Neural Comput. **9**(8), 1735–80 (1997)

10. Joachims, T.: Text categorization with support vector machines: learning with many relevant features. In: Nédellec, C., Rouveirol, C. (eds.) ECML 1998. LNCS, vol. 1398, pp. 137–142. Springer, Heidelberg (1998). https://doi.org/10.1007/BFb0026683
11. Kalchbrenner, N., Grefenstette, E., Blunsom, P.: A convolutional neural network for modelling sentences. In: ACL vol. 1, pp. 655–665. The Association for Computer Linguistics (2014)
12. Kim, Y.: Convolutional neural networks for sentence classification. In: EMNLP, pp. 1746–1751 (2014)
13. Kingma, D.P., Ba, J.: Adam: a method for stochastic optimization. CoRR abs/1412.6980 (2014). http://dblp.uni-trier.de/db/journals/corr/corr1412.html/KingmaB14
14. Lai, S., Xu, L., Liu, K., Zhao, J.: Recurrent convolutional neural networks for text classification. In: AAAI, pp. 2267–2273. AAAI Press (2015)
15. Lai, S., Xu, L., Liu, K., Zhao, J.: Recurrent convolutional neural networks for text classification. In: Proceedings of the Twenty-Ninth AAAI Conference on Artificial Intelligence, AAAI 2015, pp. 2267–2273. AAAI Press (2015). http://dl.acm.org/citation.cfm?id=2886521.2886636
16. Maas, A.L., Daly, R.E., Pham, P.T., Huang, D., Ng, A.Y., Potts, C.: Learning word vectors for sentiment analysis. In: HLT-NAACL, pp. 142–150 (2011)
17. Mikolov, T., Sutskever, I., Chen, K., Corrado, G., Dean, J.: Distributed representations of words and phrases and their compositionality. In: NIPS, pp. 3111–3119 (2013)
18. Pang, B., Lee, L., Vaithyanathan, S.: Thumbs up? Sentiment classification using machine learning techniques. In: EMNLP, pp. 79–86, July 2002
19. Song, J., Guo, Z., Gao, L., Liu, W., Zhang, D., Shen, H.T.: Hierarchical LSTM with adjusted temporal attention for video captioning. CoRR abs/1706.01231 (2017). http://arxiv.org/abs/1706.01231
20. Tang, D., Qin, B., Liu, T.: Document modeling with gated recurrent neural network for sentiment classification. In: EMNLP, pp. 1422–1432 (2015)
21. Wang, S.I., Manning, C.D.: Baselines and bigrams: Simple, good sentiment and topic classification. In: ACL vol. 2, pp. 90–94. The Association for Computer Linguistics (2012)
22. Xiao, Y., Cho, K.: Efficient character-level document classification by combining convolution and recurrent layers. CoRR abs/1602.00367 (2016)
23. Xu, K., et al.: Show, attend and tell: Neural image caption generation with visual attention. Comput. Sci., 2048–2057 (2015)
24. Yang, Z., Yang, D., Dyer, C., He, X., Smola, A.J., Hovy, E.H.: Hierarchical attention networks for document classification. In: HLT-NAACL, pp. 1480–1489 (2016)
25. Zhang, D., Wang, D.: Relation classification via recurrent neural network. CoRR abs/1508.01006 (2015)
26. Zhang, X., Zhao, J.J., LeCun, Y.: Character-level convolutional networks for text classification. In: NIPS, pp. 649–657 (2015)
27. Zhou, P., Qi, Z., Zheng, S., Xu, J., Bao, H., Xu, B.: Text classification improved by integrating bidirectional lstm with two-dimensional max pooling. In: Calzolari, N., Matsumoto, Y., Prasad, R. (eds.) COLING, pp. 3485–3495. ACL (2016). http://dblp.uni-trier.de/db/conf/coling/coling2016.html//ZhouQZXBX16
28. Zhou, P., et al.: Attention-based bidirectional long short-term memory networks for relation classification. In: ACL (2016)

Learning Restricted Regular Expressions with Interleaving from XML Data

Yeting Li[1,2], Xiaolan Zhang[1,2], Han Xu[2,3], Xiaoying Mou[1,2], and Haiming Chen[1(✉)]

[1] State Key Laboratory of Computer Science, Institute of Software, Chinese Academy of Sciences, Beijing 100190, China
{liyt,zhangxl,mouxy,chm}@ios.ac.cn
[2] University of Chinese Academy of Sciences, Beijing, China
[3] China National Science Library, Chinese Academy of Science, Beijing, China
xuhan@mail.las.ac.cn

Abstract. The presence of a schema for XML documents has numerous advantages. However, many XML documents in practice are not accompanied by a schema or by a valid schema. Therefore, it is essential to devise algorithms to learn a schema from XML documents. The fundamental task in XML schema learning is inferring restricted subclasses of regular expressions. Previous work in this direction lacks support of interleaving. In this paper, based on the analysis of real data, we first propose a new subclass of regular expressions with interleaving, named as *Extended Subclass of Regular Expressions with Interleaving (ESIRE)*. Then, based on *single occurrence automaton* and *maximum independent set*, we propose an algorithm *GenESIRE* to infer *ESIREs* from a set of given samples. Finally, we conduct a series of experiments to analyze the inference effectness of *GenESIRE*. Experimental results show that regular expressions inferred by *GenESIRE* are more precise compared with other methods, measured by different indicators.

Keywords: Schema learning · Regular expression · Interleaving

1 Introduction

XML schemas are widely used in data processing, automatic data integration, static analysis of transformation, semantic data modeling and so on [1,8,15,17–20,23]. However, many XML documents in practice are not accompanied by a schema or by a valid schema. A survey showed that XML documents with corresponding schema definitions only account for 24.8%, with the proportion of 8.9% for valid ones [14]. Hence, it is essential to learn a schema from XML documents. Besides, schema inference is also useful in schema cleaning and dealing with noise [5], where a schema is already available. Previous research has shown that the

Work supported by the National Natural Science Foundation of China under Grant Nos. 61472405.

J. C. Trujillo et al. (Eds.): ER 2018, LNCS 11157, pp. 586–593, 2018.
https://doi.org/10.1007/978-3-030-00847-5_43

essential task in schema learning is inferring regular expressions from a set of given samples [3,5,12]. In fact, in some cases these learned regular expressions can be used either as part of the schema, or the most important component of the schema inference in other cases. Therefore research on schema learning has focused on inferring regular expressions from a set of given samples.

Inference of regular expressions from a set of given samples belongs to the problem of language learning. Gold proposed a classical language learning model (*identification in the limit*) and pointed out that the superfinite class of languages could not be identifiable from positive examples only [13]. Consequently, researchers have turned to study restricted subclasses of regular expressions [22]. In this paper we focus on devising an algorithm to infer a more precise regular expression from the given samples.

Previous studies have discussed various subclasses [2,3,16,21]. Bex et al. proposed two inference algorithms *RWR* and *CRX* for SOREs and Simplified CHAREs, respectively [3,4]. Freydenberger et al. [10] gave another two algorithms *Soa2Sore* and *Soa2Chare* for them. Regular expressions inferred by latter algorithms are more precise than that inferred by the former two. However, there have been only few works that support interleaving. Two subclasess DME and ME [6] support a weaker form of interleaving without considering the sequence order of siblings. ME even does not support the union operator. They gave an inference algorithm based on *maximum clique* for DME [9]. Peng et al. [24] proposed a subclass *SIREs*. It requires each symbol can occur at most once. However, *SIREs* do not support the union operator. They presented an approximate algorithm to infer *SIREs* [24].

In this paper we propose a new subclass of regular expressions with interleaving, named as *Extended Subclass of Regular Expression with Interleaving (ESIRE)*. *ESIRE* supports union, concatenation and interleaving operators together. For example, $E_4 = (a^+b^?)\&c^*$ and $E_5 = a^?((b^+|c)d^*\&ef^?)$ are two *ESIREs*. Then we propose an algorithm to infer *ESIREs* from a set of given samples S based on *Maximum Independent Set* (MIS) and SOA. The main contributions of this paper are listed as follows.

- Based on the analysis of real-world data, we propose a useful subclass of regular expressions *ESIRE*.
- We design an inference algorithm *GenESIRE* to infer *ESIREs*.
- Experiments are conducted to study the effectiveness of *GenESIRE*, compared with other methods (*InstanceToSchema*, $Learn^+_{DME}$ and *conMiner*) on the data set *SwissProt*, using different measures. Experimental results show that the regular expressions inferred by *GenESIRE* are more precise.

This paper is organized as follows. Section 2 is the basic definitions. Section 3 gives the inference algorithm *GenESIRE*. Section 4 introduces the experiments. Conclusions are drawn in Sect. 5.

2   Preliminaries

2.1   Definitions

Definition 1 *Regular Expression with Interleaving. Let Σ be a finite alphabet. The set of all finite words over Σ is denoted by Σ^*. The empty word is denoted by ε. A regular expression with interleaving over Σ is defined inductively as follows: ε or $a \in \Sigma$ is a regular expression, for regular expressions E_1 and E_2, the disjunction $E_1|E_2$, the concatenation $E_1 \cdot E_2$, the interleaving $E_1 \& E_2$, or the Kleene-Star E_1^* is also a regular expression. The length of a regular expression E, denoted by $|E|$, is the total number of symbols and operators occurred in E. The language generated by E is defined as follows: $L(\emptyset) = \emptyset$; $L(\varepsilon) = \{\varepsilon\}$; $L(a) = \{a\}$; $L(E_1^*) = L(E)^*$; $L(E_1E_2) = L(E_1)L(E_2)$; $L(E_1|E_2) = L(E_1) \cup L(E_2)$; $L(E_1 \& E_2) = L(E_1)\&L(E_2)$. $E^?$ and E^+ are abbreviations of $E|\varepsilon$ and EE^*, respectively.*

Let $u = au_0$, $v = bv_0$ where $a, b \in \Sigma$ and $u, u_0, v, v_0 \in \Sigma^*$. Then $u\&\varepsilon = \varepsilon\&u = \{u\}$. $u\&v = \{a(u_0\&v)\} \cup \{b(u\&v_0)\}$. Strings set generated by $a\&bc$ is $\{abc, bac, bca\}$.

Definition 2 *Single Occurrence Automaton (SOA). [10] Let Σ be a finite alphabet. src and snk be distinct symbols that do not occur in Σ. A single occurrence automaton (short: SOA) over Σ is a finite directed graph $G = (V, D)$ such that*

- *src, snk $\in V$, and $V \subseteq \Sigma \cup \{src, snk\}$;*
- *src has only outgoing edges, snk has only incoming edges and every node $v \in V$ lies on a path from src to snk.*

Subclass *ISORE* is extended from *SORE* [3] considering interleaving. In this paper, a *generalized single-occurrence automaton (generalized SOA)* over Σ is defined as a directed graph in which each node $v \in V \setminus \{src, snk\}$ is an *ISORE* and all nodes are pairwise alphabet-disjoint *ISOREs*.

Definition 3 *Constraint Set (CS) and Non Constraint Set (NCS). Let S be a set of strings, POR(S) is the set of all partial order relations of each string in S. CS(S) and NCS(S) are defined as follows.*

- *$CS(S) = \{\langle x,y\rangle | \langle x,y\rangle \in POR(S) \text{ and } \langle y,x\rangle \in POR(S)\}$;*
- *$NCS(S) = \{\langle x,y\rangle | \langle x,y\rangle \in POR(S) \text{ but } \langle y,x\rangle \notin POR(S)\}$.*

Clearly, for a set of given samples S, $CS(S) \cap NCS(S) = \emptyset$. If $CS(S_1) \neq CS(S_2)$ (or $NCS(S_1) \neq NCS(S_2)$), the given sample sets (S_1 and S_2) must be different.

Definition 4 *Extended String (ES). Let Σ be a finite alphabet and $a \in \Sigma$. An ES is a finite sequence $f_1f_2 \cdots f_{n_1}$. f_i can be either the form of a^b where $b \in \{1, ?, +, *\}$, or $(e_1|e_2|\cdots|e_{m_1})^t$ where $t \in \{1, ?\}$ and e_j is the form of a^b. For example, both $a^*b^?c^+d$ and $a^*b(c|d^+)^?$ are ESs.*

Definition 5 Filter(P,s). $Filter(P, s)$ *is a function in which P is a finite set of symbols and s is a string. It returns a new string each symbol of which is computed as follows: $\pi_s(P, s_i) = s_i$ if $s_i \in P$; $\pi_s(P, s_i) = \varepsilon$ otherwise. The result is reduced by $x\varepsilon = \varepsilon x = x$. Let $P = \{b, c, r\}$ and $s = ebbdfc$. Then $s' = Filter(P, s) = bbc$.*

2.2 The New Subclass

Definition 6 Extended Subclass of Regular Expressions with Interleaving (ESIRE). *Let Σ be a finite alphabet and $a \in \Sigma$. ESIRE is a class of regular expressions defined over Σ, in which each terminal symbol can occur at most once. It consists of a finite sequence of two kinds of factors. One kind is of the form $(b_1|b_2|\cdots|b_m)^t$ where $t \in \{1, ?\}$, $m \geq 1$ and b is a or a^+. The other kind is of the form $c_1 \& c_2 \& \cdots \& c_q$ where c_i is an ES, $q \geq 2$.*

For example, $E = (a^?(b^+|c^?)\&d^?e^*)(f|g^+)^?$ is an *ESIRE*.

3   Inference Algorithm

For a set of given samples S, $L(SOA(S))$ is a minimal-inclusion generalization of S using *2T-INF* [11]. Here minimal-inclusion means that there is no other SOA A such that $S \subseteq L(A)$ but $L(A) \subset L(SOA(S))$. Finding an MIS for a graph G is a NP-hard problem. Hence we use the method $clique\_removal()$ [7] to find an approximate result. all_mis is the set containing all MISs iteratively obtained from G. $sym(A)$ is the set of all symbols occurring in A. $G.setln()$ is to assign each node a level number ln. $G.isSL(i)$ returns $true$ if i is a skip level and $false$ otherwise. $Combine(V, \text{"|"})$ (or $Combine(V, \text{"\&"})$) is to combine all elements in V with (|) (or (\&)) operator. The input is S and the output is an *ESIRE*. The main procedures and the pseudo-code of *GenESIRE* are as follows.

1. Construct a graph $G(V, D) = SOA(S)$ for S using *2T-INF* [11].
2. For each node v with a self-loop, rename it with v^+ and remove the self-loop.
3. For each non-trival strongly connected component NTSCC, call function $Repair(NTSCC, S)$. Replace the NTSCC with a new node and label it with the return value of $Repair(NTSCC, S)$. All relations with any node in NTSCC of G rebuild new relations with the new node.
4. Assign the level numbers for the new graph and compute all skip levels.
5. Nodes of each level are converted into one or more chain factors.

Algorithm Analysis. For a graph $G(V, D) = SOA(S)$, let $n = |V|$ and $m = |D|$. It costs time $O(n)$ to find all nodes with loops and $O(m + n)$ to find all NTSCCs. The time complexity of $clique\_removal()$ is $O(n^2 + m)$. For each NTSCC, computation of all_mis costs time $O(n^3 + m)$. For each mis, there is no NTSCC at all. Hence $Repair()$ only costs time $O(n^3 + m)$. The number of NTSCCs in a SOA is finite. Then computation of all_mis for all NTSCCs also costs time $O(n^3 + m)$. Assigning level numbers and computing all skip levels will be finished in time $O(m + n)$. All nodes will be converted into specific chain factors of *ESIRE* in $O(n)$. Therefore, the time complexity of *GenESIRE* is $O(n^3 + m)$.

Algorithm 1. GenESIRE(S)

Input: A set of strings S
Output: An ESIRE R

1 Construct the $G(V, D) = SOA(S)$ using *2T-INF* [11];
2 Rename each node v with loop v^+ and remove the loop; For each *NTSCC*, call algorithm *Repair(NCSCC,S)*. Update the graph G;
3 $G.setln()$; $R \leftarrow \varepsilon$; $ln \leftarrow 1$;
4 **while** $ln \leq (ln$ of $G.snk) - 1$ **do**
5     $B \leftarrow$ all nodes with "&"; $V_S \leftarrow$ all nodes without "&";
      $A \leftarrow Combine(V_S, "|")$;
6   **if** $A !=\varepsilon$ and $B==\emptyset$ and $!G.isSL(ln)$ **then**
7       $R \leftarrow R \cdot A$;
8   **else if** $A !=\varepsilon$ and $B==\emptyset$ and $G.isSL(ln)$ **then**
9       $R \leftarrow R \cdot A^?$;
10   **else if** $A==\varepsilon$ and $|B| > 1$ **then**
11   **for** *each* $b \in B$ **do**
12         $R \leftarrow R \cdot b^?$;
13   **else if** $A==\varepsilon$ and $|B| == 1$ and $!G.isSL(ln)$ **then**
14       $R \leftarrow R \cdot B_0$;
15   **else if** $A==\varepsilon$ and $|B| == 1$ and $G.isSL(ln)$ **then**
16       $R \leftarrow R \cdot B_0^?$;
17   **else if** $A !=\varepsilon$ and $B !=\emptyset$ **then**
18       $R \leftarrow R \cdot A^?$;
19   **for** *each* $b \in B$ **do**
20         $R \leftarrow R \cdot b^?$;
21     $ln = ln + 1$;
22 **return** R

---

4   Experiments and Analysis

Our data set is SwissProt[1] (protein sequence database). We extracted sets of strings for the element **Entry** from *SwissProt*. In this section, based on different indicators, we analyze the inferred results of *GenESIRE* and compare with other methods (*InstanceToSchema*[2], $Learn_{DME}^+$ [9] and *conMiner* [24]) on the data set *SwissProt*. Our experiments were conducted on a machine with 16 cores Intel Xeon CPU E5620 @ 2.40 GHz with 12M Cache, 24G RAM, OS: Windows 10.

We introduce an indicator *Combinatorial Cardinality* (*CC*) that will be used in our experiments. Let $rep = \{1, ?, *, +\}$. Consider the regular expression with interleaving E^t where $t \in rep$. *rep* is the set which illustrates the possible repetition times of E. In present paper, we consider all components that will be

---

[1] http://aiweb.cs.washington.edu/research/projects/xmltk/xmldata.
[2] http://www.xmloperator.net/i2s/.

repeated in E and name them as the *skeleton* of regular expressions. For example, the *skeleton* of $E = (a + b)^*$ is $\{\{a\}, \{b\}\}$. The *Combinatorial Cardinality* (*CC*) of E is defined as the size of the *skeleton* ($|skeleton|$).

Algorithm 2. Repair(V,S)

Input: A set of nodes V and a set of given samples S
Output: A regular expression $newRE$

1 $pattern \leftarrow V$; $S' \leftarrow \bigcup_{s \in S} Filter(pattern, s)$; Compute sets $CS(S')$, $NCS(S')$ using $POR(S')$;
2 **if** $CS(S') == \emptyset$ **then**
3   $\quad$ **return** $(Graph(CS).combine(V))^+$;

4 **else**
5 $\quad$ $G \leftarrow Graph(CS)$;
6 $\quad$ **while** $G.nodes()! = \emptyset$ **do**
7 $\quad\quad$ $v \leftarrow clique\_removal(G)$; $G \leftarrow G \setminus v$; $all\_mis.append(v)$;
8 $\quad$ **for** *each* $mis \in all\_mis$ **do**
9 $\quad\quad$ $sub\_ex \leftarrow GenESIRE(\bigcup_{s \in S} Filter(mis, s))$;
   $\quad\quad$ $RE_{mis}.append((\varepsilon \in S')? sub\_ex^? : sub\_ex)$;
10 $\quad$ $newRE \leftarrow Combine(RE_{mis}, "\&")$;
11 $\quad$ **return** $newRE$;

Definition 7 *Combinatorial Cardinality (CC)*. *Let* Σ *be the finite alphabet set.* E_i *is a regular expression with interleaving over* Σ. $a, b \in \Sigma$ *and* $u, u', v, v' \in \Sigma^*$.

- $CC(a) = CC(\varepsilon) = 1$, *where* $a \in \Sigma$;
- $CC(E^t) = CC(E)$, *where* $t \in rep$;
- $CC(E_1 | E_2 \cdots | E_n) = CC(E_1) + CC(E_2) + \cdots + CC(E_n)$;
- $CC(E_1 \cdot E_2 \cdots E_n) = CC(E_1) \times CC(E_2) \times \cdots \times CC(E_n)$;
- $CC(u \& v) = CC(a \cdot (u' \& v)) + CC(b \cdot (u \& v'))$, *where* $u = au'$ *and* $v = bv'$;
- $CC((E_1 | E_2 | \cdots | E_m) \& (E'_1 | E'_2 | \cdots | E'_n)) = CC(\bigcup_{i=1 j=1}^{i=m j=n} E_i \& E'_j)$.

For example, $E_1 = a \& b$, $E_2 = (a|b) \& (c|d)$. $CC(E_1) = 2$. $CC(E_2) = 8$. CC describes the compactness of a regular expression. Smaller the CC is, more compact the regular expression will be. Hence it is a good indicator to reflect the preciseness of regular expressions for the set of given samples S.

All inferred regular expressions are measured by different indicators (*Length of Regular Expression* ($|RE|$) and CC). The experimental results are shown in Table 1. We evaluate the performance of inference algorithm of *GenESIRE* from two aspects: *conciseness* and *preciseness*.

Conciseness. Values of $|RE|$ from Table 1 tell us that the regular expression inferred by *GenESIRE* performs as well as regular expressions inferred by other methods. Therefore, ESIRE is concise and readable in real-world applications.

Preciseness. For the regular expression inferred by *GenESIRE* on string set of *Entry*, value of *CC* is much smaller to a large extent than other methods. It reflects the results inferred by *GenESIRE* are more precise. Hence the inference algorithm is more effective.

Table 1. Results of inference using different methods on **Entry**

Sample Size	From	Element Name		$|RE|$	CC
50000	SwissProt	Entry			
Method		**Regular Expression**			
Instance ToSchema		$a_{11}^* \& a_{33}^* \& a_{10}^? \& a_{32}^* \& a_{13}^* \& a_{35}^* \& a_{12}^* \& a_{34}^* \& a_{15}^* \& a_{37}^* \& a_{14}^? \& a_{36}^* \& a_{17}^* \& a_{39}^* \&$ $a_{16}^* \& a_{38}^* \& a_{19}^* \& a_{18}^* \& a_{40}^* \& a_{20}^* \& a_{42}^* \& a_{41}^* \& a_{22}^* \& a_{21}^? \& a_{43}^* \& a_{24}^* \& a_{23}^? \& a_{26}^* \&$ $a_{25}^* \& a_{28}^* \& a_{27}^* \& a_{29}^? \& a_7^+ \& a_3^- \& a_4^+ \& a_5^- \& a_6^* \& a_7^+ \& a_8^- \& a_9^* \& a_{31}^* \& a_{30}^?$		131	$3.48*10^{55}$
learn$_{DME^+}$		$(a_{29}^?|a_6^*|a_{26}^?|a_{20}^*|a_{38}^*|a_{41}^*|a_{39}^*|a_{35}^*|a_{31}^?|a_{40}^*|a_{30}^?|a_{33}^*)\&(a_{36}^*|a_{25}^*|a_{21}^*|$ $a_{23}^*|a_{19}^*|a_{37}^*)\&(a_{42}^*|a_{32}^*|a_{17}^*|a_{34}^*)\&(a_{27}^*|a_{10}^?)\&a_{43}^*\&a_{22}^*\&a_{28}^*\&a_{24}^*\&a_9^*\&$ $a_4^*\&a_2^+\&a_1^-\&a_5^+\&a_{13}^*\&a_{18}^*\&a_{15}^*\&a_7^-\&a_3^-\&a_8^-\&a_{12}^*\&a_{16}^*\&a_{11}^*\&a_{14}^?$		136	$1.48*10^{25}$
conMiner		$a_{10}^?a_{27}^*a_{34}^*a_{25}^*a_{38}^*a_{11}^*a_{12}^*a_{18}^*a_{15}^*a_1^*a_2^+a_3^-a_4^+a_5^-a_6^*a_7^+a_8^-a_9^*a_{16}^*a_{30}^*a_{35}^*a_{41}^*$ $a_{43}^*a_{23}^*a_{20}^*a_{40}^*a_{19}^*a_{17}^?a_{26}^?a_{36}^*a_{42}^*a_{31}^?a_{24}^*a_{33}^*a_{37}^*a_{32}^*a_{39}^*a_{21}^?a_{29}^?a_{22}^?a_{28}^*a_{13}^?a_{14}^?$		128	$5.64*10^8$
GenESIRE		$a_1^+a_2^+a_3^-a_4^*a_5^-a_6^*a_7^+a_8^-a_9^*a_{16}^*(a_{10}^?\&(a_{27}^*|a_{30}^?)(a_{41}^*|a_{35}^*|a_{36}^*|a_{34}^*|$ $a_{43}^*)(a_{20}^*|a_{24}^*|a_{23}^*|a_{39}^*|a_{19}^?)(a_{33}^*|a_{21}^*|a_{17}^*|a_{37}^*|a_{25}^?)a_{29}^*a_{11}^*$ $a_{12}^*a_{18}^*a_{15}^*\&(a_{42}^*|a_{40}^*|a_{26}^?|a_{22}^?|a_{32}^*|a_{38}^*|a_{31}^?)a_{28}^*)^?a_{13}^?a_{14}^?$		140	$1.16*10^6$

5   Conclusion and Future Work

We presented a useful subclass *ESIRE* of restricted regular expressions with interleaving based on the analysis of real world data. And using SOA and MIS, we proposed an inference algorithm *GenESIRE* to infer *ESIREs*. Experimental results showed that regular expressions inferred by *GenESIRE* are more precise than other methods, measured by different indicators on real world data. Our future work is to study the extended SORE with interleaving. The inference algorithms together with the construction of automata will also be considered.

References

1. Benedikt, M., Fan, W., Geerts, F.: XPath satisfiability in the presence of DTDs. J. ACM **55**(2), 1–79 (2008)
2. Bex, G.J., Neven, F., Bussche, J.V.D.: DTDs versus XML schema: a practical Study. In: International Workshop on the Web and Databases, pp. 79–84 (2004)
3. Bex, G.J., Neven, F., Schwentick, T., Tuyls, K.: Inference of concise DTDs from XML data. In: International Conference on Very Large Data Bases, Seoul, Korea, September, pp. 115–126 (2006)
4. Bex, G.J., Neven, F., Schwentick, T., Vansummeren, S.: Inference of concise regular expressions and DTDs. ACM Trans. Database Syst. **35**(2), 1–47 (2010)
5. Bex, G.J., Neven, F., Vansummeren, S.: Inferring XML schema definitions from XML data. In: International Conference on Very Large Data Bases, University of Vienna, Austria, September, pp. 998–1009 (2007)
6. Boneva, I., Ciucanu, R., Staworko, S.: Simple schemas for unordered XML. In: International Workshop on the Web and Databases (2015)

7. Boppana, R., Halldrsson, M.M.: Approximating maximum independent set by excluding subgraphs. Bit Numer. Math. **32**(2), 180–196 (1992)

8. Che, D., Aberer, K., Özsu, M.T.: Query optimization in XML structured-document databases. VLDB J. **15**(3), 263–289 (2006)

9. Ciucanu, R., Staworko, S.: Learning schemas for unordered XML. Computer Science (2013)

10. Freydenberger, D.D., Kötzing, T.: Fast learning of restricted regular expressions and DTDs. Theory Comput. Syst. **57**(4), 1114–1158 (2015)

11. Garcia, P., Vidal, E.: Inference of k-testable languages in the strict sense and application to syntactic pattern recognition. IEEE Trans. Pattern Anal. Mach. Intell. **12**(9), 920–925 (2002)

12. Garofalakis, M., Gionis, A., Rastogi, R., Seshadri, S., Shim, K.: XTRACT: learning document type descriptors from XML document collections. Data Min. Knowl. Discov. **7**(1), 23–56 (2003)

13. Gold, E.M.: Language identification in the limit. Inf. Control. **10**(5), 447–474 (1967)

14. Grijzenhout, S., Marx, M.: The quality of the XML web. Web Semant. Sci. Serv. Agents World Wide Web **19**, 59–68 (2013)

15. Koch, C., Scherzinger, S., Schweikardt, N., Stegmaier, B.: Schema-based scheduling of event processors and buffer minimization for queries on structured data streams. In: Thirtieth International Conference on Very Large Data Bases, pp. 228–239 (2004)

16. Li, Y., Zhang, X., Peng, F., Chen, H.: Practical study of subclasses of regular expressions in DTD and XML schema. In: Li, F., Shim, K., Zheng, K., Liu, G. (eds.) APWeb 2016. LNCS, vol. 9932, pp. 368–382. Springer, Cham (2016). https://doi.org/10.1007/978-3-319-45817-5_29

17. Mani, M., Lee, D., Muntz, R.R.: Semantic data modeling using XML schemas. In: S.Kunii, H., Jajodia, S., Sølvberg, A. (eds.) ER 2001. LNCS, vol. 2224, pp. 149–163. Springer, Heidelberg (2001). https://doi.org/10.1007/3-540-45581-7_13

18. Manolescu, I., Florescu, D., Kossmann, D.: Answering XML queries on heterogeneous data sources. In: International Conference on Very Large Data Bases, pp. 241–250 (2001)

19. Martens, W., Neven, F.: Typechecking top-down uniform unranked tree transducers. In: International Conference on Database Theory, pp. 64–78 (2003)

20. Martens, W., Neven, F.: Frontiers of tractability for typechecking simple XML transformations. In: ACM SIGMOD-SIGACT-SIGART Symposium on Principles of Database Systems, pp. 23–34 (2004)

21. Martens, W., Neven, F., Schwentick, T.: Complexity of decision problems for XML schemas and chain regular expressions. SIAM J. Comput. **39**(4), 1486–1530 (2013)

22. Min, J.K., Ahn, J.Y., Chung, C.W.: Efficient extraction of schemas for XML documents. Inf. Process. Lett. **85**(1), 7–12 (2003)

23. Papakonstantinou, Y., Vianu, V.: DTD inference for views of XML data. In: Nineteenth ACM SIGMOD-SIGACT-SIGART Symposium on Principles of Database Systems, pp. 35–46 (2000)

24. Peng, F., Chen, H.: Discovering restricted regular expressions with interleaving. In: Cheng, R., Cui, B., Zhang, Z., Cai, R., Xu, J. (eds.) APWeb 2015. LNCS, vol. 9313, pp. 104–115. Springer, Cham (2015). https://doi.org/10.1007/978-3-319-25255-1_9

Applications of Conceptual Modeling

A Method to Identify Relevant Genome Data: Conceptual Modeling for the Medicine of Precision

Ana León Palacio[✉] [iD], Óscar Pastor López, and Juan Carlos Casamayor Ródenas

Research Center on Software Production Methods (PROS),
Universitat Politècnica de València, Valencia, Spain
{aleon,opastor}@pros.upv.es, jcarlos@dsic.upv.es

Abstract. The use of techniques such as Next Generation Sequencing increases our knowledge about the genomic risk of suffering a certain disease, improving our ability of providing an early diagnosis and thus an appropriate treatment for each patient. In order to provide an accurate diagnosis, clinicians must perform a search in the repositories of open data available to the research community. Nevertheless, the vast amount of heterogeneous and dispersed data sources that store information about gene-disease associations as well as their variable level of quality hinder the process of determining if the variants found in the DNA sequence of a patient's sample are clinically relevant. In this paper, we present a systematic method based on conceptual modeling and data quality management techniques to tackle the aforementioned issues with the aim of helping the genomic diagnosis of a disease. To this end, we state the most prominent problems affecting repositories of open data for genomics. Then, we use a methodological approach for identifying what we called "smart data": the relevant information hidden in the genomics data lake. Finally, in order to test and validate the proposed method, we apply it to a case study based on the clinical diagnosis of Crohn's Disease.

Keywords: Genomics · Smart data · Conceptual modeling · Data quality
Precision medicine

1 Introduction

During the last two decades, our understanding of human biology has been improved thanks to the completion of the Human Genome Project in 2003 and the use of innovative techniques to read (sequence) the DNA, such as Next Generation Sequencing (NGS). Remarkable advances have been made related to the association between changes in the DNA (variants) and the risk of suffering certain diseases. This knowledge is stored in genomic data repositories, publicly available for the research community.

In order to provide a genomic diagnosis for a disease, the clinicians take samples from a patient (blood, tissue, etc.) and determine which DNA variants are present. This process is called primary and secondary analysis, and it is based on physical and digital processes. Once the variants are identified, a research must be performed in order to find out if they are clinically relevant to make an accurate diagnosis and provide an

© Springer Nature Switzerland AG 2018
J. C. Trujillo et al. (Eds.): ER 2018, LNCS 11157, pp. 597–609, 2018.
https://doi.org/10.1007/978-3-030-00847-5_44

effective treatment. This is a manual process called *"variant curation"* or *"tertiary analysis"*, that is mainly performed by searching and reading scientific articles containing relevant population studies (See Image 1). But why the use of the public genomic repositories is not an extended practice yet?

Sample   Primary analysis   Secondary analysis   Tertiary analysis   Genetic report

Image 1. Genomic diagnosis workflow.

The use of public genomic repositories is hindered by their heterogeneity and dispersion. Moreover, due to the complexity of biological processes, the variability of sequencing techniques and the lack of consensus in laboratory practices, only part of the available information -what we call *"smart data"*- is of enough quality to be considered as reliable for genomic diagnosis. All these issues together constitute a bottleneck in the genomic diagnosis workflow. In order to tackle these problems, we propose the systematic use of data quality management techniques and conceptual modeling to extract relevant information from public repositories in order to ease the variant curation process.

In [1], the concept of *"master data"* is defined as a dataset, whatever the size is, valuable for the purpose at hand and cross functional. In the mentioned work the steps to define a data quality methodology applied to a case study related to the diagnosis of Alzheimer's Disease are presented. This work goes further: we apply the acquired knowledge by describing in depth a method to gather the relevant information – the specific master data - required by the *"variant curation"* process. We validate the method by applying it to a case study with the aim of supporting the genomic diagnosis of Crohn's Disease.

With this purpose in mind, we first review the state of the art in Sect. 2. Then, in Sect. 3 we describe our method along with the case study used as a running example. Finally, in Sect. 4 we summarize the validation results to end up with conclusions and future work in Sect. 5.

2   State of the Art

Medicine of Precision is a new paradigm whereby prevention, diagnosis and treatment of a disease is based on the individuality of each patient. One of its cornerstones is the genomic diagnosis, described in Sect. 1. When clinicians start the variant curation process they must find valuable information in the databases, as well as in the published literature. To accomplish this aim, two main problems must be faced: (i) the huge amount of available repositories and (ii) the filtering of information with a variable level of quality.

The number of genomic data sources publicly available has considerably increased in the last decade. In its last update, The NAR Online Molecular Biology Database Collection summarizes information about 1,737 repositories [2]. These databases are variable in content, scope, infrastructure and quality so a first question arises, how to determine the most suitable repositories to search for the required information? Some research has been done in order to determine the quality of a biological data source, such as the one made by the Human Variome Project [3], but it is still a proposal with no implementation.

A lot of efforts are made for the production of genomic datasets and the information required to make a genomic diagnosis is not usually available in only one data source. Frequently, several heterogeneous databases must be queried in order to join all the puzzle pieces together. But these databases were not developed with the purpose of sharing information among them. This means that it is necessary to know each under-lying data representation and face problems such us inconsistencies in the way the same concept is represented, discrepancies in the nomenclature used and information redun-dancy [4]. Moreover, due to the experimental nature of the domain, the information is susceptible of containing errors or be inaccurate [5]. These errors can be propagated through the data sources increasing the noise and forcing the clinicians to make a great effort to separate the wheat from the chaff. At this point a second question arises, how can be the most accurate and reliable data identified? The genomic data integration is currently under study, and repositories such as Ensembl [6] and DisGeNet [7] gather information from multiple sources, allowing the researchers to access related data from a single platform. Nevertheless, these platforms use a reduced number of generic repo-sitories and are focused on integrating information regardless of its quality.

Despite their contribution, none of the aforementioned research lines provide a complete solution for both questions. In our opinion, they require to be answered by using a systematic approach based on data quality management principles and concep-tual modeling. Data Quality (DQ) has been defined by Wang and Strong [8] as data that are "fitness for use". The study of data quality began in the 1990s and a wide research has been carried out in different domains. Nevertheless, research about data quality in genomics has just started so there are not sound results yet. Data Quality it is not enough by itself. Its application requires a conceptual background that will help to identify which information is required to be of high quality as well as providing structure to connect each piece of knowledge, wherever it comes from. We want to emphasize the role of Conceptual Modeling in this field – Genomic diagnosis for Medicine of Precision (MP). We firmly believe that Conceptual Modeling must drive MP, a domain where the unpre-cedented growth of data generation is continuous.

The use of conceptual models in genomics has been explored by some authors such as Paton [9], Ram [10] and Bernasconi [11]. Nevertheless, these proposals are focused on specific parts of the domain and do not explore the association between variants and diseases. In a previous work we focused on providing a conceptual model-based basic framework, where data to be managed are precisely identified and represented through the Human Genome Conceptual Model designed by the PROS Research Center[1] in UPV

---

[1]   http://www.pros.webs.upv.es/.

[12]. The Human Genome Conceptual Model is currently in its 3rd version (HGCM v3) and it maintains the essential genome information through its 5 main views (structure, transcription, variants, chromosome and metabolic routes). This paper extends our work by determining how to select the most suitable data sources and identify the information with the required quality to be used during the variant curation process.

3   A Method to Manage Relevant Genomic Data

In order to systematize the variant curation process, we propose two phases in our method: (i) Selection of reliable data sources and (ii) identification of the DNA variants according to minimum DQ criteria.

We validate and exemplify each phase of the method by using our case study, the genomic diagnosis of Crohn's Disease, a chronic inflammatory disease of the gastrointestinal tract. At the end of the process, we provide a catalog of verified and reliable variants whose presence can be checked in a patient's sample, to determine the risk of suffering Crohn's Disease.

3.1   Selection of Reliable Data Sources

In the first step of our method we must select the most adequate data sources to extract the required information. Nevertheless, the huge amount of genomic data sources and their variable characteristics makes the process a non-trivial task. According to the information provided by the *NAR Online Molecular Biology Database Collection*[2] and

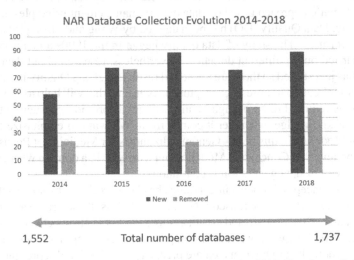

Fig. 2. Number of databases added and removed from the NAR Database Collection. A comparative from 2014 to 2018.

---

[2] https://www.oxfordjournals.org/our_journals/nar/database/c/.

the *Human Genome Variation Society Database Catalog*[3], there are over 1.700 data sources containing information about the whole spectrum of the genomic domain. Nowadays, the exact number of existing databases cannot be verified and is highly variable as can be shown in Fig. 2. An average of about 70 new databases is added to the NAR collection and 40 are removed per year. In order to identify the relevant ones, we propose three sub processes:

1. Identify the data sources based on the scope of the task we want to perform.
2. Define the minimum DQ criteria the data sources must meet.
3. Select the data sources according to the DQ criteria.

According to our case study the scope of the data sources is: human variant-disease associations and specific information about Crohn's Disease. Variants are then the relevant part of interest from our Conceptual Model of the Human Genome. Focusing on the selected scope of our task we searched in the mentioned catalogs and identified 57 candidate databases.

Then, we defined the minimum DQ criteria the data sources must meet. The definition of such criteria is guided by the concepts "data quality dimension" and "data quality

Table 1. Data quality dimensions and metrics to determine the quality of a genomic database. Each cluster has been identified by Cx, each dimension by Dx and metrics by Mx.

C1. TRUST CLUSTER: Reliability of the data source	
D1. Trustworthiness	M1. The information stored must be reviewed by a group of experts in the field
	M2. There are quality controls to assure the correctness of the submitted information
D2. Reputation	M3. The data source must be supported by a research center, association or institute with national or international relevance
C2. TEMPORAL CLUSTER: How up-to-date the database is	
D3. Timeliness	M4. There must be date stamps about when data were entered as well as when last modified
	M5. The database must provide the version number
D4. Currency	M6. The database must be active (less than 1 year from the last update)
C3. COMPLETENESS CLUSTER: Amount of available information	
D5. Completeness	M7. The database must provide at least 100% of all the minimum required attributes
C4. ACCESSIBILITY CLUSTER: Type of access and information retrieval	
D6. Availability	M8. The information must be public and freely accessible
D7. Accessibility	M9. The database must provide mechanisms to download the search results
	M10. It is highly recommended to allow the programmatic access to the information stored
C5. CONSISTENCY CLUSTER: Consistent representation of data	
D8. Consistency	M11. It is highly recommended the use of standards such as:
	• Variant names: HGVS[a]
	• Gene names: HGNC[b]
	• Diseases/Phenotypes: HPO[c]
	• Consequences: VARiO[d]

[a] http://varnomen.hgvs.org/

[b] https://www.genenames.org/

[c] https://human-phenotype-ontology.github.io/

[d] http://variationontology.org/

---

[3] http://www.hgvs.org/content/databases-tools.

metric". DQ is a multidimensional concept where a data quality dimension (DQD) is defined as a set of attributes, which can be assessed using specific metrics in order to get a measure that represents the quality. In our case, we have defined a set of metrics which covers all the common causes of low quality in a genomic data source [4]. The DQDs has been grouped in clusters according to their purpose, resulting in 5 clusters, 8 dimensions and 11 metrics. The minimum DQ criteria proposed are summarized in Table 1.

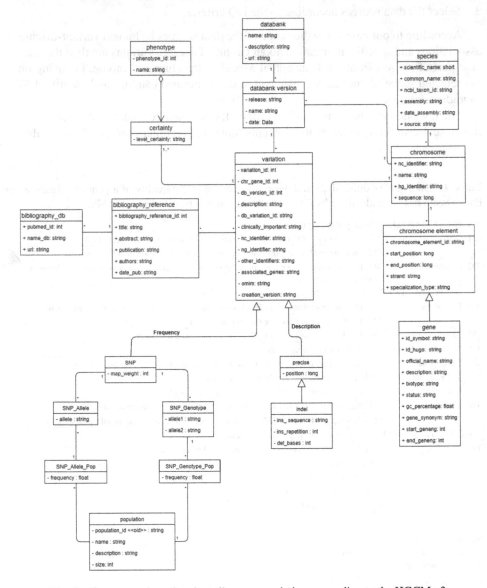

Fig. 3. Representation of variant-disease associations according to the HGCM v3.

In order to measure completeness, we must previously identify which attributes are required according to our purpose. There are no guidelines about which information is mandatory to provide a genomic diagnosis so, in our case we turn to the Human Genome Conceptual Model. By analyzing its structure, we were able to determine the main classes and attributes involved in the description of disease-variant associations, which can be seen in Fig. 3.

Table 2. Classification of the attributes required to represent a variant-disease association, according to their importance.

Class	Required	Recommended	Other
Species	assembly	scientific_name common_name	ncbi_taxon_id date_assembly source
Chromosome	nc_identifier		name sequence
Databank	name		description url
Databank Version	release	date	
Chromosome element	chromosome_element_id	start_position end_position strand	specialization type
Gene	id_symbol	official_name biotype gene_synonym start_gene_ng end_gene_ng	id_hugo description status gc_percentage
Variation	db_variation_id clinically_important		description other_identifiers associated_genes omim creation_version
Precise	position		
Indel			ins_sequence ins_repetition del_bases
SNP			map_weight
SNP Allele	allele		
SNP Allele Pop	frequency		
Population	name	size	description population_id
SNP Genotype	name		
Certainty	level of certainty		
Bibliography db	name_db		url

Following a conceptual model-based approach is key due to any data of interest has mandatorily a counterpart represented as an attribute coming from a conceptual model class.

Once the conceptual model is fixed, we can group the attributes under three categories: "required", "recommended" and "other", which are summarized in Table 2. According to the criteria established by the data quality metric M7, the database must provide information for at least 100% of the attributes in the "required" category, which are the minimum attributes needed to represent a variant-disease association according to the conceptual model.

Once the required attributes are established, the quality metrics can be applied to each of the 57 candidate databases in order to determine the relevant ones.

The selected databases are:

- **ClinVar:** The most well-known repository of relationships among human variants and phenotypes, with supporting evidence[4].
- **ENSEMBL:** A genome browser for vertebrate genomes that supports research in comparative genomics, evolution, and sequence variant and transcriptional regulation[5].
- **SNPedia:** A wiki investigating human genomics that shares information about the effects of variants in DNA, citing peer-reviewed scientific publications[6].
- **dbSNP:** A central repository for short variants[7].

The set of quality clusters proposed covers the basic criteria to assure that the databases are reliable, up-to-date, accessible, consistent and provides the required information to perform the variant curation analysis. The most common reasons why 53 databases were discarded are (i) inactivity of the database, (ii) the information is partially available for public access, (iii) lack of information corresponding to some of the "required" attributes and (iv) the reputation of the database supporters does not fit the established quality criteria. More clusters, dimensions and metrics can be added if required.

3.2 Identifying the Variants According to Minimum DQ Criteria

Once the most adequate data sources are identified, in the second step of our method we are going to filter the information to determine the most accurate and reliable data. First of all, we performed a query by disease on each database by using the APIs provided and as result 96 variants from ClinVar, 502 variants from Ensembl, 49 variants from SNPedia and 516 variants from dbSNP were obtained. After integrating the information and removing redundancies, we obtained a final dataset consisting in 517 variants. The integration process in detail as well as how the inconsistencies were solved is out of the scope of this work.

---

[4] https://www.ncbi.nlm.nih.gov/clinvar/.
[5] https://www.ensembl.org/index.html.
[6] https://www.snpedia.com/index.php/SNPedia.
[7] https://www.ncbi.nlm.nih.gov/projects/SNP/.

We focus now on the main contribution of this part of the work, related to the variant curation process: how relevant the variant is to provide a genomic diagnosis. As well as has been done in the previous step, in order to select the relevant variants, we must define the minimum DQ criteria that the information must meet. These criteria can be shown in Table 3.

Table 3. Data quality dimensions and metrics required to determine the quality of the information.

C1. TRUST CLUSTER: Reliability of the information	
D1. Trustworthiness	M1. Each variant-disease association must be supported by at least one publication
D2. Reliability	M2. There must not be conflicts in the clinical interpretation of each variant
	M3. There must not be conflicts in the information among databases
C2. COMPLETENESS CLUSTER: Amount of available information	
D5. Completeness	M4. The phenotype (disease) and the clinical significance must be clearly specified
C2. USEFULNESS CLUSTER: Relevance of the information for the task at hand	
D6. Usefulness	M5. The clinical significance of a variant must be different than "benign", "likely benign" and "likely pathogenic"
	M6. The studies provided by the bibliography must conform the next statistical requirements: • At least 500 participants • For pathogenic variants, the Odds Ratio and the Interval of Confidence must be >1 • For protective variants, the Odds Ratio must be <0.9 and the Interval of Confidence must be <1 • Replicated studies are desirable

The proposed criteria are intended to be applied in general to any type of disease. Nevertheless, it is important to take into account that under certain circumstances they can be changed. For example, if the disease under study is catalogued as a "rare disease" (1/100.00 affected people) it is very difficult to find studies involving more than 500 participants. In such cases, the criteria must be adapted to detect significant variants. Nevertheless, in such cases the state of the research is considered to be under 100%. Next, the reasons why these metrics have been selected are explained.

Trust Cluster
The Trust Cluster groups the dimensions related to the reliability of the information (Trustworthiness Dimension and Reliability Dimension). The results of the studies performed over a population about a certain disease are usually published in journals of high impact and available for the scientific community. This is the reason why when an assertion about the association between a variant and a disease is made, it is required to provide the bibliography that supports such assertion. Sometimes, the information is loaded without providing any supporting evidence which makes it difficult to be

contrasted. We consider that a disease-variant assertion is trustworthiness when there is at least one article supporting the assertion.

Another issue arises when contradictory information about the same variant is found. For example, when different users provide contradictory information about the effect (pathogenic, benign, etc.) in the same database, or when different databases provide contradictory information. When this happens, the effect of the variant cannot be considered as reliable and a deeper study must be performed in order to determine the origin of the conflicts.

Completeness Cluster

In order to determine the relevance of a variant it is required to have as much information as possible, in particular the specific name of the disease (type or subtype) and the effect of the variant over the health. Sometimes, one or both of this attributes are not provided and consequently, the relevance of the variant cannot be established.

Usefulness Cluster

This cluster groups the dimensions related to the usefulness of the information for the task at hand. In this case it is important to determine which type of effect over the health is required and the minimum criteria that the studies performed must fulfill. The list of possible effects of a variant is summarized in Table 4.

Table 4. List of terms related to the effect of the variant.

Term used	Description
Affects	The variant causes a non-disease phenotype, such as lactose intolerance
Association	The variant has been identified in a Genome-wide association study
Benign	The variant is reported to be benign
Drug response	The variant is reported to affect a drug response
Likely benign	The variant is reported to be likely benign
Likely pathogenic	The variant is reported to be likely pathogenic
Pathogenic	The variant is reported to be pathogenic
Protective	The variant is reported to decrease the risk for a particular disease
Risk factor	The variant is reported to be a risk factor for a particular disease

The term "likely" is used when the effect it is not clear but is probable. In order to provide the higher level of accuracy, the variants associated with "benign", "likely benign" or "likely pathogenic" are discarded until new evidences about the disease-variant association are published.

Due to the high variability in the experimental techniques and population studies, as well as the complexity of the mechanism of the disease, a deep analysis of the publications available must be performed in order to determine the most reliable and accurate ones. But this is a laborious process that must be guided by a set of quality criteria to help the selection of relevant publications. Following the recommendations of experts in genomic diagnosis we established the minimum criteria that the studies included in the literature associated to a variant must fulfill:

- The number of participants in the study must be up to 500.
- For protective variants, the Odds Ratio must be <0.9 and the Confidence Interval must be <1.
- For the rest of variants, the Odds Ratio and the Confidence Interval must be >1.
- It is desirable that the studies are replicated, but as this can be difficult in most cases this is not required.

The data quality criteria are applied in a certain order to ease the classification of the variants, as can be seen in Fig. 4.

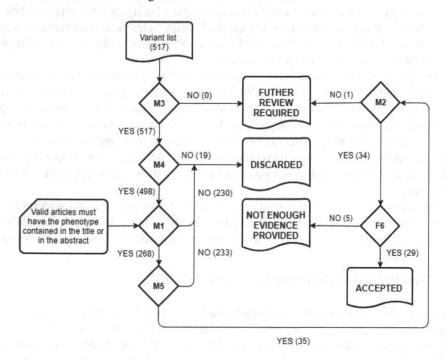

Fig. 4. Data quality workflow for the classification of relevant variants related to Crohn's disease. Quality metrics are represented as M1:M6 and the number of variants that pass each filter is specified in brackets.

At the end of the process, the variants are classified according to 4 different categories:

- *Accepted:* 29 variants have passed all the filters and can be used to provide a clinical diagnosis.
- *Not enough evidence provided:* 5 variants have associated publications but they do not conform the minimum statistical requirements.
- *Further review required:* 1 variant presents conflicts in the clinical significance.
- *Discarded:* 482 variants are not of enough quality to be used with clinical purposes.

Once the relevant variants to provide a genomic diagnostic were identified, they were validated with professionals in genomic diagnosis. In the next section, the results of the validation are presented.

4   Validation

In order to validate the method proposed, the results were reviewed by a group of geneticists and clinicians, with experience in providing genomic diagnosis. The variant curation they perform is a manual process consisting in identifying bibliography related to the disease from PubMed and reading the published articles. Some databases such as ClinVar or GWAS Central are occasionally accessed in order to obtain extra information, but they are not systematically reviewed.

The results of our method were contrasted with the information they manage to diagnose Crohn's Disease. As a result, 90% of the variants they manage were correctly identified. Besides the coincidences, we provided 8 new variants and their corresponding evidence, which are now under study to be added to their diagnosis process. These results were remarkably positive, especially because the use of the method was efficient (less total time for the identification of relevant variants), and there was a clear process behind, where traceability between the selected variants and why they were selected, was clearly determined.

The method has been applied to identify the risk of suffering other diseases such as Neuroblastoma [13] and is currently being used in the context of a project in collaboration with clinical experts in lung cancer from the "Hospital de Clínicas" in Paraguay.

5   Conclusions and Future Research Lines

In this paper, we have presented a systematic method to identify the relevant information required to ease the *"variant curation"* or *"tertiary analysis"* in order to provide an accurate genomic diagnosis. The interest of this method stems from the huge effort required to review and connect the information provided by the genomic repositories publicly available, which forces the geneticist and clinicians to read hundreds of scientific articles in order to find relevant population studies.

The presented method is supported by the use of conceptual models and data quality management techniques, and the results were validated in a real case of study related to the diagnosis of Crohn's disease. As a further work, we plan to instantiate the method with different diseases and to develop a process to keep the information updated as soon as new findings are uploaded to the public data sources.

Acknowledgements. The authors would like to thank the members of the PROS Research Centre Genome group for the fruitful discussions regarding the application of CM in the medicine field. This work has been supported by the Spanish Ministry of Science and Innovation through project DataME (ref: TIN2016-80811-P) and the Research and Development Aid Program (PAID-01-16) of the Universitat Politècnica de València under the FPI grant 2137.

References

1. León, A., Pastor, O.: From big data to smart data: a genomic information systems perspective. In: 2018 12th International Conference on Research Challenges in Information Science (RCIS), pp. 1–11. IEEE, (2018). https://doi.org/10.1109/RCIS.2018.8406658
2. Rigden, D.J., Fernández, X.M.: The 2018 Nucleic Acids Research database issue and the online molecular biology database collection. Nucleic Acids Res. 46(D1), D1–D7 (2018). https://doi.org/10.1093/nar/gkx1235
3. Vihinen, M.: Human variome project quality assessment criteria for variation databases. Hum. Mutat. 37, 549–558 (2016). https://doi.org/10.1002/humu.22976
4. León, A., Reyes, J., Burriel, V., Valverde, F.: Data quality problems when integrating genomic information. In: Link, S., Trujillo, J.C. (eds.) ER 2016. LNCS, vol. 9975, pp. 173–182. Springer, Cham (2016). https://doi.org/10.1007/978-3-319-47717-6_15
5. Müller, H., Naumann, F., Freytag, J.C.: Data quality in genome databases. In: Proceedings of Conference on Information Quality (IQ 2003), pp. 269–284 (2003). https://doi.org/10.18452/9205
6. Yates, A., et al.: Ensembl 2016. Nucleic Acids Res. 44(D1), D710–D716 (2016). https://doi.org/10.1093/nar/gkv1157
7. Piñero, J., et al.: DisGeNET: a comprehensive platform integrating information on human disease-associated genes and variants. Nucleic Acids Res. 45, D833–D839 (2017)
8. Wang, R.Y., Strong, D.M.: Beyond accuracy: what data quality means to data consumers. Manag. Inf. Syst. 12, 5–33 (1996). https://doi.org/10.1080/07421222.1996.11518099
9. Paton, N.W., et al.: Conceptual modelling of genomic information. Bioinformatics 16, 548–557 (2000). https://doi.org/10.1093/bioinformatics/16.6.548
10. Ram, S., Wei, W.: Modeling the semantics of 3D protein structures. In: Atzeni, P., Chu, W., Lu, H., Zhou, S., Ling, T.-W. (eds.) ER 2004. LNCS, vol. 3288, pp. 696–708. Springer, Heidelberg (2004). https://doi.org/10.1007/978-3-540-30464-7_52
11. Bernasconi, A., Ceri, S., Campi, A., Masseroli, M.: Conceptual modeling for genomics: building an integrated repository of open data. In: Mayr, H.C., Guizzardi, G., Ma, H., Pastor, O. (eds.) ER 2017. LNCS, vol. 10650, pp. 325–339. Springer, Cham (2017). https://doi.org/10.1007/978-3-319-69904-2_26
12. Reyes Román, J.F., Pastor, Ó., Casamayor, J.C., Valverde, F.: Applying conceptual modeling to better understand the human genome. In: Comyn-Wattiau, I., Tanaka, K., Song, I.-Y., Yamamoto, S., Saeki, M. (eds.) ER 2016. LNCS, vol. 9974, pp. 404–412. Springer, Cham (2016). https://doi.org/10.1007/978-3-319-46397-1_31
13. Burriel, V., Reyes Román, J.F., Heredia Casanoves, A., Iñiguez-Jarrín, C.E., León Palacio, A.: GeIS based on conceptual models for the risk assessment of neuroblastoma. In: 2017 11th International Conference on Research Challenges in Information Science, pp. 451–452. IEEE, Brighton (2017). https://doi.org/10.1109/rcis.2017.7956581

Experiences Applying e³ Value Modeling in a Cross-Company Study

Jennifer Horkoff[1,2]([✉]), Juho Lindman[1], Imed Hammouda[1,2,3],
and Eric Knauss[1,2]

[1] University of Gothenburg, Gothenburg, Sweden
{jennifer.horkoff,imed.hammouda,eric.knauss}@cse.gu.se,
juho.lindman@ait.gu.se
[2] Chalmers Institute of Technology, Gothenburg, Sweden
[3] Mediterranean Institute of Technology, South Mediterranean University,
Tunis, Tunisia

Abstract. Driven by business interests, (product/customer) value has become a critical topic in system and software engineering as well as enterprise planning. The conceptual modeling community has responded to this challenge with several modeling approaches, including e³ value modeling, focusing on capturing and analyzing value flows in value networks. This modeling approach has risen from practical e-commerce experiences and has been further studied in an academic context. In this experience paper, we report the advantages and disadvantages of applying e³ value modeling as part of a cross-company case study focusing on understanding the internal and external value of APIs from a strategic perspective. We found that value modeling was generally well-received and understood by the company representatives, but also found drawbacks when used in our context, including challenges in modeling internal value networks, capturing problematic or missing values, finding quantitative value measures, and showing underlying motivations for flows. Our findings can help to improve language aspects, methods and tools, and can help to guide future value analysis in similar contexts.

Keywords: Value Modeling · e³ Value Modeling
Modeling Experiences · Empirical Studies · Multi-case Studies

1 Introduction

Given the competitive nature of business, technology development and strategy has evolved to focus on value [1,2]. This focus aims to ensure that technology, processes, or interactions have a clear contribution to customer or business value. Various conceptual models, from an enterprise [3], process [4], or ecosystem perspective [5], have been developed to support value analysis.

In this work, we apply e³ value modeling in a comparative study which models the internal and external value of strategic APIs. As part of a continuous project,

© Springer Nature Switzerland AG 2018
J. C. Trujillo et al. (Eds.): ER 2018, LNCS 11157, pp. 610–625, 2018.
https://doi.org/10.1007/978-3-030-00847-5_45

we worked with five companies to understand the company context and the API of focus, capturing and validating our understanding via an e^3 value model for each case. The models were used to support a comparative analysis across cases, understanding different manifestations and configurations of API value.

Much work has focused on presenting the original e^3 value concepts and methods [6–8], and on using the language with other languages in a complementary way [9,10]. Although the e^3 framework was developed based on the e-commerce consulting experience of the authors [6], many of the examples presented in subsequent papers are illustrative, intended to show the strengths of the framework. In our case, we apply the language as outsiders to real, complex cases, reporting our experiences, in order to evaluate the benefits and drawbacks of the language in our context. Our primary research question is as follows: **RQ1.** What are the benefits and limitations of applying e^3 value modeling in our context, assessing the value of a particular technology?

In existing work, e^3 modeling is usually applied to value networks, which could also fall under the topic of ecosystem modeling or mapping (e.g., [5]). These ecosystems are typically external, involving interaction between cooperating but separate companies. In our case, many of our APIs, and the resulting API ecosystems are internal to large companies. Thus, as a sub-research question, we also ask: **RQ1.1.** What are the benefits and limitations of applying e^3 value modeling to internal value networks/ecosystems?

We found that value modeling was generally well-received and understood by the company representatives, but also found drawbacks when used in our context, including finding reciprocal value flows, problematic or missing values, finding quantitative value measures, and showing underlying motivations or flows. At least some of these findings can be attributed to the internal nature of the networks. Our findings are summarized in a list of high-level recommendations, which can be further developed to improve language aspects, methods and tools, and can help to understand differing use cases for e^3 value modeling.

As this work arises from an ongoing project, it is one of a number of complimentary papers. The initial framework for strategic API analysis, without extensive use of models, appears in [11]. Early experiences comparing goal modeling, workflow modeling and value modeling have been reported in [12]. This previous work focuses on the results of a project deliverable where value modeling was applied to only one company. Results from the same deliverable, from an API analysis perspective appear in [13]. The current work is the first to focus specifically on experiences with e^3 value modeling from a conceptual modeling perspective, using new and more extensive case data.

This paper is organized as follows: Sect. 2 briefly reviews related modeling approaches and our study context. Section 3 describes our study method, while Sect. 4 describes our findings. Section 5 answers our research questions and describes threats to validity, while Sect. 6 concludes the paper and outlines current work.

2   Background and Related Work

Value Modeling. Several value modeling frameworks have been introduced in the last 10–15 years. From an agile software/lean development perspective, value stream mapping has become wide-spread, particularly in industry [4]. These models are a type of process model with an emphasis on production and flow data. The emphasis is to look at the flow of manufacturing or products and eliminate waste or bottlenecks, optimizing processes from value-perspective. In contrast, our emphasis is on assessing the value of planned/existing technology.

From a business perspective, Osterwalder's Business Model Ontology (BMO) and the resulting Business Model Canvas emphasize the concepts of value proposition and value chain [14,15]. From an Enterprise Information Systems perspective, the conceptual accounting framework, Resource-Event-Agent (REA) has been extended by Geerts and McCarthy to include value chains, focusing on economic events [3]. In a different approach, Value Network Analysis (VNA) focuses on value conversion networks, in which tangible and intangible values are exchanged for economic and social good [16]. In contrast, e^3 value modeling focuses on the viability of value proposition, making e-commerce value propositions more concrete and amenable to analysis. The intention is to ensure that a new business idea is viable, by examining value exchanges in a value network [6,7].

Value has been accounted for in previous work to model ecosystems. Handoyo et al. focus on capturing software ecosystems via value chains, using software supply network diagrams as a foundation [5]. OMG (Object Management Group) has standardized the modeling of value in its Value Delivery Modeling Language (VDML) [17]. The metamodel has been designed to support value stream analysis, VNA, e^3 value modeling, and REA (Resource Event Agent) modeling. More recent work has focused on an ontology analysis of value ascription and value propositions [18]; however, such work has not yet been used to create or improve a usable modeling framework, complete with an abstract and concrete syntax.

e^3 **Value Modeling.** Although experiences applying any value modeling method would have been a contribution to the conceptual modeling community, we chose to apply e^3 value modeling as per [6–8]. We made this choice due to the available supporting material and tooling, the presence of a concrete graphical syntax, the relative simplicity of the language, continued interest in the language from the research community (e.g., [19,20]), and interest from our company partners based on a single case application in a previous project deliverable [12].

In more detail, e^3 models consist of actors in a value network, which may be composite; market segments, or sets of actors; value objects, which are exchanged in value exchanges from one actor to another through value ports. Value ports are grouped into value interfaces, and value activities, which create value. Examples of e^3 models, including a legend, can be found in Fig. 1. The e^3 authors combine the framework with use case maps for representations of flow, including start and stop stimulus, and AND/OR forks and joins, showing which value objects must flow as the result of a stimulus. In our case we focused on the e^3-value ontology, making minimal use of the use case map notations.

e^3 value modeling supports quantitative analysis of value. Profit sheets for each activity or actor can be created to show the monetized value streams in and out. The net value is then calculated, with positive numbers indicating the e-commerce idea is potentially feasible for that actor or activity, negative flows indicating an infeasible idea. Such calculations also allow for sensitivity analysis [8]. The authors instruct to monetize all value flows which have a currency equivalent (e.g., Payment, Fees), but also to assign economic utilities to values which are not directly economic [6].

e^3 value models have been extended (e.g., [19]) or mapped to other languages in various ways. In particular, it has been combined or used in a complementary way with various forms of goal models, e.g., [9,10], or with REA modeling [20], or with BMO [21]. However, in our study, we chose to apply the original form of value modeling as described in [6,7], and as supported by the available tooling[1].

Table 1. Case company API description

Company	API description
Company 1 (C1)	C1 offers many APIs to its physical devices as well as a through cloud services. The API of focus, a cloud API, was in the planning stages during this study. Key issues included the role of 3rd party developers in the external cloud API
Company 2 (C2)	C2 supports its distributed employees via a set of scripts automating common tasks involving company hardware. Thus the API was internal, but geographically distributed. Part of the API of focus had been developed, while the part focusing on automation was under development. Key issues include avoiding risks by limiting direct access to hardware while still providing needed functionality
Company 3 (C3)	The investigated API is mainly internal and related to the reuse of common function signatures across products. The API of focus was in partial operation. Key issues included governance, and the role of the API owner
Company 4 (C4)	C4 supports several APIs managing access to various technologies. The APIs are used internally and range from very active to near retirement. Key issues included the retirement of API functionality and governance of API change
Company 5 (C5)	C5 is working on a new API facilitating access to devices via mobile and other cloud devices. The API would be generally for internal use. Key issues include supporting differing development speeds of different components

---

[1] http://e3value.few.vu.nl/tools/.

Case Contexts. This research is carried out as part of the Chalmers Software Center (SWC)[2]. Work in the center is organized into half-year sprints, renewable as part of continuous projects. This particular work is a result of Sprint 13, occurring in the Fall of 2017, while [11–13] cover experiences from past sprints. Projects involved interested software center companies, including many of the leading and largest software companies in Northern Europe. Each company had one to three specific representatives assigned to our project. Representatives are experienced in their respective fields, with a technical background. They were generally familiar with modeling, but had only seen value modeling briefly in a previous sprint.

All project companies (C1–C5) are SWC partner companies working in the embedded systems domain. Each company selected a particular API for more in-depth work as part of the project. We describe the APIs of focus for each company in Table 1, including a brief description of their ecosystem, key issues, internal/external classification, and their level of maturity. Details are kept deliberately vague to preserve anonymity of the case companies.

The high-level goal of the research sprint was continue to develop aspects of the strategic API framework while providing analytic value in API management and strategies to our five partner companies. In this sprint, driven by company interest, we had a particular interest in API value analysis, and decided, with company partner approval, to do so via e^3 value modeling of each case. The purpose of the modeling was to (1) map the API ecosystems for each company for better internal communication and understanding, (2) understand how the APIs provided or could provide internal and external value, and (3) support a visual comparison among each API case, allowing for comparisons across companies.

3   Methodology

We created and validated the five e^3 value models through a series of company workshops, supplemented by email feedback. The primary modelers were the first two authors (academics) with frequent input from the company representatives. We summarize the process as follows.

Initial Cross-Company Workshop: This two to three hour workshop involved one to two representatives from each company. Here we solidified sprint plans, specifically to apply e^3 value modeling to a selected case from each company. A short reminder of e^3 value syntax was provided.

Optional Preparatory Meeting: Two of the five API cases provided by the companies had been studied in the previous sprint, while three companies provided cases which were new to the project. In two out of three of the new cases, we had a one hour meeting to introduce the new API case.

---

[2] https://www.software-center.se/.

Draft Model Preparation: We created draft e³ value models for the APIs in question using either our existing knowledge of the case and goal models from the previous sprint, or from preparatory meetings, documentation and diagrams provided by the companies for the new cases. The goal models were mainly used to extract the actors in the API ecosystems, with some goal model dependencies included as draft value exchanges. The draft models were sent to the companies in advance of the in-person workshops. In some cases we received feedback via email and made further changes before the workshop, and in other cases we received our first feedback during the in-person workshop.

Company Workshops: We attended an in-person workshop at the location of each company. The workshops were three hours long, except in the cases where a preparatory workshop had taken place, then they were two hours. Workshops were attended by anywhere from one to six company representatives, with an average of about three to four attendees. Attendees were selected by the main company representatives, usually those who had detailed knowledge of the API and could represent various roles (users, designers, guardians).

During the workshop the researchers again provided a short reminder of e³ value syntax, and then went over the details of the draft model, projected onto a screen or whiteboard. The participants asked questions and suggested changes, sometimes at their own initiative, and sometimes responding to researcher questions ("Is this correct?"). We made an attempt to find reciprocal value flows, often if we had flows in a single direction, we asked "Why does this actor provide this value? What do they get in return?", trying to increase model completeness. Changes and additions were noted but not made in real time, due to low real-time usability of the e³ value tool. In one case, we were able to project the model on a whiteboard and make changes over top of it.

In all workshops, in addition to changes to the model instances, we recorded general comments or discussion points about the modeling language itself, including limitations or difficulties representing certain information.

Post-Workshop Iterations: After the individual workshops, we updated the models to reflect received feedback, then sent the resulting model back to the participants via email. In some cases we received further feedback and made further iterations, while in other cases the companies had no further feedback.

Final Cross-Company Workshop: To wrap-up the sprint, we had a final, three hour cross-company workshop where we presented the e³ value models for all the companies, and discussed findings, comparing results. The company representatives had a final opportunity to correct the content of the models, to compare cases, and to discuss modeling issues and utility in a group setting.

4   Results

We describe observations from our experience and the feedback from our company partners concerning our application of e³ value modeling. First, we provide an overview of the resulting models, providing two anonymized examples, Figs. 1 and 2. We then describe positive and negative findings as found in our cases.

4.1   Overview of Models

As stated, we created five e³ value models, one for each company case. Statistics concerning the models, including counts of elements used are shown in Table 2. We provide these in order to give an idea of the content and complexity of the models. In this case our count of composite actors include all actors inside another actor, and value object is a count of the labels on value flows, thus for each counted value object there are two value ports.

Table 2. Final e³ value model statistics

Company	Actors	Composite actors	Value objects	Value activities	Value interfaces	Stimuli	Stimuli connections
C1	11	6	29	8	31	0	0
C2	15	14	27	15	35	2	2
C3	14	9	49	13	58	3	2
C4	40	38	52	8	71	12	6
C5	6	4	17	15	52	5	9

In our models, we did not name the value objects on value flows when they were obvious, in order to avoid clutter. For example, in Fig. 1, on the left, the 'Protocol' value object is labelled in the flow between the 'Smart Phone Provider' and 'Embedded Device Provider', but is not labelled again when flowing to the 'Use API' or 'Provide API activity'. We count this as a single value object 'Protocol'. Thus we have more value interfaces than value objects as the objects flow through 2–6 interfaces. In some of our models, Use Case stimuli and connections were used. We discuss their use later in this section.

We show some of the example value objects elicited in Table 3. We omit the full list of 174 value objects due to space constraints. Our selection attempts to cover the wide range of objects discovered.

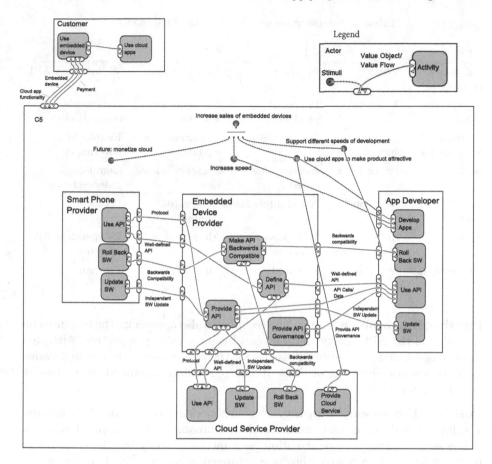

Fig. 1. e³ Value model for C5 anonymized

4.2 Findings

In this section we describe the positive and negative findings discovered when applying e³ modeling in practice. One finding presented at the end of this section is neither clearly positive or negative.

Positive: Relatively Easy to Read. Our company representatives found the value models to be generally understandable. As Table 2 shows, the models were often not simple. However, as most participants had a technical background and were generally familiar with conceptual modeling, the complexity level of e³ value models was not perceived as excessively complex to our partners.

We should note that in the previous sprint the companies were exposed to iStar goal modeling as a means of ecosystem mapping [22]. Although the models were perceived as useful, they were also complex, and several company representatives mentioned that they appreciated the comparative simplicity of e³ value models. We discuss the relationship of our findings to goal modeling in Sect. 5.

Table 3. Sample value objects from the final models

C1	C2	C3	C4	C5
Support & training	Views	Payment	New features	Well-defined API
Equipment sales	Avoid risks	Improved product	Payment for updates	Backwards compatibility
Maintenance & upgrades	Further automate	User needs	Governance of updates	Enable API governance
Services	Faster tasks	Increase dev speed	Faster/Cheaper dev	Component independence
Enriched content	New apps	Avoid duplication	Backwards compatibility	Protocol
Trust	Hide functionality	Well-proven solutions	Maintenance	Independent SW updates
Customer data	Automation	Automation	Simplicity	Payment

Positive: Emphasis on Value. The companies also appreciate the emphasis on the value of the APIs, particularly from a business value perspective. Although the partners could give technical reasons for the API projects, the strategic value of APIs was not always as obvious. It was particularly useful to see this value on a per-actor level.

Positive: Ecosystem Mapping. e^3 value models were intended to capture a value network, or e-commerce idea. More broadly, the concept of software ecosystems has gained more attention, as a means to analyzing the cooperative relationships of often complex business arrangements (e.g., [5]). Our companies were interested in ecosystem analysis from an API perspective. Due to their inclusion of actors, composite actors, and value flows between actors, e^3 value models were able to provide an overview of the API ecosystems. Even in the internal cases, our companies were large and complex enough to warrant an analysis of their internal ecosystem (see Fig. 2 for an example internal ecosystem as captured by e^3 value models).

Positive: Facilitating Comparison. As we created one value model for each case, the models were helpful in facilitating a cross-case comparison, particularly comparing the specific value provided by the APIs, the motivations for the API, and the configuration of internal and external actors.

Positive: Governance. Our partners found it helpful to see the various ways the companies managed API governance as captured in the models. We were able to capture governance practices either as activities and flows (e.g., 'Provide API Governance' from the 'Embedded Device Provider' to the 'App Developer' in Fig. 1 or as actors, activities, and flows (e.g., 'Governance Team', 'Verification (governance)' and 'Accept Changes' in Fig. 2).

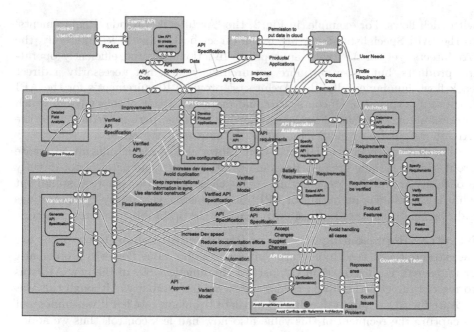

Fig. 2. e³ Value model for C3 anonymized

Negative: Quantifying Values. One of the strengths of e³ modeling is the ability to quantify and analyze net value via profit sheets. Examining the value objects collected in our models (see Table 3 and Figs. 1 and 2 for examples), we see that most are not directly economic, with the exception of various forms of 'Payment'. In order to create profit sheets, we would have had to work with our company partners to find utility functions for values objects such as 'Trust', 'Automation', and 'Component Independence'. Although theoretically possible, in practice our company contacts were not interested in monetizing otherwise quantifying these values, as this would have been a time consuming exercise with very approximate results. Furthermore, the fact that many of these value flows were internal, rather than exchanges between independent enterprises, made them particularly difficult to monetize, e.g., in Fig. 2, the monetized value of providing a 'Fixed interpretation of technical concepts'. As such, it was not feasible to perform the quantitative analysis advocated by e³ modeling.

Negative: Reciprocity in Internal Value Networks. We noted that it was relatively straightforward to capture reciprocal flows between independent actors, i.e., the companies and their customers/collaborators. For example, in Fig. 1, C5 provides an 'Embedded device' and 'Cloud functionality', while the 'Customer' provides 'Payment'. The other models have similar external flows.

However, internally, it was more difficult to find consistent and clear reciprocity between actors. Some effort was placed in the modeling sessions to elicit reciprocal flows, but often it was difficult for the representatives to come up with such flows. For example, in Fig. 2, the 'Architects' provide 'Requirements' to the 'API Specialists', but what is the direct back value flow? Indirectly, the 'Architects' get an API with appropriate functionality, and efficiently operating products, but these are more overarching values, not necessarily a direct back flow. Fundamentally, the 'Architects' provide 'Requirement's for the API because it is their job. Of course we could model an inflow of salary (from which internal actor?) but such flows are somewhat obvious, perhaps reductive to the 'Architects', and may clutter the model without analytic value.

The lack of reciprocal internal value flows also made it difficult to create profit sheets. If there are only flows in or out, obviously an actor is not directly profitable. Accounting for salary may make these calculations more possible, but what is the value of an output of API 'Requirements' vs. 'Salary'? Our findings indicate that it may not be sensible to consider the profitability of internal actors.

Negative: Representing Missing or Problematic Values. In two of the five cases, there was a need to capture that a particular value flow was problematic or undesired. For example, in C2, part of the functionality of the new API was to restrict direct access to particular hardware functionality. Although this was captured with a regular value flow 'No Risk of Problems', we felt it was important to capture the recipient of this value also now had less control, thus we added a red colored value flow 'Less Control' to the same actor. In another part of the model, we used a similar convention to capture the removed access to a particular hardware layer. The idea was to capture negative consequences and trade-offs the of the new API design. Tasks would be less dangerous and more efficient, but the user would have less direct access to and control over hardware.

In another example, in C4, the company identified desired values objects which were not fully satisfied, e.g., 'Coordinate Parallel Development Work'. In the C4 model, we identified six such desired but problematic value flows, and again colored them red. Although colors were used in [7], they were applied to scenario paths to indicate conformance with semantics, not problematic flows.

Negative: Capturing Interfaces. As we focused on modeling the value of APIs, we noted it was not obvious how an interface should be modeled in e^3 modeling. Was it an actor? A value object? An activity? Instead of dictating a solution to the companies, we followed their direction and captured the API in question in a variety of ways. In Fig. 1, the API is captured mainly via activities using the API and value flows concerning qualities of the API. In Fig. 2, the API is captured via a model supporting variants, thus it is captured as a composite actor. In most other cases the APIs were captured as actors within actors. However, APIs are interfaces, so this solution was not necessarily intuitive. Generally, it was not very clear how a technical component should be captured in e^3 value modeling, as explored further in the next point.

Negative: Human vs. Non-human Actors. One of our company contacts for C3 requested differentiating syntax for human vs. non-human actors, thus the purple boxes for non-human actors in Fig. 2. As e^3 value modeling is intended to model value networks, typically, each actor is a consumer/customer or enterprise. It is not clear if this notation is meant to support non-human, technology actors, e.g., 'Cloud Analytics' or the 'API Model' in Fig. 2. Yet these entities are core entities for our collaborating companies, and certainly they provide critical internal value, as well as potential external value via the use of APIs. Do non-human actors receive value? Does it make sense to calculate a Profit Sheet for such actors? Likely not, yet we believe these components and the value they provide should be made explicit in value models. In our case we used specially denoted actors for non-human actors, but such a solution does not necessarily work well for interfaces.

Negative: Lacking Motivation. Although the models provided an overview of actors and value flows in the internal ecosystem, we found the overall motivation for the value exchanges and the setup of the ecosystem was not obvious. For example, in Fig. 1, our original model did not include the stimuli near the top. We understood that there were actors providing API-related values and the qualities for the API, but not the motivations for the APIs. Upon examining the e³ value literature, we made use of the stimuli concept to capture the motivation behind value flow. Note that we misused this notation somewhat, as it is intended to show the start of a flow, ending in a stop stimulus. For our purposes, the flow was not important, but the motivation for the value exchanges was important. In this case, we were able to identify that the API was present to support different speeds of development, increasing the overall speed of development, and increasing sales of devices. We may have been able to capture this as further value flows, but these were not necessarily exchanges between two internal actors, but motivations for the overall network, e.g., 'Support different speeds of development' is broader than a single value exchange. Note that we did not feel the need to use this concept in the case with an external API (C1), as there the purpose was clear by looking at the external value flows. However, we felt that including stimuli as motivation improved comprehensibility in the remaining internal cases.

Unknown: Diversity of values. As can be seen in Table 3, the value objects we collected were diverse, ranging from monetary value ('Payment') to what could be considered qualities or non-functional requirements ('Backwards Compatibility', 'Simplicity', and 'Faster tasks') to functionality ('New apps', 'New features'). In eliciting value objects from our company participants, we did not strictly define the concept, or place any limitations on what they considered as value or a value flow. This diversity in value could be seen as positive or negative. On one hand, the modeling notation becomes very expressive and flexible when anything can be a value object. On the other hand, this flexibility makes it difficult to perform structured analysis, such as profit analysis. Existing work analyzing value from an ontological perspective may help to clarify the nature of value [18]; however, applying more restrictions to value modeling may make it less usable in practice. Further work, both academic and in practice, is needed.

5 Discussion

We have described the benefits and limitations of applying e^3 value modeling to assess the value of a particular technology (**RQ1**). More specifically, we have found that applying e^3 value models to internal value networks/ecosystems is useful **RQ1.1**, as it helps to elicit and understand actors, value flows and complex interactions. However, in the internal cases it was difficult to capture reciprocal flows and to analyze profitability.

One can argue that we have made negative observations because we have used e^3 value models for a purpose slightly differing from its' originally intended use: to evaluate the feasibility of an e-commerce idea, including profitability. Instead, we have been using the framework to understand the strategic value of a particular technical component, both internal and external to an organization, with an emphasis on internal value.

However, as we have reported, the application of e^3 modeling for our purposes was generally successful. We believe it could and should be used for a wider purpose than analyzing e-commerce networks. As such, some of the capabilities of the language may need to be updated. In particular, it was unfortunate that we were not able to apply any form of structured or systematic analysis to our models. To address this and other shortcomings, we suggest the following areas of expansion or future development.

1. One can argue that we should be strict when eliciting values, focusing on those which are quantifiable. However, many important value exchanges were difficult for our companies to quantify, yet were still critical in understanding the role of the APIs. Thus we recommend structured analysis for e^3 value models which supports intangible values.
2. Modeling and analysis should account for the possibility of non-reciprocal value flow, particularly in internal ecosystems and for non-human actors.
3. The e^3 language should be expanded beyond its' core to support negative or problematic flows as well as non-human, technical components or interfaces. To reduce complexity, this could be an optional extension to the language.
4. Motivation for flows should be supported, either by stimuli, as we have used it, or by other means.
5. Further reflection is needed on the nature of values and value objects, and whether differences in types of values can or should influence the language.

Ideally the above points are considered while still mainly retaining the simplicity and value-oriented expressiveness of the language. Of course, similar exercises applying e^3 value modeling in practice should be conducted, and our findings and the utility of our recommendations should be confirmed or refuted, depending on contextual factors.

Our discovered limitations may have been avoided by applying an alternative form of value modeling, as outlined in Sect. 2. For example, VNA supports analysis of intangible values in a network [16]. However, other languages are less well supported and explored, and it is not obvious their use would have provided a clear advantage. Further work should compare use of these languages in practical settings, making the context of application clear.

Threats to Validity. In terms of *Construct Validity*, the researchers and the company representatives may have misunderstood the syntax or semantics of e³ value models (e.g., misusing stimuli). We made an effort to learn from the source material, provide frequent overviews to our company contacts, and were constrained by the available tooling. Although we are not e³ experts, the first author is an experienced modeler, and the company contacts are familiar with modeling. Thus we are representative of those who may adopt e³ modeling.

Considering *Internal Validity*, our modeling process was interactive and went through many iterations, and our results were member checked during a final cross-company workshop. Group validation may suffer from group decision biases, yet one of the aims of collaborative modeling is to build consensus among teams. In future work we will aim for more individual evaluation, e.g., through surveys.

Examining *External Validity*, all case companies are located in Scandinavia and work with embedded systems; however, the companies are international. Furthermore, we have focused analysis on APIs. Applying the same modeling process to different companies with different technologies may produce differing observations. However, we mitigate this possibility by involving five different companies. We also believe most of our observations generalize to other technologies, with the possible exception of interface-specific challenges.

Finally, considering *Reliability*, it is possible that the experiences of the first author in working with and developing goal modeling (e.g., [22]), as well as the experiences of the rest of the participants having used goal modeling in the previous sprint may have affected the results. In fact, the perception of value model as generally comprehensible is in part due to a comparison with goal models. However, most modelers who are not method developers may have extensive experience with some other type of model (e.g., BPMN, UML), that may effect their perceptions of a language, e.g., noting the lack of motivation in e³ modeling. Thus we acknowledge this bias and analyze our results with this leaning in mind. In fact, the company partners had positive reactions to both goal and value modeling, in this way, we confirm the findings of previous work suggesting that goal and value modeling are complementary [9,10].

6 Conclusions

We have presented experiences in applying e³ value modeling to real cases from five companies. In doing so we provide valuable input to the language and its users, including an evaluation of language use for a differing purpose. Overall, despite reported challenges, we find e³ modeling useful for understanding an internal ecosystem, and for understanding the value of technologies such as APIs.

In our current project work, we are focusing on method transferability: developing methods for e^3 value modeling for APIs, as well as a method for value to goal modeling, which allows the modeling to be performed independently by the companies. We are also exploring various solutions for capturing interfaces in e^3 models.

Acknowledgments. Thanks to company contacts and the Chalmers Software Center for support.

References

1. Miller, P.: Explaining technology's value. Forbes May 2015
2. Kiessel, A.: How strategic is it? - assessing strategic value. Oracle February 2012
3. Geerts, G.L., McCarthy,W.E.: The ontological foundation of REA enterprise information systems. In: Annual Meeting of the American Accounting Association, Philadelphia, PA, vol. 362, pp. 127–150 (2000)
4. Rother, M., Shook, J.: Learning to see: value stream mapping to add value and eliminate MUDA. Lean Enterprise Institute, Brookline (2003)
5. Handoyo, E., Jansen, S., Brinkkemper, S.: Software ecosystem modeling: the value chains. In: Proceedings of the Fifth International Conference on Management of Emergent Digital Ecosystems, pp. 17–24. ACM (2013)
6. Gordijn, J., Akkermans, J.: Value-based requirements engineering: exploring innovative e-commerce ideas. Requir. Eng. **8**(2), 114–134 (2003)
7. Gordijn, J., Akkermans, H., Van Vliet, J.: Designing and evaluating e-business models. IEEE Intell. Syst. **16**(4), 11–17 (2001)
8. Gordijn, J.: E-business value modelling using the e3-value ontology. Value Creat. E-Bus. Model. 98–127(2004)
9. Gordijn, J., Petit, M., Wieringa, R.: Understanding business strategies of networked value constellations using goal-and value modeling. In: 14th IEEE International Conference Requirements Engineering, pp. 129–138. IEEE (2006)
10. Henkel, M., Johannesson, P., Perjons, E., Zdravkovic, J.: Value and goal driven design of e-services. In: IEEE International Conference on e-Business Engineering, ICEBE 2007, pp. 295–303. IEEE (2007)
11. Lindman, J., Hammouda, I., Horkoff, J., Knauss, E.: Emerging perspectives to API strategy. IEEE software (in press)
12. Horkoff, J., et al.: Goals, workflow, and value: case study experiences with three modeling frameworks. In: Poels, G., Gailly, F., Serral Asensio, E., Snoeck, M. (eds.) PoEM 2017. LNBIP, vol. 305, pp. 96–111. Springer, Cham (2017). https://doi.org/10.1007/978-3-319-70241-4_7
13. Horkoff, J., et al.: Modeling support for strategic API planning and analysis. In: International Conference on Software Business (ICSOB) (2018, in press)
14. Osterwalder, A., et al.: The business model ontology: A proposition in a design science approach (2004)
15. Osterwalder, A., Pigneur, Y.: Business Model Canvas (2010, Self published, Last)
16. Allee, V.: Value network analysis and value conversion of tangible and intangible assets. Journal of intellectual capital **9**(1), 5–24 (2008)
17. Object Management Group (OMG): Value Delivery Metamodel, Version 1.0. OMG Document Number formal/2015-10-05 (2015) (www.omg.org/spec/VDML/1.0/)

18. Sales, T.P., Guarino, N., Guizzardi, G., Mylopoulos, J.: An ontological analysis of value propositions. In: 2017 IEEE 21st International on Enterprise Distributed Object Computing Conference (EDOC), pp. 184–193. IEEE (2017)
19. Kundisch, D., John, T.: Business model representation incorporating real options: an extension of e3-value. In: 2012 45th Hawaii International Conference on System Science (HICSS), pp. 4456–4465. IEEE (2012)
20. Schuster, R., Motal, T.: From e3-value to REA: modeling multi-party e-business collaborations. In: IEEE Conference on Commerce and Enterprise Computing, CEC 2009, pp. 202–208. IEEE (2009)
21. Gordijn, J., Osterwalder, A., Pigneur, Y.: Comparing two business model ontologies for designing e-business models and value constellations. In: BLED 2005 (2005)
22. Dalpiaz, F., Franch, X., Horkoff, J.: iStar 2.0 language guide. arXiv preprint arXiv:1605.07767 (2016)

Event-Context-Feedback Control Through Bridge Workflows for Smart Solutions

P. Radha Krishna[1(✉)] and Kamalakar Karlapalem[2]

[1] Computer Science and Engineering Department, National Institute of
Technology, Warangal, India
prkrishna@nitw.ac.in
[2] Data Sciences and Analytics Centre, IIIT-Hyderabad, Hyderabad, India
kamal@iiit.ac.in

Abstract. The utility and impact of IoTs driven, smart solution is in their ability to react to the changes in the environment by controlling the systems managed by them. In most cases, such controlling depends on the flow of data and events from the sensors to the actionable logic of the smart solutions. The flow of data and control is typically hard-coded and the solutions provided are not flexible enough to cater to evolving requirements of smart solutions. In our solution, we make the data and control flow explicit by modeling them as workflows (with data operators) and bring in bridge workflows to comprehend events from sensor data, and pertinent (aggregated) data for the smart solutions. The smart solutions based on the feedback from sensors can instruct the smart solution controller to change some control parameters and ascertain the impact of events got from bridge workflow. Thus, the data and control loop between sensors and smart solutions and then to the smart controller is orchestrated by the bridge workflows. The detection of events, context, feedback and control actions are done by the workflows (tasks) as per the smart solution requirements. We illustrate our solution through a traffic management solution in a smart city environment.

Keywords: IoT sensors · Data and control flows · Context · Bridge workflows

1 Introduction

The Internet of Things (IoT) comprises of large numbers of sensors which are heterogeneous in nature. Data from the environment is collected by sensors, and the triggering actions are done by actuators. Smart solutions are designed to deploy certain smart services. For instance, a smart traffic advisor solution provides *driver assistance service* regarding route or facilities available, and alerts the driver based on the current location of the car derived from smartphone and road conditions. Each smart service has specific characteristics such as user requirements (need to shop for vegetables), defined comfort zones (need to avoid traffic) and ranges (time taken for the trip must be between 30 min to 1 h). Similarly, the time taken for shipment of goods is 1 h to 2 h for a smart logistics solution. These comfort parameters (also refers to the quality of service parameters) are based on latency, availability, throughput, etc. and the upper

© Springer Nature Switzerland AG 2018
J. C. Trujillo et al. (Eds.): ER 2018, LNCS 11157, pp. 626–634, 2018.
https://doi.org/10.1007/978-3-030-00847-5_46

and lower bounds on the metrics to be maintained at runtime. The service requirements may evolve over a period of time, and smart solutions should able to meet this requirement by performing/adjusting certain parameters/capabilities. In most smart solutions, the flow of data and control is typically hard-coded, and the solution provided is not flexible enough to cater to evolving requirements of smart solutions. Further, there is little vertical integration (or there is hard-coded integration) from the data source from sensors to the data and event consumption by smart solutions.

In this paper, we present a modeling approach that allows controlling the smart objects/sensors as per the requirement changes in a smart solution (Fig. 1). In our approach, smart solution controller captures this feedback and carries out adjustments as needed so that the parameters for smart services are within the range. The data that is generated by sensors and the data (mostly in aggregated form) that is required for executing smart solutions are usually different. We model data and event flows from sensors to smart solutions as workflows, where workflow tasks involve data operations. We develop bridge workflows that are used to capture the data, comprehend the events and produce (aggregated) data for the smart solutions. Moreover, the sensor data is also used to provide context information, for instance, GPS data from driver's smartphone provides context information such as the location of the vehicle and driver. The rest of the paper is organized as follows. Section 2 presents the related work. In Sect. 3, we describe our approach for designing smart solutions. In Sect. 4, we discuss the traffic advisory solution to demonstrate the proposed approach. The paper concludes with Sect. 5.

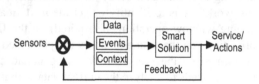

Fig. 1. A basic Smart Solution view

2 Related Work

Mervat et al. [4] discussed a data management framework for IoT that serves as a seed to build a comprehensive IoT data management solution. Narendra et al. [5] proposed data filtering algorithms to filter the sensor data required by the IoT application depending on the context when the data is generated. Zambonelli [8] presented a common set of features for IoT systems such as objects, Middleware, Services and Applications, and described software engineering concepts for developing complex IoT applications in a more systematic way. Zanella et al. [9] provided a comprehensive survey related to enabling technologies, protocols, and architecture for an urban IoT to support smart cities. Gaur et al. [1] described an architecture for Multi-Level Smart City using semantic web technologies to inference rules to combine sensor information. Most of these works focused on providing architecture details that show functional and non-functional aspects of various components and applications [6], but lacks the conceptual modeling aspects for IoT applications. In our earlier works [2, 3], we presented an ER model that allows capturing conceptual level details required for smart applications. In this work, we concentrate on data logistics and its modeling.

3 Our Solution – Bridge Workflows

Sensor devices sense the environment continuously and identify the environmental events and context. Further, the data that is captured from environment should be made accessible to smart solutions. However, sensor data that is in raw form needs to be transformed so that smart solution can utilize it for its services. Sensor data plays a critical role in understanding the sensors/smart objects, environment, activities, etc. and responding to it. **Bridge Workflows** facilitate coordination between sensor data and data required for IoT applications. They allow collection, modeling, reasoning, and distribution of context in relation to sensor data.

Figure 2 shows three workflows namely *bridge sensor data workflow*, *bridge data workflow* and *bridge smart solution workflow*. These three workflows collectively serve as Bridge workflows. *Bridge sensor data workflow* have tasks closely coupled with sensors sense the environment, capture the sensor data, transform the data (e.g., cleansing, converting from one unit of measure to another, etc.) and store the data in a data store. This data is fine-grained data which forms the basis for smart solutions. The sensor data is captured either continuously (i.e., stream) or periodically, and the data is stored accordingly. The second workflow namely *Bridge data workflow* facilitates keeping the data that is ready for the enactment of smart solutions. The first task of this sub-workflow is to identify or get the next smart solution task. Then, it performs the sliding window operation (if required) on the fine-grained sensor data, performs data operations on the data and stores the processed data in the data store. The analysis is performed on the processed data to generate events and context. In case, the smart solution needs to analyze the current data with respect to past processed data, then data from the data store is also used to generate the events and context. The events are logged into a log file.

The *Bridge Smart solution workflow* tracks each event and processes it. In addition, the workflow checks for the addition of resources to process the actions specified for the event when a condition and/or context is met. Next, required action is performed with respect to the event, and, if required, another bridge smart solution workflow is initiated. The new workflow may need data from either the same sensors or a new sensor, and accordingly it triggers first, second and/or third workflow(s). The dashed arrows in Fig. 2 indicate the inter-workflow dependency representing coordination between various tasks within the workflow or across the bridge workflows. The coordination activities include input data flow, control flow and triggering of a task according to the business processes of the sensor and smart solutions. Appropriate bridge workflows are initiated and executed for each sensor as well as smart solutions.

Bridge workflows used in this work have three major functions: (i) capturing sensor data and keeping it ready for downstream processing, (ii) handling the data logistics from sensor sources to smart solutions using data operators, and generating data, identifying events and determining the context, and (iii) processing the events to

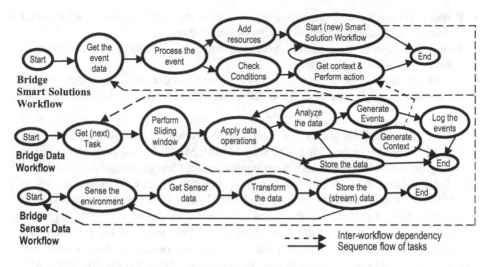

Fig. 2. Bridge workflows

orchestrate the workflows that initiate the functioning of smart solutions. MVPP^DEC
(see Sect. 3.1.) simplifies orchestration of bridge data workflows with enhanced scalability (e.g., the addition of new sensor due to evolving requirements, policy changes (such as additional security) or new technology adoption and flow of its data). Bridge smart solution workflow execution requires (Processed) Data, events (e.g., congestion event, parking full event, etc.) and context (e.g., (<CurrentLocation: Street No. 4>, <Time: 4 PM>, <Visibility: Less>)) that are required to perform a service operation (e.g., Driver Assistance service, virtual assistant service – to attend a pre-scheduled meeting) facilitated through a smart solution.

3.1   Data and Control Flows

IoT driven smart solutions are data-focused. The source of data is mostly carried from the sensors. The data flow starts from sensors and reaches smart solutions. This data flow includes events and context, *which are generated based on the sensor data values*. For instance, based on the GPS data, the density at a particular location can be determined. At the same time, if the density is beyond a specified threshold, a congestion event is raised, and the location and time together can serve as a context. So, the data is transformed into processed data, events and context. Bridge workflows facilitate the complete data and control flows. As the data is acquired from the sensors, the events are generated to control the processing of the workflows. The tasks of bridge data workflows include aggregation, generate events, determine context, etc. These tasks perform operations using a set of operators, which are similar to database operators such as select, project and join operators. Examples operations using the data operators are:

- **Select** (σ) – select the sensor (stream) data periodically for a defined time window or in a certain range of values; Constraints on incoming data.

- **Project** (Π) – select few properties (e.g., location and time data from a smartphone and speed of a vehicle) form two or more sensor data.
- **Join** ($\bowtie$) – capture the data from more than one sensor; combine event and data to generate new data and events.
- **Aggregate** (Ω) – Apply aggregate functions (e.g., count, sum, avg, etc.) on the incoming data (e.g., average rainfall during the day or certain time period).

In this work, we introduce MVPP^{DEC}, an extension of Multi-View Processing Plan (MVPP) [7], to represent the bridge data workflows which specify events (E) and context (C) as nodes, besides data (D). Like MVPP, MVPP^{DEC} supports many inputs and many outputs where each path from input to output form a bridge data workflow path for handling data, events, and context together. Figure 3 shows a sample MVPP^{DEC} for traffic advisory solution. MVPP^{DEC} is a directed graph representing various data flows. Leaf nodes at the bottom (denoted with ✿) of MVPP^{DEC} represent sensors (or group of sensors of the same type), which form the basis for generating the raw data. Root (top) nodes and non-leaf represent the processed data, events and context. The events can be either atomic or composite. Some events combined with data can generate other events, new context and/or new data. Data nodes are denoted with circles, events with dashed circles and context as filled circles. Root nodes establish the data that are required for performing necessary actions defined by smart solutions. The root nodes can take the result from its child node(s) and process it to generate events and context. We model the path from leaf to root nodes as data workflows. So, MVPP^{DEC} serve as a model that can be considered as a logical specification of data flow requirements. MVPP^{DEC} is augmented with control flows to ensure the movement of data ascertaining to quality of service requirements. Smart solution requires specification and execution of set of workflows for its fulfillment. The modeling aspects for developing a smart solution using our approach include:

1. What are the data requirements for a smart solution from the sensors?
2. Given the data requirements, is there corresponding MVPP^{DEC} which can model those requirements?
3. Given a MVPP^{DEC}, can we execute an appropriate MVPP^{DEC} workflow as and when required by the smart solution to detect events and context?

The sensed data can be used by different smart solutions. Therefore, there can be some commonality in the data needed for different smart solutions. Thus, identifying and making this common data explicit is required. The main aim of MVPP^{DEC} is to identify the portions of bridge data workflows that perform the same operations on the data, events, and context, and share the (common) intermediate results (data) between the workflows. Further, explicit representation of data flows in the form of MVPP^{DEC} helps in debugging the processing of sensor data and events generated. It also helps in materializing the intermediate nodes for faster query processing. Though we use relational algebra operators in MVPP^{DEC}, the actual implementation of these operators is done as per the storage and logical data model constructs of source data. That is why, presenting MVPP^{DEC} as a workflow with tasks provides the flexibility of supporting the operators by a library of tasks with equivalent functionality.

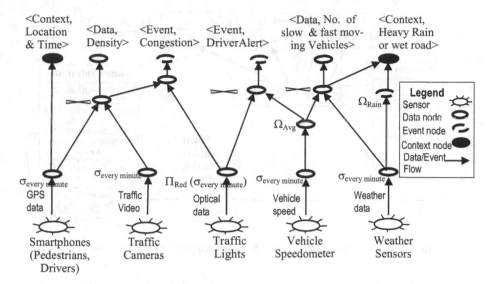

Fig. 3. An example MVPP^{DEC} for traffic advisory solution

3.2   Event-Context-Feedback Control

Smart solution management refers to the event-based and context-sensitive execution and adaptation to its overall processes. Figure 4 depicts the conceptual view of smart solution that supports requirements evolution. Sensors $S_1 \dots S_m$ senses the environment. Bridge workflows execute certain tasks ($T_1 \dots T_n$) to perform operations for smart solutions. Execution of these tasks needs data from sensors. Data transporter acquires the data from various sensors and supplies to bridge workflows' tasks. The data requirements for each task may vary, for instance, data from more than one sensor is required for a task T_1, aggregate the data from a sensor for a defined time window for task T_2, etc. Usually, there is a time delay in capturing the data from heterogeneous sensors and intimating the data to workflow tasks. Let $\lambda_1, \lambda_2, \lambda_3, \dots \lambda_m$ be the time delay from sensors $S_1, S_2, S_3, \dots S_m$ to data transporter respectively.

Bridge workflow tasks perform the operations (like aggregation) and send the data, events, and context to smart solutions with a time lag of $\mu_1, \mu_2, \dots \mu_n$. Smart solutions ($SS_1, SS_2, \dots SS_p$) execute necessary actions (say, $A_1, A_2, \dots A_k$) to perform the smart services and give feedback to Smart Solution Controller. Smart solution controller observes the service parameters obtained from the actions and monitors whether these parameters are in the defined ranges or not. This impacts the evolution requirements that need to be handled, for instance, for smooth traffic flow. Examples of evolution requirements for a traffic management solution includes time of arrival at meeting location which is rescheduled; unavailability of certain parking places due to periodic maintenance activity, traffic diversion due to a road accident, etc.

Smart solution controller checks whether these values are out of comfort zone (as per quality of service requirements) or not, and accordingly controls the parameters, and acquires their impact through events got from bridge workflows. The actions

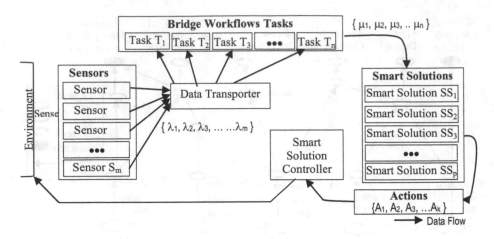

Fig. 4. A conceptual view of Smart Solution requirement evolution support

performed results in a change in the environment, and thereby necessitates requirements to evolve. When the values are out of range, smart solution controller initiates
actions such as adding additional sensors (e.g., the addition of new traffic lights in a
smart traffic management scenario). Thus, the loop between sensors and smart solutions
and then to the smart controller is orchestrated by the bridge workflows. The feedback
loop enables controlling the sensing infrastructure as well as smart solution execution
so that the services can meet the specified Service Level Agreements.

The arrival rates $\lambda_i (1 < i < m)$ and service rates $\mu_j (1 < j < n)$ are helpful in deciding
the intermediate nodes of MVPP^{DEC} that can be materialized. Suppose, a set of sensors
S generate 10 million records (i.e., λ_i) in a minute, and the smart solution needs
processed data (i.e., μ_j) of 100 records out of 10 million records. In such cases, one can
materialize the corresponding intermediates nodes so that the overall processing time is
reduced (for instance, how many rows in the materialized view that need to be processed to understand the traffic for the last five minutes). In this way, MVPP^{DEC} helps
in optimization of data workflows which fascinates faster responses to senses data.

Note that there is a time lag between initiation of a change of environment and its
impact on resetting the comfort range. Bridge workflows support these changes by
explicitly modeling them and study data and control flow design in terms of time and
effort to ensure that out of comfort range parameters get back to comfort zone within
the expected time duration. We use Event-Condition-Context-Action (EC^{2}A) rules for
specifying the behavior of smart solutions and support handling events that arise during
their enactment.

4 Smart Traffic Advisor Solution - Use Case

Consider a smart traffic advisor solution example used by a city. Figure 5 shows the
high-level view of smart traffic advisor solution. *Views* are useful to serve as a data
store and visualize the data specific to an application. MVPP^{DEC} (see Fig. 3) can be

used to materialize some of these views for fast query execution. Views can capture data not only from sensors but also from other external entities. For instance, the direction of traffic movement at traffic signals is useful in assessing the traffic condition. Event Detectors analyze the data to detect the events at runtime or patterns in the data. Some events deal with (a) control the objects, (b) alter the business process (application) or (c) both. For instance, when traffic is heavy on a specific road, a congestion event arises which controls the traffic signals (such as on/off and adjust the time for traffic lights namely red, yellow and green). On the other hand, the smart vehicles sense a congestion event (by polling the traffic data at periodic intervals) to decide on a new route that encounters less traffic for a vehicle. *EC²A Rule Manager* reacts to events and chooses appropriate actions to control the objects/business processes based on the context detected by the *Context Analyzer*. Actions correspond to the execution of EC²A rules and coordinate with the smart environment. The application/solution provides the information about specification of workflows, how to control workflows, what data is modeled, etc. and to ensure the consistency while executing entire solution with the help of bridge workflows. It also provides necessary feedback (based on the MVPP^{DEC} workflow execution results) on what additional data need to be captured and modeled, additional views to be created, exception handling, etc. Enactment of Smart city advisor solution checks for the events

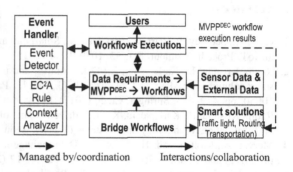

Fig. 5. Smart traffic advisor solution architecture overview

continuously from the sensors involved. An example of *Event-Condition-Context-Action (EC²A)* rule is given below:

Event: Congested Route
Condition: *Presence of high traffic densities* at multiple junctions of a road
Context: *Time, Location, Past events*
Action: Notify commuters advising alternate routes

In our example scenario, the computation of traffic density and categorizing into high, medium and low at each traffic junction can be made by the traffic camera itself (as the context location is known apriori) and transmit only density information to the next task handler. In this way, the communication traffic with regard to video transmission can be reduced. Similarly, data generated for time parameter (e.g., timestamp) can serve as both event and context. This time context (e.g., peak hours) can be used as part of an action, for instance, observe the traffic for the next 3 h.

5 Conclusion

We made explicit the process of acquiring, aggregation and presenting data as *Views* for smart solutions. Data and changes to data drive the orchestration of various tasks in the smart solutions life cycle. The external events and data-driven events together along with a particular context initiate actions that drive the fulfillment of requirements of the smart solution. In this paper, we discussed the modeling aspects for designing smart solutions. The main objective of this work is to provide a conceptual modeling approach that drives the understanding of the flow of data from sensors to the smart application. We developed bridge workflows to model data, control and process flow and presented MVPP^{DEC} that represent various data operations as part of bridge data workflow.

References

1. Gaur, A., Scotney, B., Parr, G., McClean, S.: Smart city architecture and its applications based on IoT. Procedia Comput. Sci. **52**, 1089–1094 (2015)
2. Radha Krishna, P., Karlapalem, K.: Data, control, and process flow modeling for IoT driven smart solutions. In: Mayr, H. C., Guizzardi, G., Ma, H., Pastor, O. (eds.) ER 2017. LNCS, vol. 10650, pp. 419–433. Springer, Cham (2017). https://doi.org/10.1007/978-3-319-69904-2_32
3. Radha Krishna, P., Karlapalem, K.: IoTs driven smart solutions life cycle. In: 27th Workshop on Information Technology and Systems (WITS), Seoul, South Korea (2017)
4. Mervat, A., Mohammad, H., Najah, A.: Data management for the Internet of things: design primitives and solution. Sensors **13**(11), 15582–15612 (2013)
5. Narendra, N., Ponnalagu, K., Ghose, A., Tamilselvam, S.: Goal-driven context-aware data filtering in IoT-based systems. In: 18th IEEE ITSC, pp. 2172–2179. IEEE, USA (2015)
6. Santana, E.F.Z., Chaves, A.N., Gerosa, M.A., Kon, F., Milojicic, D.S.: Software platforms for smart cities: Concepts, requirements, challenges, and a unified reference architecture. ACM Comput. Surv. **50**(6), 1–37 (2017). Article 78
7. Yang, J., Karlapalem, K., Li, Q.: Algorithms for materialized view design in data warehousing environment. In: VLDB 1997, pp. 136–145. Morgan Kaufmann, New York (1997)
8. Zambonelli, F.: Key abstractions for IoT-oriented software engineering. IEEE Softw. **34**(1), 38–45 (2017)
9. Zanella, A., Bui, N., Castellani, A., Vangelista, L., Zorzi, M.: Internet of things for smart cities. IEEE Internet Things J. **1**(1), 22–32 (2014)

Author Index

Printed in the United States
By Bookmasters

Printed in the United States
By Bookmasters